Lecture Notes in Computer Science 1631

Edited by G. Goos, J. Hartmanis and J. van Leeuwen

Springer

Berlin
Heidelberg
New York
Barcelona
Hong Kong
London
Milan
Paris
Singapore
Tokyo

Paliath Narendran   Michael Rusinowitch (Eds.)

Rewriting Techniques and Applications

10th International Conference, RTA-99
Trento, Italy, July 2-4, 1999
Proceedings

 Springer

Series Editors

Gerhard Goos, Karlsruhe University, Germany
Juris Hartmanis, Cornell University, NY, USA
Jan van Leeuwen, Utrecht University, The Netherlands

Volume Editors

Paliath Narendran
Department of Computer Science, University at Albany – SUNY
Albany, NY 12222, USA
E-mail: dran@cs.albany.edu

Michael Rusinowitch
LORIA – INRIA Lorraine
615, rue du Jardin Botanique, F-54602 Villers les Nancy, Cedex, France
E-mail: rusi@loria.fr

Cataloging-in-Publication data applied for

Die Deutsche Bibliothek - CIP-Einheitsaufnahme

Rewriting techniques and applications : 10th international conference ;
proceedings / RTA-99, Trento, Italy, July 2 - 4, 1999. Paliath Narendran ;
Michael Rusinowitch (ed.). - Berlin ; Heidelberg ; New York ; Barcelona ; Hong
Kong ; London ; Milan ; Paris ; Singapore ; Tokyo : Springer, 1999
 (Lecture notes in computer science ; Vol. 1631)
 ISBN 3-540-66201-4

CR Subject Classification (1998): F.4, F.3.2, D.3, I.2.2-3, I.1

ISSN 0302-9743
ISBN 3-540-66201-4 Springer-Verlag Berlin Heidelberg New York

Typesetting: Camera-ready by author
SPIN: 10703367   06/3142 – 5 4 3 2 1 0   Printed on acid-free paper

Preface

This volume contains the proceedings of the 10th International Conference on Rewriting Techniques and Applications held from July 2-4, 1999 in Trento, Italy, as part of the Federated Logic Conference (FLoC'99). The RTA conferences are dedicated to all aspects of term, string and graph rewriting as well as their applications such as lambda calculi, theorem-proving, functional programming, decision procedures.

The program committee selected 23 papers as well as 4 system descriptions from 53 submissions of overall high quality (46 regular papers and 7 systems descriptions). The papers cover a wide range of topics: constraint solving, termination, deduction and higher-order rewriting, graphs, complexity, tree automata, context-sensitive rewriting, string rewriting and numeration systems ... by authors from countries including: France, Germany, India, Israel, Italy, Japan, The Netherlands, Poland, Portugal, Spain, USA.

B. Courcelle (Univ. Bordeaux) and F. Otto (Univ. Kassel) presented invited talks, on graph grammars and connections between rewriting and formal language theory respectively. F. van Raamsdonk (CWI, Amsterdam) gave an invited tutorial on higher-order rewriting.

We would like to thank the external reviewers for their contribution to preparing the program and Horatiu Cirstea for his help in maintaining the web server of the program committee.

April 1999

Paliath Narendran
Michael Rusinowitch

Program Committee

Andrea Asperti (Bologna)
Rémi Gilleron (Lille)
Bernhard Gramlich (Vienna)
Jieh Hsiang (Taipei)
Richard Kennaway (Norwich)
Delia Kesner (Orsay)
Klaus Madlener (Kaiserslautern)

William McCune (Argonne)
Paliath Narendran, co-chair (Albany)
Michael Rusinowitch, co-chair (Nancy)
Klaus U. Schulz (Munich)
Geraud Senizergues (Bordeaux)
G. Sivakumar (Bombay)
Andrei Voronkov (Uppsala)

Organizing Committee

Hubert Comon (Cachan)
Nachum Dershowitz, chair (Urbana)
Harald Ganzinger (Saarbrucken)

Paliath Narendran (Albany)
Tobias Nipkow (Munich)
Yoshihito Toyama (Tatsunokuchi)

Local Organization

Andrea Asperti (Bologna)

External Reviewers

Altenkirch, T.
Anantharaman, S.
Arnold, A.
Arts, T.
Avenhaus, J.
Bonacina, M.P.
Boudet, A.
Brock, S.
Caron, A.C.
Clark, D.
Comon, H.
Contejean, E.
Courcelle, B.
Degtyarev, A.
Deiss, T.
Devienne, P.
Diekert, V.
Dougherty, D.
Durand, I.
Elliott, R.
Ferreira, M. C. F.
Finelli, M.
Glauert, J.R.W.
Gnaedig, I.
de Groote, Ph.
Guo, Q.
Hanus, M.
Hofbauer, D.
Ida, T.
Jacquemard, F.
Kepser, S.
Lamarche, F.
Lepper, I.
Lescanne, P.
Lohrey, M.
Lugiez, D.
Marchignoli, D.
Marcinkowski, J.
Matthes, R.
McNaughton, R.
Middeldorp, A.

Miller, D.
Mokkedem, A.
Monate, B.
Nadathur, G.
Naoi, T.
Narbel, P.
Niehren, J.
Nieuwenhuis, R.
Ohlebusch, E.
Ohta, Y.
van Oostrom, V.
Palmgren, E.
Perrier, G.
van Raamsdonk F.
Rehof, J.
Ringeissen, C.
Ritter, E.
Robson, M.
Rosenkrantz, D.
Scharff, C.
Seidl, H.
Sherman, D.
Sleep, R.
Staerk, R.
Steinbach, J.
Stolzenburg, F.
Strandh, R.
Talbot, J. M.
Tison, S.
Tommasi, M.
Treinen, R.
Verma, R.
Vigneron, L.
Vogt, R.
Werner, A.
Werner, B.
Zantema, H.
Zhang, H.
Zuber, W.

Table of Contents

Session 5:

Session 6:

Session 7: Invited Tutorial

Session 8: System Descriptions

Session 9:

Session 10:

Session 11: Invited Talk

Session 12:

Solved Forms for Path Ordering Constraints

Robert Nieuwenhuis and José Miguel Rivero*

Technical University of Catalonia, Dept. LSI,
Jordi Girona 1, 08034 Barcelona, Spain
roberto@lsi.upc.es rivero@lsi.upc.es

Abstract. A usual technique in symbolic constraint solving is to apply transformation rules until a *solved form* is reached for which the problem becomes simple. Ordering constraints are well-known to be reducible to (a disjunction of) solved forms, but unfortunately no polynomial algorithm deciding the satisfiability of these solved forms is known.

Here we deal with a different notion of solved form, where fundamental properties of orderings like transitivity and monotonicity are taken into account. This leads to a new family of constraint solving algorithms for the full recursive path ordering with status (RPOS), and hence as well for other path orderings like LPO, MPO, KNS and RDO, and for all possible total precedences and signatures. Apart from simplicity and elegance from the theoretical point of view, the main contribution of these algorithms is on efficiency in practice. Since guessing is minimized, and, in particular, no linear orderings between the subterms are guessed, a practical improvement in performance of several orders of magnitude over previous algorithms is obtained, as shown by our experiments.

1 Introduction

An ordering constraint is a quantifier-free first-order formula built over the binary predicate symbols '>' and '=' which, respectively, denote a given path ordering $\succ$ and a congruence $\equiv$ on ground terms. A *solution* of a constraint C is a ground substitution σ such that $C\sigma$ evaluates to true under the given $\succ$ and $\equiv$. If C has a solution it is called satisfiable. Such constraints have many interesting applications like pruning the search space in automated theorem proving [KKR90,NR95] or deciding the confluence of ordered rewrite systems [CNNR98]. They also provide powerful decidable constraint-based termination orderings $\succ_c$ for term rewriting, defined $s \succ_c t$ if $s\sigma \succ t\sigma$ for all ground σ. If $\succ$ is the lexicographic path ordering (LPO) or the recursive path ordering with status (RPOS), such $\succ_c$ subsume other path orderings like the ones of [KNS85,Les90] since all these path orderings coincide on ground terms (see [Der87]). For example, if s is $g(f(x), f(y))$ and t is $g(g(x, y), g(x, y))$, and $f \succ_{\mathcal{F}} g$ in the precedence, then $s \not\succ_{rpo} t$, but $s \succ_c t$.

The first practical applications of ordering constraints gave rise to the distinction between *fixed signature* semantics (solutions are built over a given signature

---

* Both authors partially supported by the ESPRIT Basic Research WG CCL-II.

$\mathcal{F}$) and *extended signature* semantics (new symbols are allowed to appear in solutions) [NR95]. The satisfiability problem for ordering constraints was first shown decidable for fixed signatures when $\succ$ is a total LPO [Com90] or a total RPOS [JO91]. For extended signatures, decidability was shown for LPO in [NR95] and for RPO in [Nie93]. Regarding complexity, NP algorithms for LPO (fixed and extended signatures) and RPO (extended ones) were given in [Nie93]. Very recently, an NP algorithm has been given as well for RPO under fixed signatures in [NRV98]. NP-hardness of the satisfiability problem is known, even for one single inequation, for all these cases [CT94]. All these decision procedures use at some point the fact that a constraint C can be effectively expressed as an equivalent disjunction of expressions $s_1 > t_1 \wedge \ldots \wedge s_n > t_n$, called *solved forms* in [Com90], where for each i always s_i or t_i is a variable.

In algorithms like the ones of [Com90] and [NR95], the computation of solved forms is only a first step that is followed by other exponential phases. This is not surprising, since this notion of solved form only involves a local analysis of the inequalities considered independently. In fact any constraint $s > t$ can trivially be put into the equivalent (under extended signatures) solved form $s > x \wedge x > t$, for some new variable x. This gives some intuition why this notion of solved form needs to be refined and, in particular, why transitivity through variables needs to be considered.

On the other hand, the NP algorithms of [Nie93] and [NRV98] are not very useful in practice, since they are based on a first very expensive guess of a *simple system* for C, a particular constraint S of the form $s_n \#_n s_{n-1} \#_{n-1} \cdots \#_1 s_0$, where each $\#_i$ is either $=$ or $>$, and $\{s_n, \ldots, s_1\}$ is the set of *all subterms* of C. In [Nie93] it is shown that, roughly, C is satisfiable under extended signatures if, and only if, some simple system contains one of its own solved forms and entails C. This can be checked in polynomial time, but the number of simple systems to be considered is far too large for practical usefulness. For fixed signature semantics in both LPO and RPO, this notion of simple systems is still insufficient and more guesses are needed.

In this paper we introduce some new notions of solved form, where, in addition to the closure under the classical RPOS decomposition rules, a restricted form of transitivity through variables is applied. It is proved that if C is a normal form in this sense, then it is satisfiable under extended signatures if, and only if, it has no cycle (Section 5).

For fixed signatures (Section 6) a slightly different transitivity rule is used. First, several particular cases of signatures are considered for which more efficient methods than the general one apply. The cases depend on whether (1) the smallest non-constant symbol f is unary and (2) there is at most one constant smaller than f. The following table summarizes the results. For instance, if (1) and (2) are true, then satisfiability is again equivalent to the absence of cycles. An entry 0 in the table denotes that, for some variables x, its relation with the smallest constant 0 needs to be guessed, that is, whether $x = 0$ or $x > 0$. Similarly, an ω denotes that for some variables its relation with the smallest limit ordinal term ω has to be guessed.

Results for fixed signatures		
precondition	$f \in lex$	$f \in mul$
1,2	no cycle (Section 6.1)	no cycle (Section 6.1)
1	no cycle, ω (Section 6.3)	no cycle, ω (Section 6.3)
2	no cycle (Section 6.2)	no cycle, ω (Section 6.4)
-	no cycle, ω, 0 (Section 6.4)	no cycle, ω (Section 6.4)

For the cases marked with ω the problem is split into a natural and a non-natural part. The non-natural part is dealt with by cycle detection; the subproblem of natural number constraints, i.e., constraints where f is the only non-constant symbol and all terms and solutions correspond to the natural number fragment, can then be dealt with independently. This problem is solved for the case $f \in lex$ again by a transitivity closure, but now over the natural number ordering. For the case where f is not unary and has multiset status, we rely on the existing methods for solving multiset constraints on natural numbers.

In Section 7 we comment on some implementation issues. As we will show by experimental results from an implementation in the *Saturate* system [GNN95], our methods outperform the best previous one (an improvement of [NR95], as implemented in Saturate) by several orders of magnitude for extended signatures and for fixed signatures fulfilling the requirements (1) and (2). For other fixed signatures, apart from the prohibitive methods guessing linear orderings on all subterms of the constraint, no previous algorithms were known.

2 Path Orderings

Let $\mathcal{F}$ and $\mathcal{X}$ be sets of function symbols and variables respectively, and let $\succ_{\mathcal{F}}$ be a total ordering on $\mathcal{F}$ (the *precedence*). Furthermore let $\mathcal{F}$ be the disjoint union of two sets lex and mul, the symbols with lexicographic and multiset status, respectively, and let $=_{mul}$ denote the equality of terms up to the permutation of direct arguments of symbols with multiset status.

The recursive path ordering (with status) (RPO) on ground terms is defined as follows: $s = f(s_1, \ldots, s_m) \succ_{rpo} g(t_1, \ldots, t_n) = t$ iff:

1. $s_1 \succ_{rpo} t$ or $s_1 =_{mul} t$, for some i with $1 \leq i \leq m$ or
2. $f \succ_{\mathcal{F}} g$, and $s \succ_{rpo} t_j$, for all j with $1 \leq j \leq n$ or
3. $f = g$, $f \in lex$, $\langle s_1, \ldots, s_n \rangle \succ_{rpo}^{lex} \langle t_1, \ldots, t_n \rangle$, and $s \succ_{rpo} t_j$, for all j with $1 \leq j \leq n$
4. $f = g$, $f \in mul$, $\{s_1, \ldots, s_n\} \succ_{rpo}^{mul} \{t_1, \ldots, t_n\}$

where $\langle s_1, \ldots, s_n \rangle \succ_{rpo}^{lex} \langle t_1, \ldots, t_n \rangle$ if $\exists j \leq n$ s.t. $s_j \succ_{rpo} t_j$ and $\forall i < j$ $s_i =_{mul} t_i$. Furthermore, $\succ_{rpo}^{mul}$ is the multiset extension of $\succ_{rpo}$, defined as the smallest ordering such that $S \cup \{s\} \succ_{rpo}^{mul} S' \cup \{t_1, \ldots, t_n\}$ whenever S is equal to S' up to $=_{mul}$ and $s \succ_{rpo} t_i$ for all i in $1 \ldots n$.

The *lexicographic path ordering* is the particular case of RPO where $\mathcal{F} = lex$, and the *multiset path ordering* (or RPO without status) is the particular case where $\mathcal{F} = mul$.

3 Ordering Constraints

An *RPO-ordering constraint* is a quantifier-free first-order formula built over terms in $\mathcal{T}(\mathcal{F}, \mathcal{X})$ and over the binary predicate symbols '>' and '='. A *solution* in $(\mathcal{F}, \succ_{\mathcal{F}})$ of a constraint C is a substitution σ with range $\mathcal{T}(\mathcal{F})$ and whose domain is a set of variables containing the variables of C, such that $C\sigma$ evaluates to true if $>$ and $=$ are interpreted as the RPO defined by $\succ_{rpo}$ and $=_{mul}$ respectively. Then we say that σ satisfies C in $(\mathcal{F}, \succ_{\mathcal{F}})$.

By an *extension* $(\mathcal{F}', \succ_{\mathcal{F}'})$ of $(\mathcal{F}, \succ_{\mathcal{F}})$ we mean a set of function symbols $\mathcal{F}'$ with $\mathcal{F}' \supseteq \mathcal{F}$ and a total precedence $\succ_{\mathcal{F}'}$ extending $\succ_{\mathcal{F}}$. We will call a constraint C satisfiable under *extended signatures* if there exists some extension $(\mathcal{F}', \succ_{\mathcal{F}'})$ of $(\mathcal{F}, \succ_{\mathcal{F}})$ in which C is satisfiable.

4 Solved Forms

An ordering constraint C can be equivalently expressed without negation since $s \not> t$ is equivalent to $t > s \vee t = s$, and $s \neq t$ is equivalent to $t > s \vee s > t$. Then C can be put into disjunctive normal form, and hence satisfiability has to be checked only for conjunctive constraints without negation. In the following we will deal with such conjunctions expressed by sets of equalities and inequalities between terms. The following (non-confluent) term rewrite system R operates on such sets:

$$
\begin{array}{ll}
S \cup \{\, s > f(t_1 \ldots t_n) \,\} \;\longrightarrow\; S \cup \{\, s > t_1, \ldots, s > t_n \,\} \\
\qquad\qquad \text{if } top(s) >_{\mathcal{F}} f \\[4pt]
S \cup \{\, s > t \,\} \;\longrightarrow\; S \cup S' \\
\qquad\qquad \text{if } top(s) = top(t) \in mul \text{ and } S' \in mul(s,t) \\[4pt]
S \cup \{\, f(s_1 \ldots s_n) > f(t_1 \ldots t_n) \,\} \;\longrightarrow\; S \cup \{\, s_1 = t_1, \ldots, s_{i-1} = t_{i-1},\ s_i > t_i \\
\qquad\qquad f(s_1 \ldots s_n) > t_{i+1}, \ldots, f(s_1 \ldots s_n) > t_n \,\} \\
\qquad\qquad \text{for } 1 \le i \le n \text{ if } f \in lex \\[4pt]
S \cup \{\, f(s_1 \ldots s_n) > t \,\} \;\longrightarrow\; S \cup \{\, s_i = t \,\} \\
\qquad\qquad \text{for } 2 \le i \le n \text{ if } top(t) = f \in lex \\[4pt]
S \cup \{\, f(s_1 \ldots s_n) > t \,\} \;\longrightarrow\; S \cup \{\, s_i > t \,\} \\
\qquad\qquad \text{for } 2 \le i \le n \text{ if } top(t) = f \in lex \\[4pt]
S \cup \{\, a > t \,\} \;\longrightarrow\; \bot \\
\qquad\qquad \text{if } a \text{ is a constant and } top(t) \ge_{\mathcal{F}} a \\[4pt]
S \cup \{\, f(s_1 \ldots s_n) > t \,\} \;\longrightarrow\; S \cup \{\, s_i = t \,\} \\
\qquad\qquad \text{for } 1 \le i \le n \text{ if } top(t) >_{\mathcal{F}} f \\[4pt]
S \cup \{\, f(s_1 \ldots s_n) > t \,\} \;\longrightarrow\; S \cup \{\, s_i > t \,\} \\
\qquad\qquad \text{for } 1 \le i \le n \text{ if } top(t) >_{\mathcal{F}} f \\[4pt]
S \cup \{\, s = t \,\} \;\longrightarrow\; S\sigma \\
\qquad\qquad \text{if } \sigma \in mul\_unifiers(s,t) \\[4pt]
S \cup \{\, s = t \,\} \;\longrightarrow\; \bot \\
\qquad\qquad \text{if } mul\_unifiers(s,t) = \emptyset
\end{array}
$$

In R the following notation is used:

1. If s is $f(s_1, \ldots, s_n)$ and t is $f(t_1, \ldots, t_n)$, then $mul(s, t)$ is the set of all constraints of the form
$$\{ s_{\pi(1)} = t_{\rho(1)}, \ldots, s_{\pi(i)} = t_{\rho(i)}, u_{i+1} > t_{\rho(i+1)}, \ldots, u_n > t_{\rho(n)} \}$$
for permutations π and ρ of $1 \ldots n$ and $i \neq n$ and $u_j \in \{s_{\pi(i+1)}, \ldots, s_{\pi(n)}\}$ for all $j \in i+1 \ldots n$. Note that hence the second rule of R illustrates the fact that if $S_1 \succ_{rpo}^{mul} S_2$ then, after removing the proper subset of common (w.r.t. $=_{mul}$) elements on both sides, for each element in S_2 there is a bigger one in S_1.

2. Given two terms s and t, we denote by $mul\_unifiers(s, t)$ the set of all unifiers modulo $=_{mul}$ of s and t. The term $f(t'_{\pi(1)}, \ldots, t'_{\pi(n)})$ is called a *permutation* of $f(t_1, \ldots, t_n)$ if t'_i is a permutation of t_i for $1 \leq i \leq n$ and π is a permutation of $1 \ldots n$ where $\pi = id$ if $f \in lex$. Then $mul\_unifiers(s, t) = \{ \sigma \mid \sigma = mgu(s, t')$ and t' is a permutation of $t \}$.

The termination of R can be shown by a well-founded ordering on sets S based on (i) the number of different variables in S, and (ii) on the multiset of sizes (in number of symbols) of the (in)equalities in S. See, e.g., [Com90] for details for the LPO case of the following lemma:

Lemma 1. *R is terminating, and, for each set of equalities and inequalities S, a normal form of S with respect to R is either $\bot$ or a set of inequalities of the form $s > t$ where at least one of s and t is a variable. Furthermore, S is satisfiable if, and only if, at least one of its normal forms is satisfiable.*

We now introduce our new notion of *solved form*, which, apart from being a normal form with respect to R, is also closed (in some sense) under transitivity through variables:

Definition 1. *A solved form S is a set of inequalities (and hence $S \neq \bot$) of the form $s > t$ where at least one of s and t is a variable, and such that if $s > x \in S$ and $x > t \in S$ for some variable x and non-variable s, then some normal form with respect to R of $\{s > t\}$ is a subset of S.*

Note that deciding whether some normal form of $\{s > t\}$ is a subset of S can be done in polynomial time. It suffices to check whether $s \succ_{rpo}^S t$, where $\succ_{rpo}^S$ is defined as the usual RPO extended with the case $s \succ_{rpo}^S t$ if $s > t \in S$.

Definition 2. *Let S be a set of equalities and inequalities. A solved form S' is called a solved form of S if it contains a normal form with respect to R of S.*

Lemma 2. *Let S be a set of equalities and inequalities. Then S is satisfiable if, and only if, at least one of its solved forms is satisfiable.*

For a given S, its set of solved forms can be computed by rewriting with R and closing under the transitivity rule $s > x \wedge x > t \implies s > t$. Note that this process terminates when repeated work is avoided, since at any stage the sets contain only relations between subterms of S and the number of different such sets that can be obtained is finite. In Section 7 we will comment on some non-trivial implementation issues.

5 Cycles and Satisfiability over Extended Signatures

Definition 3. *A set of inequalities S has a cycle if $\{x > t_1[x_1], x_1 > t_2[x_2], \ldots, x_{n-1} > t_n[x]\} \subseteq S$.*

Lemma 3. *Let S be a set of inequalities with a cycle. Then S is unsatisfiable.*

Definition 4. *Let S be a solved form. We define $>_v$ as the smallest transitive relation on $Vars(S)$ such that $x >_v y$ whenever $x > t[y] \in S$.*

Lemma 4. *Let S be a solved form with no cycles. Then $>_v$ is a well-founded strict partial ordering on $Vars(S)$.*

Lemma 5. *Let S be a solved form with no cycles and let x be a variable occurring in S. Furthermore, let S' be the set obtained by removing from S all inequalities where x occurs, i.e., $S' = \{s > t \mid s > t \in S \wedge x \notin Vars(s > t)\}$. Then S' is a solved form with no cycles.*

Proof. Clearly S' is still in normal form w.r.t. R and has no cycles. We now show that S' is still closed under transitivity. Suppose $s > y \in S'$ and $y > t \in S'$ for some variable $y \neq x$. Then x does not occur in $s > t$. But then the normal form of $s > t$ that belonged to S is still in S' since x does not occur in this normal form (the rules of R do not introduce any variables). □

Definition 5. *Let S be a solved form with no cycles, and let $(\mathcal{F}', \succ_{\mathcal{F}'})$ be the extension of $(\mathcal{F}, \succ_{\mathcal{F}})$ where $\mathcal{F}' = \mathcal{F} \cup \{f, 0\}$ and where $\succ_{\mathcal{F}'}$ extends $\succ_{\mathcal{F}}$ such that $g \succ_{\mathcal{F}'} f \succ_{\mathcal{F}'} 0$ for all $g \in \mathcal{F}$.*

The minimal substitution σ for S is defined by induction on $>_v$ as follows. Let $x \in Vars(S)$ and let σ_x be the partial substitution defined for the variables of S that are smaller than x w.r.t. $>_v$. Then

- *$x\sigma = 0$ if there is no inequation $x > t$ in S and*
- *$x\sigma = f(t)$ if $t = max_{rpo}\{s\sigma_x \mid x > s \in S\}$.*

Theorem 1. *Let S be a solved form over $(\mathcal{F}, \succ_{\mathcal{F}})$. Then S is satisfiable over some extension of $(\mathcal{F}, \succ_{\mathcal{F}})$ if, and only if, S has no cycle.*

Proof. By Lemma 3, S is unsatisfiable if it has some cycle. Now we show that if S has no cycle the minimal substitution σ in the extension $(\mathcal{F}', \succ_{\mathcal{F}'})$ given in Definition 5 is a solution of S.

We proceed by induction on the number k of variables in S. If $k = 0$, then S is empty and trivially satisfiable. For the induction step, let x be a variable that is maximal w.r.t. $>_v$ in S, and let S' be the conjunctive constraint obtained by removing from S all inequalities where x occurs, i.e., $S' = \{s > t \mid s > t \in S \wedge x \notin Vars(s > t)\}$. Then by Lemma 5 S' is a solved form with no cycles.

Now let σ' be σ with its domain restricted to $Vars(S) \setminus \{x\}$. Then σ' is the minimal substitution of S' and hence, by the induction hypothesis, σ' is a solution for S'.

To prove that σ is indeed a solution of S, it remains to be checked that $s\sigma \succ_{rpo} t\sigma$ for the relations $s > t$ where x appears in s or t.

There are six cases:

1. $x > t[x]$. No such inequation exists since this is a cycle.
2. $t[x]_p > x$ with $p \neq \lambda$. Then clearly $t\sigma[x\sigma]_p \succ_{rpo} x\sigma$ by the subterm property of RPO.
3. $y > t[x]$. No such inequation exists since x is maximal.
4. $x > t$ where $x \notin Vars(t)$. Then, since $x\sigma = f(t)$ where $t = max_{rpo}\{s\sigma' \mid x > s \in S\}$, clearly $x\sigma \succ_{rpo} t\sigma$.
5. $s > x$ where $x \notin Vars(s)$ and s is not a variable.
   If there is no inequation $x > t$ in S, then $x\sigma = 0$, and hence $s\sigma \succ_{rpo} x\sigma$.
   If there is at least one inequation $x > t$ in S, then $x\sigma = f(u)$ where $u = max_{rpo}\{t\sigma' \mid x > t \in S\}$. By definition of solved form, then S contains a normal form with respect to R of $s > t$ for all $x > t \in S$. Since $x \notin Vars(s > t)$, in fact S' contains this solved form and hence $s\sigma' \succ_{rpo} t\sigma'$, and $s\sigma \succ_{rpo} t\sigma$, and, in particular, $s\sigma \succ_{rpo} u$. Since $top(s) \succ_{\mathcal{F}} f$, this implies $s\sigma \succ_{rpo} f(u)$ and $s\sigma \succ_{rpo} x\sigma$.
6. $s[x]_p > y$ where $p \neq \lambda$ and $x \neq y$. We proceed by contradiction. Let $s > y$ be an inequality in S of this form such that $s\sigma \not\succ_{rpo} y\sigma$ and where y is minimal in $>_v$. Clearly $y\sigma \neq 0$; hence $y\sigma = f(u)$ where $u = max_{rpo}\{t\sigma \mid y > t \in S\}$. We now show that $s\sigma \succ_{rpo} u$, contradicting $s\sigma \not\succ_{rpo} y\sigma$, since $top(s) \in \mathcal{F}$ and hence $top(s) \succ_{\mathcal{F}'} f$.
   By definition of solved form, S contains a normal form with respect to R of $s > t$ for all $y > t \in S$. For the inequalities in this solved form, either x does not occur in them, or else they fall under one of the five previous cases, and hence σ is a solution for them, or they have the form $s|_q > z$ for some variable z with $y >_v z$, and hence σ is a solution for them by the minimality assumption for y. Hence σ is a solution of this solved form and hence $s\sigma \succ_{rpo} u$. □

6 Fixed Signatures

Let f be the smallest non-constant symbol in $\mathcal{F}$, and let 0 be the smallest constant symbol (and hence the smallest term). In this section we will first consider two restrictions:

1. there is at most one constant symbol smaller than f, and
2. f is unary

If $(\mathcal{F}, \succ_{\mathcal{F}})$ satisfies both restrictions, it will be called *well-ended*. In several further subsections we will show how these restrictions can be dropped at the expense of adding some new rules to R.

6.1 Well-ended Signatures

Note that in this setting for every ground term t, its *successor*, the smallest term bigger than t, is $f(t)$. We sometimes write $f^n(t)$ to denote the n-th successor of t.

Example 1. Consider the constraint S of the form $f(f(0)) > x \wedge f(x) > y > z > 0$. It is a solved form in the sense of the previous section, and it has no cycle. However, it is unsatisfiable over fixed signatures, since it amounts to diophantine inequations over the natural numbers: we have $2 > x \wedge x + 1 > y > z > 0$. and there is no space for both y and z between 2 and 0.

The previous example shows us that we need to reconsider the definition of solved form, since the notion of closure under transitivity used in the previous section is too weak for fixed signatures.

Definition 6. *A solved form S is a set of inequalities of the form $s > t$ where at least one of s and t is a variable, and such that if $s > x \in S$ and $x > t \in S$ for some non-variable s and variable x, then some normal form with respect to R of $\{s > f(t)\}$ is a subset of S.*

Example 2. (Example 1 continued) Closing S under the new notion of transitivity gives the following. From $f(x) > y > z$ we get $f(x) > f(z)$, which simplifies into $x > z$. This, together with $f(f(0)) > x$ gives $f(f(0)) > f(z)$ which simplifies into $f(0) > z$. This, together with $z > 0$ gives $f(0) > f(0)$ whose unique normal form is $\perp$.

Clearly, like in the previous section, a set of (in)equalities S is satisfiable if, and only if, one of its solved forms in this new sense is satisfiable.

Theorem 2. *Let $\mathcal{F}$ be a well-ended signature, and let S be a solved form (in the sense of Definition 6) over $(\mathcal{F}, \succ_{\mathcal{F}})$. Then S is satisfiable in $(\mathcal{F}, \succ_{\mathcal{F}})$ if, and only if, S has no cycles and $0 > x \notin S$.*

Proof. The proof follows the same ideas as the one of Theorem 1. Note that Lemma 5 still applies for the new notion of solved form. Of the six cases, the first four ones remain equal. The remaining two cases become:

5. $s > x$ where $x \notin Vars(s)$ and s is not a variable.
   If there is no inequation $x > t$ in S, then $x\sigma = 0$, and hence $s\sigma \succ_{rpo} x\sigma$, since $0 > x$ is not in S.
   If there is at least one inequation $x > t$ in S, then $x\sigma = f(u)$ where $u = max_{rpo}\{t\sigma' \mid x > t \in S\}$. By the new definition of solved form, then S contains a normal form with respect to R of $s > f(t)$ for all $x > t \in S$ and we conclude as in Theorem 1.
6. $s[x]_p > y$ where $p \neq \lambda$ and $x \neq y$. We proceed by contradiction. Let $s > y$ be an inequality in S of this form such that $s\sigma \nsucc_{rpo} y\sigma$ and where y is minimal in $>_v$. Clearly $y\sigma \neq 0$, since s is a non-constant term and for any such terms $s\sigma \succ_{rpo} 0$. Hence $y\sigma = f(u)$ where $u = max_{rpo}\{t\sigma \mid y > t \in S\}$. We now show that $s\sigma \succ_{rpo} f(u)$, contradicting $s\sigma \nsucc_{rpo} y\sigma$. By the new definition of solved form, S contains a normal form with respect to R of $s > f(t)$ for all $y > t \in S$, and we can conclude as in Theorem 1. □

6.2 At Most One Constant below a Non-unary f ∈ lex

If f is non-unary and lexicographic, and there is at most one constant below f, then the successor of every ground term t is $f(0, \ldots, 0, t)$. And then the satisfiability of solved forms can be decided by cycle detection as in the previous subsection, if everywhere $f(t)$ is replaced by $f(0, \ldots, 0, t)$: if $s > x \in S$ and $x > t \in S$ for some non-variable s, a variable x, and term t, then now it is required that some normal form with respect to R of $\{s > f(0, \ldots, 0, t)\}$ is a subset of S. Similarly, the minimal substitution σ for a solved form with no cycles S is now defined by taking successors like this: $x\sigma = f(0, \ldots, 0, t)$ if $t = max_{rpo}\{s\sigma_x \mid x > s \in S\}$. From this we get the following:

Theorem 3. *Let $(\mathcal{F}, \succ_{\mathcal{F}})$ be such that the smallest non-constant symbol f is in lex, and there is at most one constant symbol smaller than f. Let S be a solved form over $(\mathcal{F}, \succ_{\mathcal{F}})$. Then S is satisfiable in $(\mathcal{F}, \succ_{\mathcal{F}})$ if, and only if, S has no cycles and $0 > x \notin S$.*

6.3 More than One Constant below a Unary f

Assume the signature is ended by $g \succ_{\mathcal{F}} f \succ_{\mathcal{F}} a_1 \succ_{\mathcal{F}} \ldots \succ_{\mathcal{F}} a_n \succ_{\mathcal{F}} 0$, with $n > 0$ and where the smallest non-constant symbol f is unary. Then 0 is the smallest term, but no longer for every ground term t, its successor is $f(t)$. We have the following increasing sequence of ground terms:

$$0, \ a_n, \ \ldots, \ a_1, \ f(0), \ f(a_n), \ \ldots, \ f(a_1), \ f(f(0)), \ \ldots, \ g(0, \ldots, 0) = \omega, \ f(\omega), \ldots$$

where ω is the first limit ordinal term, that is, the smallest term ω such that $w \succ_{rpo} t$ for infinitely many terms t. Terms below ω are called *natural terms*. Clearly the transitivity notion $s > x \wedge x > t \implies s > f(t)$ applied in the Subsection 6.1 is now correct only if t is known to be non-natural. In order to know whether t is natural or not at the non-ground level, we now add three new rules to the rewrite system R that *guess* for each variable how it is related to ω. Let R_ω be the set of the following three rules:

$$
\begin{array}{lll}
S & \longrightarrow & S\{x \mapsto \omega\} \quad & \text{if } x \in vars(S) \text{ and } x > \omega \notin S \text{ and } \omega > x \notin S \\
S & \longrightarrow & S \cup \{\, x > \omega \,\} \quad & \text{if } x \in vars(S) \text{ and } x > \omega \notin S \text{ and } \omega > x \notin S \\
S & \longrightarrow & S \cup \{\, \omega > x \,\} \quad & \text{if } x \in vars(S) \text{ and } x > \omega \notin S \text{ and } \omega > x \notin S
\end{array}
$$

Now it is clear for every normal form S with respect to R_ω whether a term t with variables is natural or not: t is natural if, and only if, it contains only symbols smaller than g and variables x for which $\omega > x \in S$. Note that in practice it is not necessary to guess the relations with ω for *all* variables, as long as for all sides of an inequality it is known whether they are natural or not. For example, t is known to be natural if it contains no symbol greater than f, or $s > t \in S$ for some natural term s, etc.

Now we can adapt the notion of solved form to include transitivity only for non-natural terms, and again a set of (in)equalities S is satisfiable if, and only if, one of its solved forms in this new sense is satisfiable.

Definition 7. *A solved form S is a set of inequalities in normal form w.r.t.
$R \cup R_\omega$, such that if $s > x \in S$ and $x > t \in S$ for some non-variable term s, a
variable x and a non-natural term t, then some normal form with respect to R
of $\{s > f(t)\}$ is a subset of S.*

We now split the satisfiability problem for a solved form S in two parts:
the *natural part* S_N, and the *non-natural* one S_ω, i.e., $S_N = \{s > t \mid s >
t \in S$ and s and t are natural $\}$ and $S_\omega = S \setminus S_N$. The satisfiability of S_N can
be easily decided. It suffices to express all ground terms as their corresponding
natural number and terms $f^k(x)$ as $x + k * (n+1)$. In fact, the resulting problem of
satisfiability of diophantine inequations is in P, since it can be solved by closure
under transitivity. Note that this is precisely what was done in the previous
section at the symbolic level (and hence we will not prove its correctness again
here): close under the rule $x + k > y \wedge y > z + k' \implies x + k > z + k' + 1$,
simplifying the conclusion to get a variable at one of both sides; if no cycle or
contradiction of the form $n > n + k$ appears, then the problem is satisfiable.

Lemma 6. *Let $(\mathcal{F}, \succ_\mathcal{F})$ be such that $g \succ_\mathcal{F} f \succ_\mathcal{F} a_1 \succ_\mathcal{F} \ldots \succ_\mathcal{F} a_n \succ_\mathcal{F} 0$, where
the smallest non-constant symbol f is unary. Let S be a solved form (in the sense
of Definition 7) S over $(\mathcal{F}, \succ_\mathcal{F})$. Then $s > t \in S$ for no s that is natural and t
that is non-natural.*

Proof. Assume such an $s > t$ is in S.

If s is not a variable, then it is $s > x$ where $x > \omega \in S$. Then, by transitivity,
a solved form of $s > f(\omega)$ is in S. If s is ground, this leads to $\perp$ by R, and then
S would be $\perp$, contradicting $s > t \in S$. If s is non-ground, it is of the form $f^n(y)$
for some $n > 0$, where $\omega > y \in S$, and the unique normal form of $s > f(\omega)$ is
$y > \omega$, which leads to $\perp$ by transitivity from $\omega > y$.

If s is a variable x, then $\omega > x \in S$. Then, by transitivity, a solved form
of $\omega > f(t)$ is in S. If t is ground, this leads to $\perp$. If t is non-ground, it either
leads to $\perp$ or to a solved form where $\omega|_p > y$ for some variable y in t such that
$y > \omega \in S$, which leads to $\perp$ as before. □

Theorem 4. *Let $(\mathcal{F}, \succ_\mathcal{F})$ be such that $g \succ_\mathcal{F} f \succ_\mathcal{F} a_1 \succ_\mathcal{F} \ldots \succ_\mathcal{F} a_n \succ_\mathcal{F} 0$,
where the smallest non-constant symbol f is unary. Let S be a solved form (in
the sense of Definition 7) over $(\mathcal{F}, \succ_\mathcal{F})$. Then S is satisfiable in $(\mathcal{F}, \succ_\mathcal{F})$ if, and
only if, S_N is satisfiable and S has no cycles.*

Proof. If S is satisfiable then S_N is also satisfiable and S has no cycles. Now
assume that S_N is satisfiable and S has no cycles. Let σ_N a solution of S_N, and
we will show that it can be extended to a solution σ for the whole S, by building
the minimal solution σ for S_ω, starting from σ_N.

First, note that for the inequalities $s > t$ where s is non-natural and t is
natural, we trivially have $s\sigma \succ_{rpo} t\sigma$, since $t\sigma$ is natural, and $s\sigma$ will be non-
natural: since s either contains some symbol g with $g \succ_\mathcal{F} f$, or else some non-
natural variable x and $x\sigma \succ_{rpo} \omega$ for all such x. Hence by Lemma 6, it only
remains to check the inequalities between non-natural terms.

Note that Lemma 5 still applies for the new notion of solved form. Of the six cases of Theorem 1, again the first four ones remain equal. The last two cases are:

5. $s > x$ where $x \notin \mathcal{V}ars(s)$ and s is not a variable.

   There is at least one inequation $x > t$ in S, since $x > \omega \in S$. Then $x\sigma = f(u)$ where $u = max_{rpo}\{t\sigma' \mid x > t \in S\}$. By the new definition of solved form, then S contains a normal form with respect to R of $s > f(t)$ for all $x > t \in S$ and we conclude as in Theorem 1.

6. $s[x]_p > y$ where $p \neq \lambda$ and $x \neq y$. We proceed by contradiction. Let $s > y$ be an inequality in S of this form such that $s\sigma \not\succ_{rpo} y\sigma$ and where y is minimal in $>_v$. Since y is non-natural, $y\sigma = f(u)$ where $u = max_{rpo}\{t\sigma \mid y > t \in S\}$, and $s\sigma \succ_{rpo} f(u)$ follows as in Theorem 2. □

6.4 When the Smallest Non-constant Symbol f Is Non-unary

We now also eliminate the restriction that f is unary. We will continue with the same methodology as before, using the rules of R_ω that guess for each variable its relation to ω, and splitting the solved forms into the two independent parts: the natural and non-natural ones.

Now different approaches are needed depending on whether f has multiset or lexicographic status.

The Multiset Case If $f \in mul$, then clearly a term t is natural if, and only if, it is built from the smallest non-constant symbol f, constants smaller than f and natural variables (i.e., variables x with $\omega > x \in S$). In this sense, the multiset case is simpler than the lexicographic one, as we will see. E.g., if f is binary, and $f \succ_{\mathcal{F}} a_1 \succ_{\mathcal{F}} \ldots \succ_{\mathcal{F}} a_n \succ_{\mathcal{F}} 0$, we have the following increasing sequence of natural ground terms:

$$0, \quad a_n, \quad \ldots, \quad a_1, \quad f(0,0), \quad f(0,a_n), \quad f(a_n,a_n), \quad f(0,a_{n-1}), \quad f(a_n,a_{n-1}),$$
$$f(a_{n-1},a_{n-1}), \quad f(0,a_{n-2}), \quad \ldots, \quad f(a_1,a_1), \quad f(0,f(0,0)),\ldots$$

and if there is at most one constant below f, we have

$$0, \quad f(0,0), \quad f(0,f(0,0)), \quad f(f(0,0),f(0,0)),$$
$$f(0,f(0,f(0,0))), \quad f(f(0,0),f(0,f(0,0))), \quad \ldots$$

Here 0 is still the smallest term, and for every non-natural ground term t, its successor is $f(0,\ldots,0,t)$. The smallest non-natural term ω is $g(0,\ldots,0)$, where g is the smallest symbol bigger than f and than 0. Solving the non-natural part can hence be done as for the case of unary f, if everywhere $f(t)$ is replaced by $f(0,\ldots,0,t)$.

The definition of solved form again considers $R \cup R_\omega$, and if $s > x \in S$ and $x > t \in S$ for some non-variable s, a variable x, and a non-natural t, then now it is required that some normal form with respect to R of $\{s > f(0,\ldots,0,t)\}$ is a subset of S. Similarly, the minimal substitution σ for a solved form with no cycles S is now defined by taking successors like this: $x\sigma = f(0,\ldots,0,t)$ if $t = max_{rpo}\{s\sigma_x \mid x > s \in S\}$. From this we get the following result:

Theorem 5. *Let $(\mathcal{F}, \succ_{\mathcal{F}})$ be such that the smallest non-constant symbol f is in mul. Let S be a solved form over $(\mathcal{F}, \succ_{\mathcal{F}})$. Then S is satisfiable in $(\mathcal{F}, \succ_{\mathcal{F}})$ if, and only if, S_N is satisfiable, and S has no cycles.*

Deciding whether S_N is satisfiable amounts to solving purely natural RPO constraints, that is, constraints built only over f, 0, and possibly other constants smaller than f, and with solutions over this same signature. If $f \in lex$ this is a simple problem over the natural numbers, but for $f \in mul$ this seems not the case. Hence for the moment we propose to use the algorithm of [JO91] or the NP one of [NRV98] for S_N, which is normally a minor part of S.

The Lexicographic Case Let $(\mathcal{F}, \succ_{\mathcal{F}})$ be such that $f \succ_{\mathcal{F}} a_1 \succ_{\mathcal{F}} \ldots \succ_{\mathcal{F}} a_n \succ_{\mathcal{F}} 0$, where the smallest non-constant symbol f is in lex (the case with at most one constant below f has been treated already in Subsection 6.2). Then a term like $f(x, 0)$ can have non-natural instances even if x is instantiated with a natural term: e.g., $f(a_n, 0)$ is precisely ω, the first limit ordinal. If f is binary, we have:

$$0, \quad a_n, \quad \ldots, \quad a_1, \quad f(0,0), \quad f(0,a_n), \quad \ldots, \quad f(0,a_1), \quad f(0,f(0,0)),$$

$$f(0, f(0, a_n)), \quad \ldots, \quad f(0, f(0, f(0,0))), \quad \ldots, \quad f(a_n, 0) \ldots$$

Therefore, in order to split the constraint into its natural and non-natural parts, we need not only to know the relation between some variables x and ω, but also whether x is 0 or not. Hence the three additional rules of R given in the previous subsection become now the following four ones, which we will call $R_{\omega,0}$:

S	$\longrightarrow$ $S\{x \mapsto 0\}$	if $x \in vars(S)$
S	$\longrightarrow$ $S\{x \mapsto \omega\}$	if $x \in vars(S)$
S	$\longrightarrow$ $S \cup \{x > \omega\}$	if $x \in vars(S)$ and $x > \omega \notin S$
S	$\longrightarrow$ $S \cup \{\omega > x, x > 0\}$	if $x \in vars(S)$ and $\{\omega > x, x > 0\} \not\subseteq S$

As for the remaining rules of R given before, sometimes the application of a rule will produce a reduction to $\bot$, and in many case these situations can be foreseen. Also, guessing whether a variable x is 0 is in fact needed only if x appears in some term t that is otherwise unknown to be natural or not. Similar efficiency issues will be discussed later on.

For normal forms with respect to $R_{\omega,0}$, clearly a term is natural if, and only if, it is of the form $f(0, \ldots, 0, f(0, \ldots, 0, f(\ldots f(0, \ldots, 0, t))))$ where t is a constant smaller than f or a natural variable (i.e., a variable x with $\omega > x \in S$), and for every non-natural ground term t, its successor is $f(0, \ldots, 0, t)$. Again the results of the previous section go through if everywhere $f(t)$ is replaced by $f(0, \ldots, 0, t)$, and we get the following:

Theorem 6. *Let $(\mathcal{F}, \succ_{\mathcal{F}})$ be such that the smallest non-constant symbol f is in lex. Let S be a solved form over $(\mathcal{F}, \succ_{\mathcal{F}})$. Then S is satisfiable in $(\mathcal{F}, \succ_{\mathcal{F}})$ if, and only if, S_N is satisfiable and S has no cycles.*

And here again the satisfiability of the natural part S_N can be decided in polynomial time by translation into diophantine inequations of the form $x + k > y$ or $x > y + k$, which can be handled by transitive closure as in the previous subsection.

7 Implementation Issues

Let us first consider a practical improvement obtained by using '$\geq$' as an additional predicate in constraints. This importantly reduces the number of solved forms under consideration. For example, the two rules of R for the case $top(t) >_{\mathcal{F}} f$, can now be covered by the single one $S \cup \{ f(s_1, \ldots, s_n) > t \} \longrightarrow S \cup \{ s_i \geq t \}$. Additional rules for decomposing atoms $s \geq t$ are defined analogously to the ones for '$>$'. If a new rule $S \cup \{ s \geq t, t \geq s \} \longrightarrow S \cup \{ s = t \}$ is added as well, and the notions of cycle and transitivity, etc., are defined as expected, it is not difficult to check that all results of the previous sections go through.

7.1 Computing Solved Forms

We have seen that the constraint satisfiability problem roughly amounts to deciding the existence of a solved form with no cycle. Solved forms can be obtained by computing normal forms with respect to R, adding new inequations by application of the transitivity rule, which in turn have to be turned into normal form, etc. In order to avoid repeated work, standard methods from theorem proving or completion for closing under inference rules can, and have to be, used.

In our implementation we deal with three sets of inequalities: the *old* ones, that are in normal form and closed under transitivity, the *new* ones, that are in normal form, and an additional set T with the recently added consequences under transitivity. Initially *old* and *new* are empty, and T contains the set of input inequalities.

The working cycle consists of the following. *One* normal form w.r.t. R of each non-redundant element $s > t$ in T is added to *new*, where $s > t$ is redundant if some normal form of it is a subset of *old* $\cup$ *new*. This can be checked in polynomial time by using $s \succ_{rpo}^{S} t$, where $\succ_{rpo}^{S}$ is defined as the usual RPO extended with the case $s \succ_{rpo}^{S} t$ if $s > t \in$ *old* $\cup$ *new*. Backtracking on the choice of normal form of $s > t$ occurs when a cycle is detected at some point. If T is empty, one inequality $u > v$ of *new* is moved to *old* and all transitivity consequences between $u > v$ and *old* are put in T. If both T and *new* are empty, and there is no cycle in *old* then the constraint is satisfiable.

7.2 Some First Practical Experiments

We experimented with a Prolog implementation based on the aforementioned procedure. In order to obtain objective problem sets, we ran *Saturate* on 10 problems in first-order theorem proving. For each problem, we kept the

set of all ordering constraints generated during the run. It turned out that for only three of the ten problems, on rings, abelian groups, and embedding, respectively, a statistically significant number of non-trivial constraints were generated. In the tables below we show the results for these three problem sets comparing our *New* method with the previously best known one *Old* (an improvement of [NR95]) as it was (quite carefully) implemented in the *Saturate* system, for extended signatures. Times are in miliseconds for Sicstus Prolog 3.7.1 on a SUN Ultra 5. The problem sets and test program are available from http://www.lsi.upc.es/~roberto. For well-ended fixed signatures very similar results are obtained. For other fixed signatures, apart from the prohibitive methods guessing linear orderings on all subterms of the constraint, no previous algorithms were known.

The leftmost column *Threshold* indicates the minimum time in miliseconds required for considering a problem. For instance, the first row considers *all* problems of the set, the second row only the ones where at least one of both algorithms takes 20 ms or more, etc. Note that for harder problems the improvement ratio is higher:

Constraints from problem Rings				
Threshold	# Problems	Total Time Old	Total Time New	Improvement Ratio
0	977	62230	1350	46.10
20	370	59980	700	85.69
50	186	55410	550	100.75
100	130	52050	490	106.22
200	64	41510	280	148.25
400	26	31450	100	314.50
1000	14	25200	50	504.00

Constraints from problem Abelian Groups				
Threshold	# Problems	Total Time Old	Total Time New	Improvement Ratio
0	246	35520	590	60.20
20	99	35010	390	89.77
50	80	34480	360	95.78
100	64	33560	330	101.70
200	35	29500	260	113.46
400	20	25560	170	150.35
1000	12	21680	120	180.67

Constraints from problem Embedding				
Threshold	# Problems	Total Time Old	Total Time New	Improvement Ratio
0	814	78000	1170	66.67
20	349	76470	500	152.94
50	185	72290	180	401.61
100	117	67660	120	563.83
200	53	59080	50	1181.60
400	27	52120	30	1737.33
1000	13	40960	10	4096.00

8 Conclusions and Further Work

We have shown that, for an adequate notion of solved form, simply based on RPO decomposition and transitivity, deciding the satisfiability of path ordering constraints roughly amounts to solved form computation and cycle detection.

This leads to new algorithms that, we believe, are currently the best choice for path ordering constraint solving under all possible precedences and semantics.

Although it is not very relevant from the practical point of view, it seems quite clear that, when more carefully formulated, our algorithms can be shown to be in NP. First one guesses a rewrite derivation with R (and R_ω or $R_{\omega,0}$ if needed) into a normal form. While doing this, in order to avoid the creation of terms of exponential size, a different treatment for the equality relation is needed (see [NRV98]). For the cases where S is split into S_N and S_ω, at some point also the satisfiability of S_N has to be checked (which is in P if $f \in lex$, and requires to apply the NP algorithm of [NRV98] if $f \in mul$).

We believe that more practical algorithms can be found for purely natural multiset constraints; this is also the subject of further work.

References

[CNNR98] Hubert Comon, Paliath Narendran, Robert Nieuwenhuis, and Michael Rusi-nowitch. Decision problems in ordered rewriting. In *13th IEEE Symp. Logic in Comp. Sc. (LICS)*, pages 410–422, Indianapolis, USA, 1998.

[Com90] Hubert Comon. Solving symbolic ordering constraints. *International Journal of Foundations of Computer Science*, 1(4):387–411, 1990.

[CT94] Hubert Comon and Ralf Treinen. Ordering Constraints on Trees. In *Proc. CAAP*, LNCS 787, Edinburgh, Scotland, Springer-Verlag.

[Der87] Nachum Dershowitz. Termination of rewriting. *Journal of Symbolic Computation*, 3:69–116, 1987.

[GNN95] Harald Ganzinger, Robert Nieuwenhuis, and Pilar Nivela. The Saturate System, 1995. (see www.mpi-sb.mpg.de/SATURATE/Saturate.html).

[JO91] J-P. Jouannaud and M. Okada. Satisfiability of systems of ordinal notations with the subterm property is decidable. In Proc *18th ICALP*, LNCS 510, Madrid, Spain, July 16–20 1991. Springer-Verlag.

[KKR90] Claude Kirchner, Hélène Kirchner, and Michaël Rusinowitch. Deduction with symbolic constraints. *Revue Française d'Intelligence Artificielle*, 4(3):9–52, 1990.

[KNS85] Deepak Kapur, Paliath Narendran, and G. Sivakumar. A path ordering for proving termination for term rewriting systems. In *Proc. of 10th ICALP*, LNCS 185, pages 173–185, Germany, 1985. Springer-Verlag.

[Les90] Pierre Lescanne. On the recursive decomposition ordering with lexicographical status and other related orderings. *J. Aut. Reasoning*, 6(1):39–49, 1990.

[Nie93] Robert Nieuwenhuis. Simple LPO constraint solving methods. *Information Processing Letters*, 47:65–69, August 1993.

[NR95] Robert Nieuwenhuis and Albert Rubio. Theorem Proving with Ordering and Equality Constrained Clauses. *J. Symbolic Comp.*, 19(4):321–351, 1995.

[NRV98] P. Narendran, M. Rusinowitch, and R. Verma. RPO constraint solving is in NP. In *CSL 98*, Brno, Czech Republic, August 23–28, 1998. Abstract at http://www.dbai.tuwien.ac.at/CSL98.

Jeopardy

Nachum Dershowitz[1] and Subrata Mitra[2]

[1] Department of Computer Science
Tel-Aviv University
Ramat Aviv, Tel-Aviv 69978, Israel
email: nachumd@cs.tau.ac.il
[2] Enterprise Component Technology
162 36th "A" Cross
3rd Main, 7th Block
Jayanagar, Bangalore 560 082, India

> Jeopardy? Isn't that a game show?
> —Faye Kellerman, *Prayers for the Dead*

Abstract. We consider functions defined by ground-convergent left-linear rewrite systems. By restricting the depth of left sides and disallowing defined symbols at the top of right sides, we obtain an algorithm for function inversion.

1 Motivation

It is thought that some ancient cultures employed a solar calendar with a fixed year-length of 360 days and a simple scheme of 12 equal-length months. Imagine a 0-based version of such a calendar. Given a date $\langle d, m, y \rangle$ consisting of a year number y, month number m and day number d, it is trivial to calculate the number of elapsed days since the onset of the calendar on date $\langle 0, 0, 0 \rangle$:

$$n(d, m, y) = 360 \times y + 30 \times m + d \qquad (1)$$

To facilitate conversion of dates between calendars (see [Dershowitz and Reingold, 1997]), one also needs to compute the inverse of n to find the date $\langle d, m, y \rangle$ corresponding to a given number of elapsed days N. The appropriate function is not all that trivial:

$$n^-(N) = \langle N \bmod 30, \lfloor (N \bmod 360)/30 \rfloor, \lfloor N/360 \rfloor \rangle \qquad (2)$$

The ideal of logic programming suggests that one should only need to specify the function n and simply let the programming language do the "dirty work" and solve for $\langle d, m, y \rangle$, given any N. That is, we want the machine to determine the appropriate question for a given answer, as in the popular game "Jeopardy". Narrowing (or any other complete semantic-unification procedure), given a goal $n(d, m, y) =^? 400$ (and some appropriate definitions of operations on natural numbers) would yield the sought-after solution $d = 10$, $m = 1$, $y = 1$. Unfortunately, it would also find numerous undesirable solutions, such as $d = 40$, $m = 12$, $y = 0$. Worse, unadulterated narrowing will continue forever seeking additional, nonexistent solutions for $y > 1$. To eliminate the undesired "solutions", one would have to add the missing parts of the specification in the form of constraints:

$$n(d, m, y) =^? N, \; d < 30 =^? T, \; m < 12 =^? T$$

To preclude nontermination, one could add "failure-causing" rules:

$$
\begin{aligned}
s(x) =^? s(y) &\to x =^? y \\
s(x) =^? 0 &\to F \\
0 =^? s(y) &\to F
\end{aligned}
\tag{3}
$$

Applied eagerly (as suggested in [Dershowitz and Plaisted, 1988]), these rules prune unsatisfiable inversion goals.

Thus, we are interested in the problem of solving sets of equations of the form $t = N$, where t is an arbitrary term containing defined function symbols, constructors (that is, undefined function symbols and constants), and variables, while N is a *value*, by which we mean a term containing constructors only, without defined symbols or variables. We will describe a broad class of functions that can be inverted in this manner. Failure rules, like (3), which hold in general for constructors in convergent systems, are built into the algorithm.

More generally, semantic matching is the process of generating a basis set of substitutions that, when applied to a "pattern" term, gives a term equal (in some theory) to a given "target" term. In other words, matching is the special ("one-way") case of (semantic) unification in which one of the two terms to be unified is ground (variable-free). Matching algorithms are required for pattern application in functional languages and have potential uses in logic-based languages. For example, given the usual definitions for *append* and *reverse* on lists, it is natural to implicitly define a predicate for checking if a list is a palindrome in the following manner:

$$
\begin{aligned}
palindrome(append(x, reverse(x))) &= T \\
palindrome(append(x, a : reverse(x))) &= T
\end{aligned}
\tag{4}
$$

To use such definitions within a functional pattern-directed language, it is necessary to match patterns of the form $append(x, reverse(x))$ against values like 1: 2: 2: 1: ϵ. To perform such matches—which is not possible in current functional languages—an inversion algorithm is required.

We restrict ourselves to equational theories that are presented as (rather typical) functional programs in the form of ground-convergent left-linear rewrite systems. Though for arbitrary linear systems, matching is unsolvable, by placing (not wholly unrealistic) syntactic restrictions on the sides of rules (left sides are restricted in depth and right sides may not have arbitrary defined symbols at the top), we can show termination of our inversion algorithm. We do not actually require sufficient completeness (only convergence), so some ground terms may have non-constructor normal forms.

For calendar computation from scratch, we also need a program for (unary) "natural arithmetic", such as

$$
\begin{aligned}
x + 0 &\to x & s(x) < s(y) &\to x < y \\
x + s(y) &\to s(x + y) & 0 < s(y) &\to T \\
& & 0 < 0 &\to F \\
x \times 0 &\to 0 & s(x) < 0 &\to F \\
x \times s(y) &\to (x \times y) + x
\end{aligned}
\tag{5}
$$

Though this program does not meet our criteria, it can be massaged into shape; see Section 8.

On the other hand, the following (somewhat peculiar) gound-convergent version (with standard abbreviating conventions) of multiplication does satisfy the requirements we impose for invertibility:

$$
\begin{aligned}
x \times 0 &\to 0 & & \\
0 \times x &\to 0 & x + 0 &\to x \\
sx \times sy &\to s(x \times sy + y) & x + sy &\to s(x + y)
\end{aligned}
\tag{6}
$$

Together with a definition of squaring:

$$
x^2 \to x \times x
\tag{7}
$$

it allows us to compute square-roots by solving goals like $x^2 \to^? s^{100}0$.

In Section 6, we give an algorithm for inversion when certain syntactic conditions, enumerated in Section 5, are fulfilled. The algorithm, the correctness of which is proved in Section 7, is based on the more generally valid goal transformation rules of Section 4. From the theoretical point of view, we are interested in probing the borderline between decidability and undecidability, so the necessity of the conditions is also shown. Prior work is summarized in Section 3 and future work is suggested in Section 8. First some preliminaries.

2 Nomenclature

We use standard notation and terminology for concepts in rewriting [Dershowitz and Jouannaud, 1990]. In particular, $s \to^! t$ means that the (first-order) term s rewrites (in zero or more steps using the system under question) to the normal-form (i.e. unrewritable term) t. For our purposes, any function symbol or constant that appears at the root of a left side of a rule is *defined*, while all others are *constructors*. A *constructor* term (or context) is composed of constructors and variables; a ground constructor term (context) is a *value* (context). A rule defining a symbol f will be called an f-rule.

A system is *left-linear* if no variable appears more than once on the left side; it is *linear* if no variable appears on either side more than once. The *depth* $\|t\|$ of a term t is the number of symbols in the longest path of its tree representation. This means that constants and variables have depth 1. A variable is *shallow* in a term if it does not appear below depth 1.

A matching *(sub)goal* takes the form $s \to^? t$, where s and t share no variables, and has a *solution* σ, assigning terms to variables in s, if and only if $s\sigma \to^! t\tau$, for some (ancillary) substitution τ of terms for variables in t. Clearly, if t itself is not in normal form, the goal has no solutions. There may of course be more than one solution to a goal. We need not compute them all: We can ignore more specific solutions than ones we do compute (e.g. $x \mapsto sz$ subsumes $x \mapsto sss0$); we can ignore solutions that are not normalized, since they must be equal to normal-form solutions (e.g. $x \mapsto 0 \times z$ is covered by $x \mapsto 0$ for the theory of multiplication).

In inversion problems $s =^? N$, the term N is a (variable-free) value. If the rewrite system R is ground convergent, then an equation $s =^? N$ has a ground solution γ in the equational theory of R if, and only if, γ is equivalent to some solution of the inversion goal $s \to^? N$. Hence, to find all solutions σ to $s =^? N$, one can look for a complete set of solutions to $s \to^? N$.

3 Background

Any complete procedure for semantic matching with respect to an arbitrary theory cannot always terminate, even if the theory is presented as a finite, linear, convergent system for two reasons: matchability for some such theories is undecidable [Heibrunner and Hölldobler, 1987]; some have no finite set of most general unifiers [Fages and Huet, 1983].

Semantic unification in a theory supplied with a finite *ground convergent* (i.e. confluent for ground terms and terminating) rewrite system is known to be computable in the following special cases:

- Every non-ground right side is a variable [Hullot, 1980].
- Every non-ground right side is a constructor term [Dershowitz *et al.*, 1992].
- Every non-ground right side is a proper subterm of its left side [Narendran, Pfenning and Statman, 1997].
- Every non-ground right side is either a constructor term or a proper subterm of its left side [Mitra, 1994].
- Every right side is composed of constructors and proper subterms of its left side [Mitra, 1994].
- All variables are shallow on the left side [Christian, 1992].
- The system is linear and every variable that appears on both sides is shallow on both sides (convergence is unnecessary) [Nieuwenhuis, 1998].
- The system is linear and the right side of every f-rule is either a constructor term or a proper subterm of the left side, except for at most one right side that may be a value context with a single subterm $g(\ldots, r_i, \ldots)$, where every r_i is either a variable or a value [Dershowitz and Mitra, 1992].
- The system is linear and the right side of every f-rule is a constructor term, except for at most one right side that may be a constructor context with a single subterm $g(\ldots, r_i, \ldots)$, where every r_i is either a variable or a value [Mitra, 1994].

Only the last two special cases are powerful enough to capture even simple recursive functions like addition.

Under the assumption of ground convergence, one need only consider innermost computations, in which rules are always applied to terms whose proper subterms are in normal form. For that reason, one need only compute normal-form solutions; they are equivalent in the equational theory R to all others. For left-linear systems, semantic matching can be simpler than unification, since we know the shape of the normal form of the instantiated pattern. In prior work [Dershowitz et al., 1992], we showed decidability of semantic matching for certain variable-preserving or left-linear ground convergent systems satisfying a (noncomputable) semantic condition, called "decreasingness". Let $[t]$ represent the depth (or other measure for which proper subterms are smaller than their superterms) of the (unique) normal form of t for a given system. The point is that:

Theorem 1. *For a ground convergent rewrite system, the normal-form solutions σ to an inversion goal $s \to^? N$ are bounded in depth, i.e. $\|x\sigma\| \leq \|N\|$, if for all defined symbols f and ground terms $[f(\ldots, t, \ldots)] \geq [t]$.*

Proof. If $s\sigma \to^! N$, then $[s\sigma] = [N]$. Hence, $\|x\sigma\| = [x\sigma] \leq [s\sigma] = [N]$, since $x\sigma$ is a normal form.

It follows that, for such systems, the (finite) set of (normal-form) solutions to inversion goals can be computed in finite time. The problem is that the condition precludes "erasing" rules that have a variable on the left that is not carried over to the right side. To get around this obstacle, one can distinguish between decreasing and non-decreasing defined symbols:

Theorem 2 ([Dershowitz et al., 1992]). *The inversion problem is computable for a left-linear ground convergent rewrite system if no right side has a defined symbol at its root, nor a defined symbol that appears below a function symbol f that is "decreasing", in the sense that there are ground terms for which $[f(\ldots, t, \ldots)] < [t]$.*

In System (5), $\times$ and $<$ are decreasing (e.g. $[s^4 0 \times 0] = [0] = 1 < 5 = [s^4 0]$), but one rule has a defined symbol at the top right. Indeed, the goal $0 \times x \to^? 0$ has infinitely many solutions $s^n 0$, and there is no more general term $s^i y$ for which $0 \times s^i y \to^! 0$. This is because $0 \times y = 0$ is only an inductive, but not an equational, theorem of (5). For this reason, we need to use convoluted rules for multiplication, as in (6).

4 Complete Inversion

For finite equational theories, satisfiability problems are recursively enumerable, and general-purpose semantic-unification procedures have been extensively studied. If we restrict ourselves to ground convergent rewrite systems that are left-linear, then the following transformation rules constitute a complete procedure for inversion of goals $s \to^? N$:

Decompose:
$$\frac{f(s_1,\ldots,s_n) \to^? f(t_1,\ldots,t_n)}{s_1 \to^? t_1, \ldots, s_n \to^? t_n}$$

Mutate:
$$\frac{f(s_1,\ldots,s_n) \to^? t}{s_1 \to^? l_1\rho, \ldots, s_n \to^? l_n\rho}$$
ρ is a solution to $r \to^? t$; $f(l_1,\ldots,l_n) \to r$ is a (renamed) rule in R

Eliminate:
$$\frac{x \to^? t}{x \equiv^? t}$$
x is a variable

Ignore:
$$\frac{s \to^? x}{}$$
x is a variable

When Decompose and Mutate can both be applied to the same subgoal, both alternatives must be explored. The rules are applied until all that remains is a set of syntactic unification goals of the form $x_j \equiv^? t_j$. Any solution to the latter is a solution to the original goal. Variables in the left half of goals are never instantiated, giving a "basic" strategy.

We need to show that this system of goal transformations has the following properties:

Soundness Given a goal $s \to^? t$, if the procedure produces a solution σ, then $s\sigma \to \cdots \to t\tau$, for some substitution τ.

Completeness Given a goal $s \to^? t$, if $s\sigma \to^! t\tau$ for some σ and τ, the procedure will produce a solution μ such that there is a substitution ρ for which $R \models x\mu\rho = x\sigma$, for each variable x appearing in s.

Soundness should be clear except for the Ignore rule, for which we need the following lemmata:

Lemma 1. *For any left-linear rewrite system, if $s' \to^? t'$ is a current subgoal and x is a variable in t', then the set of current subgoals contains only that one occurrence of x.*

Proof. In the right half of the initial goal $s \to^? N$, there are no variables. When Decomposing $f(s_1,\ldots,s_n) \to^? f(t_1,\ldots,t_n)$, any variable that appeared once on the right is now in only one t_i. When Mutating, first $r \to^? t$ is solved, at which point the terms l_i do not yet appear in any subgoal. Later, when solving $s_i \to^? l_i\rho$, any variable y in $l_i\rho$ is either from the rule, in which case it is alone, since R is left-linear, or it was introduced by ρ, in which case it derives from eliminated goals $x_j \equiv^? t_j$ and appeared alone in the t_j.

Lemma 2. *If a variable x only appears in the right half of a subgoal $s \to^? x$, that subgoal may be deleted (by Ignore).*

Proof. For each solution σ, the ancillary substitution τ includes $x \mapsto s\sigma$ to satisfy the eliminated subgoal.

Were the system not left-linear, then there would be variables y in the right half of more than one goal. That would require an additional rule to transform a goal $f(s_1, \ldots, s_n) \to^? y$ into $y \equiv^? f(y_1, \ldots, y_n), s_1 \to^? y_1, \ldots, s_n \to^? y_n$, for new variables y_i.

The proof of completeness is by induction on the number of steps in any innermost normalizing derivation $s\sigma \to^! t\tau$, and, secondarily, on the depth of s.

1. If it has zero steps, then $s\sigma = t\tau$ and Decomposition and Eliminate will find a solution at least as general as σ.
2. If t is a variable, then Ignore generates the most general (trivial) solution.
3. If s is a variable x and σ provides a normal form N for it, then by confluence, $N \to^! t$ only if N is t, and Eliminate generates the right solution. If x also has a normal form (from a prior goal) looking like u, then t and u must unify.
4. Otherwise, Decomposition proceeds until a smaller subgoal that requires a rewrite at the top arises. Consider such a derivation

$$f(s_1, \ldots, s_n)\sigma \to \cdots \to f(l_1, \ldots, l_n)\rho = l\rho \overset{\text{top}}{\to} r\rho \to^! t\tau$$

Mutation does the trick by first finding (something at least as general as) the substitution ρ for $r \to^? t$; the solution σ is then found from the $s_i \to^? l_i\rho$ subgoals.

5 Constructing Systems

In the next section we will prove that the following syntactic requirements suffice for solvability of matching goals:

(A) Each left side is of depth at most 3.
(B) If a right side is a variable or constant, then the left side of that rule is of depth at most 2.
(C) Whenever a right side is not a variable, it is headed by a constructor.

A left-linear, ground convergent system satisfying these three conditions will be called a *constructing system*. Only condition (C) is severe in practice.

Now we show that each of the above three restrictions is necessary: If one drops the requirement of left-linearity, we get undecidability using a rule $E(x, x) \to T$ to reduce an arbitrary unification problem $s =^? t$ to $E(s, t) \to^? T$.

In the remaining cases, we reduce unification in the theory of addition and multiplication over natural numbers (which is undecidable per Hilbert's Tenth Problem) to matching problems, for which we use the constructing rules in (6).

If one violates (C) and allows defined right-root symbols, we have the following counterexample:

$$\begin{aligned} f0 &\to 0 \\ E(0,0) &\to 0 \\ E(sx, sy) &\to fE(x,y) \end{aligned} \tag{8}$$

By induction it is easy to see that

$$E(x, y) \to^! 0 \text{ if and only if } x, y \to^! s^n 0 \text{ for some } n$$

Therefore, a goal of the form $E(t, u) \to^? 0$ would, in general, be unsolvable for arbitrary terms t and u involving $+$ and $\times$.

The following system illustrates the problem when a left side is of depth 3 and the right side of depth 1:

$$
\begin{array}{ll}
f0 \to 0 & E(0,0) \to s0 \\
fs0 \to 0 & E(sx, sy) \to sf E(x, y)
\end{array}
\tag{9}
$$

for which

$$E(x, y) \to^! s0 \text{ if and only if } x, y \to^! s^n 0 \text{ for some } n$$

Finally, we relax Condition (A) and allow depths greater than 2 below the left root. Consider:

$$
\begin{array}{ll}
fsx \to s0 & E(0,0) \to ss0 \\
G(ss0, ssx) \to sssx & E(sx, sy) \to sf G(E(x, y), ss0)
\end{array}
\tag{10}
$$

Since $fG(ss0, ss0) \to fsss0 \to s0$,

$$E(x, y) \to^! ss0 \text{ if and only if } x, y \to^! s^n 0 \text{ for some } n$$

and matching is undecidable.

6   Computable Inversion

We present in this section algorithm for function inversion for any theory presented by a constructing system R. For instance, the following is a constructing system for inserting a number in its correct place in an ordered list, and therefore it has a computable inverse:

$$
\begin{array}{ll}
min(x, 0) \to 0 & max(x, 0) \to x \\
min(0, x) \to 0 & max(0, x) \to x \\
min(sx, sy) \to s(min(x, y)) & max(sx, sy) \to s(max(x, y)) \\
\multicolumn{2}{c}{insert(x, nil) \to x : nil} \\
\multicolumn{2}{c}{insert(x, y : z) \to min(x, y) : insert(max(x, y), z)}
\end{array}
\tag{11}
$$

At each stage of the algorithm, we have a set of subgoals. We (don't care) non-deterministically choose one, $s \to^? t$, and consider the following cases.

1. If s and t are identical, the subgoal may be removed.
2. If t is a variable, just remove this subgoal.
3. Suppose s is a variable x:
   (a) If x is already bound to a term u, then bind it instead to the most general unification of u and t.

(b) If u and t are not unifiable, fail.

(c) If x is unbound but appears in t, fail.

(d) If x does not appear in t, add the binding $x \mapsto t$.

4. If neither $s = f(s_1, \ldots, s_n)$ nor $t = g(t_1, \ldots, t_m)$ is a variable, try *both* of the following:

(a) if $f = g$ (and $m = n$), replace the current goal with the multiset of goals, $s_1 \to^? t_1, \ldots, s_n \to^? t_n$.

(b) For each rule $f(l_1, \ldots, l_n) \to r$ in R (with all its variables renamed apart from those in the goal), do one of the following:

i. If r and t are identical, then replace the goal with subgoals $s_1 \to^? l_1, \ldots, s_n \to^? l_n$.

ii. If r is a variable x, then replace the goal with subgoals $s_1 \to^? l_1 \rho, \ldots, s_n \to^? l_n \rho$, where ρ is $x \mapsto t$.

iii. If r is headed by a constructor that is not g, fail this path.

iv. If g is a constructor, first recursively solve the subgoals $r_1 \to^? t_1, \ldots, r_m \to^? t_m$ in succession. For each solution ρ to the variables of these m subgoals, solve the new subgoals $s_1 \to^? l_1 \rho, \ldots, s_n \to^? l_n \rho$.

7 Correctness

To prove the correctness of the above inversion algorithm, we need to establish its soundness, completeness, and termination. It goes without saying that it can be exponential in cost.

Soundness will be easy, since each step of the algorithm is an application of one or more transformation rules of Section 4. For completeness, we need only check that every transformation that might lead to a solution is attempted by the algorithm.

For termination, we will use the bag (multiset) extension of the lexicographic measure $\langle \|t\|, \|s\| \rangle$ of a subgoal $s \to^? t$. We will also need the following invariant, which we show by computational induction:

For each mapping $x \mapsto u$ of a solution to a goal $s \to^? t$, we have $\|u\| \leq \|t\|$.

Indeed, it is because most general solutions are bounded in size that the inversion problem is decidable for constructing systems.

Let the current goal $s \to^? t$ be called G. Let X signify the set of variables and γ, the current partial solution. We use $\equiv$ for identity of terms.

Consider each step of the algorithm in turn:

1: *If $s \equiv t$, remove G.*

This is sound, since it is a composite of Decompose and Ignore and results in the trivial solution. There is no need to Mutate for completeness, since s must be normalized if t is.

2: *If $t \in X$, remove G.*

This is just an application of Ignore and covers all its applicable cases. It, too, results in the trivial solution.

3b,3c: *If $s = x \in X$, but $s\gamma \equiv^? t$ is unsatisfiable, fail.*

In these cases, the syntactic goal $s \equiv^? t$ will fail.

3a,3d: *If $s = x \in X$ and $mgu(s\gamma, t)$ exists, remove G and replace $x \mapsto u$ in γ with $mgu(u, t)$.*

This is Eliminate combined with syntactic unification and covers the only successful case of Elimination.
Since the most general unifier of linear terms with disjoint variables (see Lemma 1) is bounded in depth by the maximum of their depths, all substitutions satisfy the invariant.

4a: *If $s = f(s_1, \ldots, s_n)$ and $t = f(\ldots, t_i, \ldots)$, replace G with subgoals $s_i \to^? t_i$, $i = 1, \ldots, n$.*

This is just Decompose and covers all cases needed for completeness that are not included in Step 1.
Termination follows from the fact that each subgoal is smaller, since $\|t_i\| < \|t\|$. By induction all solutions are bounded by $\|t\| - 1$.

4(b)i: *If $s = f(s_1, \ldots, s_n)$ and $f(\ldots, l_j, \ldots) \to t \in R$, replace G with subgoals $s_i \to^? l_i$, $i = 1, \ldots, n$.*

This is Mutate, when $r \equiv t$, and the subgoal $r \to^? t$ is trivial, yielding the identity substitution for ρ.
If r is a constant, then, by assumption (B), $\|l_i\| = 1 = \|r\| = \|t\|$ and solutions are bounded by $\|t\|$. Since $\|s_i\| < \|s\|$, the subgoals are smaller.
If r is not a constant, then, by assumption (A), $\|l_i\| \leq 2 \leq \|r\| = \|t\|$.

4(b)ii: *If $s = f(s_1, \ldots, s_n)$, $f(\ldots, l_i, \ldots) \to x \in R$ and $x \in X$, replace G with subgoals $s_i \to^? l_i\{x \mapsto t\}$, $i = 1, \ldots, n$.*

This is Mutate for "collapsing" rules ($r \in X$).
By assumption (B), l_i is either a constant or a variable (possibly x). Thus, $\|l_i\{x \mapsto t\}\|$ is either 1 or $\|t\|$. In any case, the new subgoals are smaller by virtue of their shallower left half s_i.

4(b)iii: *If $t = g(\ldots, t_j, \ldots)$, but $l \to c(\ldots, r_k, \ldots) \in R$ has constructor $c \neq g$, this choice path fails.*

Mutation cannot succeed in this case, since $c(\ldots, r_k, \ldots)\sigma \to \cdots \to g(\ldots, t_j, \ldots)\tau$ is impossible.

4(b)iv: *If $s = f(s_1, \ldots, s_n)$, $t = c(t_1, \ldots, t_m)$, $f(\ldots, l_i, \ldots) \to c(\ldots, r_j, \ldots) \in R$, and c is a constructor, solve the set $s_i \to^? l_i\rho$ $(i = 1, \ldots, n)$, for each solution ρ of the $r_j \to^? t_j$ $(j = 1, \ldots, m)$.*

This is the only remaining case for Mutate, since assumption (C) is that the root of a non-variable right side is a constructor.

The $r_j \to^? t_j$ have shallower right halves, so the computation of ρ terminates.

By assumption (A), $\|l_i\| \leq 2$ and l_i can contribute at most 1 to the depth $\|l_i\rho\|$. By induction, $\|x\rho\| \leq \|t_j\| \leq \|t\| - 1$, so $\|l_i\rho\| \leq \|t\|$, and the second set of subgoals is smaller by virtue of their left halves. Also, their solution is bounded by $\|t\|$.

8 Extensions

A "symbolic definition" of the form

$$f(x_1, \ldots, x_n) \to e$$

which just defines a non-recursive function f that does not appear elsewhere (in e or R), like the calendar rule (1) for n or squaring rule $x^2 \to x \times x$ (7), can always be added to a decidable system, since those symbols can be immediately eliminated from any goal containing them.

Theorem 2 was refined in [Aguzzi and Modigliani, 1994] with a notion of "positional" increase. By using positional information, it is possible to handle certain systems that do not have leading constructors on the right-hand sides of rules. For example, the usual definition of $\times$ can be used instead of the three multiplication rules in (6). With such an extension, we should be able to handle insertion sort by adding $sort(\epsilon) \to \epsilon$ and $sort(x : y) \to insert(x, sort(y))$ to (11).

Looking at our proof of the correctness of the algorithm, it should be clear that one can allow left sides of depth $d > 3$ for rules such that the normal form of the right side r has no path of length less than $d + 1$. Checking normal forms of all instances is not a syntactic condition, but testing for ground reducibility is [Plaisted, 1985]. So we can replace Conditions (A) and (B) with the following:

(*) *For each rule $l \to r$, we have $\|l\| < \ell + 2$, where ℓ is the length of the shortest path from the root of r to a ground reducible position.*

Since the ground-normal form of any term is of depth at least 1, this condition guarantees that the subgoal $r \to^? t$ will result in variable bindings that are no deeper than $\|t\| - d$.

Returning to the calendar, the rules for $<$ needed to constrain the length of a month and a year violate even this condition. The calendar code, however, does not require solutions to arbitrary inequalities, only to inequalities of the form $x < N$, for ground term N. These cannot be handled by tabulated functions, with rules like $2 < 12 \to T$ (where 12 is just an abbreviation for $s^{12}0$), since the left side would be too deep. Instead, we can rephrase a constraint $x < N \to^? T$ as $x < N \to^? T^N \top$ and use:

$$
\begin{array}{ll}
sx < sy \to T(x < y) & Lsx \to TLx \\
0 < sx \to TLx & L0 \to \top
\end{array}
\tag{12}
$$

in lieu of the rules for inequality in (5).

A more interesting situation is posed by months of unequal length, as in many archaic calendars having epagomenal days, which we treat as a thirteenth month of only five days (making for a 365-day year). That requires the constraint

$$m < 13 \wedge d < 30 \wedge (m < 12 \vee d < 5)$$

for which we use the goals

$$m < 13 \rightarrow^? T^{13}\top, \ d < 30 \rightarrow^? T^{30}\top, \ (m < 12 \vee d < 5) \rightarrow^? T^{17}\top$$

and compute disjunction as follows:

$$
\begin{array}{ll}
Tx \vee y \rightarrow T(x \vee y) & \top \vee y \rightarrow \top \\
y \vee Tx \rightarrow T(x \vee y) & y \vee \top \rightarrow \top
\end{array}
\tag{13}
$$

For example,

$$12 < 12 \vee 4 < 5 \rightarrow \cdots \rightarrow T^{12}(0 < 0) \vee T^4(0 < 1)$$

$$\rightarrow T^{12}(0 < 0) \vee T^5 L0 \rightarrow T^{12}(0 < 0) \vee T^5\top$$

$$\rightarrow \cdots \rightarrow T^{17}(0 < 0 \vee \top) \rightarrow T^{17}\top$$

For non-left-linear systems, simple inversion goals of the form $f(\ldots, c_i, \ldots) \rightarrow^? N$, where the c_i are constructor terms, can be solved, provided no right side has a defined symbol at its root, nor a defined symbol that appears below a "decreasing" symbol [Mitra, 1994]. Syntactic criteria for this case are also possible.

In [Dershowitz and Mitra, 1992], we considered systems with potentially infinitely many solutions, which were captured as indexed terms, along the lines of [Comon, 1992]. (See Section 3.) Such an approach should work for some inversion problems with unbounded solutions.

The following constructing system for differentiation illustrates some of the subtleties of inversion problems:

$$
\begin{array}{ll}
Dt \rightarrow s0 & \\
D0 \rightarrow 0 & D(x + y) \rightarrow Dx + Dy \\
Dsx \rightarrow Dx + 0 & D(x \times y) \rightarrow x \times Dy + y \times Dx
\end{array}
\tag{14}
$$

The third rule has $+0$ to ensure that the right side is headed by the constructor $+$ (vis-a-vis this system, at least, it is a constructor). If we include our constructing rules for $\times$, inverting the goal $Dz \rightarrow^! t + t$ yields the indefinite integral $z \mapsto t \times t$, but, in the absence of simplifying rules for addition, we do not get more general solutions. And one cannot add addition, without turning $+$ into a defined function, at which point (14) would no longer be constructing.

Perhaps results (e.g. [Jacquemard, 1996]) on regularity of the normalizable terms, which have bearing on derivability, can help decide matching.

Finally, our algorithm can serve as the basis of a program to compile a logic program for computing the inverse of a given functional program.

Acknowledgement

We thank Claude Kirchner for his indispensable interest and gracious hospitality. A preliminary version of the inversion algorithm was presented at the Fifth Workshop on Logic, Language, Information and Computation (July 1998, São Paulo, Brazil).

References

1. Aguzzi, G., Modigliani, U.: A criterion to decide the semantic matching problem. Proc. Intl. Conf. on Logic and Algebra, Italy (1995).
2. Christian, J.: Some termination criteria for narrowing and E-narrowing. Proc. 11th Intl. Conf. on Automated Deduction, Saratoga Springs, NY. Lecture Notes in Artificial Intelligence **607**:582–588, Springer-Verlag, Berlin (1992).
3. Comon, H.: On unification of terms with integer exponents. Tech. Rep. 770, Université de Paris-Sud, Laboratoire de Réchérche en Informatique (1992).
4. Comon, H., Haberstrau, M., Jouannaud, J.P.: Decidable problems in shallow equational theories. Tech. Rep. 718, Université de Paris-Sud, Laboratoire de Réchérche en Informatique (1991).
5. Dershowitz, N., Jouannaud, J.-P.: Rewrite systems. In J. van Leeuwen, ed., Handbook of Theoretical Computer Science B: 243–320, North-Holland, Amsterdam (1990).
6. Dershowitz, N., Mitra, S.: Higher-order and semantic unification. Proc. 13th Conf. on Foundations of Software Technology and Theoretical Computer Science, Bombay, India. Lecture Notes in Artificial Intelligence **761**:139–150, Springer-Verlag, Berlin (1993).
7. Dershowitz, N., Mitra, S., Sivakumar., G.: Decidable matching for convergent systems (Preliminary version). Proc. 11th Conference on Automated Deduction, Saratoga Springs, NY. Lecture Notes in Artificial Intelligence **607**:589–602, Springer-Verlag, Berlin (1992).
8. Dershowitz, N., Plaisted, D. A.: Equational programming. In: J. E. Hayes, D. Michie, J. Richards, eds., Machine Intelligence 11: The logic and acquisition of knowledge, 21–56. Oxford Press, Oxford (1988).
9. Dershowitz, N., Reingold, E. M.: Calendrical Calculations. Cambridge University Press, Cambridge (1997).
10. Fages, F., Huet, G.: Unification and matching in equational theories. Proc. 8th Colloq. on Trees in Algebra and Programming, L'Aquila, Italy. Lecture Notes in Computer Science **159**:205–220, Springer-Verlag, Berlin (1983).
11. Heilbrunner, S., Hölldobler, S.: The undecidability of the unification and matching problem for canonical theories. Acta Informatica **24**(2):157–171 (1987).
12. Hullot, J.-M.: Canonical forms and unification. Proc. 5th Intl. Conf. on Automated Deduction, Les Arcs, France, Lecture Notes in Computer Science **87**:318–334, Springer-Verlag, Berlin (1980).
13. Jacquemard, F.: Decidable approximations of term rewriting systems. Proc. 7th Intl. Conf. on Rewriting Techniques and Applications, New Brunswick, NJ. Lecture Notes in Computer Science **1103**:362–376, Springer-Verlag, Berlin (1966).
14. Mitra, S.: Semantic unification for convergent systems. Ph.D. thesis, Dept. of Computer Science, University of Illinois, Urbana, IL, Tech. Rep. UIUCDCS-R-94-1855 (1994).

15. Narendran, P., Pfenning, F., Statman, R.: On the unification problem for Cartesian closed categories. J. Symbolic Logic **62**(2):636–647 (1997).
16. Nieuwenhuis, R.: Decidability and complexity analysis by basic paramodulation. Information and Computation **147**:1–21 (1998).
17. Plaisted, D. A.: Semantic confluence tests and completion methods. Information and Computation **65**:182–215 (1985).

Strategic Pattern Matching

Eelco Visser*

Department of Computer Science, Universiteit Utrecht
P.O. Box 80089, 3508 TB Utrecht, The Netherlands
http://www.cs.uu.nl/~visser/, visser@acm.org

Abstract. Stratego is a language for the specification of transformation rules and strategies for applying them. The basic actions of transformations are matching and building instantiations of first-order term patterns. The language supports concise formulation of generic and data type-specific term traversals. One of the unusual features of Stratego is the separation of scope from matching, allowing sharing of variables through traversals. The combination of first-order patterns with strategies forms an expressive formalism for pattern matching. In this paper we discuss three examples of *strategic pattern matching*: (1) *Contextual rules* allow matching and replacement of a pattern at an arbitrary depth of a subterm of the root pattern. (2) *Recursive patterns* can be used to characterize concisely the structure of languages that form a restriction of a larger language. (3) *Overlays* serve to hide the representation of a language in another (more generic) language. These techniques are illustrated by means of specifications in Stratego.

1 Introduction

First-order terms are used to represent data structures in term rewriting systems, functional and logic programming languages. First-order patterns are used to decompose such terms by simultaneously recognizing a structure and binding variables to subterms, which would otherwise be expressed by nested conditional expressions that test tags and select subterms. However, first-order patterns are not treated as first-class citizens and their use poses limitations on modularity and reuse: no abstraction over patterns is provided because they may occur only in the left-hand side of a rewrite rule, the arms of a case, or the heads of clauses; pattern matching is at odds with abstract data types because it exposes the data representation; a first-order pattern can only span a fixed distance from the root of the pattern to its leafs, which makes it necessary to define recursive traversals of a data structure separately from the pattern to get all needed information.

For these reasons, enhancements of the basic pattern matching features have been implemented or considered for several languages. For example, *list matching*

* This paper was written while the author was employed by the Pacific Software Research Center, Oregon Graduate Institute, Portland, Oregon, USA. This work was supported, in part, by the US Air Force Materiel Command under contract F19628-93-C-0069.

in ASF+SDF [7] is used to divide a list in multiple sublists possibly separated by element patterns. *Associative-commutative (AC) matching* in OBJ, Maude [5] and ELAN [3] supports the treatment of lists as multi-sets. *Higher-order matching* in λProlog [17] allows the matching of subterms at a variable depth. *Views* for Haskell, as proposed in [24], provide a way to view a data structure using different patterns than are used to represent them. Each of these techniques provides a mix of structure recognition, variable binding, *term traversal*, and *transformation*. For instance, in list (AC, higher-order) matching a term is first transformed by application of the associative and identity laws (associative and commutative laws, $\beta\eta$-conversion) and then matched against the given pattern. Matching a view pattern involves the transformation of the underlying data structure to the view data type.

This paper shows how the *rewriting strategies* paradigm [3, 5, 14, 22, 23] provides a general framework for describing and implementing such pattern matching combinations. Rewriting strategies are programs that determine the order in which rewriting rules are applied. One of the important aspects of strategies is the definition of term traversals to find subterms to which rules can be applied. Here the application of such traversals in the definition of patterns *at the level of individual rules* is considered.

This paper explores *strategic pattern matching* in Stratego [22, 23], a language for the specification of program transformation systems. Stratego is a layer of syntactic abstractions on top of System S, a core language for the definition of rewriting strategies. The basic actions of System S are matching and building instantiations of first-order term patterns. The language supports concise formulation of generic and data type specific term traversals. One of the unusual features of System S is the separation of scope from matching, allowing sharing of variables through traversals. The combination of first-order patterns with strategies forms an expressive formalism for pattern matching.

The next section gives a brief overview of System S and Stratego. The following sections discuss three applications of strategic pattern matching illustrated by means of specifications in Stratego: (1) Contextual rules allow matching and replacement of a subterm at an arbitrary depth with respect to the root of a pattern. Section 3 shows how contextual patterns are used in a concise specification of a type checker. (2) Recursive patterns can be used as predicates to characterize concisely the structure of languages that form a subset of a larger language. Section 4 illustrates the idea by means of a characterization of conjunctive and disjunctive normal forms as a restriction of propositional formulae. Section 5 applies the same technique to characterize the embedding of AsFix, the abstract syntax representation of ASF+SDF, into ATerms, a universal data type. (3) Overlays are pseudo-constructors that abstract from an underlying (complex) representation using real constructors. They can be used to overlay a language on top of another more generic representation language. Section 5 defines overlays for AsFix to hide the details of its embedding in ATerms. Related work is discussed in Section 6 and some conclusions are drawn in Section 7.

2 Rewriting Strategies

This section introduces System S, a calculus for the definition of tree transformations, and Stratego, a specification language providing syntactic abstractions for System S expressions. For an operational semantics see [22, 23].

2.1 System S

System S is a hierarchy of operators for expressing term transformations. The first level provides control constructs for sequential non-deterministic programming, the second level introduces combinators for term traversal and the third level defines operators for binding variables and for matching and building terms.

First-order terms are expressions over the grammar

$$t := x \mid C(t1,\ldots,tn) \mid [t1,\ldots,tn] \mid (t1,\ldots,tn)$$

where x ranges over variables and C over constructors. The arity and types of constructors are declared in signatures. The notation $[t1,\ldots,tn]$ abbreviates the list $Cons(t1,\ldots,Cons(tn,Nil))$. Transformations in System S are applied to ground terms, i.e., terms withouth variables.

Level 1: Sequential Non-deterministic Programming Strategies are programs that attempt to transform ground terms into ground terms, at which they may succeed or fail. In case of success the result of such an attempt is a transformed term. In case of failure the result is an indication of the failure. Strategies can be combined into new strategies by means of the following operators: The *identity* strategy id leaves the subject term unchanged and always succeeds. The *failure* strategy fail always fails. The *sequential composition* s1; s2 first attempts to apply s1 to the subject term and, if that succeeds, applies s2 to the result. The *non-deterministic choice* s1 + s2 attempts to apply either s1 or s2. It succeeds if either succeeds and it fails if both fail; the order in which s1 and s2 are tried is unspecified. The *deterministic choice* s1 <+ s2 attempts to apply either s1 or s2, in that order. The *recursive closure* rec x(s) attempts to apply s, where at each occurence of the variable x in s, the strategy rec x(s) is applied. The *test* strategy test(s) tries to apply s. It succeeds if s succeeds, and reverts the subject term to the original term. It fails if s fails. The *negation* not(s) succeeds (with the identity transformation) if s fails and fails if s succeeds. Two examples of strategies defined with these operators are try and repeat in Figure 1.

Level 2: Term Traversal The Level 1 constructs apply transformations to the root of a term. In order to apply transformations throughout a term it is necessary to traverse it. For this purpose, System S provides the following operators: For each n-ary constructor C the *congruence* operator $C(s1,\ldots,sn)$ is defined. It applies to terms of the form $C(t1,\ldots,tn)$ and applies si to ti for $1 <= i <= n$. An example of the use of congruences is the operator map(s) in Figure 1 that applies s to each element of a list.

```
module traversals
imports lists
strategies
  try(s)     = s <+ id              map(s)    = rec x(Nil + Cons(s, x))
  repeat(s) = rec x(try(s; x))      list(s)   = rec x(Nil + Cons(s, x))
  topdown   = rec x(s; all(x))      alltd(s)  = rec x(s <+ all(x))
  bottomup  = rec x(all(x); s)      oncetd(s) = rec x(s <+ one(x))
  downup(s) = rec x(s; all(x); s)   sometd(s) = rec x(s <+ some(x))
  onebu(s)  = rec x(one(x) <+ s)    somebu(s) = rec x(some(x) <+ s)
  downup2(s1, s2) = rec x(s1; all(x); s2)
```

Fig. 1. Specification of several generic term traversal strategies.

Congruences can be used to define traversals over specific data structures. Specification of generic traversals (e.g., pre- or post-order over arbitrary structures) requires more generic operators. The operator all(s) applies s to all children of a constructor application C(t1,...,tn). In particular, all(s) is the identity on constants (constructor applications without children). The strategy one(s) applies s to one child of a constructor application C(t1,...,tn); it is precisely the failure strategy on constants. The strategy some(s) applies s to some of the children of a constructor application C(t1,...,tn), i.e., to at least one and as many as possible. Like one(s), some(s) fails on constants.

Figure 1 defines various traversals based on these operators. For instance, oncetd(s) tries to find *one* application of s somewhere in the term starting at the root working its way down; s <+ one(x) first attempts to apply s, if that fails an application of s is (recursively) attempted at one of the children of the subject term. If no application is found the traversal fails. Compare this to the traversal alltd(s), which finds *all* outermost applications of s and never fails.

Level 3: Match, Build and Variable Binding The operators introduced thus far are useful for repeatedly applying transformation rules throughout a term. Actual transformation rules are constructed by means of pattern matching and building of pattern instantiations.

A match ?t succeeds if the pattern term t matches the subject term. As a side-effect, any variables in t are bound to the corresponding subterms of the subject term. If a variable was already bound before the match, then the binding only succeeds if the terms are the same. This enables non-linear pattern matching, so that a match such as ?F(x, x) succeeds only if the two arguments of F in the subject term are equal. This non-linear behaviour can also arise accross other operations. For example, the two consecutive matches ?F(x, y); ?F(y, x) succeed exactly when the two arguments of F are equal. Once a variable is bound it cannot be unbound.

A build !t replaces the subject term with the instantiation of the pattern t using the current bindings of terms to variables in t. A scope {x1,...,xn: s} makes the variables xi local to the strategy s. This means that bindings to these variables outside the scope are undone when entering the scope and are restored after leaving it. The operation where(s) applies the strategy s to the subject term. If successful, it restores the original subject term, keeping only the newly obtained bindings to variables.

2.2 Stratego

The specification language Stratego provides syntactic abstractions for System S expressions. A specification consists of a collection of modules that define signatures, transformation rules and strategy definitions.

A signature declares the sorts and operations (constructors) that make up the structure of the language(s) being transformed. An example signature is shown in Figure 2. A strategy definition f(x1,...,xn) = s introduces a new strategy operator f parameterized with strategies x1 through xn and with body s. Such definitions cannot be recursive, i.e., they cannot refer (directly or indirectly) to the operator being defined. All recursion must be expressed explicitly by means of the recursion operator rec. Labeled transformation rules are abbreviations of a particular form of strategy definitions. A conditional rule L : l -> r where s with label L, left-hand side l, right-hand side r, and condition s denotes a strategy definition L = {x1,...,xn: ?l; where(s); !r}. Here, the body of the rule first matches the left-hand side, and then attempts to satisfy the condition s. If that succeeds, then it builds the right-hand side r. The rule is enclosed in a scope that makes all term variables xi occurring in l, s and r local to the rule. If more than one definition is provided with the same name, e.g., f(xs) = s1 and f(xs) = s2, this is equivalent to a single definition with the sum of the original bodies as body, i.e., f(xs) = s1 + s2.

The following definitions provide a useful shorthand. The notation <s> t denotes !t; s, i.e., the strategy that builds the term t and then applies s to it. The notation s => t denotes s; ?t, i.e., the strategy that applies s to the current subject term and then matches the result against t. The combined notation <s> t => t' thus denotes (!t; s); ?t'. The <s> t notation can also be used in a build expression. For example, the strategy expression !F(<s> t, t') corresponds to {x: <s> t => x; !F(x,t')}, where x is a new variable.

This paper is about three programming idioms and the syntactic abstractions to support them. Recursive patterns are an idiom that is directly supported by Stratego as introduced above. The syntax of Stratego has been extended for contexts and overlays to provide more concise syntax for these idioms. However, these syntax extensions are implemented without extending System S.

2.3 Implementation

The Stratego compiler translates a specification to a C program that reads a term, applies the specified transformation to it, and, if succesful, outputs the transformed term. The compiler first translates a specification to a System S expression, which is then translated to a list of abstract machine instructions. The instructions are implemented in C. The run-time system is based on the ATerm library [19]. The compiler is implemented in Stratego itself.

```
module pico-syntax
imports list-basic
signature
  sorts Program Decl Stat Expr Type Id
  operations
    Block   : List(Decl) * Stat -> Program
    Decl    : Id * Type  -> Decl
    Natural : Type                    Plus  : Expr * Expr -> Expr
    String  : Type                    Minus : Expr * Expr -> Expr
    Skip    : Stat                    Conc  : Expr * Expr -> Expr
    Assign  : Id * Expr -> Stat       Var   : Id -> Expr
    Seq     : Stat * Stat -> Stat     Int   : Int -> Expr
    If      : Expr * Stat * Stat -> Stat  Str : String -> Expr
    While   : Expr * Stat -> Stat     Id    : String -> Id
```

Fig. 2. Abstract syntax of Pico.

3 Contexts

This section describes *contextual patterns*, i.e., patterns that relate some bit of information from the root pattern to a subterm at variable depth. This is illustrated by the specification of a type checker for the toy language Pico [7]. Heering [11] gives a concise specification of such a typechecker using a combination of second-order matching to relate variable declarations and their occurrences in a program and an abstract interpretation style of type checking.

Pico is a small imperative while-language. It has expressions ranging over natural number and string values and the usual statement combinators. A program consists of a block, which contains a list of variable declarations and a statement. Variable declarations associate a type (Natural or String) with a variable identifier. The abstract syntax of Pico is defined in Figure 2.

A program is statically correct if variables are used consistently with their declarations. Conventionally, type checkers are defined as a predicate that traverses the program carrying the declarations and checking the correctness of expressions and statements. In the abstract interpretation style of [11] first all variable occurrences are replaced by their types (using an injection Tp of types into identifiers), then consistent combinations of such typed expressions and statements are reduced to simpler forms. For example, let variables "a" and "b" have type String, the expression Conc(Var(Id("a")),Var(Id("b"))) is first transformed into Conc(Var(Tp(String)),Var(Tp(String))), which then reduces to Var(Tp(String)). If the program is correct it will reduce to a block with a skip statement. However, if the program contains type errors, residuals of this error will remain in the result of the type checking procedure and point to the offending parts of the program. For example, the program

```
Block([Decl(Id("a"), Natural), Decl(Id("b"), String)],
  While(Var(Id("a")), Assign(Id("b"), Plus(Var(Id("a")), Var(Id("a")))))))
```

reduces to

```
module pico-typecheck
imports pico-syntax traversals
signature
  operations
    Tp : Type -> Id
rules
  InlTp : Block(ds[Decl(Id(x), t)], s[Id(x)]) -> Block(ds, s[Tp(t)])
  IntTp : Int(n) -> Var(Tp(Natural))
  StrTp : Str(s) -> Var(Tp(String))
  Check : Seq(Skip, s) -> s
  Check : Seq(s, Skip) -> s
  Check : Assign(Tp(t), Var(Tp(t))) -> Skip
  Check : If(Var(Tp(Natural)), s1, s2) -> Seq(s2, s3)
  Check : While(Var(Tp(Natural)), s1) -> s1
  Check : Plus(Var(Tp(Natural)), Var(Tp(Natural))) -> Var(Tp(Natural))
  Check : Minus(Var(Tp(Natural)), Var(Tp(Natural))) -> Var(Tp(Natural))
  Check : Conc(Var(Tp(String)), Var(Tp(String))) -> Var(Tp(String))
strategies
  typecheck = downup2(repeat(InlTp + IntTp + StrTp), repeat(Check))
```

Fig. 3. Type checking rules and strategy for Pico.

```
Block([Decl(Id("a"),Natural), Decl(Id("b"), String)],
  Assign(Tp(String), Var(Tp(Natural))))
```

making clear that the assignment statement is not type correct.

A specification of this approach is shown in Figure 3. The `typecheck` strategy declares a `downup2` traversal over the program. On the way down identifiers and constants are replaced by their types by means of rules `InlTp`, `IntTp` and `StrTp`. On the way up well-typed expressions and statements are reduced to simpler forms by the `Check` rules. Distribution of type information over a program is achieved by means of the contextual rule `InlTp`. The sub-pattern `ds[Decl(Id(x), t)]` is a context that matches one instance of the pattern `Decl(Id(x), t)` as a subterm of the `ds` argument of the `Block` pattern. The sub-patterns `s[Id(x)]` in the left-hand side and `s[Tp(t)]` in the right-hand side form a context that replaces one occurrence of `Id(x)` somewhere in the statements by `Tp(t)`, where `x` and `t` are determined by the match in the `ds` context.

Contextual rules are implemented by translation to primitive constructs. A context `x[t]`, occurring on the left-hand side only, corresponds to a traversal over the term matching `x` trying to find a match of the pattern `t`. A context `x[l]` in the left-hand side and `x[r]` in the right-hand side corresponds to a traversal over the term matching `x` that replaces an occurrence of `l` by the corresponding instantiation of `r`. Thus, a first attempt at implementation of rule `InlTp` is:

```
InlTp : Block(ds, s) -> Block(ds, s')
        where <oncetd(?Decl(Id(x), t))> ds;
              <oncetd(?Id(x); !Tp(t))> s => s'
```

The first clause in the condition makes a traversal over the declarations finding a declaration. The second clause traverses the statements replacing an occurrence of the identifier in the declaration by its type.

However, this does not achieve the desired effect. If the first traversal finds a declaration for which there are no (more) occurrences of the identifier in the statements, then the second traversal will fail, even if there are other declarations for which it would succeed. In other words the first traversal needs to backtrack to find other declarations if the second traversal fails. This is achieved by inlining the second traversal in the first, as follows:

```
InlTp : Block(ds, s) -> Block(ds, s')
        where <oncetd(?Decl(Id(x), t);
                        where(<oncetd(?Id(x); !Tp(t))> s => s')))> ds
```

The where clause of this rule computes a new value s'. The outer traversal walks over the declarations. When a declaration for an identifier Id(x) is found the inner traversal walks over the statements and replaces one occurrence of Id(x) with its type Tp(t) from the declaration. If no occurrence of the identifier is found in the statements the outer loop continues to search for another declaration. If no declaration and matching identifier occurrence in the statements can be found, the rule fails.

4 Recursive Patterns

This section treats *recursive patterns*, i.e., patterns that describe recursive structure as opposed to the fixed structure described by first-order patterns. The idiom of recursive patterns is illustrated by the specification of language restrictions. Recursive patterns are also useful tools in program analysis.

A signature generates a language of terms. A language restriction is a subset of a language. Restrictions are not always syntactic, i.e., do not correspond to the language generated by a subsignature, but can require constructs only to be used in certain combinations. Examples of language restrictions abound in language processing: (1) The set of normal forms with respect to a set of rewrite rules is a restriction. The rewrite rules give an operational method for obtaining the normal form of a term, but they do not describe the *structure* of the normal forms. (2) A core language reflects the computational kernel of a language. Again, the transformation that translates a program in the complete language to a core language program does not define the structure of core language programs. (3) The intermediate languages produced by the stages of a compiler are often restrictions of a common language. Subsequent stages introduce lower-level features. The combination of all constructs might not form a valid language. (4) Languages embedded in a generic representation format. The generic format allows a wide range of expressions, only a few of those are expressions in the embedded language.

Language restrictions are often dealt with informally. A component of a language processor assumes its input to be in a certain form that is not defined anywhere. Descriptions of language restrictions separate from the transformations

```
module prop
signature
  sorts Prop
  operations
    Atom : String -> Prop    And : Prop * Prop -> Prop
    Not  : Prop -> Prop      Or  : Prop * Prop -> Prop
strategies
  conj(s) = rec x(And(x, x) <+ s)
  disj(s) = rec x(Or(x, x)  <+ s)
  conj-nf = conj(disj(Atom(id) + Not(Atom(id))))
  disj-nf = disj(conj(Atom(id) + Not(Atom(id))))
```

Fig. 4. Characterization of conjunctive and disjunctive normal forms.

that produce them are useful for documentation (what restriction is consumed or produced by this language processor) and validation (check that the input or output of a processor conforms to the restriction). Strategies support the concise description of language restrictions by means of *recursive patterns*. A recursive pattern is a strategy that describes the structure of a set of terms by means of recursion and congruences. This technique is illustrated by two examples: disjunctive and conjunctive normal forms of propositional formulae and, in the next section, the embedding of AsFix in ATerms.

As a first example, consider a language of propositional formulae constructed from atoms (proposition letters) with negation, conjunction and disjunction. The signature describing the abstract syntax of this language is shown in Figure 4.

A formula is in conjunctive normal form if it is a conjunction of disjunctions of atoms or negated atoms. Likewise, a formula is in disjunctive normal form if it is a disjunction of conjunctions of atoms or negated atoms. These restrictions can be characterized concisely by means of the recursive patterns in Figure 4. Given some strategy s that characterizes formulae in some form, the strategy `rec x(And(x, x) <+ s)` describes conjunctions of the form `And(And(...,  ...), And(..., ...))` with leaves of the form s. Thus, the operators `conj(s)` and `disj(s)`, describe conjuncts and disjuncts of s's, respectively. Hence, the combination `conj(disj(s))` describes conjuncts of disjuncts of s's. Unfolding the definition of `conj` and `disj` in `conj-nf` gives:

```
conj-nf = rec x(And(x, x) + rec y(Or(y, y) + Not(Atom(id)) + Atom(id)))
```

So `conj-nf` and `disj-nf` describe conjunctive and disjunctive normal forms, respectively.

5 Overlays

This section introduces *overlay patterns*, i.e., patterns composed with pseudo-constructors that abstract from a concrete representation with real constructors. Overlays are first-class citizens in the sense that all operations that apply to

normal constructors, i.e., matching, building and congruence, apply to overlays
as well. Furthermore, overlays can be defined in terms of other overlays, allowing
a hierarchy of abstractions. The technique is illustrated by the definition of
overlays for the representation of AsFix constructs in ATerms. These are applied
in another example of recursive patterns to describe the restriction of ATerms
to AsFix expressions.

5.1   ATerms and AsFix

The Annotated Term Format or ATerms [19] is a universal data type designed
for representing data types in a generic manner for the purpose of data exchange,
generic manipulation and persistent storage of data. Figure 5 gives the signature
of ATerms. An ATerm is either an application (Appl) of an AFun to a list of terms,
or a list (AList) of terms. An AFun is either an integer (Int), quoted string (Str)
or an unquoted symbol (Sym). For example, the ATerm

```
Appl(Sym("And"), [Appl(Sym("Atom"), [Appl(Str("a"), [])]),
                  Appl(Sym("Atom"), [Appl(Str("b"), [])])])
```

is an encoding of the propositional formula And(Atom("a"), Atom("b")).

AsFix is the abstract syntax for the algebraic specification formalism ASF+
SDF [7]. It is used as the intermediate representation for language processors
such as a term rewriting compiler and a pretty-printer generator. The AsFix
representation of a specification consists of a signature and a list of conditional
equations over typed first-order terms. Here only unconditional equations over
first-order terms are considered.

One of the characteristics of AsFix is its encoding of syntactic information.
In ASF+SDF constructors are defined by means of a context-free production
that declares its mix-fix syntax and the sorts of its arguments. In AsFix both
the sort information and the syntactic information is retained. For example, the
production E "+" E -> E is represented by the AsFix expression

```
Prod([Sort("E"), Lit("+"), Sort("E")], Sort("E"))
```

Productions p of this form are used as the constructor 'names' in applications
of the form App(p, [a1,...,an]). Below a precise definition of AsFix is given.

AsFix expressions can be represented as ATerms. For instance, the production
above is represented by the ATerm

```
Appl(Sym("Prod"), [AList([Appl(Sym("Sort"), [Appl(Str("E"), [])]),
                          Appl(Sym("Lit"),  [Appl(Str("+"), [])]),
                          Appl(Sym("Sort"), [Appl(Str("E"), [])])]),
                   Appl(Sym("Sort"), [Appl(Str("E"), [])])])
```

This representation allows easy exchange, persistency and generic manipulation
of AsFix expressions. However, the representation has two problems: (1) Since
the ATerm format is a universal datatype, not every ATerm is a valid AsFix
expression. (2) Since the ATerm format is bulky, specifying operations on AsFix
using pattern matching on the ATerm representation is rather tedious. The first

```
module aterms
signature
  sorts AFun ATerm
  operations
    Int   : Int -> AFun        Appl  : AFun * List(ATerm) -> ATerm
    Str   : String -> AFun     AList : List(ATerm) -> ATerm
    Sym   : String -> AFun
```

Fig. 5. ATerm signature.

problem is solved by definining a recursive pattern that characterizes the ATerms that are valid AsFix expressions. This recursive pattern can be used to validate input to language processors. The second problem is solved by defining overlays that abstract from the concrete ATerm representation of AsFix expressions, while still maintaining that representation under the hood.

5.2 Overlays for AsFix

Overlays are abstractions of term patterns. An overlay definition C(x1, ..., xn) = pat introduces a new constructor C with n arguments that is an abbreviation of the pattern pat. This new constructor can be used in all places where the pattern pat can be used, i.e., in match patterns, build patterns and congruences. An expression ?t (!t) with an occurrence of C(t1,...,tn) denotes the expression ?t' (!t'), where t' is obtained by replacing C(t1,...,tn) by pat[t1/x1,...,tn/xn] in t. A congruence expression C(s1,...,sn) denotes the instantiation of the congruence derived from the pattern t, with the strategies si substituted for the variables xi.

Figure 6 defines overlays for the constructs of AsFix. For example, the overlay

```
Prod(as, r) = Appl(Sym("Prod"), [AList(as), r])
```

defines an abstraction for the ATerm pattern encoding an AsFix production. Using these overlays the complicated ATerm above can be written as

```
Prod([Sort("E"), Lit("+"), Sort("E")], Sort("E"))
```

Overlays can now be used in the recursive pattern that characterizes the restriction of ATerms to AsFix. The patterns asfix-... in Figure 6 describe the syntactic categories of AsFix expresions using the congruences — expressions such as Lit(string) and App(asfix-prod, list(x)) — that are derived from the overlays. The pattern asfix-prod defines a production as a Prod with a list of asfix-sorts as first argument and an asfix-sort as second argument. The pattern asfix-term defines an AsFix term as a literal, a typed variable, or an application of a production to a list of terms. The pattern asfix-equ defines an equation as an Equation with a non-variable term as left-hand side and a term as right-hand side.

The recursive patterns only describes 'raw' AsFix expressions and do not check that the argument sorts in the Prod of an application correspond to the

```
module asfix
imports aterms traversals strings
overlays
  Sort(n)        = Appl(Sym("Sort"), [Appl(Str(n), [])])
  Lit(l)         = Appl(Sym("Lit"), [Appl(Str(l), [])])
  Prod(as, r)    = Appl(Sym("Prod"), [AList(as), r])
  App(p, as)     = Appl(Sym("App"), [p, AList(as)])
  Var(s, n)      = Appl(Sym("Var"), [s, n])
  Equation(l, r) = Appl(Sym("Equation"), [l, r])
strategies
  asfix-sort = Sort(string) + Lit(string)
  asfix-prod = Prod(list(asfix-sort), asfix-sort)
  asfix-term = rec x(Lit(string) +
                     Var(asfix-sort, string) +
                     App(asfix-prod, list(x)))
  asfix-equ  = Equation(asfix-term; not(Var(id, id)), asfix-term)
  asfix-eqs  = list(asfix-equ)
rules
  Check1 : Var(x, _) -> x
  Check2 : App(Prod(args, res), args) -> res
  Check3 : Equation(srt, srt) -> Equation(srt, srt)
strategies
  typecheck-term = rec x(Lit(id) + Check1 + App(id, list(x)); Check2)
  typecheck-eqs = list(Equation(typecheck-term, typecheck-term); Check2)
```

Fig. 6. AsFix: overlays, recursive pattern and type checker.

sorts of the actual arguments of the application. To test this a typechecker in an abstract interpretation style similar to that of the Pico typechecker in Section 3 is defined in Figure 6. Only now there is no need to distribute type information, since terms are already annotated with their types.

The process of defining overlays to hide the underlying representation can be repeated, e.g., to define on top of the AsFix abstractions another layer to describe patterns for a specific instantiation of AsFix terms. For instance, take the SDF productions E "+" E -> E and E "*" E -> E. The following overlays define shorthands for AsFix terms using these productions:

```
overlays
  BinexpOp(o)    = Prod([Sort("E"), Lit(o), Sort("E")], Sort("E"))
  Binexp(l, o, r) = App(Binexp(o), [l, Lit(o), r])
  Plus(l, r)     = Binexp(l, "+", r)
  Mul(l, r)      = Binexp(l, "*", r)
```

These overlays allow the 'domain-specific' transformation rule

```
Distr : Mul(x, Plus(y, z)) -> Plus(Mul(x, y), Mul(x, z))
```

Although this rule is written at the level of the embedded language of expressions, they are applied at the level of the underlying ATerm representation.

6 Related Work

Programmable rewriting strategies originate in theorem proving tactics and were first introduced in rewriting in the specification language ELAN [13]. In the algebraic specification formalism Maude [5] strategies can be defined by the user as meta-level specifications. System S and Stratego were developed in [14] (sequential non-deterministic programming and generic term traversal) and [22, 23] (breaking down rewrite rules in matching and building term patterns). ELAN supports congruences and recursive equations, which should support definition of recursive patterns. Overlays and contexts are not supported either by ELAN or by Maude. See [22, 23] for more details on the relation between these languages.

A wide range of languages introduce enhanced pattern matching features. A brief and necessarily incomplete overview follows:

The transformation languages Dora [10] and TXL [6, 15] are examples of languages with some ad-hoc combinations of traversal and pattern matching.

Context patterns can be implemented by means of higher-order matching in λProlog [17]. A higher-order pattern F(t) instantiates the function variable F such that application to t yields the term that is matched. Heering [11] gives an example of second-order matching that we discussed in Section 3. Mohnen's context patterns for Haskell [16] are similar to higher-order matching in λProlog. Sellink and Verhoef [20] show how list matching can be used to implement shallow contexts (that can find statements in a list of statements, but at a fixed nesting depth) used for transforming COBOL programs. Stratego contexts provide the additional possibility of specifying the traversal to be used. This implies that restrictions on the structure of the context can be imposed and that more than one replacement can be done.

Aiken and Murphy [1] describe a language of regular tree expressions for program analysis. Their language is very similar to the recursive patterns in this paper, but is restricted to recognition only.

An overlay is an abbreviation for three abstractions: a match abstraction, a build abstraction and a congruence abstraction. The pattern templates for SML of Aitken and Reppy [2] define two abstractions: a match abstraction and a build abstraction. Congruences are not supported in SML. Another difference with templates is that templates need to be linear. In a definition C(x) = t, the variable x can occur only once in t. Overlays do not need to be linear.

A view type in Wadler's proposal for views in Haskell [24] presents an alternative view to a representation data type by means of a pair of conversion functions that translate between the representation type and the view type. Views are more general than overlays and templates, in that they allow rearrangement of the underlying pattern. However, this added expressivity turns into a disadvantage if one considers pattern matching. Overlays are abstractions that do not result in a loss of efficiency, while views can require an arbitrary transformation. View-like transformations are of course expressible in Stratego. Thompson's lawful types [21] for Miranda are similar to views.

Another problem with general views is that it can destroy equational reasoning [4]. Several proposals [4, 8, 9, 18] repair this by only allowing views in match expressions and not in build expressions. Values in the underlying data representation should be constructed by means of functions. Erwig's active patterns [8] could be considered as functions that inspect and transform their argument and then bind subterms to variables or fail. Fähndrich and Boyland [9] syntactically restrict the patterns used in pattern abstractions such that pattern matching becomes statically checkable.

7 Conclusions

This paper presented three examples of strategic pattern matching: contexts, recursive patterns and overlays. These idioms provide concise specification of expressive patterns that enhance standard first-order patterns. Their definition follows naturally from the features of System S; for some of the techniques new syntactic abstractions were added to Stratego, but no new System S constructs were needed. The key features that enable this expressivity are: (1) ability to abstract over pattern matching (where abstraction over building is a common feature of many languages), and (2) the separation of variable scope and matching, which enables the communication of variable bindings over other operations, and (3) generic term traversals through `all`, `some` and `one`.

The techniques described in this paper have been applied in the specification of the (bootstrapped) Stratego compiler, in an optimizer for RML [23], and in a specification of the warm fusion transformation for functional programs [12]. Future work includes: the application of these techniques in other program transformations; the development of more abstractions for concise specification of program transformations; and the optimization of strategies, in particular traversal fusion, which is important for the optimization of contextual rules.

Acknowledgements The author thanks Andrew Tolmach, Patty Johann and the referees for comments on drafts of this paper.

References

1. A. Aiken and B. Murphy. Implementing regular tree expressions. In *Functional Programming and Computer Architecture (FPCA'91)*, pages 427–447, Aug. 1991.
2. W. E. Aitken and J. H. Reppy. Abstract value constructors. In *ACM SIGPLAN Workshop on ML and its Applications*, pages 1–11, San Francisco, Cal., June 1992.
3. P. Borovanský, C. Kirchner, and H. Kirchner. Controlling rewriting by rewriting. In J. Meseguer, editor, *Proceedings of the First International Workshop on Rewriting Logic and its Applications*, volume 4 of *Electronic Notes in Theoretical Computer Science*, Asilomar, Pacific Grove, CA, September 1996. Elsevier.
4. F. W. Burton and R. D. Cameron. Pattern matching with abstract data types. *Journal of Functional Programming*, 3(2):171–190, April 1993.
5. M. Clavel, S. Eker, P. Lincoln, and J. Meseguer. Principles of Maude. In J. Meseguer, editor, *Proceedings of the First International Workshop on Rewriting Logic and its Applications*, volume 4 of *Electronic Notes in Theoretical Computer Science*, pages 65–89, Asilomar, Pacific Grove, CA, September 1996. Elsevier.

6. J. R. Cordy, I. H. Carmichael, and R. Halliday. *The TXL Programming Language, Version 8*, Apr. 1995.
7. A. Van Deursen, J. Heering, and P. Klint, editors. *Language Prototyping. An Algebraic Specification Approach*, volume 5 of *AMAST Series in Computing*. World Scientific, Singapore, September 1996.
8. M. Erwig. Active patterns. In *Implementation of Functional Languages*, volume 1268 of *Lecture Notes in Computer Science*, pages 21–40, 1996.
9. M. Fähndrich and J. Boyland. Statically checkable pattern abstractions. In *International Conference on Functional Programming (ICFP'97)*, pages 75–84, Amsterdam, The Netherlands, June 1997. ACM SIGPLAN.
10. C. D. Farnum. *Pattern-Based Languages for Prototyping of Compiler Optimizers*. PhD thesis, University of California, Berkeley, 1990. Technical Report CSD-90-608.
11. J. Heering. Second-order term rewriting specification of static semantics: An exercise. In Van Deursen et al. [7], chapter 8, pages 295–305.
12. P. Johann and E. Visser. Warm fusion in Stratego. A case study in generation of program transformation systems. Technical report, Department of Computer Science, Universiteit Utrecht, 1999. http://www.cs.uu.nl/~visser/stratego/.
13. C. Kirchner, H. Kirchner, and M. Vittek. Implementing computational systems with constraints. In P. Kanellakis, J.-L. Lassez, and V. Saraswat, editors, *Proceedings of the first Workshop on Principles and Practice of Constraint Programming*, pages 166–175, Providence R.I., USA, 1993. Brown University.
14. B. Luttik and E. Visser. Specification of rewriting strategies. In M. P. A. Sellink, editor, *2nd International Workshop on the Theory and Practice of Algebraic Specifications (ASF+SDF'97)*, Electronic Workshops in Computing, Berlin, November 1997. Springer-Verlag.
15. A. Malton. The denotational semantics of a functional tree-manipulation language. *Computer Languages*, 19(3):157–168, 1993.
16. M. Mohnen. Context patterns, part ii. In *Implementation of Functional Languages*, pages 338–357, 1997.
17. G. Nadathur and D. Miller. An overview of λProlog. In R. A. Kowalski, editor, *Logic Programming. Proceedings of the Fifth International Conference and Symposium*, volume 1, pages 810–827, Cambridge, Mass., USA, 1988. MIT Press.
18. C. Okasaki. Views for Standard ML. In *SIGPLAN Workshop on ML*, pages 14–23, Baltimore, Maryland, USA, September 1998.
19. P. A. Olivier and H. A. de Jong. Efficient annotated terms. Technical report, Programming Research Group, University of Amsterdam, August 1998.
20. M. P. A. Sellink and C. Verhoef. Native patterns. In M. Blaha, A. Quilici, and C. Verhoef, editors, *Proceedings of the 5-th Working Conference on Reverse Engineering (WCRE'98)*, pages 89–103, Honolulu, Hawaii, USA, October 1998.
21. S. Thompson. Laws in Miranda. In *ACM Symposium on Lisp and Functional Programming*, pages 1–12. ACM, August 1986.
22. E. Visser and Z.-e.-A. Benaissa. A core language for rewriting. In C. Kirchner and H. Kirchner, editors, *Second International Workshop on Rewriting Logic and its Applications (WRLA'98)*, Electronic Notes in Theoretical Computer Science, Pont-à-Mousson, France, September 1–4 1998. Elsevier.
23. E. Visser, Z.-e.-A. Benaissa, and A. Tolmach. Building program optimizers with rewriting strategies. In *International Conference on Functional Programming (ICFP'98)*, pages 13–26, Baltimore, Maryland, September 1998. ACM.
24. P. Wadler. Views: A way for pattern matching to cohabit with data abstraction. In *ACM Symposium on Principles of Programming Languages*, pages 307–313, Munich, January 1987. ACM.

On the Strong Normalisation of Natural Deduction with Permutation-Conversions

Philippe de Groote

LORIA UMR n° 7503 – INRIA
Campus Scientifique, B.P. 239
54506 Vandœuvre lès Nancy Cedex – France
e-mail: Philippe.de.Groote@loria.fr

Abstract. We present a modular proof of the strong normalisation of intuitionistic logic with permutation-conversions. This proof is based on the notions of negative translation and CPS-simulation.

1 Introduction

Natural deduction systems provide a notion of proof that is more compact (or, quoting Girard [6], more *primitive*) than that of sequent calculi. In particular, natural deduction is better adapted to the study of proof-normalisation procedures. This is true, at least, for the intuitionistic systems, where proof-normalisation expresses the computational content of the logic. Nevertheless, even in the intuitionistic case, the treatments of disjunction and existential quantification are problematic. This is due to the fact that the elimination rules of these connectives introduce arbitrary formulas as their conclusions. Consequently, in order to satisfy the subformula property, the so-called permutation-conversions are needed.

Strong normalisation proofs for intuitionistic logic [12] are more intricate in the presence of permutation-conversions. For instance, the proofs given in textbooks such as [6] and [14] do not take permutation-conversions into account.[1] In this paper, we revisit this problem and present a simple proof of the strong normalisation of intuitionistic logic with permutation-conversions. This proof, which is inspired by a similar proof in [4], has several advantages:

- It is modular and, therefore, easily adaptable to other systems. Indeed, the problem related to the interaction between permutation- and detour-conversions is avoided (see Lemma 16, in Section 5).
- It is based on a continuation-passing-style interpretation of intuitionistic logic, which sheds light on the computational content of the several conversion rules. In particular, it shows that the computational content of permutation-conversions is nil.

---

[1] In [6], extending the proof to permutation-conversions is left to the reader while in [14], the technique that is used is not adapted to the case of permutation-conversions.

– It is based on an arithmetisable translation of intuitionistic logic into the simply typed λ-calculus. Consequently, when combined with an arithmetisable proof of strong normalisation of the simply typed λ-calculus (see [3], for instance), it yields a completely arithmetisable proof of the strong normalisation of intuitionistic logic. This must be contrasted with the proof in [14], which is based on an interpretation into higher-order Heyting arithmetic.

The paper is organised as follows.

Section 2 is an introduction to the proof-theory of intuitionistic positive propositional logic (**IPPL**). In particular, we define the notions of detour- and permutation-conversions by means of an associated λ-calculus ($\lambda^{\to\wedge\vee}$). Our presentation is essentially inspired by [14].

In Section 3, we establish the strong normalisation of **IPPL** with respect to permutation-conversions. The proof consists simply in assigning a norm to the untyped terms of $\lambda^{\to\wedge\vee}$, and showing that this norm strictly decreases by permutation conversion.

Section 4 provides a negative translation of **IPPL** into the implicative fragment of intuitionistic logic. At the level of the proofs, this negative translation corresponds to a CPS-simulation of $\lambda^{\to\wedge\vee}$ into the simply typed λ-calculus. We show that this CPS-simulation (slightly modified) commutes with the relations of detour-conversion and β-reduction, from which we conclude that $\lambda^{\to\wedge\vee}$ is strongly normalisable with respect to detour-conversions.

In Section 5, we collect the results of the two previous sections in order to show that $\lambda^{\to\wedge\vee}$ is strongly normalisable with respect to detour- and permutation-conversions mixed together. To this end, we show that the modified CPS-translation of Section 4 interprets the relation of permutation-conversion as equality. This means that we have found a negative translation commuting with the permutation-conversions, answering a problem raised by Mints [9].

Our proof may be easily adapted to full intuitionistic propositional calculus (by adding negation), and to first-order intuitionistic logic (by adding quantifiers). We do not present this extension here, for the lack of space.

2   Intuitionistic positive propositional logic

2.1   Natural deduction

The formulas of intuitionistic positive propositional logic (**IPPL**) are built up from an alphabet of atomic propositions $\mathcal{A}$ and the connectives →, ∧, and ∨ according to the following grammar:

$$\mathcal{F} \quad ::= \quad \mathcal{A} \mid \mathcal{F} \to \mathcal{F} \mid \mathcal{F} \wedge \mathcal{F} \mid \mathcal{F} \vee \mathcal{F}$$

Following Gentzen [5], the meaning of the connectives is specified by the introduction and elimination rules of the following natural deduction system (where a bracketed formula corresponds to a hypothesis that may be discarded when applying the rule).

$$
\frac{\begin{array}{c}[\alpha]\\ \vdots\\ \beta\end{array}}{\alpha \to \beta} \qquad \frac{\alpha \to \beta \quad \alpha}{\beta}
$$

$$
\frac{\alpha \quad \beta}{\alpha \wedge \beta} \qquad \frac{\alpha \wedge \beta}{\alpha} \qquad \frac{\alpha \wedge \beta}{\beta}
$$

$$
\frac{\alpha}{\alpha \vee \beta} \qquad \frac{\beta}{\alpha \vee \beta} \qquad \frac{\alpha \vee \beta \quad \overset{[\alpha]}{\gamma} \quad \overset{[\beta]}{\gamma}}{\gamma}
$$

As observed by Prawitz [11], an introduction immediately followed by an elimination corresponds to a *detour* that can be (locally) eliminated. Consider, for instance, the case of implication:

$$
\frac{\dfrac{\begin{array}{c}[\alpha]\\ \vdots\ \Pi_1\\ \beta\end{array}}{\alpha \to \beta}\ \text{(Intro.)} \qquad \begin{array}{c}\vdots\ \Pi_2\\ \alpha\end{array}}{\beta}\ \text{(Elim.)}
$$

Π_1 is a proof of β under the hypothesis α. On the other hand, Π_2 is a proof of α. Consequently, one may obtain a direct proof of β by grafting Π_2 at every place where α occurs in Π_1 as a hypothesis discarded by Rule (Elim.):

$$
\begin{array}{c}
\vdots\ \Pi_2\\
\alpha\\
\vdots\ \Pi_1\\
\beta
\end{array}
$$

When such local reduction steps allow any proof to be transformed into a proof without detour, one says that the given natural deduction system satisfies the *normalisation property*. Moreover, when this property holds independently of the strategy that is used in applying the reduction steps, one says that the system satisfies the *strong normalisation property*.

2.2 Natural deduction as a term calculus

As well-known, there exists a correspondence between natural deduction systems and typed λ-calculi, namely, the Curry-Howard isomorphism [2,7]. This correspondence, which is described in the table below, allows natural deduction proofs to be denoted by terms.

Natural deduction	λ-calculus
propositions	types
connectives	type constructors
proofs	terms
introduction rules	term constructors
elimination rules	term destructors
active hypothesis	free variables
discarded hypothesis	bound variables

In the case of **IPPL**, the corresponding term calculus is the simply typed λ-calculus with product and coproduct, which we will call $\lambda^{\to\wedge\vee}$. In particular, introduction and elimination rules for conjunction correspond to pairing and projection functions, while introduction and elimination rules for disjunction correspond to injection and case analysis functions. The syntax is described by the following grammar, where $\mathcal{X}$ is a set of variables:

$$\mathcal{T} ::= \mathcal{X} \mid \lambda\mathcal{X}.\mathcal{T} \mid (\mathcal{T}\,\mathcal{T}) \mid \mathbf{p}(\mathcal{T},\mathcal{T}) \mid \mathbf{p}_1\,\mathcal{T} \mid \mathbf{p}_2\,\mathcal{T} \mid \mathbf{k}_1\,\mathcal{T} \mid \mathbf{k}_2\,\mathcal{T} \mid \mathbf{D}_{\mathcal{X},\mathcal{X}}(\mathcal{T},\mathcal{T},\mathcal{T})$$

and the typing rules are as follows:

$$\frac{\begin{array}{c}[x:\alpha]\\ M:\beta\end{array}}{\lambda x.\,M:\alpha\to\beta} \qquad \frac{M:\alpha\to\beta \quad N:\alpha}{M\,N:\beta}$$

$$\frac{M:\alpha \quad N:\beta}{\mathbf{p}(M,N):\alpha\wedge\beta} \qquad \frac{M:\alpha\wedge\beta}{\mathbf{p}_1\,M:\alpha} \qquad \frac{M:\alpha\wedge\beta}{\mathbf{p}_2\,M:\beta}$$

$$\frac{M:\alpha}{\mathbf{k}_1\,M:\alpha\vee\beta} \qquad \frac{M:\beta}{\mathbf{k}_2\,M:\alpha\vee\beta} \qquad \frac{M:\alpha\vee\beta \quad N:\gamma \quad O:\gamma}{\mathbf{D}_{x,y}(M,N,O):\gamma}\quad {\scriptstyle [x:\alpha]\ [y:\beta]}$$

2.3 Detour-conversion rules

The Curry-Howard isomorphism is not only a simple matter of notation. Its deep meaning lies in the relation existing between proof normalisation and λ-term evaluation. Indeed an introduction immediately followed by an elimination corresponds to a term destructor applied to term constructor. Consequently, Prawitz's detour elimination steps amount to evaluation steps. For instance, the detour elimination step for implication corresponds exactly to the familiar notion of β-reduction:

$$\frac{\dfrac{\begin{array}{c}[x:\alpha]\\ \vdots\ \Pi_1\\ M:\beta\end{array}}{\lambda x.\,M:\alpha\to\beta} \quad \begin{array}{c}\vdots\ \Pi_2\\ N:\alpha\end{array}}{(\lambda x.\,M)\,N:\beta} \quad\longrightarrow\quad \begin{array}{c}\vdots\ \Pi_2\\ N:\alpha\\ \vdots\ \Pi_1\\ M[x:=N]:\beta\end{array}$$

It is therefore possible to specify the detour elimination steps as simple rewriting rules between λ-terms. Following [14], these rules are called *detour-conversion* rules. We use "$\to_D$" to denote the corresponding one-step reduction relation between λ-terms and, following [1], we use "$\overset{+}{\to}_D$" and "$\twoheadrightarrow_D$" to denote, repectively, the transitive closure and the transitive, reflexive closure of "$\to_D$".

Definition 1. (Detour-conversions)

1. $(\lambda x.\, M)\, N \to_D M[x{:=}N]$
2. $\mathbf{p}_1\, \mathbf{p}(M, N) \to_D M$
3. $\mathbf{p}_2\, \mathbf{p}(M, N) \to_D N$
4. $\mathbf{D}_{x,y}(\mathbf{k}_1\, M, N, O) \to_D N[x{:=}M]$
5. $\mathbf{D}_{x,y}(\mathbf{k}_2\, M, N, O) \to_D O[y{:=}M]$ ∎

It must be clear that the Curry-Howard isomorphism allows one to reduce proof normalisation problems to λ-term normalisation problems. In particular, the strong normalisation of intuitionistic implicative logic corresponds to the strong normalisation of the simply typed λ-calculus [1] (which is, by the way, the only normalisation result we assume in this paper). There is, however, a slight difference that must be stressed. The grammar given in Section 2.2 defines untyped λ-terms, some of which do not correspond to natural deduction proofs. Consequently, the question whether the untyped λ-terms satisfy some normalisation property has no direct equivalent in the logical setting.

2.4 Permutation-conversion rules

In the disjunction free fragment of **IPPL**, the normal proofs (i.e., the proofs without detour) satisfy the subformula property. This means that if Π is a normal proof of a formula α under a set of hypotheses Γ, then each formula occurring in Π is a subformula of a formula in $\Gamma \cup \{\alpha\}$. In the presence of disjunction, the detour-conversions of Definition 1 are no longer sufficient to guarantee the subformula property. Consider the following example:

$$
\cfrac{\alpha \vee \beta \qquad \cfrac{\begin{array}{c}[\alpha,\gamma]\\ \vdots\ \Pi_1\\ \delta\end{array}}{\gamma \to \delta}\ \text{(Intro.)} \qquad \cfrac{\begin{array}{c}[\beta,\gamma]\\ \vdots\ \Pi_2\\ \delta\end{array}}{\gamma \to \delta}\ \text{(Intro.)}}{\cfrac{\gamma \to \delta \qquad\qquad\qquad \begin{array}{c}\vdots\ \Pi\\ \gamma\end{array}}{\delta}\ \text{(Elim.)}}
$$

A priori there is no reason why γ and $\gamma \to \delta$ would be subformulas of δ or of any hypothesis from which δ is derived. This is due to the fact that there are introduction rules followed by an elimination rule. Indeed, one would like to reduce the above example as follows:

$$
\begin{array}{cc}
& \vdots\ \Pi \qquad\qquad \vdots\ \Pi \\
[\alpha]\ \dot{\gamma} \qquad\qquad [\beta]\ \dot{\gamma} \\
\vdots \qquad\qquad \vdots\ \Pi_1 \qquad\quad \vdots\ \Pi_2 \\
\alpha \vee \beta \quad\quad \delta \qquad\qquad\quad \delta \\
\hline
\delta
\end{array}
$$

However, such a reduction is not possible by applying only the detour-conversion rules of Definition 1 because, in the above example, the elimination rule does not *immediately* follow the introduction rules. For this reason, some other conversion rules are needed, the so-called *permutation-conversion*. For instance, to reduce the above example, we need the following rule:

$$
\cfrac{\cfrac{\alpha \vee \beta \quad [\alpha]\;\vdots\;\gamma \to \delta \quad [\beta]\;\vdots\;\gamma \to \delta}{\gamma \to \delta} \quad \vdots\;\gamma\;\Pi}{\delta}\;\Pi \quad\longrightarrow\quad \cfrac{\alpha \vee \beta \quad \cfrac{[\alpha]\;\vdots\;\gamma \to \delta \;\; \vdots\;\gamma\;\Pi}{\delta} \quad \cfrac{[\beta]\;\vdots\;\gamma \to \delta \;\; \vdots\;\gamma\;\Pi}{\delta}}{\delta}
$$

The above conversion, which concerns the implication elimination rule, obeys a general scheme:

$$
\cfrac{\cfrac{\alpha \vee \beta \quad [\alpha]\;\vdots\;\gamma \quad [\beta]\;\vdots\;\gamma}{\gamma} \quad \vdots}{\delta}\;(\text{Elim.}) \quad\longrightarrow\quad \cfrac{\alpha \vee \beta \quad \cfrac{[\alpha]\;\gamma\;\vdots}{\delta}(\text{Elim.}) \quad \cfrac{[\beta]\;\gamma\;\vdots}{\delta}(\text{Elim.})}{\delta}
$$

All the permutation-conversions may be obtained, from the above scheme, by replacing Rule (Elim.) by the different elimination rules of **IPPL**. Of course, it is also possible to express these permutation-conversions as rewriting rules between λ-terms. This is achieved in the following definition.

Definition 2. (Permutation-conversions)

1. $\mathbf{D}_{x,y}(M,N,O)\,P \to_P \mathbf{D}_{x,y}(M,N\,P,O\,P)$
2. $\mathbf{p}_1\,\mathbf{D}_{x,y}(M,N,O) \to_P \mathbf{D}_{x,y}(M,\mathbf{p}_1\,N,\mathbf{p}_1\,O)$
3. $\mathbf{p}_2\,\mathbf{D}_{x,y}(M,N,O) \to_P \mathbf{D}_{x,y}(M,\mathbf{p}_2\,N,\mathbf{p}_2\,O)$
4. $\mathbf{D}_{u,v}(\mathbf{D}_{x,y}(M,N,O),P,Q) \to_P \mathbf{D}_{x,y}(M,\mathbf{D}_{u,v}(N,P,Q),\mathbf{D}_{u,v}(O,P,Q))$ ■

We are now in a position to state precisely the question addressed by the present paper: how can we give a modular proof of the strong normalisation of **IPPL** with respect to both the detour- and permutation-conversions? or, equivalently, how can we prove that the typed λ-terms of Section 2.2 satisfy the strong normalisation property with respect to the reduction relation induced by the union of the rewriting systems of definitions 1 and 2?

3   Strong Normalisation of permutation-conversions

In this section, we establish the strong normalisation of $\lambda^{\to\wedge\vee}$ with respect to permutation-conversions. The proof consists simply in assigning a norm to the λ-terms and then in proving that this norm (which we call the permutation degree) is decreasing under the reduction relation $\to_P$.

Definition 3. (Permutation degree)

1. $|x| = 1$
2. $|\lambda x.\, M| = |M|$
3. $|M\, N| = |M| + \#M \times |N|$
4. $|\mathbf{p}(M, N)| = |M| + |N|$
5. $|\mathbf{p}_1\, M| = |M| + \#M$
6. $|\mathbf{p}_2\, M| = |M| + \#M$
7. $|\mathbf{k}_1\, M| = |M|$
8. $|\mathbf{k}_2\, M| = |M|$
9. $|\mathbf{D}_{x,y}(M, N, O)| = |M| + \#M \times (|N| + |O|)$

where:

10. $\#x = 1$
11. $\#\lambda x.\, M = 1$
12. $\#M\, N = \#M$
13. $\#\mathbf{p}(M, N) = 1$
14. $\#\mathbf{p}_1\, M = \#M$
15. $\#\mathbf{p}_2\, M = \#M$
16. $\#\mathbf{k}_1\, M = 1$
17. $\#\mathbf{k}_2\, M = 1$
18. $\#\mathbf{D}_{x,y}(M, N, O) = 2 \times \#M \times (\#N + \#O)$ ■

Lemma 4. *Let M and N be two λ-terms of $\lambda^{\to\wedge\vee}$ such that $M \to_P N$. Then $\#M = \#N$.*

Proof. Let $C[\,]$ be any context, i.e., a λ-term with a hole. It is straightforward that $\#C[M] = \#C[N]$ whenever $\#M = \#N$. Hence it remains to show that $\#$ is invariant under each rewriting rule of Definition 2.

$$\#\mathbf{D}_{x,y}(M, N, O)\, P$$
$$= \#\mathbf{D}_{x,y}(M, N, O)$$
$$= 2 \times \#M \times (\#N + \#O)$$
$$= 2 \times \#M \times (\#N\, P + \#O\, P)$$
$$= \#\mathbf{D}_{x,y}(M, N\, P, O\, P)$$
$$\#\mathbf{p}_i\, \mathbf{D}_{x,y}(M, N, O)$$
$$= \#\mathbf{D}_{x,y}(M, N, O)$$
$$= 2 \times \#M \times (\#N + \#O)$$
$$= 2 \times \#M \times (\#\mathbf{p}_i\, N + \#\mathbf{p}_i\, O)$$
$$= \#\mathbf{D}_{x,y}(M, \mathbf{p}_i\, N, \mathbf{p}_i\, O)$$

$$\#\mathbf{D}_{u,v}(\mathbf{D}_{x,y}(M,N,O),P,Q)$$
$$= 2 \times \#\mathbf{D}_{x,y}(M,N,O) \times (\#P + \#Q)$$
$$= 4 \times \#M \times (\#N + \#O) \times (\#P + \#Q)$$
$$= 2 \times \#M \times (2 \times \#N \times (\#P + \#Q) + 2 \times \#O \times (\#P + \#Q))$$
$$= 2 \times \#M \times (\#\mathbf{D}_{u,v}(N,P,Q) + \#\mathbf{D}_{u,v}(O,P,Q))$$
$$= \#\mathbf{D}_{x,y}(M,\mathbf{D}_{u,v}(N,P,Q),\mathbf{D}_{u,v}(O,P,Q))$$

$\square$

Lemma 5. *Let M and N be two λ-terms of $\lambda^{\to\wedge\vee}$ such that $M \to_P N$. Then $|M| > |N|$.*

Proof. The proof is similar to that of the previous lemma. We show that the permutation degree is strictly decreasing under the rewriting rules of Definition 2.

$$|\mathbf{D}_{x,y}(M,N,O)\,P|$$
$$= |\mathbf{D}_{x,y}(M,N,O)| + \#\mathbf{D}_{x,y}(M,N,O) \times |P|$$
$$= |M| + \#M \times (|N| + |O|) + 2 \times \#M \times (\#N + \#O) \times |P|$$
$$> |M| + \#M \times (|N| + |O|) + \#M \times (\#N + \#O) \times |P|$$
$$= |M| + \#M \times (|N| + \#N \times |P| + |O| + \#O \times |P|)$$
$$= |M| + \#M \times (|N\,P| + |O\,P|)$$
$$= |\mathbf{D}_{x,y}(M,N\,P,O\,P)|$$

$$|\mathbf{p}_i\,\mathbf{D}_{x,y}(M,N,O)|$$
$$= |\mathbf{D}_{x,y}(M,N,O)| + \#\mathbf{D}_{x,y}(M,N,O)$$
$$= |M| + \#M \times (|N| + |O|) + 2 \times \#M \times (\#N + \#O)$$
$$> |M| + \#M \times (|N| + |O|) + \#M \times (\#N + \#O)$$
$$= |M| + \#M \times (|N| + \#N + |O| + \#O)$$
$$= |M| + \#M \times (|\mathbf{p}_i\,N| + |\mathbf{p}_i\,O|)$$
$$= |\mathbf{D}_{x,y}(M,\mathbf{p}_i\,N,\mathbf{p}_i\,O)|$$

$$|\mathbf{D}_{u,v}(\mathbf{D}_{x,y}(M,N,O),P,Q)|$$
$$= |\mathbf{D}_{x,y}(M,N,O)| + \#\mathbf{D}_{x,y}(M,N,O) \times (|P| + |Q|)$$
$$= |M| + \#M \times (|N| + |O|) + 2 \times \#M \times (\#N + \#O) \times (|P| + |Q|)$$
$$> |M| + \#M \times (|N| + |O|) + \#M \times (\#N + \#O) \times (|P| + |Q|)$$
$$= |M| + \#M \times (|N| + \#N \times (|P| + |Q|) + |O| + \#O \times (|P| + |Q|))$$
$$= |M| + \#M \times (|\mathbf{D}_{u,v}(N,P,Q)| + |\mathbf{D}_{u,v}(O,P,Q)|)$$
$$= |\mathbf{D}_{x,y}(M,\mathbf{D}_{u,v}(N,P,Q),\mathbf{D}_{u,v}(O,P,Q))|$$

$\square$

We immediately obtain the expected strong normalisation result from the above lemma.

Proposition 6. *$\lambda^{\to\wedge\vee}$ is strongly normalisable with respect to permutation-conversions.* $\square$

Remark that this proposition also holds for the untyped terms. This fact confirms that the permutation-conversions do not have a real computational meaning. The fact that they are needed to obtain the subformula property may be seen as a defect of the syntax.

4 Negative translation and CPS-simulation

We now establish the strong normalisation of $\lambda^{\to\wedge\vee}$ with respect to detour-conversions. To this end we interpret **IPPL** into intuitionistic implicative logic by means of a negative translation. This corresponds to a translation of $\lambda^{\to\wedge\vee}$ into the simply typed λ-calculus. This translation must satisfy two requirements. On the one hand, it must provide a simulation of the detour-conversions. On the other hand it must be compatible with the permutation-conversions in a sense that will be explained. In order to satisfy the first requirement, the negative translation we use is a generalisation of the one used by Meyer and Wand in the implicative case [8], i.e., a generalisation of the translation induced by Plotkin's call-by-name CPS-translation [10].

Definition 7. (Negative translation) The negative translation $\overline{\alpha}$ of any formula α is defined as:
$$\overline{\alpha} = \sim\sim\alpha^{\circ}$$
where
$$\sim\alpha = \alpha \to o$$
for some distinguished atomic proposition o (that is not used elsewhere), and where:

1. $a^{\circ} = a$
2. $(\alpha \to \beta)^{\circ} = \overline{\alpha} \to \overline{\beta}$
3. $(\alpha \wedge \beta)^{\circ} = \sim(\overline{\alpha} \to \sim\overline{\beta})$
4. $(\alpha \vee \beta)^{\circ} = \sim\overline{\alpha} \to \sim\sim\overline{\beta}$ ∎

Then we accomodate Plotkin's call-by-name simulation to the case of $\lambda^{\to\wedge\vee}$.

Definition 8. (CPS-translation)

1. $\overline{x} = \lambda k. x\,k$
2. $\overline{\lambda x. M} = \lambda k. k\,(\lambda x. \overline{M})$
3. $\overline{(M\,N)} = \lambda k. \overline{M}\,(\lambda m. m\,\overline{N}\,k)$
4. $\overline{\mathbf{p}(M,N)} = \lambda k. k\,(\lambda p. p\,\overline{M}\,\overline{N})$
5. $\overline{\mathbf{p}_1\,M} = \lambda k. \overline{M}\,(\lambda p. p\,(\lambda i. \lambda j. i\,k))$
6. $\overline{\mathbf{p}_2\,M} = \lambda k. \overline{M}\,(\lambda p. p\,(\lambda i. \lambda j. j\,k))$
7. $\overline{\mathbf{k}_1\,M} = \lambda k. k\,(\lambda i. \lambda j. i\,\overline{M})$
8. $\overline{\mathbf{k}_2\,M} = \lambda k. k\,(\lambda i. \lambda j. j\,\overline{M})$
9. $\overline{\mathbf{D}_{x,y}(M,N,O)} = \lambda k. \overline{M}\,(\lambda m. m\,(\lambda x. \overline{N}\,k)\,(\lambda y. \overline{O}\,k))$

where k, m, p, i and j are fresh variables. ∎

We now prove that the translations of Definition 7 and 8 commute with the typing relation.

Proposition 9. *Let M be a λ-term of $\lambda^{\to\wedge\vee}$ typable with type α under a set of declarations Γ. Then $\overline{M}$ is a λ-term of the simply typed λ-calculus, typable with type $\overline{\alpha}$ under the set of declarations $\overline{\Gamma}$.*

Proof. See Appendix A. □

The translation of Definition 8 does not map normal forms to normal forms. This is due to the so-called administrative redexes that are introduced by the translation. The modified translation below circumvents this problem.

Definition 10. (Modified CPS-translation) The modified CPS-translation $\overline{\overline{M}}$ of any λ-term M of $\lambda^{\rightarrow\wedge\vee}$ is defined as:

$$\overline{\overline{M}} = \lambda k. (M : k)$$

where k is a fresh variable, and where the infix operator ":" obeys the following definition:

1. $x : K = x\,K$
2. $\lambda x. M : K = K\,(\lambda x. \overline{\overline{M}})$
3. $(M\,N) : K = M : \lambda m.\,m\,\overline{\overline{N}}\,K$
4. $\mathbf{p}(M,N) : K = K\,(\lambda p.\,p\,\overline{\overline{M}}\,\overline{\overline{N}})$
5. $\mathbf{p}_1\,M : K = M : \lambda p.\,p\,(\lambda i.\,\lambda j.\,i\,K)$
6. $\mathbf{p}_2\,M : K = M : \lambda p.\,p\,(\lambda i.\,\lambda j.\,j\,K)$
7. $\mathbf{k}_1\,M : K = K\,(\lambda i.\,\lambda j.\,i\,\overline{\overline{M}})$
8. $\mathbf{k}_2\,M : K = K\,(\lambda i.\,\lambda j.\,j\,\overline{\overline{M}})$
9. $\mathbf{D}_{x,y}(M,N,O) : K = M : \lambda m.\,m\,(\lambda x.\,(N : K))\,(\lambda y.\,(O : K))$

where m, p, i and j are fresh variables, and where, in Clause 9, x and y do not occur free in K. Remark that this last condition is not restrictive since it may always be satisfied by renaming. ■

As expected, the modified translation is a β-reduced form of the CPS-translation.

Lemma 11. *Let M and K be terms of $\lambda^{\rightarrow\wedge\vee}$. Then:*

1. $\overline{M} \twoheadrightarrow_\beta \overline{\overline{M}}$,
2. $\overline{M}\,K \twoheadrightarrow_\beta M : K$.

Proof. We proceed by induction on the structure of M. Property 1 is the property of interest, while Property 2 is needed to make the induction work. □

From this lemma, we get the analogue of Proposition 9 for the modified translation.

Proposition 12. *Let M be a λ-term of $\lambda^{\rightarrow\wedge\vee}$ typable with type α under a set of declarations Γ. Then $\overline{\overline{M}}$ is a λ-term of the simply typed λ-calculus, typable with type $\overline{\alpha}$ under the set of declarations $\overline{\Gamma}$.*

Proof. The proposition follows from Proposition 9, Lemma 11, and the subject reduction property of the simply typed λ-calculus.

The modified CPS-translation allows the detour-conversions of $\lambda^{\to\wedge\vee}$ to be simulated by β-reduction. This is established by the next lemmas.

Lemma 13. *Let M and N be λ-terms of $\lambda^{\to\wedge\vee}$ and K be a simple λ-term. Then:*

1. $(M : K)[x := \overline{\overline{N}}] \twoheadrightarrow_\beta (M[x := N]) : (K[x := \overline{\overline{N}}])$,
2. $\overline{\overline{M}}[x := \overline{\overline{N}}] \twoheadrightarrow_\beta \overline{\overline{M[x := N]}}$.

Proof. Property 2 is a direct consequence of Property 1, which is established by a straightforward induction on the structure of M. ☐

Lemma 14. *Let M and N be two λ-terms of $\lambda^{\to\wedge\vee}$ such that $M \to_D N$. Then:*

1. $M : K \overset{+}{\to}_\beta N : K$, *for any simple λ-term K,*
2. $\overline{\overline{M}} \overset{+}{\to}_\beta \overline{\overline{N}}$.

Proof. Property 2 may be established as a direct consequence of Property 1. Then proving that $C[M] : K \overset{+}{\to}_\beta C[N] : K$ whenever $M : K \overset{+}{\to}_\beta N : K$ consists in a straightforward induction on the structure of the context $C[\,]$. Hence it remains to establish Property 1 for the five rewriting rules of Definition 1.

$$
\begin{aligned}
((\lambda x.\, M)\, N) : K \;&=\; (\lambda x.\, M) : \lambda m.\, m\, \overline{\overline{N}}\, K \\
&=\; (\lambda m.\, m\, \overline{\overline{N}}\, K)\, (\lambda x.\, \overline{\overline{M}}) \\
&\to_\beta\; (\lambda x.\, \overline{\overline{M}})\, \overline{\overline{N}}\, K \\
&\to_\beta\; \overline{\overline{M}}[x := \overline{\overline{N}}]\, K \\
&\twoheadrightarrow_\beta\; \overline{\overline{M[x := N]}}\, K \\
&\to_\beta\; (M[x := N]) : K
\end{aligned}
$$

$$
\begin{aligned}
(\mathbf{p}_i\, \mathbf{p}(M, N)) : K \;&=\; \mathbf{p}(M_1, M_2) : \lambda p.\, p\, (\lambda j_1.\, \lambda j_2.\, j_i\, K) \\
&=\; (\lambda p.\, p\, (\lambda j_1.\, \lambda j_2.\, j_i\, K))\, (\lambda p.\, p\, \overline{\overline{M}}_1\, \overline{\overline{M}}_2) \\
&\to_\beta\; (\lambda p.\, p\, \overline{\overline{M}}_1\, \overline{\overline{M}}_2)\, (\lambda j_1.\, \lambda j_2.\, j_i\, K) \\
&\to_\beta\; (\lambda j_1.\, \lambda j_2.\, j_i\, K)\, \overline{\overline{M}}_1\, \overline{\overline{M}}_2 \\
&\overset{+}{\to}_\beta\; \overline{\overline{M}}_i\, K \\
&\to_\beta\; M_i : K
\end{aligned}
$$

$$
\begin{aligned}
\mathbf{D}_{x_1, x_2}(\mathbf{k}_i\, M, N_1, N_2) : K \;&=\; (\mathbf{k}_i\, M) : \lambda m.\, m\, (\lambda x_1.\, (N_1 : K))\, (\lambda x_2.\, (N_2 : K)) \\
&=\; (\lambda m.\, m\, (\lambda x_1.\, (N_1 : K))\, (\lambda x_2.\, (N_2 : K)))\, (\lambda j_1.\, \lambda j_2.\, j_i\, \overline{\overline{M}}) \\
&\to_\beta\; (\lambda j_1.\, \lambda j_2.\, j_i\, \overline{\overline{M}})\, (\lambda x_1.\, (N_1 : K))\, (\lambda x_2.\, (N_2 : K)) \\
&\overset{+}{\to}_\beta\; (\lambda x_i.\, (N_i : K))\, \overline{\overline{M}} \\
&\to_\beta\; (N_i : K)[x_i := \overline{\overline{M}}] \\
&\twoheadrightarrow_\beta\; (N_i[x_i := M]) : K
\end{aligned}
$$

☐

The above lemma allows any sequence of detour-conversion steps to be simulated by a longer sequence of β-reduction steps in the simply typed λ-calculus. Therefore, since the simply typed λ-calculus is strongly β-normalisable, we immediately obtain the following proposition.

Proposition 15. $\lambda^{\to \wedge \vee}$ *is strongly normalisable with respect to detour-conversions.* $\square$

5 Strong normalisation

In this section we prove that $\lambda^{\to \wedge \vee}$ is strongly normalisable with respect to both detour- and permutation-conversion. This is not a direct consequence of Propositions 6 and 15 because detour-conversions can create permutation-redexes and, conversely, permutation conversions can create detour-redexes. Therefore we first show that the modified CPS-translation maps the relation of permutation-conversion to syntactic equality.

Lemma 16. *Let M and N be two λ-terms of $\lambda^{\to \wedge \vee}$ such that $M \to_P N$. Then:*

1. $M : K = N : K$, *for any simple λ-term K,*
2. $\overline{\overline{M}} = \overline{\overline{N}}$.

Proof. Property 2 is a direct consequence of Property 1. To show that $C[M] : K = C[N] : K$ whenever $M : K = N : K$ consists in a routine induction on the structure of the context $C[\,]$. It remains to establish Property 1 for the four rewriting rules of Definition 2.

$$
\begin{aligned}
(\mathbf{D}_{x,y}&(M,N,O)\,P) : K \\
&= \mathbf{D}_{x,y}(M,N,O) : \lambda m.\, m\, \overline{\overline{P}}\, K \\
&= M : \lambda m.\, m\, (\lambda x.\, (N : \lambda m.\, m\, \overline{\overline{P}}\, K))\, (\lambda y.\, (O : \lambda m.\, m\, \overline{\overline{P}}\, K)) \\
&= M : \lambda m.\, m\, (\lambda x.\, (N\, P : K))\, (\lambda y.\, (O\, P : K)) \\
&= \mathbf{D}_{x,y}(M, N\, P, O\, P) : K \\
(\mathbf{p}_i\, \mathbf{D}_{x,y}&(M,N,O)) : K \\
&= \mathbf{D}_{x,y}(M,N,O) : \lambda p.\, p\, (\lambda j_1.\, \lambda j_2.\, j_i\, K) \\
&= M : \lambda m.\, m\, (\lambda x.\, (N : \lambda p.\, p\, (\lambda j_1.\, \lambda j_2.\, j_i\, K))) \\
&\qquad\qquad\quad (\lambda y.\, (O : \lambda p.\, p\, (\lambda j_1.\, \lambda j_2.\, j_i\, K))) \\
&= M : \lambda m.\, m\, (\lambda x.\, (\mathbf{p}_i\, N : K))\, (\lambda y.\, (\mathbf{p}_i\, O : K)) \\
&= \mathbf{D}_{x,y}(M, \mathbf{p}_i\, N, \mathbf{p}_i\, O) : K \\
\mathbf{D}_{u,v}&(\mathbf{D}_{x,y}(M,N,O), P, Q) : K \\
&= \mathbf{D}_{x,y}(M,N,O) : \lambda m.\, m\, (\lambda u.\, (P : K))\, (\lambda v.\, (Q : K)) \\
&= M : \lambda m.\, m\, (\lambda x.\, (N : \lambda m.\, m\, (\lambda u.\, (P : K))\, (\lambda v.\, (Q : K)))) \\
&\qquad\qquad\quad (\lambda y.\, (O : \lambda m.\, m\, (\lambda u.\, (P : K))\, (\lambda v.\, (Q : K)))) \\
&= M : \lambda m.\, m\, (\lambda x.\, (\mathbf{D}_{u,v}(N, P, Q) : K))\, (\lambda y.\, (\mathbf{D}_{u,v}(O, P, Q) : K)) \\
&= \mathbf{D}_{x,y}(M, \mathbf{D}_{u,v}(N, P, Q), \mathbf{D}_{u,v}(O, P, Q)) : K
\end{aligned}
$$

$\square$

We may now prove the main result.

Theorem 17. $\lambda^{\to\wedge\vee}$ *is strongly normalisable with respect to the reduction relation induced by the union of the detour- and permutation-conversions.*

Proof. Suppose it is not the case. Then there would exist an infinite sequence of detour- and permutation-conversion steps starting from a typable term (say, M) of $\lambda^{\to\wedge\vee}$. If this infinite sequence contains infinitely many detour-conversion steps, there must exist, by Lemmas 14 and 16 an infinite sequence of β-reduction steps starting from $\overline{M}$. But this, by Proposition 12, would contradict the strong normalisation of the simply typed λ-calculus. Hence the infinite sequence may contain only a finite number of detour conversion steps. But then, it would contain an infinite sequence of consecutive permutation-conversion steps, which contradicts Proposition 6. $\square$

References

1. H.P. Barendregt. *The lambda calculus, its syntax and semantics*. North-Holland, revised edition, 1984.
2. H.B. Curry and R. Feys. *Combinatory Logic, Vol. I*. North-Holland, 1958.
3. Ph. de Groote. The conservation theorem revisited. In M. Bezem and J.F. Groote, editors, *Proceedings of the International Conference on Typed Lambda Calculi and Applications*, pages 163–178. Lecture Notes in Computer Science, 664, Springer Verlag, 1993.
4. Ph. de Groote. A simple calculus of exception handling. In M. Dezani and G. Plotkin, editors, *Second International Conference on Typed Lambda Calculi and Applications, TLCA'95*, volume 902 of *Lecture Notes in Computer Science*, pages 201–215. Springer Verlag, 1995.
5. G. Gentzen. *Recherches sur la déduction logique (Untersuchungen über das logische schliessen)*. Presses Universitaires de France, 1955. Traduction et commentaire par R. Feys et J. Ladrière.
6. J.-Y. Girard. *Proof Theory and Logical Complexity*. Bibliopolis, 1987.
7. W.A. Howard. The formulae-as-types notion of construction. In J. P. Seldin and J. R. Hindley, editors, *to H. B. Curry: Essays on Combinatory Logic, Lambda Calculus and Formalism*, pages 479–490. Academic Press, 1980.
8. A. Meyer and M. Wand. Continuation semantics in typed lambda-calculi (summary). In R. Parikh, editor, *Logics of Programs*, pages 219–224. Lecture Notes in Computer Science, 193, Springer Verlag, 1985.
9. G. Mints. Private communication, 1997.
10. G. D. Plotkin. Call-by-name, call-by-value and the λ-calculus. *Theoretical Computer Science*, 1:125–159, 1975.
11. D. Prawitz. *Natural Deduction, A Proof-Theoretical Study*. Almqvist & Wiksell, Stockholm, 1965.
12. D. Prawitz. Ideas and results in proof-theory. In J.E. Fenstad, editor, *Proceedings of the Second Scandinavian Logic Symposium*, pages 237–309. North-Holland, 1971.
13. A. Troelstra and D. van Dalen. *Constructivism in Mathematics*, volume I. North-Holland, 1988.
14. A. Troelstra and D. van Dalen. *Constructivism in Mathematics*, volume II. North-Holland, 1988.

A Proof of Proposition 9

Variable

$$\dfrac{x : \sim\sim a \quad k : \sim a}{\dfrac{x\,k : o}{\lambda k.\,x\,k : \sim\sim a}}$$

Abstraction

$$[x : \overline{\alpha}]$$
$$\vdots$$
$$\dfrac{\overline{M} : \overline{\beta}}{}$$

$$\dfrac{k : \sim(\overline{\alpha} \to \overline{\beta}) \quad \lambda x.\,\overline{M} : \overline{\alpha} \to \overline{\beta}}{\dfrac{k\,(\lambda x.\,\overline{M}) : o}{\lambda k.\,k\,(\lambda x.\,\overline{M}) : \sim\sim(\overline{\alpha} \to \overline{\beta})}}$$

Application

$$\dfrac{\overline{M} : \sim\sim(\overline{\alpha} \to \overline{\beta}) \quad \dfrac{\dfrac{m : \overline{\alpha} \to \overline{\beta} \quad \overline{N} : \overline{\alpha}}{m\,\overline{N} : \overline{\beta}} \quad k : \sim\beta^{\circ}}{\dfrac{m\,\overline{N}\,k : o}{\lambda m.\,m\,\overline{N}\,k : \sim(\overline{\alpha} \to \overline{\beta})}}}{\dfrac{\overline{M}\,(\lambda m.\,m\,\overline{N}\,k) : o}{\lambda k.\,\overline{M}\,(\lambda m.\,m\,\overline{N}\,k) : \overline{\beta}}}$$

Pairing

$$\dfrac{k : \sim\sim(\overline{\alpha} \to \sim\overline{\beta}) \quad \dfrac{\dfrac{p : \overline{\alpha} \to \sim\overline{\beta} \quad \overline{M} : \overline{\alpha}}{p\,\overline{M} : \sim\overline{\beta}} \quad \overline{N} : \overline{\beta}}{\dfrac{p\,\overline{M}\,\overline{N} : o}{\lambda p.\,p\,\overline{M}\,\overline{N} : \sim(\overline{\alpha} \to \sim\overline{\beta})}}}{\dfrac{k\,(\lambda p.\,p\,\overline{M}\,\overline{N}) : o}{\lambda k.\,k\,(\lambda p.\,p\,\overline{M}\,\overline{N}) : \sim\sim\sim(\overline{\alpha} \to \sim\overline{\beta})}}$$

Left projection

$$\dfrac{\overline{M} : \sim\sim\sim(\overline{\alpha} \to \sim\overline{\beta}) \quad \dfrac{p : \sim(\overline{\alpha} \to \sim\overline{\beta}) \quad \dfrac{\dfrac{\dfrac{i : \overline{\alpha} \quad k : \sim\alpha^{\circ}}{i\,k : o}}{\lambda j.\,i\,k : \sim\overline{\beta}}}{\lambda i.\,\lambda j.\,i\,k : \overline{\alpha} \to \sim\overline{\beta}}}{\dfrac{p\,(\lambda i.\,\lambda j.\,i\,k) : o}{\lambda p.\,p\,(\lambda i.\,\lambda j.\,i\,k) : \sim\sim(\overline{\alpha} \to \sim\overline{\beta})}}}{\dfrac{\overline{M}\,(\lambda p.\,p\,(\lambda i.\,\lambda j.\,i\,k)) : o}{\lambda k.\,\overline{M}\,(\lambda p.\,p\,(\lambda i.\,\lambda j.\,i\,k)) : \overline{\alpha}}}$$

Right projection

$$\dfrac{\dfrac{\dfrac{\dfrac{j:\overline{\beta} \quad k:\sim\beta^{\circ}}{jk:o}}{\lambda j.\,jk:\sim\overline{\beta}}}{p:\sim(\overline{\alpha}\to\sim\overline{\beta}) \quad \dfrac{\lambda i.\,\lambda j.\,jk:\overline{\alpha}\to\sim\overline{\beta}}{}}{\dfrac{\overline{M}:\sim\sim\sim(\overline{\alpha}\to\sim\overline{\beta}) \quad \dfrac{p\,(\lambda i.\,\lambda j.\,jk):o}{\lambda p.\,p\,(\lambda i.\,\lambda j.\,jk):\sim\sim(\overline{\alpha}\to\sim\overline{\beta})}}{\dfrac{\overline{M}\,(\lambda p.\,p\,(\lambda i.\,\lambda j.\,jk)):o}{\lambda k.\,\overline{M}\,(\lambda p.\,p\,(\lambda i.\,\lambda j.\,jk)):\overline{\beta}}}}$$

Left injection

$$\dfrac{k:\sim(\sim\overline{\alpha}\to\sim\sim\overline{\beta}) \quad \dfrac{\dfrac{\dfrac{i:\sim\overline{\alpha} \quad \overline{M}:\overline{\alpha}}{i\,\overline{M}:o}}{\lambda j.\,i\,\overline{M}:\sim\sim\overline{\beta}}}{\lambda i.\,\lambda j.\,i\,\overline{M}:\sim\overline{\alpha}\to\sim\sim\overline{\beta}}}{\dfrac{k\,(\lambda i.\,\lambda j.\,i\,\overline{M}):o}{\lambda k.\,k\,(\lambda i.\,\lambda j.\,i\,\overline{M}):\sim\sim(\sim\overline{\alpha}\to\sim\sim\overline{\beta})}}$$

Right injection

$$\dfrac{k:\sim(\sim\overline{\alpha}\to\sim\sim\overline{\beta}) \quad \dfrac{\dfrac{\dfrac{j:\sim\overline{\beta} \quad \overline{M}:\overline{\beta}}{j\,\overline{M}:o}}{\lambda j.\,j\,\overline{M}:\sim\sim\overline{\beta}}}{\lambda i.\,\lambda j.\,j\,\overline{M}:\sim\overline{\alpha}\to\sim\sim\overline{\beta}}}{\dfrac{k\,(\lambda i.\,\lambda j.\,j\,\overline{M}):o}{\lambda k.\,k\,(\lambda i.\,\lambda j.\,j\,\overline{M}):\sim\sim(\sim\overline{\alpha}\to\sim\sim\overline{\beta})}}$$

Case analysis

$$\dfrac{\overline{M}:\sim\sim(\sim\overline{\alpha}\to\sim\sim\overline{\beta}) \quad \dfrac{\dfrac{m:\sim\overline{\alpha}\to\sim\sim\overline{\beta} \quad \dfrac{\begin{array}{c}[x:\overline{\alpha}]\\ \vdots\\ \overline{N}:\overline{\gamma} \quad k:\sim\gamma^{\circ}\\ \hline \overline{N}\,k:o\end{array}}{\lambda x.\,\overline{N}\,k:\sim\overline{\alpha}}}{m\,(\lambda x.\,\overline{N}\,k):\sim\sim\overline{\beta}} \quad \dfrac{\begin{array}{c}[y:\overline{\beta}]\\ \vdots\\ \overline{O}:\overline{\gamma} \quad k:\sim\gamma^{\circ}\\ \hline \overline{O}\,k:o\end{array}}{\lambda y.\,\overline{O}\,k:\sim\overline{\beta}}}{\dfrac{m\,(\lambda x.\,\overline{N}\,k)\,(\lambda y.\,\overline{O}\,k):o}{\lambda m.\,m\,(\lambda x.\,\overline{N}\,k)\,(\lambda y.\,\overline{O}\,k):\sim(\sim\overline{\alpha}\to\sim\sim\overline{\beta})}}}{\dfrac{\overline{M}\,(\lambda m.\,m\,(\lambda x.\,\overline{N}\,k)\,(\lambda y.\,\overline{O}\,k)):o}{\lambda k.\,\overline{M}\,(\lambda m.\,m\,(\lambda x.\,\overline{N}\,k)\,(\lambda y.\,\overline{O}\,k)):\overline{\gamma}}}$$

Normalisation in Weakly Orthogonal Rewriting

Vincent van Oostrom[1][2]

[1] Universiteit Utrecht, Faculteit der Wijsbegeerte
P.O. Box 80126, 3508 TC Utrecht, The Netherlands
oostrom@phil.uu.nl
[2] CWI, P.O. Box 94079, 1090 GB Amsterdam, The Netherlands

Abstract. A rewrite sequence is said to be outermost-fair if every outermost redex occurrence is eventually eliminated. Outermost-fair rewriting is known to be (head-)normalising for almost orthogonal rewrite systems. We study (head-)normalisation for the larger class of weakly orthogonal rewrite systems. (Infinitary) normalisation is established and a counterexample against head-normalisation is given.

1 Introduction

The term $f(a)$ in the term rewrite system $\{a \rightarrow a, f(x) \rightarrow b\}$ can be rewritten to normal form b, but is also the starting point of the infinite rewrite sequence $f(a) \rightarrow f(a) \rightarrow \ldots$. It is then of interest to design a *normalising* strategy, i.e. a restriction on rewriting which guarantees to reach a normal form if one can be reached. How to design a normalising strategy? Observe that in the example the normal form b was reached by contracting the redex closest to the root, a so-called outermost redex. An *outermost* strategy restricts rewriting to contraction of outermost redexes. The idea of this strategy is that outermost redexes cannot 'disappear' hence must be contracted in order to reach a normal form. There are two problems with this idea:

1. Consider the TRS $\{a \rightarrow a, b \rightarrow c, f(x, c) \rightarrow d\}$. Then $f(a, b) \rightarrow f(a, b)$ by contraction of the outermost redex a. However, $f(a, b) \rightarrow f(a, c) \rightarrow d$ is a reduction to normal form.
2. Consider the TRS $\{a \rightarrow b, f(a) \rightarrow f(a)\}$. Then $f(a) \rightarrow f(a)$ by contraction of the $f(a)$-redex However, $f(a) \rightarrow f(b)$ is a reduction to normal form.

The problem in (1) is that the outermost strategy which alway contracts the redex a is not fair in the sense that the other outermost redex b persists forever. Hence we will restrict attention to *fair* outermost strategies. The problem in (2) is that the outermost redex $f(a)$ has overlap with the non-outermost redex a, and contracting them gives rise to distinct terms ($f(a)$ and $f(b)$ respectively). Hence we will restrict attention to *weakly orthogonal* term rewrite systems, where overlap may occur but does not give rise to distinct terms (see Example 2). These two restrictions suffice to establish our first result.

Theorem 1. Outermost-fair strategies are normalising for weakly orthogonal term rewrite systems.

Theorem 1 closes the gap between normalisation for almost orthogonal (see Def. 2) term rewrite systems [O'D77,Raa96] and non-normalisation for rewrite systems having overlap between a step making 'progress towards the normal form' and a step making 'no(t as much) progress', see e.g. (2) above and [Raa97].

In lazy functional programming (weak) head-normal forms, not normal forms, are computed [Pey87]. After computing a head-normal form, one zooms in on the subterms and the process repeats itself. In this way it is possible to compute (with) potentially infinite data structures, e.g. the list of prime numbers. The obvious question then is: are outermost-fair strategies head-normalising?

By the same proof method as that of Theorem 1, we obtain our next result, generalising the first-order result of [Mid97].

Theorem 2. *Outermost-fair strategies are head-normalising for almost orthogonal term rewrite systems.*

Surprisingly, head-normalisation cannot be extended from almost to weak orthogonality as witnessed by Counterexample 1. Nevertheless, we do have:

Theorem 3. *Outermost-fair strategies are infinitary normalising for weakly orthogonal term rewrite systems.*

That is, infinite normal forms will be found if they exist; a result which usually is obtained as a corollary of head-normalisation. Before presenting the outline of the paper, let's first present a summary of the results:

almost orthogonal	TRS	PRS	weakly orthogonal	TRS and PRS
normalising	+ [O'D77]	+ [Raa96]	normalising	+ Thm 1
∞-normalising	+ [Mid97]	+ Thm 2	∞-normalising	+ Thm 3
head-normalising	+ [Mid97]	+ Thm 2	head-normalising	− Counterex 1

By 'term rewrite system' in this paper we mean 'fully extended higher-order pattern rewrite system' (PRS, see e.g. [Raa97]). For the sake of exposition only definitions for the first-order case are presented, but replacing them by their higher-order equivalent preserves the results. For example, Theorem 1 generalises the normalisation result obtained in [Raa97] from almost to weakly orthogonal PRSs. By 'TRS' a first-order term rewrite system is meant.

Section 2 introduces the *contribution* relation induced by a rewrite step. In Section 3, which is the (technical) heart of the paper and based on a (re)combination of ideas from [SR90,GK94,Mid97] and [Oos97], three notions are formalised based on the contribution relation: *essentiality*, a *measure*, and the *projection*. It is shown that any rewrite sequence can be transformed into an *essential* one, i.e. a sequence where every step contributes to the result. Rewrite sequences will be *measured* by, roughly, counting the number of essential steps. It will be shown that both the transformation and the *projection* of the rewrite sequence over another one will decrease the measure. Readers not interested in technicalities can skip to Sections 4 and 5 where our main results and its corollaries are presented.

We assume the reader to be familiar with term rewriting, e.g. [BN98]. For background information on outermost-fair rewriting and its normalisation the reader should consult [Mid97] and [Raa97] or the paper by Van Raamsdonk on higher-order rewriting in this volume.

2 Terms and Trees

In this section some notations are fixed and the contribution relation induced by a term rewrite system is defined. It is important to note that throughout this paper **all term rewrite systems are (assumed to be) left-linear.**

We use s, t, r, l to range over *terms*, which are built from *variables* (x, y, z) and *function symbols* (f, g, h, a, b, c). The *rewrite (step)* relation generated by a TRS $\mathcal{R}$ is denoted by $\rightarrow_{\mathcal{R}}$. We use u, v to range over rewrite steps, and d, e to range over rewrite sequences. A *collapsing* rule is a rule whose right-hand side is a single variable. We use $\preceq$ to denote the usual *prefix* relation on *positions* (o, p, q). It is a well-founded partial order with least element ε, the *root*. If $o \preceq p$ then we say o is *outside/above* p and p is *inside/below* o. We use σ, τ to range over substitutions.

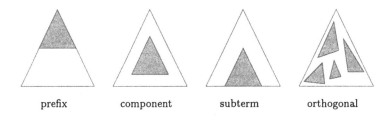

| prefix | component | subterm | orthogonal |

Fig. 1. Term parts.

Definition 1. 1. *A linear term* encompassed *[BN98, p. 166] by a term s is a* component *of s, where we (may) assume that s contains no variables.*
2. *By an* occurrence *of an object, e.g. a component, we formally mean a pair consisting of a term and the object in that term. Such a pair is called a marked* term *and the object its* mark. *Notions for terms are extended to notions for marked terms via the first projection. A* head-component *or* prefix *is a component occurring at the head (root).*
3. *A* redex *is a pair consisting of a component occurrence of the left-hand side of some rule, and the rule itself. We identify redexes with their induced step.*

Performing the obvious replacements in the definition of a notion for ordinary rewriting yields the corresponding *head*-notion, e.g. a *head*-redex is a redex having a head-component.

Example 1. Some components of $g(f(a, a))$ are the component x at any position, three components $g(x)$, $g(f(a, x))$, and $g(f(a, a))$ all occurring at the root, $f(x, y)$ and $f(x, a)$ at position 1, and a at both 11 and 12. Some non-components of $g(f(a, a))$ are $f(x, x)$ and the 'component' consisting of the two occurrences of a. Among these only x, $g(x)$, $g(f(a, x))$, and $g(f(a, a))$ are prefixes.

We assume the reader to be familiar with the correspondence between terms and trees. We introduce two relations on trees induced by term rewrite steps: the *copy* and the *contribution* relation. The idea of the copy relation is to relate the 'context' part and the 'substitution' part of a rewrite step to 'themselves'. On top of this, the contribution relation relates the 'destroyed' part to the 'created' part. This is abstractly illustrated in Figure 2. We define the relations by means of two

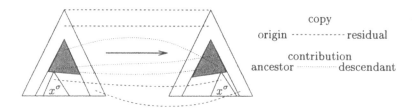

Fig. 2. Abstract copy (dashed) and contribution (dashed and dotted).

examples, since the formal definitions are cumbersome and largely irrelevant. On

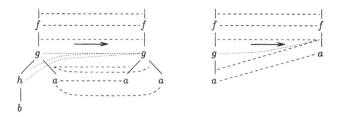

Fig. 3. Concrete copy (dashed) and contribution (dashed and dotted).

the left in Figure 3 the step $f(g(h(b),a)) \rightarrow f(g(a,a))$ due to the rewrite rule $g(h(x),y) \rightarrow g(y,y)$ is displayed. Both vertices g and h and the edge between them contribute to g. On the right the step $f(g(a)) \rightarrow f(a)$ due to the collapsing rewrite rule $g(x) \rightarrow x$ is shown. The vertex g contributes to the edge between f and a. In both steps f and a and the edges connected to them are copied.

Observe that we are minimalistic in the sense that the edges connecting the component of a redex to its surroundings do not belong to the component, so are copied and not destroyed/created by the rewrite step. Note furthermore that the copy relation is 'type-preserving' in the sense that it relates vertices to vertices and edges to edges only. Unfortunately, this is not true for the contribution relation, but this seems unavoidable if one wants to deal with collapsing rules in a way respecting Lévy labelling [Lév78], cf. also [BKV98].[1]

The copy and contribution relation for redexes are defined via their compo-nents, on which the relation must be a tree isomorphism in case of the copy relation. The contribution and copy relation induced by a rewrite sequence are

---

[1] The higher-order relations can be retrieved from [Raa97,Oos94] and [Oos97].

obtained via the relational composition of the corresponding relations induced by the constituting steps. The empty rewrite sequence induces the identity relation both for contribution and for copy.

If some occurrence of an *object*, i.e. a position, prefix, component, redex, or set of those, is related by the contribution (copy) relation to another such occurrence, then the former is called an *ancestor* (*origin*) of the latter (*along the rewrite sequence*) and the latter a *descendant* (*residual*) of the former (*after the rewrite sequence*). By *the* ancestor of a rewrite sequence, the ancestor of the whole final term in the initial term is meant. Let $\mathcal{O}$ be any object. An $\mathcal{O}$ *sequence*/$\mathcal{O}$ *chain* is a marked rewrite sequence such that every mark is an $\mathcal{O}$ and is the ancestor/origin of the mark of its successor (if any).

Definition 2. *A left-linear term rewrite system is* orthogonal, *if it has no critical pairs [Nip91]. If all critical pairs are* trivial, *i.e. of the form* $\langle s, s \rangle$, *(and arise from head-steps) the term rewrite system is* weakly *(almost) orthogonal.*

Example 2. Combinatory logic is orthogonal. Parallel or $\{or(x, T) \to T, or(T, x) \to T, or(F, F) \to F\}$, is almost orthogonal, since the only critical pair $\langle T, T \rangle$ is trivial and arises from overlap at the head: $T \leftarrow or(T, T) \to T$. Prede/successor $\{p(s(x)) \to x, s(p(x)) \to x\}$, is weakly orthogonal since its left-hand sides are linear, and it has critical peaks $s(x) \leftarrow s(p(s(x))) \to s(x)$ and $p(x) \leftarrow p(s(p(x))) \to p(x)$ giving rise to trivial critical pairs $\langle s(x), s(x) \rangle$ and $\langle p(x), p(x) \rangle$.

3 Essentiality, a Measure, and the Projection

Suppose one is interested in only a component of the final term of a rewrite sequence. Our goal will then be to *extract* from the sequence the steps which are *essential* for the component while *discarding* the inessential ones. In this section we show how this can be done in case the component one is interested in is a prefix. After that we introduce two concepts auxiliary to the proof of Theorem 1, but interesting in their own right: *measure* and *projection*. Our notion of essentiality and its properties are an adaptation of [GK94] to the non-orthogonal case. We illustrate it by means of an example.

Example 3. Consider a TRS having rules $\{a \to b, f(x, y) \to g(x, x)\}$ and suppose one is interested in the whole final term $g(b, b)$ of the rewrite sequence $f(a, a) \to f(b, a) \to f(b, b) \to g(b, b)$. The third step $f(b, b) \to g(b, b)$ is essential since it creates g by means of the rule $f(x, y) \to g(x, x)$. The second step $f(b, a) \to f(b, b)$ is not essential since the second argument of f is erased in the third step, hence this step should be discarded. The first step $f(a, a) \to f(b, a)$ is essential, since it contributes to both b's occurring in the final term; the first step creates b via the rule $a \to b$, then b is copied once in the second step, and finally copied twice in the third step. The extracted sequence will be $f(a, a) \to f(b, a) \to g(b, b)$.

Suppose now we are interested only in the head symbol g in the final term of the rewrite sequence $f(a, e) \to g(a, a) \to g(b, a)$. The second step $g(a, a) \to g(b, a)$ is not essential since we are not interested in the b, hence this step should be discarded. On the other hand, the first step $f(a,e) \rightarrow g(a,a)$ is essential since it contributes to g; g is created in the first step by means of the rule $f(x,y) \rightarrow g(x,x)$, and just copied in the second step.

Suppose that the first rewrite step in a sequence does not contribute to a given component in the final term, whereas all the others do. This implies that the step is *orthogonal* to the ancestor of the component, and in particular to all the rewrite steps in the sequence. The classical [Ros35] idea is then to use the fact that orthogonal rewrite steps commute [Bou85,Oos94] to *permute* [Lév78] the non-contributing rewrite step towards the tail of the rewrite sequence.

Example 4. Consider the rewrite sequence $f(a,e) \rightarrow f(b,e) \rightarrow g(b,b)$ in the TRS of Example 3. If we are interested in g in the final term, then the second step does, but the first step does not contribute to it, hence they should be permuted. After 'permutation' we have $f(a,e) \rightarrow g(a,a) \rightarrow g(b,a) \rightarrow g(b,b)$, where the last two steps do not contribute to g hence we take the origin of g along them, resulting in the sequence $f(a,e) \rightarrow g(a,a)$.

The example shows that permuting redexes which are *nested*, i.e. one is inside an argument of the other, may lead to multiplication; the example witnesses duplication of the a-redex. We deal with this phenomenon by introducing a simultaneous rewrite relation $\multimap\!\!\rightarrow$ which allows to perform several *orthogonal* steps simultaneously, e.g. $g(a,a) \multimap\!\!\rightarrow g(b,b)$ by contracting the two orthogonal a-redexes simultaneously.

Definition 3. *Two sets of positions are* non-overlapping *if they are disjoint. Two redexes, a set of redexes, and a set of rewrite steps are* non-overlapping *if their components are, every pair of distinct elements is, and its set of redexes is, respectively. Since we assume all (underlying) rules to be left-linear we often use* orthogonal *instead of* non-overlapping *(see Figure 1).*

One can give an inductive definition of performing a non-overlapping set of rewrite steps in one go in the style of Tait & Martin-Löf [Bar84]. For our purposes it is more convenient to do this via the possible *developments* [CR36] of the set.

Definition 4. *1. A development of a non-overlapping set of rewrite steps is*
   (a) the empty rewrite sequence if the set is empty, and
   (b) an arbitrary rewrite step from the set, followed by a development of the residual of the set after that rewrite step otherwise.
2. The existence of a development of an orthogonal set of rewrite steps from s to t is denoted by $s \multimap\!\!\rightarrow t$, possibly subscripted with the set. $\multimap\!\!\rightarrow$ is the simultaneous rewrite relation and we employ U, V, W to range over simultaneous rewrite steps, and D, E, F to range over simultaneous rewrite sequences. To indicate that the rewrite steps are not nested we write $-\!|\!|\!\rightarrow$ instead of $\multimap\!\!\rightarrow$. $-\!|\!|\!\rightarrow$ is the parallel *rewrite relation induced by the term rewrite system [Hue80].*
3. The contribution and copy relation are lifted to simultaneous rewrite steps via the corresponding relations induced by any development of the set.

The induced relations are independent of the particular development by (a minor variation on) the *finite developments theorem* (FD) [HL91,Klo80,Oos94], which expresses that all developments of an orthogonal set of rewrite steps are finite, end in the same term and induce the same copy and contribution relation.

Definition 5. *Let $U;V$ be a rewrite sequence. $U;V$ is permutable if all redexes in V are residuals after U of some orthogonal set of redexes. Let V' be the minimal such set. The result of permuting $U;V$ is $V';U'$, where U' is the set of all residuals of U after V'.*

Observe that a step in U contributes to an object in the final term if and only if at least one of its residuals in U' does so. Moreover, a step in V contributes to an object in the final term if and only if its origin in V' does so.

Example 5. Consider the simultaneous rewrite sequence $f(a, e) \twoheadrightarrow f(b, e) \twoheadrightarrow g(b, b)$ in the TRS of Example 3. Since the head-redex contracted in the second step is a residual of 'itself' in the initial term, the steps are permutable giving rise to the sequence $f(a, e) \twoheadrightarrow g(a, a) \twoheadrightarrow g(b, b)$. Note that both a-redexes contracted in the second simultaneous step are residuals of the a-redex in the initial term.

Now we show that if one is interested in the rewrite steps contributing to a prefix of the final term of a sequence, then the non-contributing steps can all be discarded by permuting them towards the tail of the rewrite sequence.

Definition 6. *A position in a term in a rewrite sequence is* essential *for a set of positions in the final term, if it contributes to it. A set of positions, a redex, a single rewrite step, and a simultaneous rewrite step are* (in)essential *if every (no) element is, its component is, its redex is, and all its rewrite steps are, respectively. A rewrite sequence/development is* essential *if all its steps are.*

Lemma 1. *A single rewrite step is either essential or inessential.*

Proof. The contribution relation is defined such that if some position in its left-hand side is related to say o, all positions in its left-hand side are. $\square$

So, any rewrite step u in an ordinary rewrite sequence can be labelled as essential (u_e) or inessential ($u_{\bar{e}}$). Simultaneous steps may contain both essential and inessential steps, but by FD it follows that any such step U factors as $U_e; U_{\bar{e}}$, where the subscript e ($\bar{e}$) indicates that all constituting rewrite steps are (in)essential. It remains to factor rewrite sequences. For this we employ a technical result expressing monotonicity of ancestors. But first we remark that there are simultaneous steps which are both essential and inessential: empty simultaneous steps. We assume they are silently removed.

Lemma 2. *The ancestor of a prefix is a prefix.*[2]

---

[2] Replacing 'prefix' by 'component' the lemma still holds for TRSs, but fails for PRSs, e.g. in the λ-calculus for the component (uz) w.r.t. $(\lambda y.x(yz))u \rightarrow_\beta x(uz)$.

Proof. By a case analysis on the definition of the contribution relation.   □

From this it follows that marking the final term of a rewrite sequence with a prefix, uniquely induces a prefix sequence.

Lemma 3. *A simultaneous prefix sequence D factors as D_e; $D_{\bar{e}}$ (cf. [Ros35,GK94,Mel97]).*

Proof. It clearly suffices to prove that any subsequence of the form $U_{\bar{e}}$; U'_e factors as V_e; $V_{\bar{e}}$. By definition, all rewrite steps in $U_{\bar{e}}$ are inessential hence they are all below the prefix and do not 'touch' the prefix when performed. Since U'_e is essential all steps in it are part of this prefix, hence the origin of U'_e is 'itself', hence orthogonal to $U_{\bar{e}}$, hence the steps are permutable.   □

Observe that if the prefix of the final term is the whole final term, then the sequence $D_{\bar{e}}$ is empty, i.e. the sequence was transformed into an essential one!

3.1   A Measure

The proof of Theorem 1 will be by induction, using a measure on simultaneous rewrite sequences inspired by [SR90]. It is the lexicographic product of the length of the sequence and the maximal lengths of essential developments of its simultaneous steps, from the tail to the head of the sequence.

Definition 7. *Let $D = U_1; U_2; \ldots; U_n$ be a simultaneous prefix sequence. The measure $\mu(D)$ of D is the n-tuple $(l_n, \ldots, l_1)$, where l_i is defined as the maximal length of an essential development of U_i. Tuples are compared first by their length and then by their successive elements (in the natural order). This yields a well-founded order which is denoted by $<$.*

Example 6. The measure of the simultaneous prefix sequence $f(a, a, a)$ $\twoheadrightarrow$ $f(b, b, a)$ $\twoheadrightarrow$ $g(b, b, b)$ induced by the prefix $g(b, b, y)$ of its final term, in the TRS $\{f(x, b, z) \rightarrow g(z, z, z), a \rightarrow b\}$, is the tuple $(3, 1)$. Since the sequence has length 2, the tuple has 2 elements. The second step of the sequence contracts an a- and an f-redex, both of which are essential. An essential development of maximal length is $f(b, b, a) \rightarrow g(a, a, a) \rightarrow g(b, a, a) \rightarrow g(b, b, a)^3$ explaining the first element, 3, of the tuple. The first step contracts two a-redexes, of which only the rightmost is essential. Hence its development has length 1, explaining the second element.

Our measure differs in two respects from that of [SR90]. Firstly, we shifted from parallel rewriting ($\twoheadrightarrow_U$) and measuring the size (of U), to simultaneous rewriting ($\twoheadrightarrow_U$) and measuring the maximal length of a development (of U). This is the obvious thing to do when trying to lift results from first- to higher-order. Secondly, we count the maximal length of essential developments only, not of arbitrary developments. The reason is that in the higher-order case residuals of essential steps may get nested inside inessential ones. Hence the length of an arbitrary development may depend on inessential steps, which is undesirable and

---

3 $g(b, b, a) \rightarrow g(b, b, b)$ is inessential though the redex is a residual of an essential one.

measuring by it would render the next lemma invalid. This was the motivation for introducing the contribution relation. First remark that removing empty steps decreases the length hence the measure.

Lemma 4. *If $D_e; D_{\bar{e}}$ is a factorisation of D, then $\mu(D) \geq \mu(D_e)$.*

Proof. By enriching the proof of Lemma 3 remarking that the measure of an inessential step is 0 and if $U_{\bar{e}}; U_e$ permutes into $V_e; V_{\bar{e}}$, then $\mu(U_e) = \mu(V_e)$. □

3.2 The Projection

Our main proof technique is based on *projecting* a rewrite sequence to some *result* over arbitrary rewrite steps, thereby (eventually) decreasing its measure. The projection is the weakly orthogonal projection as defined in [Oos94,Raa97] combined with discarding inessential steps on the fly.

Definition 8. *A set of* results *(see [GK94,Mel98] for ample motivation) is a set of prefixes closed under (inverse) rewrite steps below it. Moreover for a prefix sequence to a result with initial prefix O and final prefix P, if some step u overlaps O, then u overlaps an ancestor of some essential step.*

Lemma 5. *Consider a prefix sequence to a result, with an initial term which is not a result, having prefix O. Then some redex is entirely in O.*

Let D be a simultaneous prefix sequence to a result, with initial prefix O and final prefix P. Let e be a rewrite sequence coinitial with D of length m. The *projection* of D over e is defined by lexicographic induction on the pair $(\mu(D), m)$. By Lemma 4, D factors as $D_e; D_{\bar{e}}$. By closure of results under *expansions*, i.e. inverse steps, below P, D_e is a simultaneous prefix sequence to a result, such that $\mu(D) \geq \mu(D_e)$ by Lemma 4. That is, for our induction it is harmless to assume that simultaneous prefix sequences to a result are essential, so we assume it.

If e is empty the projection is just D. Suppose $e = v; e_v$. As in [Mid97], two cases are distinguished. In ($\perp$) the measure does not change, in ($\not{\perp}$) it decreases.

($\perp$) If v does not overlap O, then the standard orthogonal projection [Klo80], can be applied. More precisely, one can construct by repeated application of FD a simultaneous prefix sequence D', and a rewrite sequence e', such that $D; e'$ and $v; D'$ start and end in the same term, where e' consists of steps below P. Hence D' is an (essential simultaneous prefix) sequence to a result, by closure of results under rewriting below the prefix. Moreover, $\mu(D) = \mu(D')$. Repeat the construction on D' and e_v.

($\not{\perp}$) If v does overlap O, then by the definition of result it overlaps the ancestor of some step in D, hence D is non-empty say $D = U; D_U$.

    ($\perp$)' If v does not overlap U, then applying FD yields simultaneous steps U' and V' such that $U; V'$ and $v; U'$ start and end in the same term and V' overlaps the ancestor of P along D_U. That is, V' develops as $v'; V'_{v'}$ for some v' overlapping the ancestor of P. Hence repeating the construction on these we are again in case ($\not{\perp}$), resulting in a sequence D'_U such that $\mu(D_U) > \mu(D'_U)$. Applying the construction once again on e_v and $U'; D'_U$ yields the desired D' such that $\mu(D) > \mu(U'; D'_U) \geq \mu(D')$.

($\not{L}$)' If v does overlap U, then U can be developed as $u; U_u$ such that u overlaps v. By weak orthogonality u and v result in the same term. By FD $\mu(D) > \mu(U_u; D_U)$, so the construction can be repeated on e_v and $U_u; D_U$ yielding the desired D' such that $\mu(D) > \mu(U_u; D_U) \geq \mu(D')$.

4 Normalisation

In this section Theorem 1 is proven and some of its corollaries are discussed. First we define the concepts involved: strategy, outermost-fair and normalising.

Definition 9. *A subrelation of an abstract rewrite system is called a* strategy. *It is* →*-normalising if it is terminating and its normal forms coincide with those of the abstract rewrite system* →. *A one-step/parallel term* strategy for a term rewrite system $\mathcal{R}$ *is a restriction on the set of allowed ordinary/parallel steps. Notions for abstract rewrite system strategies extend to notions for term strategies via their underlying abstract rewrite system. A term strategy for a term rewrite system* $\mathcal{R}$ *is* normalising *if it is* →$_\mathcal{R}$*-normalising. It is* outermost-fair *if it does not allow an infinite outermost redex chain.*

Example 7. 1. The rewrite sequence $f(a, a) \to f(a, b) \to f(a, a) \to f(a, b) \to$ ... in the TRS $\{f(x, b) \to f(x, a), a \to b\}$ is outermost-fair. Observe that the leftmost a gives rise to an infinite chain. Although infinitely many redexes in the chain are outermost, the chain itself is not outermost since infinitely many redexes in it are non-outermost as well.
  2. From the rewrite sequence $f(a) \to f(a) \to$ ... in the TRS $\{f(x) \to f(x), a \to a\}$, it cannot be determined whether it is outermost-fair or not. One needs to know which redexes are contracted; it is outermost-fair only if infinitely often the first rule is applied.

Theorem 1. Outermost-fair strategies are normalising for weakly orthogonal term rewrite systems.

Proof. To start, observe that the set of normal forms is a set of results. Closure under rewriting and expansion is trivial since the prefix is the whole term and there's nothing below it. A redex which contributes to a normal form but has no overlap with any of the essential steps, would have a residual in the normal form which is impossible. Now suppose for a contradiction that for some term s it holds that:

  1. a(n essential) rewrite sequence D from s to normal form z exists, and
  2. an infinite outermost-fair rewrite sequence e from s exists.

Since e is infinite, it is of the form $v; e_v$ for some rewrite step v and infinite outermost-fair rewrite sequence e_v. Projecting D over v yields by the above a rewrite sequence D' such that $\mu(D) \geq \mu(D')$. If we can prove that by repeated projection the measure eventually strictly decreases, then we are done by well-foundedness of the measure and the observation that only the empty sequence has the minimal measure ().

Suppose to the contrary that from some moment on, the measure doesn't decrease anymore. Then the projection is orthogonal each time, and the prefix O of the initial term remains the same by projection. By Lemma 5, O contains some redex and since O is a prefix there's also an outermost redex u at some position $o \in O$.[4] By outermost-fairness this step must be eliminated sometime in e. This can happen in two ways.

(overlap) Some redex v overlapping with u is contracted, but then contracting u instead of v yields the same result by weak orthogonality and gives rise to case ($\not\perp$) in the projection, decreasing the measure.

(above) Some redex v at a position p above u is created, but then $p \prec o$. By well-foundedness of the prefix order eventually case (overlap) applies.   □

By doing some proof-hacking it is possible to ϵ-relax the weak orthogonality requirement to left-linearity and *biclosedness*, where a term rewrite system is called *biclosed* if for every critical pair $\langle s, t \rangle$ both s rewrites to t and t rewrites to s in a number of steps.

Corollary 1. *1. Parallel outermost strategies are normalising for weakly orthogonal term rewrite systems ([O'D77,Raa97]).*

2. Normalisation and termination coincide for non-erasing weakly orthogonal term rewrite systems ([Chu41,Klo80,Gra96]), where a step $s \to t$ is non-erasing if every position in s has a descendant in t.[5]

3. The computable, history-free and sequential strategy S_ω [AM96]. is normalising for weakly orthogonal TRSs.

4. Needed strategies are (hyper) normalising for orthogonal term rewrite systems [HL91]. A strategy is needed if the redexes it selects contribute to the normal form, and hyper normalisation of a strategy S is normalisation of any strategy which always eventually performs an S-step.

5. Every rewrite sequence in a term rewrite system can be transformed into a standard one, where a rewrite sequence is standard if for any step in the sequence, no step properly outside all its descendants is performed later ([GLM92]).

Proof. 1. The parallel outermost strategy is outermost fair.

2. Due to non-erasingness the ancestor of any rewrite sequence is always the whole term. Hence ($\not\perp$) always applies in the proof of Theorem 1.

3. From Theorem 1 and [AM96, Theorem 6.7].

4. An easy consequence of FD is that for orthogonal term rewrite systems, whether a position contributes to its (unique if any) normal form does not depend on the particular rewrite sequence to the normal form. Hence all redexes in the ancestor of a rewrite sequence to normal form are needed and the

---

[4] Beware: u may be partially below O! E.g. in the sequence $f(b, a) \to f(b, b) \to_2 b$ in the TRS $\{a \to b, f(b, x) \to_1 x, f(x, b) \to_2 x\}$, the redex a is in the initial prefix, but the outermost redex $f(b, a)$ is only partially so (b is below the prefix!).

[5] For TRSs this definition of non-erasingness is equivalent to the standard one, requiring all variables in the left-hand side of a rule to occur in its right-hand side.

others are not. Moreover, from Lemma 5 we know that the ancestor of any non-empty rewrite sequence contains some (needed) redex. Combining these observations it is clear that case ($\nmid$) must and can always be applied (eventually, in the case of hyper normalisation) in the proof of Theorem 1, when projecting over a sequence generated by a needed strategy (cf. [BKV98]).

5. Take the component consisting of the whole final term of a rewrite sequence d and repeat the following until the first item fails.
   (a) Consider an outermost position in the component to which some step in the sequence contributes.
   (b) Select the first such step and take the component of its redex
   Call the origin in the initial term of the last step selected by this procedure a *first* step of d. By the selection procedure any first step u is essential and no step outside its descendants is performed in the sequence. By the former property, projecting d over u results in a rewrite sequence e, such that $\mu(d) > \mu(e)$. By the latter property, u is not duplicated until contracted hence d and e are permutation equivalent [Lév78]. By the decrease in measure repeated selection/projection terminates. The resulting sequence of steps $(u \ldots)$ is standard by the properties of the selection which are invariant under permutation equivalence. □

The proof of Theorem 1 can be strengthened such that the weak orthogonality requirement in the second item, can be relaxed, allowing critical pairs satisfying the *CPC'* critical pair criterion of [Gra96, Definition 3.4.29], providing an alternative proof of his result. We conjecture that left-linearity can be dropped as well (see [Gra96] for the first-order case).

The third item already appears as [AM96, Corollary 7.2] with the same proof, i.e. based on Theorem 1 of the present paper which at that moment was thought to have been proven by Van Raamsdonk. Unfortunately a counterexample against the proof method was found, and at that moment the result had to be restricted to almost orthogonal TRSs (see [Raa96,Raa97]).

A consequence of the fourth item is normalisation of leftmost-outermost strategies for left-normal [Klo80] higher-order rewrite systems like the λ-calculus.

The procedure in the last item is closely related to the procedures lmc and STD described in [Klo80] and [GLM92]. The difference is that we rely on properties of contribution for its correctness and on our measure for its termination.

5   Head-normalisation

In [Mid97] it was shown that for almost orthogonal TRSs outermost-fair strategies are head-normalising (there called root-normalizing). Unfortunately this property fails for the slightly larger class of weakly orthogonal systems.

Counterexample 1. Consider the term $f(g(a, a))$ in the TRS:

$$a \to b$$
$$f(g(a, x)) \to f(g(b, x))$$
$$g(b, x) \to g(x, x)$$

There is only one critical pair $\langle f(g(b, x)), f(g(b, x)) \rangle$ (arising from overlap between the first two rules). Since the critical pair is trivial, the TRS is weakly orthogonal. Now on the one hand from $f(g(a, a))$ the head-normal form $f(g(b, b))$ *can* be reached, e.g. $f(g(a, a)) \to f(g(a, b)) \to f(g(b, b))$ by applying the first rule twice. But on the other hand the infinite parallel outermost rewrite sequence $f(g(a, a)) \to f(g(b, a)) \to f(g(a, a)) \to \ldots$ *does not* reach a head-normal form.

The counterexample shows that the head-normal forms do not constitute a set of results in the weakly orthogonal case (cf. [Mel98]).

Lemma 6 (coherence). *([BN98, Lemma 9.3.10], [Raa96, Lemma 5.3.2]) Let $s = l^\sigma$ be an instance of a left-hand side l in an almost orthogonal term rewrite system. Suppose $s \to t$ by a non-head step u. Then u does not overlap l.*

Theorem 2. *Outermost-fair strategies are head-normalising for almost orthogonal term rewrite systems.*

Proof. To replay the proof of Theorem 1 it suffices to show that the set of head-normal forms is a set of results. Closure under rewriting below the head is trivial. For a proof of closure under expansion, suppose that $s \to_{\bar{\varepsilon}} z$ and z is a result. If s were not in head-normal form it would rewrite by some non-head step to a head-redex l^σ. By confluence and coherence, z and l^σ would have a common reduct, via non-head steps, of the form l^τ contradicting the assumption that z is in head-normal form. Like for normal forms, a redex contributing to a head-normal form but having no overlap with the essential steps, would have a residual overlapping at the head with the head-normal form. □

For the higher-order case this is a new result, although it is the obvious combination of the results of [Mid97] and [Raa97]. Finally, we show that outermost-fair/parallel outermost strategies are infinitary normalising, a result which is usually proven via head-normalisation [Mid97]. We refer to that paper for a precise definition; one can think of generating the (infinite) list of prime numbers. Due to Counterexample 1 the proof via head-normalisation is blocked. [6]

Theorem 3. *Outermost-fair strategies are infinitary normalising for weakly orthogonal term rewrite systems.*

Proof (sketch). Suppose d is an infinite rewrite sequence from a finite term s to infinite normal form z. This implies that any finite prefix Q of z remains 'untoched' after a finite initial segment of d. Let $q_1, q_2, \ldots$ be an enumeration of all positions of z in breadth-first order. Applying factorisation with respect to the successive prefixes yields a sequence $D = D_1; D_2; \ldots$ from s to z such that D_i is essential for q_i, for every $i \geq 0$. Define $\mu(d) = (\mu(D_1), \mu(D_2), \ldots)$, i.e. the infinite sequence consisting of the measures of the essential subsequences. We show that by projecting D over an outermost-fair sequence an ever growing

---

[6] The counterexample does not contradict infinite normalisation, since the term $f(g(b, b))$ does not have a(n infinite) normal form.

prefix of the measure will be filled by ()'s, establishing the result. Since s is finite, initially the set of minimal positions which have measure different from () is finite, and 'covers' every infinite branch of z. Let q be the first one. Replaying the proof of Theorem 1, we may assume that eventually a redex contributing to q is contracted. This 'decreases' the measure of the sequence corresponding to q, which entails that an ever growing prefix of the measure will consist of ()'s. □

As in [Mid97], our proof is restricted to normalisation by strongly converging sequences (leaving an ever growing prefix 'untouched'). Consider the weakly orthogonal TRS $\{h^i(a) \to h^{i+1}(a) \mid i \geq 0\}$. The infinite sequence $a \to h(a) \to h(h(a)) \dots$ is Cauchy-converging and its limit is the infinite normal form $z = h(h(h(\dots)))$. z may be reached by repeated application of the rule $a \to h(a)$, but also by always applying the 'largest' possible rule. Call these sequences d and e. d is strongly converging, but e is not since it is a head-sequence. (Both notions of convergence coincide for finite TRSs.) Note that $\mu(e)$ is not well-defined since the prefix of e consisting of steps contributing to the root is e itself, so infinite. Hence, our proof (projecting e over d) doesn't work if e is only Cauchy-converging. Still, we conjecture that outermost-fair rewriting is infinitary Cauchy-normalising for weakly orthogonal term rewrite systems.

6 Conclusion

We believe that the contribution relation and our proof methods based upon it are fundamental for the syntactic study of (left-linear term) rewrite systems. For example, we expect the extraction procedure of [Lév78] to be amenable to an analysis generalising the analysis of factorisation in this paper. Furthermore, we hope to employ our methods to prove normalisation results for strategies in the realm of explicit substitution/proof normalisation.

Acknowledgements I would like to thank Jan Willem Klop, Aart Middeldorp, Femke van Raamsdonk and Roel de Vrijer for inspiring discussions and the anonymous referees for useful comments.

References

[AM96] Sergio Antoy and Aart Middeldorp. A sequential reduction strategy. *Theoretical Computer Science*, 165(1):75–95, 1996.

[Bar84] H.P. Barendregt. *The Lambda Calculus, Its Syntax and Semantics*, volume 103 of *Studies in Logic and the Foundations of Mathematics*. North-Holland, Amsterdam, revised edition, 1984.

[BKV98] Inge Bethke, Jan Willem Klop, and Roel de Vrijer. Descendants and origins in term rewriting. IR 458, Vrije Universiteit, Amsterdam, December 1998.

[BN98] Franz Baader and Tobias Nipkow. *Term Rewriting and All That*. Cambridge University Press, February 1998.

[Bou85] G. Boudol. Computational semantics of term rewriting systems. In M. Nivat and J.C. Reynolds, editors, *Algebraic Methods in Semantics*, pages 169–236. Cambridge University Press, 1985.

[Chu41] Alonzo Church. *The Calculi of Lambda-Conversion.* Princeton University Press, Princeton, New Jersey, 1941. Second Printing 1951.

[CR36] Alonzo Church and J.B. Rosser. Some properties of conversion. *Transactions of the American Mathematical Society,* 39:472–482, January to June 1936.

[GK94] John Glauert and Zurab Khasidashvili. Relative normalization in orthogonal expression reduction systems. In *CTRS'94,* volume 968 of *Lecture Notes in Computer Science,* pages 144–165. Springer, 1994.

[GLM92] Georges Gonthier, Jean-Jacques Lévy, and Paul-André Melliès. An abstract standardisation theorem. In *Proceedings of the 7th IEEE Symposium on Logic in Computer Science,* pages 72–81, 1992.

[Gra96] Bernhard Gramlich. *Termination and Confluence Properties of Structured Rewrite Systems.* PhD thesis, Universität Kaiserslautern, 19. Januar 1996.

[HL91] Gérard Huet and Jean-Jacques Lévy. Computations in orthogonal rewriting systems, I. In Jean-Louis Lassez and Gordon Plotkin, editors, *Computational Logic: Essays in Honor of Alan Robinson.* MIT Press, 1991.

[Hue80] G. Huet. Confluent reductions: Abstract properties and applications to term rewriting systems. *Journal of the ACM,* 27(4):797–821, October 1980.

[Klo80] J.W. Klop. *Combinatory Reduction Systems.* PhD thesis, Rijksuniversiteit Utrecht, June 1980. Mathematical Centre Tracts 127.

[Lév78] Jean-Jacques Lévy. *Réductions correctes et optimales dans le λ-calcul.* Thèse de doctorat d'etat, Université Paris VII, 1978.

[Mel97] Paul-André Melliès. Axiomatic rewriting theory III, a factorisation theorem in rewriting theory. In *CTCS'97,* volume 1290 of *Lecture Notes in Computer Science,* pages 49–68. Springer, 1997.

[Mel98] Paul-André Melliès. Axiomatic rewriting theory IV, a stability theorem in rewriting theory. In *Proceedings of the 14th IEEE Symposium on Logic in Computer Science,* pages 287–298, 1998.

[Mid97] Aart Middeldorp. Call by need computations to root-stable form. In *Proceedings of the 24th Annual ACM SIGPLAN-SIGACT Symposium on Principles of Programming Languages, Paris,* pages 94–105, 1997.

[Nip91] Tobias Nipkow. Higher-order critical pairs. In *Proceedings of the 6th IEEE Symposium on Logic in Computer Science,* pages 342–349, 1991.

[O'D77] M.J. O'Donnell. *Computing in Systems Described by Equations,* volume 58 of *Lecture Notes in Computer Science.* Springer Verlag, 1977.

[Oos94] Vincent van Oostrom. *Confluence for Abstract and Higher-Order Rewriting.* PhD thesis, Vrije Universiteit, Amsterdam, March 1994.

[Oos97] Vincent van Oostrom. Finite family developments. In *RTA'97,* volume 1232 of *Lecture Notes in Computer Science,* pages 308–322. Springer, 1997.

[Pey87] Simon L. Peyton Jones. *The Implementation of Functional Programming Languages.* Prentice-Hall International, 1987.

[Raa96] Femke van Raamsdonk. *Confluence and Normalisation for Higher-Order Rewriting.* PhD thesis, Vrije Universiteit, Amsterdam, May 1996.

[Raa97] Femke van Raamsdonk. Outermost-fair rewriting. In *TLCA'97,* volume 1210 of *Lecture Notes in Computer Science,* pages 284–299. Springer, 1997.

[Ros35] J.B. Rosser. A mathematical logic without variables I. *Annals of Mathematics,* 36:127–150, 1935.

[SR90] R.C. Sekar and I.V. Ramakrishnan. Programming in equational logic: Beyond strong sequentiality. In *Proceedings of the 5th IEEE Symposium on Logic in Computer Science,* pages 230–241, 1990.

Strong Normalization of Proof Nets
Modulo Structural Congruences*

Roberto Di Cosmo[1] and Stefano Guerrini[2,**]

[1] DMI-LIENS (CNRS URA 1347) Ecole Normale Supérieure
45, Rue d'Ulm - 75230 Paris France. Email:dicosmo@ens.fr
[2] Dipartimento di Scienze dell'Informazione - Università di Roma I, 'La Sapienza'
Via Salaria, 113 - I-00198, Roma, Italy - Email: guerrini@dsi.uniroma1.it

Abstract. This paper proposes a notion of reduction for the *proof nets* of Linear Logic modulo an equivalence relation on the *contraction links*, that essentially amounts to consider the contraction as an associative commutative binary operator that can float freely in and out of proof net *boxes*. The need for such a system comes, on one side, from the desire to make proof nets an even more parallel syntax for Linear Logic, and on the other side from the application of proof nets to λ-calculus with or without explicit substitutions, which needs a notion of reduction more flexible than those present in the literature. The main result of the paper is that this relaxed notion of rewriting is still strongly normalizing.

Keywords: Proof Nets. Linear Logic. Strong Normalization.

1 Introduction

In his seminal paper [6], Girard proposed proof nets as a *parallel syntax* for Linear Logic, where uninteresting permutations in the order of application of logical rules are de-sequentialised and collapsed. Nevertheless, in the presence of exponentials, that are necessary to translate λ-terms into proof nets, the traditional presentation of proof nets turns out to be inadequate: too many inessential details concerning the order of application of independent structural rules (*e.g.*, contraction) are still present.

When using proof nets to simulate λ-calculus, this redundancy already gets in the way, so that it is necessary to consider an extended notion of reduction, or a special version of proof nets with an *n*ary structural link and a brute force normalization procedure. But if one tries to simulate the behavior of explicit substitutions, then one is really forced to consider contraction links as a sort of associative-commutative operator.

Looking carefully at these difficulties, one can see that what is really needed is an extension of the notion of reduction on proof nets where the order of application of the contraction rules, and the relative order of contraction rules and box formation rules is

---

* Partially supported by: Italian MURST project *Tecniche Formali per la Specifica, l'Analisi, la Verifica, la Sintesi e la Trasformazione di Sistemi Software*; UE TMR Network *LINEAR: Linear Logic in Computer Science.*
** Written while at the Department of Computer Science of Queen Mary and Westfield College, London, UK supported by an EPSRC grant.

abstracted away. This can be done by defining an equivalence relation over regular proof nets that essentially amounts to consider the contraction as an associative-commutative binary operator that can float freely in and out of proof net *boxes*, and define a notion of reduction on the corresponding equivalence classes. Both λ-calculus and systems of explicit substitution can be very easily simulated in such a system. Also, this system allows to abstract away all the uninteresting permutations in the order of application of *structural* rules, which are de-sequentialised and collapsed into the same equivalence class. Yet, up to now, it was unknown whether such an extension would enjoy the same good properties as proof nets, and first of all, strong normalization. The main result of the paper is that this relaxed notion of rewriting is still strongly normalizing.

In the following, we shall first recall the traditional definition of proof nets and of their reduction, as well as the systems proposed by Danos and Regnier [4] to simulate λ-calculus, and by Di Cosmo and Kesner [5] to simulate a calculus with explicit substitution. Then, we shall define our equivalence relation and prove our main theorem.

1.1 Linear Logic and Proof Nets

Let us recall some classical notions from Linear Logic. We shall consider Multiplicative Exponential Linear Logic (MELL) without constants, *i.e.*, the fragment of Linear Logic whose formulas are: $\mathcal{F} ::= a \mid \mathcal{F} \otimes \mathcal{F} \mid \mathcal{F} \otimes \mathcal{F} \mid !\mathcal{F} \mid ?\mathcal{F}$, where a ranges over a non-empty set of *atoms* $\mathcal{A}$ that is the sum of two disjoint subsets $\mathcal{P}$ and $\mathcal{P}^\perp$, corresponding to the *positive* atoms p and to the *negative* atoms $p^\perp$ respectively. In particular, $p^\perp$ is named the *linear negation* of p, and vice versa. Linear negation extends to every formula A by means of the following De Morgan equations: $(A \otimes B)^\perp = A^\perp \otimes B^\perp$, $(?A)^\perp = !A^\perp$, $A^{\perp\perp} = A$. The connectives $\otimes$ (tensor) and $\otimes$ (par) are the *multiplicatives*; the connectives ! (of-course) and ? (why-not) are the *exponentials*. For the definition of the sequent calculus of Linear Logic, we refer the reader to [6].

One of the advantages of MELL is the availability of a graph-like representation of proofs that is highly non-sequential, that is, which is often able to forget the order in which some rules are used in a sequent calculus derivation, when this order is irrelevant. This representation is known as Proof Nets.

A (MELL) proof net is a finite (hyper)graph whose vertices are occurrences of MELL formulas (in the following, we shall often write 'formula' for 'occurrence of formula') and whose (hyper)edges, named *links*, correspond to connections between the active formulas of some rule of the sequent calculus of MELL. The formulas below a link are the *conclusions* of the link; the formulas above a link are its *premises*.

Fig. 1 gives the inductive rules for the construction of proof nets. As usual Γ, $?\Gamma$ and Δ stand for sets of formulas—in this case, sets of conclusions of the net above them—in particular, $?\Gamma$ denotes a set of ?-formulas. The rule *axiom* is the base case: a proof net formed of a unique link of type ax. The rules *par*, *contraction*, *dereliction* and *weakening* add a new link of the corresponding type to a previously constructed proof net. The rules *tensor* and *cut* add a new link and merge two (distinct) proof nets. Finally, the *promotion* rule promotes a formula A to $!A$. In order to apply that rule, we need a proof net M whose conclusions but A are of type ?. As a result, promotion encloses M into a *box* whose conclusions are the promoted formula $!A$ and a copy of each ?-

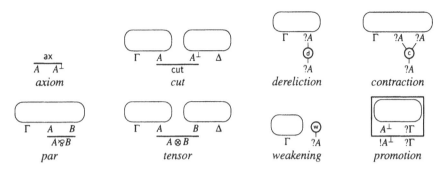

Fig. 1. Proof Nets.

conclusion of M. The conclusion !A is the *principal port* of the box; the conclusions in ?Γ are its *auxiliary ports*.

Boxes force a strong constraint on the *sequentialization* of a proof net (*i.e.*, on the construction of a proof net by application of rules in Fig. 1): in any possible sequentialization of a proof net that contains a box B, no rule corresponding to a link below a conclusion of B can be applied before the complete sequentialization of B. However, the notion of box is crucial for the definition of proof net cut-elimination. In fact, because of the side condition on promotion (recall that all the auxiliary premises of a box must be of type ?), we have to keep track of the context that allowed the promotion of A (again, for a more detailed analysis, refer to [6]).

Remark 1. A proof net M is a (hyper)graph, so it does not contain any explicit information on the ways in which it can be sequentialized (*e.g.*, think at the strings of some context free language; the strings do not contain any information on their derivations in the context free grammar of the language). Therefore, let us assume to have a (hyper)graph M formed of formulas and links—such (hyper)graphs are known as *proof structures*. The problem 'is the proof structure M a proof net?' is clearly decidable, *e.g.*, take the brute force approach that tries ordering links in all the possible ways. The so called *correctness criteria* characterize proof nets with no explicit reference to the rules in Fig. 1. For instance, the Danos-Regnier criterion states that M is a proof net when all the *switches* of M are trees (a switch is a graph obtained by collapsing some boxes and by removing some edges). For a detailed discussion of correctness criteria and of their complexity, see [3, 7].

The rewriting rules in Fig. 2 define the cut-elimination procedure for proof nets. In fact, each cut-elimination rule in Fig. 2 transforms a proof net into a proof net (see [6]). In Fig. 2, a link between instances of the same set of formulas means that there is a link between each pair/triple of corresponding formulas in that sets.

Definition 1 (PN). *Proof Nets is the smallest set of (hyper)graphs closed by the rules in Fig. 1.* PN *is the rewriting system defined on Proof Nets by the rules in Fig. 2.*

In the following, $M \in$ PN will denote that M is a proof net. Moreover, since we shall consider several variants of proof net reduction, this will also mean that M reduces according to the rules of PN.

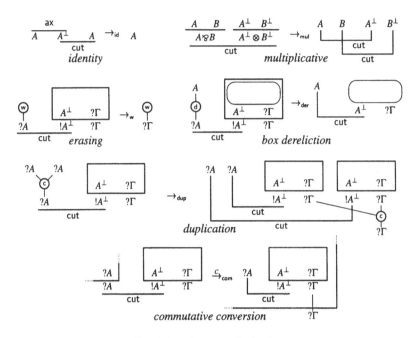

Fig. 2. Proof net cut-elimination.

Theorem 1. PN *is strongly normalizing and confluent (Church-Rosser). As a consequence,* PN *has the unique normal form property.*

Strong normalization (SN) was proved by Girard in [6] (Girard's proof of SN uses the *candidats de réductibilité*; a completely syntactical proof of SN can be found in Joinet's thesis [8]); the Church-Rosser property (CR) was proved by Danos in [2].

Henceforth, let us write nf(N), for the normal form of $N \in$ PN. More generally, since all the reduction systems that we shall analyze will be derived from PN and will be named by sub/superscripted variants of PN, $N \in$ PN$_x^y$ will denote that N reduces according to the rules of PN$_x^y$ and nf$_x^y(N)$ will denote its normal form (if any).

2 Survey and our proposal

2.1 Simulating the λ-calculus: collapsed structural links

When simulating the β-reduction of λ-calculus in PN, the rigidity of the exponential links makes things difficult: the net translation of a term t does not always reduce exactly to the translation of the reduct term s, due to the different shape of the contraction trees in the translation. This is quite annoying, to the point that the first really satisfactory proof of simulation can be found in [4], where Danos and Regnier introduce a system where all exponential links are collapsed into one single $n$ary link.

Usual proof nets are mapped into those proposed in [4] by a transformation μ that pushes contraction and dereliction out of all boxes and contracts them together. Fig. 3

describes μ by applying it to an example; see the mapping on the left. The root of the exponential tree in the example is not the premise of a contraction and is not above the auxiliary port of a box. The collapsed link of type ? that replaces the tree preserves the branches of the tree and the number of boxes that they cross. Every weakening link is replaced by a new link of type × that introduces a special (crossed) occurrence $A^\times$ of the formula A. Every formula $A^\times$ marks a *weakening branch* of the ?-link. A *?-weakening tree* is a ?-link connected to weakening branches only; it is the translation of an exponential tree whose leaves are all weakening links. A *?-weakening* is a ?-weakening tree formed of one weakening branch only; it corresponds to the translation of an exponential tree formed of a weakening link only (*e.g.*, see the mapping on the right in Fig. 3). The introduction of the weakening branches is due to technical reasons; the rationale is that we want to keep track of all the erasing rules required by the reduction. The ×-link is not present in [4], where weakening branches are simply erased.

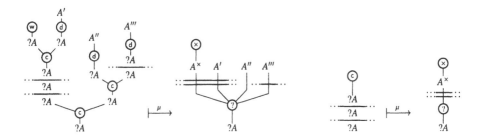

Fig. 3. Collapsing an exponential tree into a ?-link.

Definition 2 (PN$_C$). *Let* PN$_C$ *be the set of the proof nets where contractions and exponential crossings at the auxiliary doors of boxes collapse into a unique nary link of type* ?, *and all the exponential reductions but erasing are collapsed into a unique* exponential *reduction step that performs unboxing, duplication and box inclusion, as shown by the example in Fig. 4.*

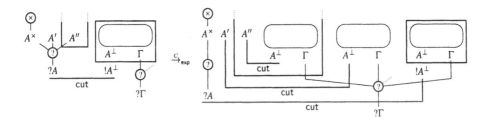

Fig. 4. The exponential rule of PN$_C$.

The exponential rule of PN_C introduces a ?-weakening cut for every weakening branch of the ?-link in the redex. In order to erase the corresponding boxes, that cuts must be explicitly eliminated by means of an erasing rule. The erasing rule of PN_C is the obvious translation of the erasing rule of PN: on the left-hand side, replace the weakening link by a ?-weakening and the auxiliary port crossings by ?-link branches; on the right-hand side, transform each branch into a weakening branch by putting a ×-link above its leaf. When the ?-link in the redex is a ?-weakening tree with n branches, the exponential rule degenerates into a *weakening duplication* that creates n copies of the box in the redex and splits the cut into n ?-weakening cuts. In particular, when the tree is a ?-weakening (*i.e.*, $n = 1$), the left-hand side and right-hand side would coincide; therefore, in order to not introduce trivial reduction loops, the exponential rule does not apply to a ?-weakening cut; the only rule that applies to that cuts is erasing. In [4], the absence of weakening branches corresponds to an exponential rule in which the ?-weakening cuts introduced by our version of the rule are automatically eliminated.

Remark 2 (No exponential axioms). The transformation μ is not defined for the proof nets that contain exponential axioms (*i.e.*, $!A, ?A^\perp$ axioms). From the point of view of provability, this is not a problem, for it is well-known that each proof net can be η-expanded into another one with the same conclusions that contains atomic axioms only (*i.e.*, $p, p^\perp$ axioms only). But, for a detailed analysis of proof net reduction and of its relations with λ-calculus, that unrestricted η-expansion is unacceptable. Therefore, let us constrain η-expansion to exponential axioms. Namely, the η_e-expansion replaces each $!A, ?A^\perp$ axiom with a box containing the axiom $A, A^\perp$ and a dereliction link from $A^\perp$ to $?A^\perp$. Every reduction of $M \in PN$ is simulated by a reduction of its η_e-expansion, and similarly for $M \in PN_C$. Therefore and w.l.o.g., in the following, we shall restrict PN to the case without exponential axioms. In this way, $\mu : PN \to PN_C$ is total.

Proposition 1. *Let $M \in PN$. For every $r : \mu(M) \xrightarrow{C} P$, there is a non-empty $\rho : M \to^* N$ s.t. $P = \mu(N)$. Therefore, PN_C is SN and CR, and $nf_C(\mu(M)) = \mu(nf(M))$.*

The obvious limitation of this approach is that its reduction is too coarse grained: it really performs in one single step all the duplication, erasure and unboxing operations involved in a β-reduction step for the λ-calculus. For this reason, if one wants to study finer reductions on the λ-terms, like the ones involved in handling explicit substitutions, this system turns out to be inadequate: it throws out the baby with the bath water.

2.2 Simulating explicit substitutions: fusion and splitting of contraction links

In [5], the limitations of both PN and PN_C are recognised, and another system is proposed, where it is possible to fuse two *n*ary contraction links together (see the *fusion* rule in Fig. 5) and where the irrelevance of the order of contraction and box formation is taken into account via a reduction rule that allows to push some contractions inside a box (see the push rule in Fig. 5).

This approach is less coarse grained, and it was the first solution for interpreting explicit substitutions in PN, but it still suffers from a certain rigidity of the extended reductions, that makes the translation from λ-calculus with explicit substitutions into PN cumbersome (while the propagation of the substitutions is faithfully mirrored, the translation of a cut forces all the duplications to be performed at once).

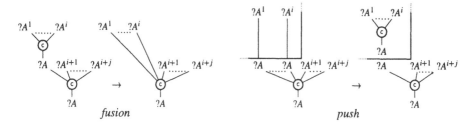

Fig. 5. Fusion and push.

2.3 Our approach: rewriting modulo an equivalence relation

If one looks carefully at the previous approaches, one really finds out that they are both trying to handle contraction links as associative-commutative operators freely floating in and out of boxes: Danos and Regnier work on a representative of the *AC* (associative-commutative) equivalence class which is obtained by collapsing all the trees of exponential links and pushing them outside of all boxes; Di Cosmo and Kesner allow a finer control on how to collapse and push in or out of boxes the contraction links.

The limitations of the previous approaches clearly point out the need of a more flexible system, which accepts explicitly the associative-commutative nature of the contraction operator, allowing a finer control of duplication and propagation of substitutions in the nets. For this reason, we introduce an equivalence relation $\sim$ on Proof Nets and define reduction on the corresponding equivalence classes.

Definition 3 (PN$_{AC}$). *The equivalence relation $\sim$, named AC, is the context closure of the graph equivalences in Fig. 6. Let us extend the reduction of* PN *to the equivalence classes of Proof Nets as* $M \xrightarrow{AC} N$ *iff* $\exists M', N' : M \sim M' \to N' \sim N$. *We shall write* PN$_{AC}$ *for Proof Nets equipped with this new reduction.*

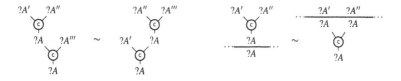

Fig. 6. *AC* congruence.

That extension of PN preserves the normal forms, as shown by the next proposition, which proves indeed that PN$_{AC}$ is a fine analysis of PN$_C$.

Proposition 2. *For every* $M, N \in$ PN$_{AC}$, $M \sim N$ *iff* $\mu(M) = \mu(N)$. *Then, let* $M \xrightarrow{AC}^* N$.

1. *There are* $\mu(M) \xrightarrow{C}^* P$ *and* $\mu(N) \xrightarrow{C}^* P$.
2. $\mathsf{nf}_C(\mu(M)) = \mathsf{nf}_C(\mu(N))$ *and* $\mathsf{nf}(M) \sim \mathsf{nf}(N)$.

3 Main results

The main result of the paper is that PN_{AC} is strongly normalizing and has the unique normal form property (modulo AC).

Theorem 2. *Let $M \in PN_C$.*

1. Let $M \xrightarrow{C}^{*} N$ with N cut-free. Then $N \sim nf(M)$.
2. Every reduction of M is finite.

The first item is a trivial consequence of Proposition 2 (a particular case of it). The proof of strong normalization is by reduction to termination of PN_C.

3.1 Overview of the proof technique

The key point in relating PN_{AC} to PN_C is the study of the so-called *persistent paths*, an invariant introduced by Geometry of Interaction. Persistent paths capture the intuitive idea that every connection (path) between the nodes of a reduct N of M is the deformation of some connection (path) between the nodes of M (see [4]). In fact, along the reduction of M certain connections are broken (*e.g.*, take the path between A and $B^{\perp}$ in the multiplicative rule), while others *persist*; in particular, the paths that persist after every reduction yield the normal form. Geometry of Interaction is an algebraic formulation of the previous notion of path deformation, even if the idea 'reduction as path composition' was already implicit in Lévy labelled λ-calculus. For a survey on the relations between persistent paths, Lévy's labels and Geometry of Interaction *regular paths* see [1].

Persistent paths will be defined and studied in section 4. There, we shall assign a norm to every $M \in PN_{AC}$ in terms of the persistent paths of $\mu(M)$ (actually, in terms of the persistent paths that do not collapse). That norm is decreased by the reductions of PN_{AC} with a correspondence in PN_C, while it is left unchanged by duplication and commutative conversion. In section 5, we shall analyze the transformations that simulate duplication and commutative conversion in PN_C. That analysis will lead us to define a second norm (section 5.4) that is decreased by every one-step reduction.

Unfortunately, the previous proof schema does not work if directly applied to PN_C and PN_{AC}. In fact, in order to fully exploit it, we must tackle two technical difficulties.

The first problem is connected with duplication: we need a way to count the number of box duplications in a reduction. For that purpose, instead of resorting to some measure defined on the whole reduction, we exploit the presence of weakening. Namely, using weakening, we define a proof structure $T^{\checkmark}$, a *tick* (see section 5.3), that reduces to the empty net and s.t. the proof structure $M^{\checkmark}$ obtained by inserting a tick into each box of M is a proof net. Since each box duplication duplicates a tick, the number of boxes duplicated in a reduction is equal to the number of new ticks in the result.

The second problem is that ticks might disappear along the reduction because of an erasing rule. Thus, in order to preserve our counting device, we have to delay garbage collection until the end of the computation (indeed, this approach simplifies other technical parts also). Namely, let us denote by $M \xrightarrow[\neg w]{AC}^{*} N$ a reduction that does not contain erasing rules and by $PN_{AC}^{\neg w}$ the restriction of PN_{AC} to that non-erasing reduction.

Lemma 1. *For every $M \in \mathsf{PN}_{AC}$, if $M \xrightarrow{AC}{}^* N$ then $M \xrightarrow[\neg w]{AC}{}^* P \xrightarrow{AC}{}^*_w N$. Therefore, PN_{AC} is terminating iff $\mathsf{PN}_{AC}^{\neg w}$ is terminating.*

Henceforth, we shall restrict to the study of $\mathsf{PN}_{AC}^{\neg w}$ and of the corresponding system $\mathsf{PN}_C^{\neg w}$, *i.e.*, PN_C restricted to the non-erasing reduction $\xrightarrow[\neg w]{AC}$. That analysis will conclude with the proof of strong normalization of $\mathsf{PN}_{AC}^{\neg w}$ (Lemma 13) that, by Lemma 1, proves the strong normalization of PN_{AC} as well.

4 Paths in $\mathsf{PN}_C^{\neg w}$

A *path* in a proof net M is an undirected path in the graph of M that, crossing any link but axiom and cut, moves from a premise to the conclusion of the link and that, crossing an axiom/cut, moves from one conclusion/premise of the axiom/cut to the other conclusion/premise.

Let M be a proof net. We shall denote by $\Phi(M)$ the set of its paths and we shall write $\psi \sqsubseteq \phi$ to denote that ψ is a subpath of ϕ. Remarkably, when M is in normal form, $\Phi(M)$ is finite and is the set of the *elementary paths* of M (a path is elementary when it does not cross any cut); instead, when M contains cuts, the paths of M may loop and $\Phi(M)$ may be infinite.

4.1 Persistent and permanent paths

After a reduction step, paths deform or even vanish, so there is a natural notion of *residual* of a path along a proof net reduction: as in [4], this notion can be captured by associating to every $r : M \xrightarrow[\neg w]{C} N$, a function $\bar{r} : \Phi(N) \xrightarrow[\neg w]{C} \Phi(M)$ that maps a path of N to its *ancestor* in M. The notion of residual extends to a reduction $\rho = r_0 r_1 \ldots r_k$ by function composition, *i.e.*, $\bar{\rho} = \bar{r_0} \cdot \bar{r_1} \cdot \ldots \cdot \bar{r_k}$.

We remark that $\bar{\rho}$ is total; that is, for $\rho : M \xrightarrow[\neg w]{C}{}^* N$, every path $\phi \in \Phi(N)$ is the deformation of some path in M. Moreover, every deformed path ϕ results from the contraction to a node of some subpath of $\rho(\phi)$; therefore, either ϕ is essentially the same as $\bar{\rho}(\phi)$, or $|\phi| < |\bar{\rho}(\phi)|$. However, $\bar{\rho}$ is not onto. In fact, a path of M disappears in the following cases:

1. The path contracts to a connection between the premises of a cut that is then reduced along ρ (*e.g.*, the path between $A \mathbin{⅋} B$ and $A^\perp \otimes B^\perp$ in Fig. 7).
2. The execution of a multiplicative or exponential cut disconnects the path. For instance, take the dashed path in the right-hand side of Fig. 7.

The two cases above correspond to two completely different phenomena. In the first case, the path disappears enclosed into a longer path that eventually contracts to a formula. In the second case, the reduction *splits* the path. Thus, in the first case, we can say that the path *persists* along the reduction, as a trace of it is still present in the resulting proof net; in the second case, the path has no image in the result.

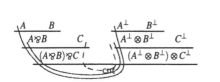 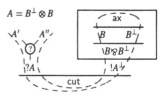

Fig. 7. Paths.

Definition 4 (persistent paths). *Let* $\rho : M \xrightarrow{C}{}_{\neg w}^* N$. *A path* $\phi \in \Phi(M)$ *is* ρ-*persistent when there is* $\psi \in \Phi(N)$ *s.t.* $\phi \sqsubseteq \bar{\rho}(\psi)$. *The* ρ-*persistent path* ϕ *is said* ρ-*permanent when* $\phi = \bar{\rho}(\psi)$ *for some* $\psi \in \Phi(N)$. *A path of M is* persistent, *or* permanent, *when it is* ρ-*persistent, or* ρ-*permanent, for every reduction* ρ *of M.*

Henceforth, $\Psi(M)$ will denote the set of the permanent paths of M and $\Psi_{\sqsubseteq}(M)$ will denote the set of its persistent paths. By definition, $\Psi_{\sqsubseteq}(M)$ is a superset of the closure by subpaths of $\Psi(M)$; further, we shall prove that $\Psi_{\sqsubseteq}(M)$ is that closure, see Lemma 4.

Lemma 2. *Let* $M \in \mathrm{PN}_C^{\neg w}$. *Every occurrence of formula in M is persistent.*

Therefore, the set of the persistent paths is not empty. Indeed, it is readily seen that every path corresponding to a redex (*i.e.*, every cut pair A, $A^{\perp}$) is persistent. Moreover, every *virtual redex*, *i.e.*, every path that along some reduction will eventually reduce to a cut pair, is persistent, see [4] and [1].

4.2 Folding and unfolding of permanent paths

The permanent paths of a proof net M are the connections of M that are invariant under any reduction. So we expect that $\Psi(M)$ be an image of $\mathrm{nf}_C^{\neg w}(M)$; that is, we expect $\Psi(M) = \bar{\rho}(\Phi(\mathrm{nf}_C^{\neg w}(M)))$, for any normalizing reduction ρ. However, that equivalence is not immediate. In fact, though $\mathrm{PN}_C^{\neg w}$ has the unique normal form property, two distinct reductions might build the same path of $\mathrm{nf}_C^{\neg w}(M)$ by combining different paths of M.

Lemma 3. *Let* $M \in \mathrm{PN}_C^{\neg w}$. *For every* $r_1 : M \xrightarrow{C}{}_{\neg w} M_1$ *and* $r_2 : M \xrightarrow{C}{}_{\neg w} M_1$, *there exist* $\rho_1 : M_1 \xrightarrow{C}{}_{\neg w}^* N$ *and* $\rho_2 : M_2 \xrightarrow{C}{}_{\neg w}^* N$, *s.t.* $\overline{r_1\rho_1} = \overline{r_2\rho_2}$.

Proposition 3. *Let* $N = \mathrm{nf}_C^{\neg w}(M)$. *There is a canonical map* $\mathrm{fold}_M : \Phi(N) \to \Phi(M)$ *s.t.* $\mathrm{fold}_M = \rho$, *for every* $\rho : M \xrightarrow{C}{}_{\neg w}^* N$. *Moreover,* $\Psi(M) = \mathrm{fold}_M(\Phi(N))$.

The previous proposition proves the soundness of the definition of permanent paths. Moreover, let $\rho : M \xrightarrow{C}{}_{\neg w}^* N$; it proves that the restriction of $\bar{\rho}$ to permanent paths is an onto map $\hat{\rho} : \Psi(N) \xrightarrow{C}{}_{\neg w}^* \Psi(M)$ (this is a consequence of $\Psi(M) = \mathrm{fold}_M(\Phi(P)) = \bar{\rho} \cdot \mathrm{fold}_N(\Phi(P))$, where $P = \mathrm{nf}_C^{\neg w}(M) = \mathrm{nf}_C^{\neg w}(N)$). We stress that $\bar{\rho}(\Psi(N)) = \Psi(M)$ is not a trivial consequence of the definition of permanent paths, as that definition trivially implies $\bar{\rho}(\Psi(N)) \supseteq \Psi(M)$ only. Finally, as a corollary of Proposition 3, we get that every persistent path can be prolongated to a permanent path.

Lemma 4. *For every* $\phi \in \Psi_{\sqsubseteq}(M)$, *there is* $\psi \in \Psi(M)$ *s.t.* $\phi \sqsubseteq \psi$.

The *unfolding* of $\phi \in \Phi(M)$ is the set of its residuals in the normal form, *i.e.*,

$$\mathsf{unfold}_M(\phi) = \{\psi \in \Phi(\mathsf{nf}_C^{\neg w}(M)) \mid \mathsf{fold}_M(\psi) = \phi\} = \mathsf{fold}_M^{-1}(\phi)$$

The *cardinality* of a path is the cardinality of its unfolding, *i.e.*,

$$\#(\phi) = |\mathsf{unfold}_M(\phi)|$$

By definition, $\#(\phi) > 0$ iff $\phi \in \Psi(M)$. Thus, $\sum\{\#(\phi) \mid \phi \in \Phi(M)\} = \sum\{\#(\phi) \mid \phi \in \Psi(M)\} = |\Phi(\mathsf{nf}_C^{\neg w}(M)|$; that is another way to express the combinatorial fact that no finite reduction creates an infinite number of residuals (*i.e.*, $\#(\phi)$ is always finite).

4.3 The norm of $\mathsf{PN}_C^{\neg w}$

In the reduction of $\mathsf{PN}_C^{\neg w}$ we have two distinct phenomena. On one side, exponential reductions tend to unfold permanent paths, increasing their number; on the other side, every reduction reduces the length of some permanent path. The previous considerations summarize in the following lemma (as usual, $|\phi|$ denotes the length of the path ϕ, while $\overline{\rho}^{-1}(\phi) = \{\psi \mid \overline{\rho}(\psi) = \phi\}$).

Lemma 5. *Let* $\rho : M \xrightarrow[\neg w]{C}{}^* N$. *For every* $\phi \in \Psi(M)$,

1. $\#(\phi) = \sum\{\#(\psi) \mid \psi \in \overline{\rho}^{-1}(\phi)\}$;
2. $|\phi| \geqslant |\psi|$, *for every* $\psi \in \overline{\rho}^{-1}(\phi)$.
3. *Moreover, if* ρ *is not empty and is not a sequence of weakening duplications, then* $|\phi| > |\psi|$ *for some* $\phi \in \Psi(M)$.

Let us equip $\mathsf{PN}_C^{\neg w}$ with the following norm:

$$\|M\|_C^{\phi} = \sum\{\#(\phi) \cdot |\phi| \mid \phi \in \Phi(M)\} = \sum\{\#(\phi) \cdot |\phi| \mid \phi \in \Psi(M)\}$$

We remark that, since $\Psi(M)$ is finite, $\|M\|_C^{\phi}$ is well-defined (*i.e.*, it is finite).

Lemma 6. *For every* $\rho : M \xrightarrow[\neg w]{C}{}^* N$, $\|N\|_C^{\phi} \leqslant \|M\|_C^{\phi}$. *Moreover, when* ρ *is not empty and is not a sequence of weakening duplications,* $\|N\|_C^{\phi} < \|M\|_C^{\phi}$.

5 Relating $\mathsf{PN}_{AC}^{\neg w}$ to $\mathsf{PN}_C^{\neg w}$

The grain of the reduction in $\mathsf{PN}_{AC}^{\neg w}$ is finer than in $\mathsf{PN}_C^{\neg w}$. In particular, the commutative conversion and the duplication rule have no correspondence in $\mathsf{PN}_C^{\neg w}$; moreover, in $\mathsf{PN}_{AC}^{\neg w}$ we reduce modulo AC. For the part of $\mathsf{PN}_{AC}^{\neg w}$ with a direct correspondence in PN_C the situation is clear: since $M \xrightarrow[\neg w]{AC} N$ implies $\mu(M) \xrightarrow[\neg w]{C} \mu(N)$, this part of the system is strongly normalizing and

$$\|M\|_{AC}^{\phi} = \|\mu(M)\|_C^{\phi} \qquad \text{for } M \in \mathsf{PN}_{AC}^{\neg w}$$

seems the natural candidate for expressing that property. For the remaining part of $\mathsf{PN}_{AC}^{\neg w}$, let us analyze each rule separately.

5.1 Commutative conversion

When $r : M \xrightarrow{AC}_{\text{com}} N$, $\mu(N)$ and $\mu(M)$ are equal but for some boxes of $\mu(M)$ that have been moved inside some other box of $\mu(N)$, see Fig. 8.

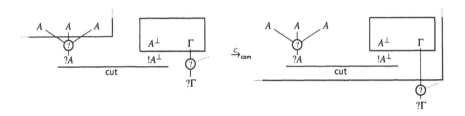

Fig. 8. Commutative conversion in PN_C.

Lemma 7. *Let* $r : M \xrightarrow{AC}_{\text{com}} N$.

1. $\text{nf}_C^{\neg w}(\mu(M)) = \text{nf}_C^{\neg w}(\mu(N))$;
2. $\text{fold}_{\mu(M)} = \text{fold}_{\mu(N)}$ *and* $\Psi(\mu(M)) = \Psi(\mu(N))$;
3. $\|M\|_{AC}^{\phi} = \|N\|_{AC}^{\phi}$.

Therefore, the commutative conversion induced on PN_C preserves normal forms and persistent paths. Moreover, though it does not decrease the norm on paths, it is readily seen that we cannot have an infinite sequence of commutative conversions.

Definition 5 (depth). *The depth of an !-link, and then of the corresponding box, is the number of boxes that encapsulate it. The depth* $\partial(M)$ *of a proof net* M *is the sum of the depths of its !-links.*

Let $\text{n}^!(M)$ be the number of !-links in M. We define

$$\|M\|^! = \text{n}^!(M)^2 - \partial(M)$$

Lemma 8. *For any* $M \in \text{PN}_{AC}^{\neg w}$.

1. $\|M\|^! \geqslant 0$.
2. *If* $r : M \xrightarrow{AC}_{\text{com}} N$, *then* $\|N\|^! < \|M\|^!$.

5.2 Duplication

This is the trickiest case. Fig. 9 illustrates by means of an example the transformation $\delta_r : \mu(M) \xrightarrow{AC}_{\text{dup}} \mu(N)$ corresponding to $r : M \xrightarrow{C}_{\text{dup}} N$. In that example, we assume that the contraction c in the redex r join two exponential subtrees whose leaves are A' and A'', A''', respectively; that two sets of leaves are the premises of the two new instances of c in $\mu(N)$. As every rule in PN_C, δ_r defines a map $\overline{\delta_r} : \Phi(\mu(N)) \rightarrow \Phi(\mu(M))$.

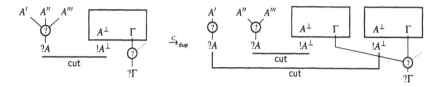

Fig. 9. Duplication in PN$_C$.

Lemma 9. *Let* $r : M \xrightarrow{AC}_{dup} N$.

1. $\mathsf{nf}_C^{-w}(\mu(M)) = \mathsf{nf}_C^{-w}(\mu(N))$;
2. $\mathsf{fold}_{\mu(M)} = \overline{\delta}_r \cdot \mathsf{fold}_{\mu(N)}$ *and* $\Psi(\mu(M)) = \overline{\delta}_r(\Psi(\mu(N)))$;
3. $\|M\|_{AC}^\phi = \|N\|_{AC}^\phi$.

5.3 Ticked proof nets

Usually, the proof that duplication is terminating exploits the fact that, in a sequence of duplications, no box is duplicated twice by the same contraction link—this is the intuitive idea; formally, we should reason in terms of residuals. However, since we assume to know that PN$_C^{-w}$ is strongly normalizing, we can resort to a technical trick.

Duplication does not decrease the length of any permanent path. So, in order to prove that it is terminating, we need a measure of the unfolding that it causes. The remark that duplication tends to increase the number of persistent paths seems unfruitful: unfortunately, there are $M \xrightarrow{AC}_{dup} N$ for which $|\Psi(\mu(M))| = |\Psi(\mu(N))|$. For instance, the proof net M in Fig. 10 reduces to an axiom; so the path ϕ drawn in the figure is the only non-empty permanent path of M. The path ϕ contains two occurrences of the path ψ (rooted at $!(A \otimes A^\perp)$) that loops inside the box, *i.e.*, $\phi = \phi_0 \psi \phi_1 \psi \phi_2$. After $M \xrightarrow{C}_{com} N$, the residual of ϕ is $\phi' = \phi_0' \psi' \phi_1' \psi'' \phi_2'$, where ψ' and ψ'' are residuals of ψ that loop inside two distinct boxes of N. In other words, instead of duplicating some permanent path, the duplication in M unfolds the loop described by the unique permanent path in the proof net. The situation would be different if the box B in M would contain a permanent path: that path would be duplicated by the duplication of B.

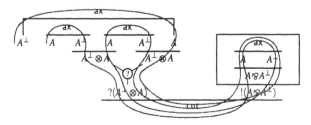

Fig. 10. Unfolding the loop of a permanent path.

Lep p be any atomic formula. A *tick* of PN_C is a proof structure $T^\checkmark$ as that in Fig. 11; and $\mu(T^\checkmark)$ is a tick of PN_{AC}. A tick is not a proof net but, for every $N \in PN_C$, the proof structure $M = N \cup T^\checkmark$ obtained by attaching the tick $T^\checkmark$ to N is a proof net (*i.e.*, $M \in PN_C$); moreover, $M \xrightarrow{AC}_w N$, by contraction of the weakening cut in $T^\checkmark$. Therefore, let N be the interior of the most external box of some proof net; by replacing M for N, we get a ticked box $B \in PN_C$. Then, by recursive application of this ticking procedure to the boxes in B, we eventually get a proof net whose boxes are all ticked.

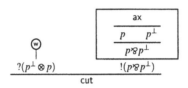

Fig. 11. A tick.

Definition 6 ($PN_{AC}^\checkmark$). *A box contains (at least) a tick when its interior is a proof net $B \cup T^\checkmark$ and $T^\checkmark$ is a tick. A proof net of PN_{AC} is ticked when each of its boxes contains a tick. Let us denote by $PN_{AC}^\checkmark$ the set of the ticked proof nets of PN_{AC}. We say that $M^\checkmark \in PN_{AC}^\checkmark$ is a ticking of $M \in PN_{AC}$ when M can be obtained from $M^\checkmark$ by erasing some of its ticks.*

The set of the ticked proof nets $PN_{AC}^\checkmark$ is closed by reduction, *i.e.*, for any $M^\checkmark \in PN_{AC}^\checkmark$ and any $\rho : M^\checkmark \xrightarrow{AC}_{\neg w}{}^* N^\checkmark$, $N^\checkmark \in PN_{AC}^\checkmark$. In the following, $M^\checkmark$ will always denote some ticking of $M \in PN_C$ (by the way, there exists at least one $M^\checkmark$ for every M). By definition, $M^\checkmark \xrightarrow{AC}_w{}^* M$, for any $M^\checkmark$.

Lemma 10. *The $\xrightarrow{AC}_{\neg w}$-reduction of $M \in PN_{AC}$ is terminating iff the $\xrightarrow{AC}_{\neg w}$-reduction of any $M^\checkmark$ is terminating.*

By Lemma 10, strong normalization of PN_{AC} reduces to that of $PN_{AC}^\checkmark$. Moreover, as the ticks of $M^\checkmark$ are permanent, duplication is not a problem in $PN_{AC}^\checkmark$. In fact, let $n^\checkmark(M)$ be the number of ticks in M. For any $M \in PN_{AC}^{\neg w}$, we define

$$\|M\|_{AC}^\checkmark = \|\mu(M)\|_C^\checkmark \quad \text{where} \quad \|P\|_C^\checkmark = n^\checkmark(nf_C^{\neg w}(P)) - n^\checkmark(P) \quad \text{for } P \in PN_C^{\neg w}$$

Lemma 11. *For any $M^\checkmark \in PN_{AC}^\checkmark$.*

1. $\|M^\checkmark\|_{AC}^\checkmark \geqslant 0$.
2. *If $r : M^\checkmark \xrightarrow{AC}_{\neg w} N^\checkmark$, then $\|N^\checkmark\|_{AC}^\checkmark \leqslant \|M^\checkmark\|_{AC}^\checkmark$; moreover, $\|N^\checkmark\|_{AC}^\checkmark < \|M^\checkmark\|_{AC}^\checkmark$, when r is a duplication.*

5.4 The norm of $PN_{AC}^{\checkmark}$

Let $M \in PN_{AC}^{\neg w}$, we take

$$\langle\!\langle M \rangle\!\rangle_{AC} = \langle\; \|M\|_{AC}^{\phi} + \|M\|_{AC}^{\checkmark}, \|M\|^{!}\;\rangle$$

with the lexicographic ordering, *i.e.*, $\preccurlyeq$ is the reflexive closure of $\langle a_1, b_1 \rangle \prec \langle a_2, b_2, \rangle$ iff $(a_1 < a_2)$ or $(a_1 = a_2 \wedge b_1 < b_2)$. By definition, $\langle 0, 0 \rangle \preccurlyeq \langle\!\langle M \rangle\!\rangle_{AC}$.

We remark that, for any $M, N \in PN_{AC}^{\neg w}$, $M \sim N$ implies $\langle\!\langle M \rangle\!\rangle_{AC} = \langle\!\langle N \rangle\!\rangle_{AC}$.

Lemma 12. *Let $M^{\checkmark} \in PN_{AC}^{\checkmark}$. For every $r : M^{\checkmark} \xrightarrow[\neg w]{AC} N^{\checkmark}$, $\langle\!\langle N^{\checkmark} \rangle\!\rangle_{AC} \prec \langle\!\langle M^{\checkmark} \rangle\!\rangle_{AC}$.*

Lemma 13. $PN_{AC}^{\neg w}$ *is strongly normalizing.*

6 Conclusions and future work

We have presented here for the first time a proof of strong normalization for Multiplicative Exponential Linear Logic's Proof Nets with an associative-commutative contraction free to float in and out of proof boxes. This is interesting for several reasons.

First, this is another significant application of the *normalization by persistent paths* slogan which can be found in Girard's Geometry of Interaction. But also, now that we know that we can rearrange contraction trees as we like during a reduction of a proof net, and still have the strong normalization property, we can go back to analyse how the classical β-reduction of the lambda calculus, or the more refined reductions of calculi with explicit substitutions are simulated in our system. We expect not only to be able to provide a much simpler simulation than the ones in the literature, but also to extract from PN_{AC} a calculus of explicit substitutions with good properties.

References

[1] A. Asperti, V. Danos, C. Laneve, and L. Regnier. Paths in the lambda-calculus: three years of communications without understanding. In *Proceedings, Ninth Annual IEEE Symposium on Logic in Computer Science (LICS)*, pages 426–436, Paris, France, 1994.

[2] V. Danos. *Une Application de la Logique Linéaire à l'Ètude des Processus de Normalisation (principalement du λ-calcul)*. PhD Thesis, Université Paris 7, Paris, June 1990.

[3] V. Danos and L. Regnier. The structure of multiplicatives. *Archive for Mathematical Logic*, 28:181–203, 1989.

[4] V. Danos and L. Regnier. Proof-nets and the Hilbert space. In J.-Y. Girard, Y. Lafont, and L. Regnier, editors, *Advances in Linear Logic*, pages 307–328. Cambridge University Press, 1995. Proceedings of the Workshop on Linear Logic, Ithaca, New York, June 1993.

[5] R. Di Cosmo and D. Kesner. Strong normalization of explicit substitutions via cut elimination in proof nets. In *Proceedings, Twelfth Annual IEEE Symposium on Logic in Computer Science (LICS)*, pages 35–46, Warsaw, Poland, 1997. Full Paper available as http://www.dmi.ens.fr/~dicosmo/Pub/esll.ps.gz.

[6] J.-Y. Girard. Linear logic. *Theoretical Comput. Sci.*, 50(1):1–102, 1987.

[7] S. Guerrini. Correctness of multiplicative proof nets is linear. In *Proceedings, Fourteenth Annual IEEE Symposium on Logic in Computer Science (LICS)*, Trento, Italy, 1999.

[8] J.-B. Joinet. *Étude de la normalisation du calcul des séquents classique à travers la logique linéaire*. PhD Thesis, Université Paris 7, Paris, 1993.

Hierarchical Graph Decompositions Defined by Grammars and Logical Formulas[*]

Bruno Courcelle

LaBRI, Université Bordeaux I,
351 cours de la Libération
33405 Talence, France
email: courcell@labri.u-bordeaux.fr

Abstract

A context-free grammar specifies a set of words, and for each of these words one or more derivation trees. Such a tree represents the (or a) hierarchical structure of the corresponding word and is the input of algorithms like those used in a compiler.

Context-free graph grammars have been defined and studied which generate sets of finite graphs (and hypergraphs). The generated objects also have derivation trees which are useful for example as input to drawing algorithms. Furthermore, many NP-complete graph properties can be decided in linear time for graphs generated by such grammars, and given by the derivation trees. This is the case in particular for properties expressible in Monadic Second-order logic, via a suitable representation of graphs by relational structures. A Monadic Second-order formula representing such a property can be compiled into a deterministic finite-state tree automaton that verifies the property by traversing the derivation tree.

There are other connections between context-free graph grammars and Monadic Second-order logic. In particular, every context-free set of graphs is the image of the set of finite binary trees under a transformation of trees into graphs expressible in Monadic Second-order logic. (Similarly context-free languages are the images under rational transductions of certain languages encoding trees). In certain cases the *parsing* mapping, i.e. the transformation of the graph into one of its derivation trees, is expressible in Monadic Second-order logic.
However, (but not surprizingly by the algorithmic results recalled above) context-free graph grammars suffer a severe limitation: no single grammar can generate the set of all finite graphs, or even the set of all finite planar graphs.

We know from graph theory that graphs have certain canonical hierarchical decompositions. One of them is the decomposition of a graph into a forest over

---

[*] This research was supported by the European Community Training and Mobility in Research network GETGRATS.

its 3-connected components, another one is the *modular decomposition*. These two hierarchical decompositions are based respectively on graph gluing operations and substitutions of graphs for vertices, which fit within the framework of context-free graph grammars. But since the basic blocks are not finite, they cannot be handled by context-free graph grammars since these grammars must have finitely many production rules.

We are interested in situations where such hierarchical graph decompositions can be defined in Monadic Second-order logic. (In some sense, we aim at replacing grammars by logic.) This the case for the decomposition in 3-connected blocks, and also for the modular decomposition under the additional assumption that the graph is given with some linear order of its set of vertices. (Any two linear orders yield the same decomposition since modular decomposition is unique; we do not know how to do th e construction without an auxiliary linear order.)

After reviewing these known results, we will present the new notion of modular decomposition for *finite undirected hypergraphs of unbounded rank* (we call rank the maximal size of a hyperedge). Such hypergraphs can be handled as bipartite graphs, but their modular decompositions are not the ones for graphs via this representation. The trees representing them can be defined in Monadic Second-order logic without needing any auxiliary ordering.

A hypergraph is *convex* if there exists a linear order on the vertices such that all hyperedges are intervals. If it is *prime* (i.e., is a basic compo nent of the modular decomposition), then only two linear orders witness its convexity (one and its opposite). They can be defined by Monadic Second-order formulas.

We deduce from this construction that if a set of finite convex hypergraphs has a decidable Monadic Second-order theory then, considered as a set of bipartite graphs, it is a subset of a context-free set. This provides us with a new case of validity of a conjecture made by Seese that structures having a decidable Monadic Second-order theory are definable from trees by Monadic Second-order formulas.

For a survey on Monadic Second-order logic, see:

B. Courcelle, *The expression of graph properties and graph transformations in monadic second-order logic*, Chapter 5 of the *Handbook of graph grammars and computing by graph transformations, Vol. 1 : Foundations*, G. Rozenberg ed., World Scientific, 1997, pp. 313-400.

For other references, see:
http://dept-info.u-bordeaux.fr/~courcell/ActSci.html

Undecidability of the $\exists^*\forall^*$ Part of the Theory of Ground Term Algebra Modulo an AC Symbol [*]

Jerzy Marcinkowski

jma@tcs.uni.wroc.pl

Institute of Computer Science
University of Wrocław,
ul. Przesmyckiego 20
51-165 Wrocław, Poland

Abstract. We show that the $\exists^*\forall^*$ part of the equational theory modulo an AC symbol is undecidable. This solves the open problem 25 from the RTA list ([DJK91],[DJK93],[DJK95])

1 Introduction

Formulae built of terms and the equality predicate are one of the most natural objects in rewriting. One of the most natural ways of modeling the semantic behavior of real objects is considering additional equality axioms between terms. The most natural set of equality axioms is AC what means the associativity axiom: $f(x, f(y, z)) = f(f(x, y), z)$ and the commutativity axiom: $f(x, y) = f(y, x)$. Validity of an equational formula modulo AC symbol is in general an undecidable problem. This paper is not the first attempt to demarcate the decidability/undecidability border, that is to find out what are the classes of the "simple" formulas, for which there exists an algorithm deciding validity.

We consider a finite signature S of function symbols, containing also some AC symbols. We also assume that there is at least one constant in the signature, so the set of ground terms over S is not empty. Then we consider the first order equational theory of the ground terms algebra over S: the only relational symbol of the theory is the equality, the function symbols are the symbols in S, and the variables range over the set of ground terms.

The measure of the complexity of a formula is the number of alternations of quantifiers in the prenex form. On the undecidability side of the border it was proved in [T90] and [T92] that the Σ_3 part the theory is undecidable (this part contains the formulae whose quantifier prefix is of the form $\exists^*\forall^*\exists^*$). On the decidability side, it is known that the Σ_1 part (existential formulas) is decidable [C93]. Also several papers (including [LM93] and [F93]) were written about the decidability of some special cases of so called *AC complement problem*, which itself is a special case of the Σ_2 part of the theory. Decidability of

---
[*] Research supported by the Polish KBN grant 8T11C02913

the whole Σ_2 part was stated as an open problem in [T90] and then on the RTA list of open problems ([DJK91],[DJK93],[DJK95]). In this paper we present the negative solution of the problem.

The rest of the paper is organized as follows: in Section 2 we prove undecidability of the existential-universal (Σ_2) part of the theory of an AC idempotent symbol: we assume there is a function symbol in the signature which is not only commutative and associative but also idempotent, what means that it satisfies the axiom $f(x,x) = x$. In Section 3 we present our main result: undecidability of the $\exists^*\forall^*$ theory of an AC symbol. In Section 3 we prove the result of Section 2 for the smallest possible signature. Finally, in Section 4 we show that the main result holds also if infinite terms are allowed.

2 $\exists^*\forall^*$ theory of one idempotent AC symbol

Let us start from a very simple case of the equational theory of an idempotent AC symbol. In [T90] and [T92] Treinen shows that the $\exists^*\forall^*\exists^*$ part of the theory is undecidable. We would like to begin this technical part showing the undecidability of the $\exists^*\forall^*$ part of the theory. It will be a good introduction to the methods used in the following sections.

Let $\Pi = \{< l_1, r_1 >, < l_2, r_2 > \ldots < l_k, r_k >\}$ be an instance of the Post Correspondence Problem. This means that each l_i and r_i is a nonempty word over some finite alphabet (we assume it is $\{a,b\}$). Let us remind that a nonempty word $w = j_1 j_2 \ldots i_{j_s}$ over the alphabet $\{1, 2, \ldots k\}$ is called *solution of Π* if $l_{j_1} l_{j_2} \ldots l_{j_s} = r_{j_1} r_{j_2} \ldots r_{j_s}$, and that the existence of a solution is an undecidable property of Π.

We consider the signature consisting of the binary AC symbol $+$, two unary function symbols a and b, and the function symbol h of arity 2. The only constant is c.

The words l_i, r_i can be naturally understood as unary contexts built over the signature $\{a,b\}$: for example the word abb means for us the same as the context $a(b(b(X)))$.

Let us define:

$$\chi_1(x) =$$

$$\forall w, w_1, w_2$$
$$(h(w_1, w_2) + w = x \wedge (w_1 \neq w_2 \vee w_1 = w_2 = c)) \Rightarrow$$

$$[\, h(l_1(w_1), r_1(w_2)) + w + h(w_1, w_2) = x \vee$$
$$h(l_2(w_1), r_2(w_2)) + w + h(w_1, w_2) = x \vee$$

$$\ldots$$

$$\vee\, h(l_k(w_1), r_k(w_2)) + w + h(w_1, w_2) = x) \,]$$

$$\chi_2(x) =$$

$$\exists s \ h(c,c) + s = x$$

And
$$\chi = \exists x \ \chi_1(x) \wedge \chi_2(x)$$

Obviously χ is an $\exists^*\forall^*$ formula.

Theorem 1. *1. Formula χ is valid if and only if Π is solvable.*
2. The $\exists^*\forall^*$ part of the equational theory modulo an idempotent AC symbol is undecidable.

Of course (ii) follows form (i). The proof of (i) is left for the reader as an easy exercise. Let us however explain the meaning of the formulas above. The existentially quantified variable x is understood as a "set". Formula χ_2 says that the pair ε, ε of words, encoded as $h(c,c)$ is "in x". This happens to be the "first pair of the solution of the PCP instance Π" but what is important here is that this pair is the initialization of some process, whose termination is undecidable. Formula $\chi_1(x)$ says that if $h(w_1, w_2)$ is in x then one of the possible next configurations of the process is also in x (unless $h(w_1, w_2)$ encodes the final configuration of the process). Since we consider first order terms, the set represented by x is finite. So it can only exist if the process terminates (similar proof can be given also if we accept infinite terms, see Section 5 for details). The technical problem is how to say "is in x" by a universal formula. To express the fact that $y \in x$ we need a "witness" w, such that $w + y = x$. But using such a witness may lead to a $\exists^*\forall^*\exists^*$ formula: *There exists such an x that the initial configuration is in x and for every configuration y and every witness w if $y + w = x$, what means if y is in x, then there exists a witness v such that $v + u_1 = x$ or $v + u_2 = x$ or $\ldots v + u_k = x$, where $u_1, u_2, \ldots u_k$ are possible configurations reachable from y in one step.* In this section we could go around this difficulty by reusing the witness. Thanks to the idempotency v can be built as $w + h(w_1, w_2)$. This is not the case in the following sections where we consider a non-idempotent AC symbol.

3 $\exists^*\forall^*$ theory of an AC symbol

In this section we prove:

Theorem 2. *The $\exists^*\forall^*$ equational theory modulo an AC symbol is undecidable.*

Let us define:

$$\phi_1(t_l, t_r, s_l, s_r) =$$

$$(s_l = l_1(t_l) \wedge s_r = r_1(t_r)) \vee (s_l = l_2(t_l) \wedge s_r = r_2(t_r)) \vee \ldots \vee (s_l = l_k(t_l) \wedge s_r = r_k(t_r))$$

and
$$\phi_2(x) =$$

$\forall w, z$

$w + z = x \wedge w$ is of the form $f(w_1, w_2, w_3, w_4) \Rightarrow$
w is of the form $f(t_l, t_r, t, f(r_l, r_r, r, v))$ or of the form $f(s, s, c, c)$ with $s \neq c$

$\phi_2(x)$ can be written as a universal formula:

$\forall w, w_1, w_2, w_3, w_4, w_5$
$\quad \neg(x = w + f(w_1, w_2, w_3, a(w_4))) \wedge$
$\quad \neg(x = w + f(w_1, w_2, w_3, b(w_4))) \wedge$
$\quad \neg(x = w + f(w_1, w_2, w_3, w_4 + w_5)) \wedge$
$\quad \neg[x = w + f(w_1, w_2, w_3, c) \wedge (w_3 \neq c \vee w_1 \neq w_2 \vee w_1 = c)]$

Define:
$$\phi_3(x) =$$
$\forall y, s_l, s_r, t, r_l, r_r, w, v$
$\quad f(s_l, s_r, t, f(r_l, r_r, w, v)) + y = x \Rightarrow$
$\quad \phi_1(s_l, s_r, r_l, r_r) \wedge w + f(r_l, r_r, w, v) = t$

and:
$$\phi_4(x) =$$
$\exists s_1, s_2 \quad x = s_1 + f(c, c, s_1, s_2)$

Now, define ϕ as

$$\exists x \; \phi_2(x) \wedge \phi_3(x) \wedge \phi_4(x)$$

ϕ is clearly an $\exists^* \forall^*$ formula.

Lemma 1. *If Π is solvable then ϕ is valid.*

Proof:
If Π is solvable then there exists a finite sequence $x_0, y_0, x_1, y_1 \ldots x_l, y_l$ of terms such that $x_0 = y_0 = c$ and $\phi_1(x_i, y_i, x_{i+1}, y_{i+1})$ holds for each $i = 0, 1, \ldots l - 1$ and that $x_l = y_l$.
Define $t_l = f(x_l, y_l, c, c)$. When t_j is defined for some $j > 0$ as $f(s_1, s_2, s_3, s_4)$ define t_{j-1} as
$\quad f(x_{j-1}, y_{j-1}, s_3 + f(s_1, s_2, s_3, s_4), f(s_1, s_2, s_3, s_4))$
Then define $x = t_0 + t_1 + \ldots + t_l + c$
Notice that if t_j is $f(s_1, s_2, s_3, s_4)$ for some $j \geq 0$ then $s_1 = x_j$ and $s_2 = y_j$. If $x = w + f(s_1, s_2, s_3, s_4)$ for some s_1, s_2, s_3, s_4 then $f(s_1, s_2, s_3, s_4)$ is t_j for some j and so we can check directly that ϕ_2 and ϕ_3 hold for x.

To prove that also ϕ_4 holds first notice that $t_0 = f(c, c, s_3, s_4)$ for some s_3 and s_4. Then use induction to show that if t_j is $f(r_1, r_2, r_3, r_4)$ then $r_3 = c + t_l + \ldots + t_{j+1}$. So $s_3 = c + t_l + \ldots + t_1$ and $x = t_0 + s_3$. $\quad\square$

Lemma 2. *If ϕ is valid then Π is solvable.*

Suppose ϕ is valid and let x be such term that $\phi_2(x) \wedge \phi_3(x) \wedge \phi_4(x)$ holds.
Take such s_1 and s_2 that $x = s_1 + f(c, c, s_1, s_2)$. They exist since $\phi_4(x)$ holds. Define t_0 as $f(c, c, s_1, s_2)$. Now, if t_j is defined for some j and t_j is of the form $f(z_1, z_2, z_3, f(w_1, w_2, w_3, w_4))$ then define t_{j+1} as $f(w_1, w_2, w_3, w_4)$.

Lemma 3. *For every $i \geq 0$, if t_i is defined as $f(z_1, z_2, z_3, z_4)$ then*

1. *Either $t_i + z_3 = x$ or there exists w such that $w + t_i + z_3 = x$*
2. *$z_1 = z_2$ or t_{i+1} is defined.*
3. *Suppose t_{i+1} is defined as $f(u_1, u_2, u_3, u_4)$ for some terms u_1, u_2, u_3, u_4. Then $\phi_1(z_1, z_2, u_1, u_2)$ holds.*
4. *If t_{i+1} is defined as $f(u_1, u_2, u_3, u_4)$ for some terms u_1, u_2, u_3, u_4 then u_1 is larger than z_1.*

Proof of Lemma 3:
Notice that for given i claim (ii) follows from (i) (since $\phi_2(x)$ is valid). Claim (iii) follows from (i) and (ii) (since $\phi_3(x)$ is valid). Claim (iv) follows from (iii).
If i is 0 then claim (i) follows from ϕ_4.
Suppose that the Lemma holds for some $i - 1$ and that t_i is defined.
Let $t_{i-1} = f(z_1, z_2, z_3, f(w_1, w_2, w_3, w_4))$. By hypothesis either $t_{i-1} + z_3 = x$ or there exists w such that $w + t_{i-1} + z_3 = x$. Since $z_3 = f(w_1, w_2, w_3, w_4) + w_3$ we get that either $t_{i-1} + f(w_1, w_2, w_3, w_4) + w_3 = x$ or $w + t_{i-1} + f(w_1, w_2, w_3, w_4) + w_3 = x$ $\quad\square$
Now, notice that by Lemma 3 (i) for every defined t_i there exists w such that $w + t_i = x$. But for given x there are only finitely many such terms v that there exists w such that $x = v + w$. On the other hand, if $i \neq j$ and t_i and t_j are defined then they are different (this is by Lemma 3 (iv)). That implies that there exists l such that $t_l = f(z_1, z_2, z_3, z_4)$ is defined but t_{l+1} is not. By Lemma 3 (ii) this implies that $z_1 = z_2$. Consider the sequence $r_1^0, r_2^0, \ldots r_1^l, r_2^l$ of the first and second arguments of $t_0, t_1, \ldots t_l$ respectively. By Lemma 3 (iii) $\phi_1(r_1^i, r_2^i, r_1^{i+1}, r_2^{i+1})$ holds for each $i < l$. Since $r_1^0 = r_2^0 = c$ and $r_1^l = r_2^l$ this sequence is a solution of Π. $\quad\square$
Theorem 2 follows now from Lemma 1, Lemma 2 and from undecidability of the Post Correspondence Problem.

4 The simplest possible signature

Now we are going to show that Theorem 2 holds also if we restrict the signature so that it contains only the binary AC symbol $+$, a unary function symbol g and a constant c. This is the simplest case in which undecidability can be conjectured.

As noticed in [T90] without g (that is if we only have the AC symbol and some number of constants in the signature) the theory is decidable for the same reasons as Presburger arithmetic is (one can proceed here as in [MY78]).

In the proof of Theorem 2 we decided to use the Post Correspondence Problem as the one to which we reduce our problem. This was mainly an esthetic choice. A Turing machine, for example, could work as well. In this case f in formula ϕ_1 should be of arity 5: instead of the two Post words we would encode the state of the finite control, the tape to the left of the head and the tape to the right of the head. Technically this choice would not change anything, just the notations would be a little bit more complicated. Another possible choice could be a machine with two counters.

The first trouble that we have in this section with the Post Correspondence Problem is that if we want to encode it like in formula ϕ_1 then we need two different monadic function symbols a and b: PCP for words over an alphabet containing only one symbol is decidable. To go around this difficulty we will encode words in $\{a, b\}^*$ as numbers:

Definition 1. *For a given word $w \in \{a, b\}^*$ let $c(w)$ (or code of w) be the natural number (in decimal notation) achieved by replacing all the symbols a of w by 1, and all the symbols b of w by 2.*

The following obvious lemma states the property of the encoding c which will be useful in our construction:

Lemma 4. *If w, l are words over $\{a, b\}$ then $c(wl) = c(l) + 10^{|l|} c(w)$.*

define $\psi_1^i(x, y, z, t)$ as the formula:
$$z = x + x + \ldots + x + c(l_i) \wedge t = y + y + \ldots y + c(r_i)$$
where x is added $10^{|l_i|}$ times and y is added $10^{|r_i|}$ times.

Now we are ready to write the formula ψ_1, which is a counterpart of the formula ϕ_1 from the previous section:

$$\psi_1(x, y, z, t) =$$

$$\psi_1^1(x, y, z, t) \vee \psi_1^2(x, y, z, t) \vee \ldots \vee \psi_1^l(x, y, z, t)$$

In order to write the formula ψ_2, the counterpart of ϕ_2 we need a trick to get rid of the arity 4 function symbol. We can use $+$ instead, thanks to the associativity it has any arity we need. The problem is that, due to commutativity we forget the order of the arguments then. Informally:

$$\psi_2(x) =$$

$\forall w, z$
    $w + z = x \wedge w$ is of the form $g(u) \implies$
    u is of the form $gggg(u_1) + ggg(u_2) + gg(u_3) + g(u_4)$
    where

none of u_1, u_2, u_3, u_4 has g in the root, and
either u_4 is of the form $gggg(v_1) + ggg(v_2) + gg(v_3) + g(v_4)$
or $u_4 = u_3 = c$ and $u_1 = u_2 \neq c$

Like ϕ_2, also ψ_2 can be written as a universal formula, but one must really be patient here:

$\forall w, z, u_1, u_2, u_3, u_4, u_5, v_1, v_2, v_3, v_4, v_5, v_6$

(1) $\neg(g(c + u_1) + z = x) \wedge$

(2) $\neg(g(g(u_1)) + z = x) \wedge$

(3) $\neg(g(g(u_1) + g(u_2)) + z = x) \wedge$

(4) $\neg(g(g(u_1) + g(u_2) + g(u_3)) + z = x) \wedge$

(5) $\neg(g(u_1 + u_2 + u_3 + u_4 + u_5) + z = x) \wedge$

(6) $\neg(g(ggggg(u_1) + u_2) + z = x) \wedge$

(7) $\neg(g(gggg(u_1) + gggg(u_2) + u_3) + z = x) \wedge$

(8) $\neg(g(ggg(u_1) + ggg(u_2) + ggg(u_3) + u_4) + z = x) \wedge$

(9) $\neg(g(gg(u_1) + gg(u_2) + gg(u_3) + gg(u_4) + z = x) \wedge$

(10) $\neg[[g(g(u_1) + g(u_2) + u_3) + z = x] \wedge$
      $[u_1 = v_1 + v_2 \vee u_1 = c] \wedge [u_2 = v_1 + v_2 \vee u_2 = c]] \wedge$

(11) $\neg[[g(gg(u_1) + gg(u_2) + u_3) + z = x] \wedge$
      $[u_1 = v_1 + v_2 \vee u_1 = c] \wedge [u_2 = v_1 + v_2 \vee u_2 = c]] \wedge$

(12) $\neg[[g(ggg(u_1) + ggg(u_2) + u_3) + z = x] \wedge$
      $[u_1 = v_1 + v_2 \vee u_1 = c] \wedge [u_2 = v_1 + v_2 \vee u_2 = c]] \wedge$

(13) $\neg[g(g(u_1) + gg(u_2) + ggg(u_3) + gggg(c)) + z = x \wedge (u_1 \neq u_2 \vee u_3 \neq c)] \wedge$

(14) $\neg(g(g(u_1) + gg(u_2) + ggg(u_3) + gggg(c + v_1) + z = x) \wedge$

(15) $\neg(g(g(u_1) + gg(u_2) + ggg(u_3) + gggg(g(v_1)) + z = x) \wedge$

(16) $\neg(g(g(u_1) + gg(u_2) + ggg(u_3) + gggg(g(v_1) + g(v_2)) + z = x) \wedge$

(17) $\neg(g(g(u_1) + gg(u_2) + ggg(u_3) + gggg(g(v_1) + g(v_2) + g(v_3))) + z = x) \wedge$

(18) $\neg(g(g(u_1) + gg(u_2) + ggg(u_3) +$
      $+ gggg(g(v_1) + g(v_2) + g(v_3) + g(v_4) + g(v_5))) + z = x) \wedge$

(19) $\neg(g(g(u_1) + gg(u_2) + ggg(u_3) + gggg(ggggg(v_1) + v_2)) + z = x) \wedge$

(20) $\neg(g(g(u_1) + gg(u_2) + ggg(u_3) + gggg(gggg(v_1) + gggg(v_2) + v_3)) + z = x) \wedge$

(21) $\neg(g(g(u_1) + gg(u_2) + ggg(u_3) +$
      $+ gggg(ggg(v_1) + ggg(v_2) + ggg(v_3) + v_4)) + z = x) \wedge$

(22) $\neg(g(g(u_1) + gg(u_2) + ggg(u_3) +$
      $+ gggg(gg(v_1) + gg(v_2) + gg(v_3) + gg(v_4))) + z = x) \wedge$

(23) $\neg[[g(g(u_1) + gg(u_2) + ggg(u_3) + gggg(g(v_1) + g(v_2) + v_3)) + z = x] \wedge$
      $[v_1 = v_5 + v_6 \vee v_1 = c] \wedge [v_2 = v_5 + v_6 \vee v_2 = c]] \wedge$

(24) $\neg[[g(g(u_1) + gg(u_2) + ggg(u_3) + gggg(gg(v_1) + gg(v_2) + v_3)) + z = x] \wedge$
      $[v_1 = v_5 + v_6 \vee v_1 = c] \wedge [v_2 = v_5 + v_6 \vee v_2 = c]] \wedge$

(25) $\neg[[g(g(u_1) + gg(u_2) + ggg(u_3) +$
      $+ gggg(ggg(v_1) + ggg(v_2) + v_3)) + z = x] \wedge$
      $[v_1 = v_5 + v_6 \vee v_1 = c] \wedge [v_2 = v_5 + v_6 \vee v_2 = c]]$

The first line of the formula says that if $g(u) + z = x$ then u does not have c as a summand. The lines from (2) to (4) say that such u is a sum of at least 4 summands. The fifth line says that u is not a sum of 5 summands or more. So here we already know that there are exactly 4 summands in such u, and all of them have g in the root. The sixth line says that no summand in u is of the form $ggggg(v)$. At this point we know that u is a sum of four summands, each of them of one of the forms $g(v), gg(v), ggg(v)$ or $gggg(v)$, where v does not start from g. Since the formula is universal we cannot say now *for each of the four forms there is a summand in u which has this form.* Instead we say, in lines (7)-(12), *There is at most one summand of each of those forms.* To be more precise, in line (7) we say *There is at most one summand of the form $gggg(v)$, where v is any term.* In line (8) we say *there are at most two summands of the form $ggg(v)$, where v is any term* (we know that one of them has also the form $gggg(v)$) and in line (9) we say *there are at most three summands of the form $gg(v)$, where v is any term.* But still we must exclude the possibility that there is more than one summand of each of the forms $g(v), gg(v)$ and $ggg(v)$. This is done in the lines (10)-(12).

At this point only u_4 from the informal definition of ψ_2 needs to be described. Line (13) says that if u_4 is c then also u_3 is c and u_1 and u_2 are equal. In the lines (14)-(19) we repeat the trick from lines (1)-(6) to ensure that u_4 is a sum of four summands, each of them of the form $g(v), gg(v), ggg(v)$ or $gggg(v)$, where v does not start from g. Then, in the lines (20)-(22) we repeat the trick from lines (7)-(9) to ensure that u_4 is of the form $g(v_1) + gg(v_2) + ggg(v_3) + gggg(v_4)$, where none of the $v_1, \ldots v_4$ begins with g. Finally, in the lines (23)-(25) we repeat, for u_4, the trick from lines (10)-(12).

Now we are ready to write ψ_3, the counterpart of ϕ_3:

$\psi_3(x) =$

$\forall\, w, v, w_1, w_2, w_3, v_1, v_2, v_3, v_4$
$\quad x = w + g(g(w_1) + gg(w_2) + ggg(w_3) +$
$\qquad\qquad +gggg(g(v_1) + gg(v_2) + ggg(v_3) + gggg(v_4))) \;\Rightarrow$
$\quad \psi_1(w_1, w_2, v_1, v_2) \wedge w_3 = v_3 + g(g(v_1) + gg(v_2) + ggg(v_3) + gggg(v_4))$

And ψ_4, the counterpart of ϕ_4:

$\psi_4(x) =$

$\exists s_1, s_2 \;\; x = s_1 + g(g(\mathbf{c}(l_1) + gg(\mathbf{c}(r_1) + ggg(s_1) + gggg(s_2))$

Notice that we could not postulate the existence "in x" of a term with the codes $\mathbf{c}(\varepsilon)$ in two first positions (as it was done in ϕ_4). This is because $\mathbf{c}(\varepsilon)$ is zero, and we only know how to count positive natural numbers. That is why we use a bit different version of Post Correspondence Problem. The following Lemma is an obvious consequence of the undecidability of the standard version of the Post Correspondence Problem:

Lemma 5. *The existence of a solution*

$$l_{i_1} l_{i_2} \ldots l_{i_m} = r_{i_1} r_{i_2} \ldots r_{i_m}$$

of an instance Π of the Post Correspondence Problem remains undecidable even if we require that $i_1 = 1$.

Finally, we write formula ψ. It is not very hard to guess that ψ is:

$$\exists x \; \psi_2(x) \wedge \psi_3(x) \wedge \psi_4(x)$$

Clearly, ψ is a $\exists^* \forall^*$ formula.

Now undecidablity of the $\exists^* \forall^*$ part of the theory over the signature with only single monadic function symbol and single constant follows from:

Lemma 6. *ψ is valid if and only if Π solvable.*

To prove the lemma one can simply repeat the proofs of Lemmas 1 and 2, with the obvious notational changes.

5 Infinite terms

In this section we assume that the quantification ranges over (possibly) infinite terms. It turns out that, with only some minor modifications, we can repeat also for this case the result and method of Section 3.

Let us start from the remark, that one can imagine two different definitions of what equality modulo an AC symbols means. First possibility is that we consider two infinite terms equal only if their equality can be proved in a finite number of AC-steps. Second possibility is that we allow infinite number of AC-steps. The proof below works for both the cases.

The main difference between the situation in this section and the one in Section 3 is that we cannot write here: *There exists x such that the initial configuration is in x, and together with a nonterminal configuration y the set x contains one of the configurations reachable from y in one step*. If we allow infinite terms then x as required by the formula exists even if the process does not terminate. Instead the formula should be: *There exists x such that the initial configuration is in x, such that together with every configuration y the set x contains all the configurations reachable from y in one step and such that no terminal configuration is in x.*

To write this formula one can for example consider the signature with the function symbol f of arity $k + 3$, where k is the number of pairs in the PCP. The formula θ_1 will be:

$$\theta_1(w, v, w_1, v_1, w_2, v_2, \ldots w_k, v_k) =$$

$$w_1 = l_1(w) \wedge v_1 = r_1(v) \wedge w_2 = l_2(w) \wedge v_2 = r_2(v) \wedge \ldots w_k = l_k(w) \wedge v_k = r_k(v)$$

We are going to give only an informal description of θ_2 . The reader who understood Section 3 and 4 can easily imagine how to write it formally as a universal formula.

$\theta_2(x) =$

$\forall w, u$
    $w + u = x \wedge w$ is of the form $f(y_l, y_r, y, y_1, y_2, \ldots y_k) \Rightarrow$
    $(y_l \neq y_r$ or $y_l = y_r = c)$ and each of $y_1, y_2, \ldots y_k$ is of the form $f(v_l, v_r, v, v_1, v_2, \ldots v_k)$

$\theta_3(x) =$

$\forall w, y_l, y_r, y, v_l^1, v_r^1, v^1, v_1^1, v_2^1, \ldots v_k^1, v_l^2, v_r^2, v^2, v_1^2, v_2^2, \ldots v_k^2 \ldots v_l^k, v_r^k, v^k, v_1^k, v_2^k, \ldots v_k^k$
    $x = w + f(y_l, y_r, y,$
$$f(v_l^1, v_r^1, v^1, v_1^1, v_2^1, \ldots v_k^1),$$
$$f(v_l^2, v_r^2, v^2, v_1^2, v_2^2, \ldots v_k^2),$$
$$\ldots$$
$$f(v_l^k, v_r^k, v^k, v_1^k, v_2^k, \ldots v_k^k)) \implies$$

$$[y = f(v_l^1, v_r^1, v^1, v_1^1, v_2^1, \ldots v_k^1) + v^1 +$$
$$+ f(v_l^2, v_r^2, v^2, v_1^2, v_2^2, \ldots v_k^2) + v^2 + \ldots$$
$$\ldots + f(v_l^k, v_r^k, v^k, v_1^k, v_2^k, \ldots v_k^k) + v^k \wedge$$
$$\theta_1(y_l, y_r, v_l^1, v_r^1, v_l^2, v_r^2, \ldots v_l^k, v_r^k)]$$

$\theta_4(x) =$

$\exists s, s_1, s_2, \ldots s_k \ \ x = s + f(c, c, s, s_1, s_2, \ldots s_k)$

and:

$\theta = \exists x \ \theta_2(x) \wedge \theta_3(x) \wedge \theta_4(x)$

Now, θ is false if and only if Π is solvable.

6   Acknowledgment

Many thanks to ToMasz Wierzbicki for the discussion which has opened my eyes.
    I also would like to thank the anonymous referees who found many typos and small bugs in the submitted version.

References

[C93] H. Comon, *Complete axiomatizations of some quotient term algebras* Theoretical Computer Science, 118(2), September 1993,

[DJK91] N. Dershowitz, J-P Jouannaud, J. W. Klop, *Problems in Rewriting*, Proceedings of 4 RTA, 1991, (Springer LNCS vol. 448), pp 445-456,

[DJK93] N. Dershowitz, J-P Jouannaud, J. W. Klop, *More Problems in Rewriting*, Proceedings of 5th RTA, 1993, (Springer LNCS vol. 690) pp. 468-487,

[DJK95] N. Dershowitz, J-P Jouannaud, J. W. Klop, *Problems in Rewriting III*, Proceedings of RTA 95, (Springer LNCS vol. 914) pp 457-471,

[F93] M. Fernandez, *AC-Complement Problems: Validity and Negation Elimination* Proceedings of 5th RTA, 1993, (Springer LNCS vol. 690) pp. 358-373,

[LM93] D. Lugiez and J.-L. Moysset. *Complement problems and tree automata in AC-like theories* Proceedings of the Symposium on Theoretical Aspects of Computer Science 1993, (Springer LNCS vol. 665) pp. 515-524,

[MY78] M. Machtey, P.Young, *An Introduction to the General Theory of Algorithms*, Elsevier 1978,

[T90] R. Treinen, *A New Method for Undecidability Proofs of First Order Theories* in Proceedings of the Tenth Conference on Foundations of Software Technology and Theoretical Computer Science, (Springer LNCS 472) volume 472 , 1990,

[T92] R. Treinen, *A new method for undecidability proofs of first order theories* Journal of Symbolic Computation 14(5) November 1992 pp 437-458,

Deciding the Satisfiability of Quantifier Free Formulae on One-Step Rewriting[*]

Anne-Cécile Caron, Franck Seynhaeve, Sophie Tison, and Marc Tommasi

LIFL, Bât M3, Université Lille 1, F59655 Villeneuve d'Ascq cedex, France
phone: (+33) 320416178, fax: (+33) 320436566
email: lastname@lifl.fr, Web: http://www.lifl.fr/~lastname.

Abstract. We consider quantifier free formulae of a first order theory without functions and with predicates x *rewrites to y in one step* with given rewrite systems. Variables are interpreted in the set of finite trees. The full theory is undecidable [Tre96] and recent results [STT97], [Mar97], [Vor97] have strengthened the undecidability result to formulae with small prefixes ($\exists^*\forall^*$) and very restricted classes of rewriting systems (*e.g.* linear, shallow and convergent in [STTT98]). Decidability of the positive existential fragment has been shown in [NPR97].
We give a decision procedure for positive and negative existential formulae in the case when the rewrite systems are quasi-shallow, that is all variables in the rewrite rules occur at depth one.
Our result extends to formulae with equalities and memberships relations of the form $x \in \mathcal{L}$ where $\mathcal{L}$ is a recognizable set of terms.

Introduction

The theory of one-step rewriting for a given rewrite system R and signature $\mathcal{F}$ is the first-order theory of the following structure: its universe consists of all $\mathcal{F}$-ground terms, and its only predicate is the relation "x rewrites to y in one step by R". The structure contains no function symbols and no equality. In [Tre96] it has been shown undecidable. This result has been refined in many recent papers [STT97,Mar97,Vor97]. Even in the case of very short fragments and systems with strong restrictions, the undecidability result holds. For instance, the $\exists^*\forall^*$-fragment for the class of *linear shallow and convergent* rewrite systems in [STTT98] is undecidable.

Decidability of the existential fragment is an open problem but the positive existential fragment has been shown to be decidable by Niehren *et al* in [NPR97]. The elegant solution presented in [NPR97] uses a deep result of Schmidt-Schauss on second order unification [SS96]: a positive existential formula of the theory of one-step rewriting is solution-equivalent to a stratified second order unification problem. In [Jac96], Jacquemard gives a translation for any formula of the positive existential fragment of one-step rewriting into a formula of a decidable

---

[*] Partially supported by The Esprit working group CCL II (22457), and "GDR AMI" Groupement De Recherche 1116 du CNRS.

extension of the weak monadic theory of two successors. In this latter result, restrictions on rewrite system are necessary: in every rule, lengths of the positions of all occurrences of the same variable must be equal.

Niehren *et al* results are also a first step towards the study of connections between context unification and one step rewriting. Context unification ([Com91,SS94,Lev96]) is unification of first-order terms with context variables that range over terms with one hole, and its decision is an open problem (RTA List of Open Problems [RTA]). Solving a rewriting constraint consists in finding a context under which the rewrite relation applies. A rewriting constraint $x \rightarrow^{rul} y$ is simply expressed by a context unification constraint saying that there exists a common context C and two variables z_l and z_r such that z_l and z_r respectively match the left hand side and the right hand side of the rule rul and $x = C[z_l]$ and $y = C[z_r]$.

Recent developments [NTT98] have shown that a weak extension of one step rewriting that allows to express relative orders for rewritings is sufficient to encode any stratified context unification problem. Surprisingly, the positive fragment of one step rewriting is a difficult problem, and our result is also interesting for context unification because it addresses problems with negation.

Our result concerns relations "x rewrites to y in one step by r" ($x \rightarrow^r y$) where r is a rule of a finite quasi-shallow rewrite system. Here, quasi-shallow means that variables in both left-hand side and right-hand side of rules occur at depth exactly one. Contrary to shallow rewrite systems, finite closed terms of unbounded height are allowed in quasi-shallow rules but collapsing rules of the form $f(x_1, \ldots, x_n) \rightarrow x$ are not allowed.

In this paper we give a decision procedure based on inference rules to solve existential formulae with equality and all boolean operators for quasi-shallow rewrite systems. Let us sketch our method.

Solved forms we want to obtain are formulae without any rewriting constraints. So, the aim of the algorithm, described by a set of inference rules and a control, is to eliminate them. Given a formula with n variables, the algorithm tries to find n terms satisfying the formula in the following way. At each step, it first guesses a root symbol for each variable, that is a substitution of the form $x = f(x_1, \ldots, x_p)$ is applied to each variable. Thus new variables $x_1, \ldots, x_p$ are added to the problem. Second, the algorithm tries to satisfy each rewrite relation $f(x_1, \ldots, x_p) \rightarrow^r g(y_1, \ldots, y_m)$ either at the root position or below. Both cases involve adding in the problem equalities or differences or rewrite relations between the new variables. In the case of a non rewrite relation $f(x_1, \ldots, x_p) \not\rightarrow^r g(y_1, \ldots, y_m)$, a dual treatment is performed: the algorithm tries to contradict the rule r instead of satisfying it.

Let us consider the following example.

$$P_0 \equiv x \rightarrow^r y \land y \not\rightarrow^r x \land x \neq y$$

with the rewrite rule $r \equiv f(\alpha, \beta) \rightarrow f(a, \alpha)$. First, guess root symbols for x and y. For instance $x = f(x_1, x_2)$ and $y = f(y_1, y_2)$. We obtain

$$P_1 \equiv f(x_1, x_2) \rightarrow^r f(y_1, y_2) \land f(y_1, y_2) \not\rightarrow^r f(x_1, x_2),$$

where x_i and y_i are fresh variables. For the positive part of P_1, either the rewriting applies at the root position, and therefore we delete a rewriting constraint and add $x_1 = y_2 \wedge y_1 = a$, or not and propagate the constraint through one of the components: e.g. we transform $f(x_1, x_2) \rightarrow^r f(y_1, y_2)$ into $x_1 \rightarrow^r y_1 \wedge x_2 = y_2$. For the negative part of P_1, we must check that the rewriting does not apply at the root position, i.e. $x_1 \neq a$ or $y_1 \neq x_2$ and furthermore we must also check that it does not apply below. Since we have $x \neq y$ in P_0, we can distinguish three cases and nondeterministically add one of the three constraints:

$$y_1 \not\rightarrow^r x_1 \wedge y_2 \neq x_2 \qquad y_2 \not\rightarrow^r x_2 \wedge y_1 \neq x_1 \qquad y_1 \neq x_1 \wedge y_2 \neq x_2$$

After that, we obtain a new problem and the number of rewrite constraints has not increased. To ensure termination, we prove that the part of the problem that is not concerned with any rewriting can be solved independently (because of the shalowness of rewrite systems). So, the number of different problems, when restricted to variables involved in rewriting constraints is bounded and this is sufficient to obtain termination.

On the contrary, when $x = y$ in P_0, we obtain a constraint $f(x_1, x_2) \not\rightarrow^r f(x_1, x_2)$, and must say that $x_1 \not\rightarrow^r x_1$ AND $x_2 \not\rightarrow^r x_2$. Therefore, the number of rewritings increases. But an important property of quasi-shallow rewrite systems is that the set of solutions of constraints of the form $x \not\rightarrow^r x$ are $\mathsf{REC}_{\neq}$-languages, i.e. languages recognizable by tree automata with constraints between brothers ([BT92]). The decision procedure we propose also deal with membership constraints of the form $x \in L$ where L some given $\mathsf{REC}_{\neq}$-language. Hence, we are able to transform any constraint $x \not\rightarrow^r x$ into a membership constraint and thus we avoid increasing the number of rewritings in the transformation steps.

A precompilation of the problem given in input allows us to build a unique $\mathsf{REC}_{\neq}$-automaton to control any membership constraint that may be generated during the execution of our algorithm. Managing membership constraints also increases the expressive power of the language considered and leads to simpler proofs and constructions (e.g. to take care of ground terms).

The expressivity of the fragment addressed here is relatively poor regarding classical properties useful in term rewriting theory. For instance, formulae can express confluence only for a bounded number of steps (due to the restriction "one-step") and for quasi-shallow rewrite systems. Nevertheless, undecidability results and connections to context unification show that solving one step rewriting formulae is difficult but interesting by its own.

Shallowness of rewrite systems is very important to get this decidability result. Even if some technical difficulties are now hidden in the use of $\mathsf{REC}_{\neq}$-automata, the result seems not surprising, and difficult to extend, even in the positive case, to non quasi-shallow rewrite systems. But we hope, with the help of "good strategies" for the application of our set of inference rules, that our methods could be adapted for non shallow rewrite systems. It seems that this difficulty arises as soon as collapsing rules are allowed.

Full paper [CSTT99] is available on www.grappa.univ-lille3.fr/~tommasi.

1 Preliminaries

Terms and term rewriting systems Let $\mathcal{F}$ be a finite ranked alphabet, let $\mathcal{X}$ be a set of variables. Let $\mathcal{F}_n$ be the set of n-ary symbols in $\mathcal{F}$. The set $T(\mathcal{F}, \mathcal{X})$ denotes the set of terms and $T(\mathcal{F})$ denotes the set of ground terms.

We denote $\overrightarrow{x_n}$ a sequence of n variables $x_1, \ldots, x_n$ and $\widetilde{\alpha_n}$ a sequence of n terms $\alpha_1, \ldots, \alpha_n$ such that each α_i is either a variable or a ground term. Let us note $\overrightarrow{x_n} \cap \overrightarrow{y_{n'}} = \emptyset$ if $\{x_1, \ldots, x_n\} \cap \{y_1, \ldots, y_{n'}\} = \emptyset$. For a sake of brevity, we drop the index n of $\overrightarrow{x_n}$ and $\widetilde{\alpha_n}$ when n is clear from the context.

A rewrite system $\mathcal{R}$ is quasi-shallow if for each rule $l \to r$, the left-hand side l and the right-hand side r are of the form $f(\widetilde{\alpha_n})$ where n is positive or null. In other words, variables occur exactly at depth one. Let us note that l or r may be a constant.

We denote by $t \to_{\mathcal{R}} t'$ if term t rewrites to t' with the system $\mathcal{R}$. In this paper, we fix a rewrite system and we drop $\mathcal{R}$ in the index of the rewrite relation. We also write $t \to^r t'$ (respectively $t \to^r_\bullet t'$) when t rewrites to t' with the rule r (respectively at the root position of t and t').

Tree Automata with Comparisons between Brother Terms Tree automata with comparisons between brothers ($\mathsf{REC}_{\neq}$-automata) have been introduced by Bogaert and Tison [BT92]. They strictly increase the expressive power of finite tree automata. Rules of $\mathsf{REC}_{\neq}$-automata can test with constraint expressions equalities or differences between brother terms.

A constraint expression is a boolean combination of equations $x_i = x_j$, (so that inequations $x_i \neq x_j$ are allowed) or sign $\top$ (null constraint), where x_i and x_j are variables. The set of constraint expressions will be denoted by CE and CE_n is the set of constraint expressions over at most n variables.

A tuple of terms $t_1, \ldots, t_n$ satisfies a constraint expression c if and only if c holds when every variable x_i is substituted by the corresponding term t_i in the sequence.

Definition 1. *An automaton $\mathcal{A}$ with comparisons between brothers, briefly $\mathsf{REC}_{\neq}$-automaton, is a tuple $(\mathcal{F}, \mathcal{S}, \mathcal{S}_f, \Delta)$, where $\mathcal{S}$ is a finite set of states, $\mathcal{S}_f \subseteq \mathcal{S}$ is a set of final states and $\Delta \subseteq \bigcup_i (\mathcal{F}_i \times CE_i \times \mathcal{S}^{i+1})$ is a set of rules (a rule $(f, c, s_1, \ldots, s_n, s)$ will be denoted $f(s_1, \ldots, s_n)[c] \to s$).*

Let $f \in \mathcal{F}_n$ and $t_1, \ldots, t_n$ be terms of $T(\mathcal{F})$. The relation $\xrightarrow{*}_{\mathcal{A}}$ is defined as follows:

$$f(t_1, \ldots . t_n) \xrightarrow{*}_{\mathcal{A}} s \text{ if and only if}$$
$$\left(\begin{array}{l} \exists f(s_1, \ldots, s_n)[c] \to s \in \Delta \text{ such that } \forall i, t_i \xrightarrow{*}_{\mathcal{A}} s_i \\ \text{and } t_1, \ldots, t_n \text{ satisfy the constraint } c \end{array} \right)$$

Let s be a state of $\mathcal{S}$. We denote by $L_{\mathcal{A}}(s)$ the set of terms t such that $t \xrightarrow{*}_{\mathcal{A}} s$. A tree $t \in T(\mathcal{F})$ is said to be accepted by $\mathcal{A}$ if there exists a final state s_f such that $t \in \mathcal{L}_{\mathcal{A}}(s_f)$. The language $\mathcal{L}(\mathcal{A})$ recognized by $\mathcal{A}$ is the set of accepted terms.

Clearly, regular tree languages are $REC_{\neq}$-recognizable. The class $REC_{\neq}$ of sets of tree languages recognized by the class of $REC_{\neq}$-automata is closed by all boolean operations (intersection, union and complement). Moreover, the emptiness problem — is a language in $REC_{\neq}$ empty ? — and the finiteness problem — is a language in $REC_{\neq}$ finite ? — are decidable. All proofs for these properties are constructive. By construction, the proof of finiteness allows us to enumerate members of a finite $REC_{\neq}$ tree language. Therefore, one can compute the cardinality of such a language. The reader is reported to [BT92] and [CDG$^+$97] for more details on tree automata with equality tests.

Example 1. The set of well-balanced trees over $\{a, f\}$ is recognized by $\mathcal{A} = (\{a, f\}, \{s, s_f\}, \{s_f\}, \Delta)$ with Δ

$$a \to s_f \ , \qquad f(s_f, s_f)[x_1 = x_2] \to s_f \ ,$$
$$f(s_f, s_f)[x_1 \neq x_2] \to s_p \ , \qquad f(s_p, s_f)[\top] \to s_p \ ,$$
$$f(s_f, s_p)[\top] \to s_p \ , \qquad f(s_p, s_p)[\top] \to s_p \ .$$

One-step rewriting logic Let $\mathcal{R}$ be quasi-shallow rewrite systems and $\mathcal{L}$ be a set $\{L_1, \ldots, L_q\}$ of $REC_{\neq}$-languages. We focus on the satisfiability problem of the existential fragment of the theory $\mathcal{T}_{\mathcal{R}, \mathcal{L}}$ whose underlying language is defined as follows. (In $\mathcal{T}_{\mathcal{R}, \mathcal{L}}$ rewrite systems and $REC_{\neq}$-languages are given as parameters.)

Atomic formulae consist in equalities, membership, or rewrite relations $\to$ between variables. Closure under boolean operators of atomic formulae defines formulae of the language.

$$at ::= x \to y \mid x = y \mid x \in L_i$$
$$\phi ::= \phi_1 \wedge \phi_2 \mid \phi_1 \vee \phi_2 \mid \neg \phi \mid at$$

We interpret variables in the set of ground terms over the ranked alphabet $\mathcal{F}$. If σ is an interpretation, then $x = y$ holds if and only if $x\sigma = y\sigma$, $x \in L_i$ holds if and only if $x\sigma$ belongs to L_i and $x \to y$ holds if and only if $x\sigma$ rewrites in $y\sigma$ in one step with the rewrite system $\mathcal{R}$. A solution σ of ϕ is an interpretation of the variables that satisfies ϕ.

Theorem 1. *Let $\mathcal{R}$ be a rewrite system and $\{L_1, \ldots, L_n\}$ be $REC_{\neq}$-languages. The existential fragment of the theory of $\mathcal{T}_{\mathcal{R}, \mathcal{L}}$ of one-step rewriting with membership constraints and equality is decidable.*

We present an algorithm and sketch the proof in Section 2. The complete proof appears in [CSTT99].

For a sake of clarity, we consider formulae using the predicate $\to^r$ — i.e. *rewrites with the rule r,* rather than $\to$. The latter presentation is more convenient for the exposition of our algorithm and is equivalent to the former — w.r.t. the decision problem we are interested in, since $x \to y$ if and only if $\bigvee_{r \in \mathcal{R}} x \to^r y$.

Finally if formulae are written in a disjunctive normal form, $\phi = \bigvee_{i \in I} \phi_i$, then ϕ is satisfiable if and only if at least one of the ϕ_i is satisfiable. So we will consider formulae without disjunction in the rest of this paper.

We use letters P, Q, P', Q' to denote formulae without disjunction, also called *problems*, in this latter syntax.

The set of the variables occurring in P is denoted by $\mathsf{Var}(P)$. For convenience, we consider sometime a problem P as a set of atomic formulas (instead of a conjunction). For instance, we denote by $\mathsf{Card}(P)$ the number of (different) atomic formulas of P; or we will say that $P \subset Q$ if there exists P' such that Q is syntactically equal to $P \wedge P'$.

Non-rewrite relations and $\mathsf{REC}_{\neq}$-automata The decision procedure must deal with expressions of the form $x \not\rightarrow^r x$ and we show in this section that we can get rid of them using tree automata. We will replace rewrite constraints $x \not\rightarrow^r x$ by membership constraints. Membership constraints will also be useful for the treatment of ground parts of rewrite rules.

This transformation of rewrite constraints $x \not\rightarrow^r x$ into membership constraints leads to the definition of new $\mathsf{REC}_{\neq}$-languages from $\mathcal{T}_{\mathcal{R},\mathcal{L}}$'s definition viewpoint.

Let us recall we are given a set of rewrite rules $\mathcal{R}$ and a set of $\mathsf{REC}_{\neq}$-languages $L_1, \ldots, L_q$. Rewrite rules may be built with ground terms like $g(a)$ for instance in the rule $f(g(a), x) \rightarrow f(x, x)$. We denote by $L_{q+1}, \ldots, L_{q'}$ singleton sets containing a ground term occurring in the set of rules $\mathcal{R}$.

We also consider languages L_r that are defined by $\{t \in T(\mathcal{F}) \mid t \not\rightarrow^r t\}$ and we will prove that languages L_r are $\mathsf{REC}_{\neq}$-recognizable. Languages L_r are also numbered in the following way $L_{q'+1}, \ldots, L_n$ so that every $\mathsf{REC}_{\neq}$-language considered here belongs to the set $\{L_1, \ldots, L_n\}$.

For the rest of this paper, languages $\{L_1, \ldots, L_n\}$ are now fixed.

In order to prove that languages $\{t \in T(\mathcal{F}) \mid t \not\rightarrow^r t\}$ are $\mathsf{REC}_{\neq}$-recognizable, we first show that tree languages defined by $\{t \in T(\mathcal{F}) \mid t \rightarrow^r t\}$ are $\mathsf{REC}_{\neq}$-recognizable. Then we use closure properties of automata to get the result.

Proposition 1. *Given a quasi-shallow rewrite system $\mathcal{R}$, languages $\{t \in T(\mathcal{F}) \mid t \not\rightarrow^r t\}$ where $r \in \mathcal{R}$ are $\mathsf{REC}_{\neq}$-recognizable.*

Proof. Basically, the set $\{t \in T(\mathcal{F}) \mid t \rightarrow^r t\}$ is also defined by the set of terms that encompass the most general unifier of r's left-hand side and right-hand side. Because rules of the rewrite system are quasi-shallow, the most general unifier is also quasi-shallow. Hence, such a construction can be done with (complete and deterministic) automata with equality tests between brothers. For instance, let us consider a rule $f(x, y, y, z) \rightarrow f(g(a), y, z, y)$. The set $\{t \in T(\mathcal{F}) \mid t \rightarrow^r t\}$ is the set of terms that emcompass $f(g(a), x, x, x)$. Hence, one can use closure under complementation and determinization to build a deterministic $\mathsf{REC}_{\neq}$-automaton for each language $L_r = \{t \in T(\mathcal{F}) \mid t \not\rightarrow^r t\}$ associated with a rewrite rule r. $\square$

Product automaton In a natural way, conjunctions of non-rewrite relations $x \not\to^r x \wedge x \not\to^{r'} x$ must be transformed into membership constraints $x \in L_r \cap L_{r'}$. It will be convenient to consider a unique automaton to handle all possible conjunctions. A classical construction of a product automaton is suited to do that.

We just introduce the construction in these preliminaries and point out the properties of product automata. Let us consider complete and determinitic $REC_{\neq}$-automata $\mathcal{A}_1, \ldots, \mathcal{A}_n$ which respectively recognize languages $L_1, \ldots, L_n$. Each automaton is given by a tuple $(\mathcal{F}, \mathcal{S}_i, \mathcal{S}_{f_i}, \Delta_i)$ and we suppose that the set of states are pairwise disjoint. The product automaton $\mathcal{A}$ is defined by $(\mathcal{F}, \mathcal{S}, \mathcal{S}, \Delta)$ where $\mathcal{S} = \mathcal{S}_1 \times \mathcal{S}_2 \times \cdots \times \mathcal{S}_n$ and Δ is the set of rules of the form:

$$f(s^1, \ldots, s^m)[C] \to s$$

where C is a n-tuple of constraint expressions $(c_1, \ldots, c_n)$, every s^j is a n-tuple $(s_1^j, \ldots, s_n^j)$, $s = (s_1, \ldots, s_n)$ and $f(s_i^1, \ldots, s_i^m)[c_i] \to s_i \in \Delta_i$.

Note that the product automaton is complete and deterministic since each automaton $\mathcal{A}_i$ is.

The set of final states is useless for our study and we only are interested in the sets of terms $L_{\mathcal{A}}(s) = \{t \in T(\mathcal{F}) \mid t \xrightarrow{*}_{\mathcal{A}} s\}$. Clearly if $s = (s_1, \ldots, s_n)$ and s_i is final for the automaton $\mathcal{A}_i$, then any term t such that $t \xrightarrow{*}_{\mathcal{A}} s$ belongs to L_i, and thus $L_{\mathcal{A}}(s) \subseteq L_i$. Because the product automaton built here is complete and deterministic, the opposite property also holds. If $s = (s_1, \ldots, s_n)$ and s_i is not final for the automaton $\mathcal{A}_i$, then any term t such that $t \xrightarrow{*}_{\mathcal{A}} s$ does not belong to L_i, and thus $L_{\mathcal{A}}(s) \cap L_i = \emptyset$.

2 Decision algorithm

We first build the product automaton $\mathcal{A}$ according to Section 1 corresponding to fixed rewrite system $\mathcal{R}$ and languages $\{L_1, \ldots, L_n\}$.

Example 2. Let us consider a signature with exactly one binary symbol f and one constant a. Let $\mathcal{R}$ be the rewrite system with only one rule $r \equiv f(\alpha, \beta) \to f(a, \alpha)$. Let $\mathcal{L}$ be the set $\{L_1, L_2\}$ where $L_1 = \{t \mid t \not\to^r t\} = \{a\}$ and L_2 is the set of well-balanced trees over f and a. The product automaton will be, in a more compact way than the one that should be issued by the construction of Section 1, $\mathcal{A} = (\{a, f\}, \{s_a, s_f, s_p\}, \{s_a, s_f, s_p\}, \Delta)$ with Δ

$$a \to s_a \; ; \qquad f(s_a, s_a)[\top] \to s_f \; ;$$
$$f(s_f, s_f)[x_1 = x_2] \to s_f \; ; \qquad f(s_f, s_f)[x_1 \neq x_2] \to s_p \; ;$$
$$f(s, s')[\top] \to s_p \qquad \text{for every } (s, s') \neq (s_f, s_f).$$

We have $L_{\mathcal{A}}(s_a) = L_1$ and $L_{\mathcal{A}}(s_a) \cup L_{\mathcal{A}}(s_f) = L_2$.

Definition 2 (Input problems). *An input problem P is such that neither $x \not\to^r x$, nor $x \notin L$, nor $x \in L_i$ occurs in P and such that $x \neq y$ occurs for every couple of distinct variables x and y and for every variable x, there exists some state s such that $x \in L_{\mathcal{A}}(s)$ occurs in P.*

To obtain a problem Q' that fulfills such properties from any problem Q of the existential fragment of $\mathcal{T}_{\mathcal{R},\mathcal{L}}$, we proceed with nondeterministic transformations:

1. For any constraint $x \not\to^r x$ in P, choose a state s such that $L_A(s) \subseteq L_r$ and replace $x \not\to^r x$ by a membership constraint $x \in L_A(s)$,
2. For any couple of distinct variables x, y such that $x \neq y$ is not in P, either add $x \neq y$ or substitute x by y.
3. For any membership constraint $x \notin L_i$, choose s such that $L_A(s) \cap L_i = \emptyset$ and replace $x \notin L_i$ by $x \in L_A(s)$.
4. For any membership constraint $x \in L_i$, choose s such that $L_A(s) \subseteq L_i$ and replace $x \in L_i$ by $x \in L_A(s)$.
5. For any variable x, choose a state s and add $x \in L_A(s)$ to P.

Clearly, there is a finite number of choices for every transformation, because the set of states is finite (So, Q is satisfiable if and only if at least one of the outcome of the tranformation is.)

When L_r is not empty, there are states $s_{r_1}, \ldots, s_{r_k}$ in the product automaton A such that $\bigcup_{i \in 1 \leq k} L_A(s_{r_i}) = L_r$. In the first step we select one of them. (If L_r is empty, then there is no such a state and the problem is unsatisfiable). In the second step we guess for every couple of variables whether they are equal or not. Three last steps are similar to the first one.

Example 3 (continued).
Let P be $x \to^r y \wedge y \not\to^r z \wedge x' \in L_2$. A possible input problem obtained from P, if we let $y = z$, is P_0: $x \to^r y \wedge y \in L_A(s_a) \wedge x \in L_A(s_f) \wedge x \neq y \wedge x' \in L_A(s_f) \wedge x' \neq x \wedge x' \neq y$.

The decision algorithm takes in input two parameters: i) an input problem P; ii) a list of problems initially empty that will memorize problems. It nondeterministically transforms problems according to a set of inference rules. Transformation essentially modifies the rewriting part of problems, that is atomic formulae whose variables occurs in a rewriting constraint, and other part remain unchanged. When no clash is discovered, termination is based either on the emptiness of the rewriting part or on an equivalence relation $\cong$ on problems that avoid infinite computations.

We use several notations we define now.

Definition 3. *Let P be a problem.*

- *The* rewriting part *of P is $P^R = P^{RR} \wedge P^{R\in} \wedge P^{R\neq}$ where*
  - *P^{RR} is the conjunction of atomic formulas $x \to^r y$, $x \not\to^r y$, $x \to^r x$ occurring in P.*
  - *$P^{R\in}$ is the conjunction of atomic formulas $x \in L_A(s)$ occurring in P such that $x \in Var(P^{RR})$.*
  - *$P^{R\neq}$ is the conjunction of atomic formulas $x \neq y$ occurring in P such that $x, y \in Var(P^{RR})$.*
- *The* difference part *of P, denoted by $P^{\neq}$, is the conjunction all of atomic formulas $x \neq y$ occurring in P.*

- *The* membership part of P, denoted by $P^{\in}$, *is the conjunction of atomic formulas* $x \in L_A(s)$ *occurring in* P.

A clash can be discovered because there are too many differences, with respect to membership constraints. In our example, two distinct variables cannot belong to the set $L_A(s)$.

Definition 4. *Let P be an existential problem of $\mathcal{T}_{\mathcal{R},\mathcal{L}}$.*

$$D_P(s) = \mathsf{Card}(\{x \in \mathsf{Var}(P) \mid (x \in L_A(s)) \text{ is in } P\})$$

Intuitively, $D_P(s)$ is the minimal number of different ground terms which must be in $L_A(s)$ if we want to satisfy the constraints of the form $x \in L_A(s)$.

Example 4 (continued). Given input problem P_0, we have $P_0^R \equiv x \to^r y \wedge x \in L_A(s_f) \wedge y \in L_A(s_a) \wedge x \neq y$, $P^{\neq} \equiv x' \neq x \wedge x' \neq y \wedge x \neq y$, $P^{\in} \equiv x \in L_A(s_f) \wedge y \in L_A(s_a) \wedge x' \in L_A(s_f)$ and for instance $D_P(s_a)$ equals to 1.

Definition 5. *Let Q and P be two problems. We denote $Q \cong P$ if and only if there exists a variable renaming ρ, i.e. an injective mapping from $\mathsf{Var}(Q)$ into $\mathsf{Var}(P)$, such that $\rho(Q^R) = P^R$.*

Algorithm

1: **DecShallow**(P, PP) : Boolean
    {P is a problem, PP a list of problems}
2: **if** P contains $\perp$ **then**
3:     Return FAIL
4: **else if** there exists a state $s \in S$ such that $D_P(s) > \mathsf{Card}(L_A(s))$ **then**
5:     Return FAIL
6: **else if** $Q \cong P$ with $Q \in PP$ **then**
7:     Return FAIL
8: **else if** P^R is empty **then**
9:     Return TRUE
10: **else**
11:     Push P^R on PP.
        {Guess a head symbol for each variable in P^R.}
12:     Let $f_1, \ldots, f_n$ in $\mathcal{F}^n$
13:     Let $P^{(0)}$ be $P^R[x_1/f_1(\overrightarrow{x^1}), \ldots, x_n/f_n(\overrightarrow{x^n})]$
        with $\{x_1, \ldots, x_n\} = \mathsf{Var}(P^R)$ and $\overrightarrow{x^i}$ is a sequence of new variables.
        {Guess differences and equalities between variables}
14:     Guess a partition on variables $\mathsf{Var}(P^{(0)}) = \biguplus_i \pi_i$ and let $P^{(1)}$ be the problem obtained by
15:         Fix a variable x_{π_i} in each class π_i
16:         For each i substitute every x in π_i by x_{π_i}
17:         For each i, j $i \neq j$, add $x_{\pi_i} \neq x_{\pi_j}$
        {Propagate or delete differences.}

18:    Let $P^{(2)}$ be the closure of $P^{(1)}$ under $\mathcal{S}_{diff}$.
    {Propagate or delete membership constraints.}

19:    Let $P^{(3)}$ be the closure of $P^{(2)}$ under $\mathcal{S}_{member}$.
    {Propagate or delete rewritings}

20:    Let $P^{(4)}$ be the closure of $P^{(3)}$ under $\mathcal{S}_{gen}$, $\mathcal{S}_{apply}$.
    {Simplify the problem with equalities and differences.}

21:    Let $P^{(5)}$ be the closure of $P^{(4)}$ under $\mathcal{S}_{equal}$.
    {Recursive call with the new problem.}

22:    Return DecShallow(P', PP)

23: **end if**

The inference rules ($\mathcal{S}_{equal}$, $\mathcal{S}_{diff}$, $\mathcal{S}_{gen}$, $\mathcal{S}_{apply}$, $\mathcal{S}_{member}$) are given for quasi-shallow rewrite systems. They can easily be adapted for any rewrite system. On the contrary, termination is solely proved for quasi-shallow rewrite systems.

In order to explain this algorithm, we distinguish termination tests and the rest of the procedure called the inference part.

Inference Let us focus on Lines 12-22 to study problem transformations.

In lines 12-13, a root symbol is guessed for each variable in the rewriting part, and new variables are introduced. Note that we are only interested in transformations of the rewriting part of the problem and we get rid of membership and differences parts.

For any pair x, y of those new variables, we guess in Line 14 whether $x = y$ or $x \neq y$. After, we proceed in such a way predicates $\neq$, $\not\to^r$, $\in$ or $\to^r$ only involve variables and not terms of the form $f(x_1, \ldots, x_n)$.

- Considering atomic formulae $f(\overrightarrow{x}) \neq g(\overrightarrow{y})$, differences must be propagated (if $f = g$) or deleted (if $f \neq g$) to the corresponding new variables in $\overrightarrow{x}$ and $\overrightarrow{y}$. This is done in Line 18 with the system $\mathcal{S}_{diff}$.
- In Line 19, a similar treatment is done for membership constraints. It just consists in following the automaton rules.
- In Line 20, rewrite (resp. non-rewrite) relations are deleted or propagated. This is done with the two systems $\mathcal{S}_{gen}$ and $\mathcal{S}_{apply}$. Recall that $t \to^r_\bullet t'$ if t rewrites to t' by rule r at the root position of t and t'.

Of course, this step introduces new equalities and new differences between variables because some rewrite rules may be non linear. Hence a simplification of the problem inferred in Line 21 must be done. Since a guess has been done in Line 14 concerning equalities or differences between new variables, the simplification just consists in two rules, a clash or a substitution, see the system $\mathcal{S}_{equal}$.

Termination tests Termination is controlled in Lines 2-9. We give an informal presentation through four cases:

1. Clash (Line 2): the problem contains the symbol $\bot$. This is inferred from simple contradictions, *e.g.* $x = y$ and $x \neq y$. Clashes can also be discovered because a non-rewrite relation cannot be satisfied (see for example rule (19)).

When a clash is inferred, the meaning is that choices that have been done do not lead to a solution.

2. Too many differences (Line 4): recall that the quantity $D_P(s)$ denotes the number of distinct variables whose interpretation must belong to $L_A(s)$.

3. Loop (Line 6): the current problem is "solution-equivalent" to a problem inferred before—*i.e.* in PP. This is the main test necessary to obtain termination. Two problems P and Q are solution-equivalent if they contain the same (non-)rewritings and the same membership constraints up to some variable renaming in their rewriting part.

4. Solution (Line 8): There is no more rewriting nor non-rewriting constraints in the problem. The problem obtained only contains difference constraints and membership constraints. The test Line 4 ensures that the conjunction of memberships and differences is satisfiable.

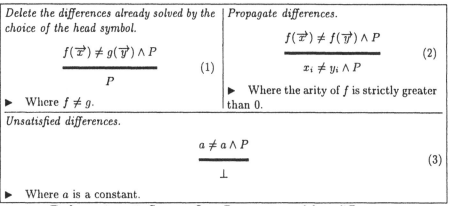

Rule system 1: System $\mathcal{S}_{diff}$. Propagate or delete differences.

3   Concluding Remarks

Extensions Restriction on rewrite systems implies that no unification constraint of the form $x = t$ where $t \in T(\mathcal{F}, \mathcal{X})$ occurs in problems inferred along a computation. This property shows that the satisfiability is only depending on membership and rewriting constraints. When we relax restrictions on rewrite systems, this property is lost, and thus we cannot prove termination. (By the way, with some slight modification of the inference system, one can design a semi-algorithm).

For instance, let us consider $P = \{x = h(x_1, x_2) \; ; \; x \to^r x_2 \; ; \; x_1 \to^{r'} y\}$ with $r = f(\alpha_1, \alpha_2) \to \alpha_2$ and $r' = f(\alpha_1, \alpha_2) \to g(\alpha_1, \alpha_2)$. Due to equalities introduced between terms at different height, we can generate a sequence of problems whose size is increasing (w.r.t. our ordering). We do not succeed in

Propagate or delete a membership constraint. $$\frac{f(\overrightarrow{x}) \in L_{\mathcal{A}}(s) \wedge P}{\bigwedge_i x_i \in L_{\mathcal{A}}(s_i) \wedge P} \quad (4)$$ ▶ Where $f(s_1, \ldots s_n) \to s \in \Delta$.	First clash. No rule in the automaton. $$\frac{f(\overrightarrow{x}) \in L_{\mathcal{A}}(s) \wedge P}{\perp} \quad (5)$$ ▶ If there is no rule $f(s_1, \ldots s_n) \to s \in \Delta$.
Second clash. The automaton is determinisitic. $$\frac{f(\overrightarrow{x}) \in L_{\mathcal{A}}(s) \wedge f(\overrightarrow{x}) \in L_{\mathcal{A}}(s') \wedge P}{\perp} \quad (6)$$ ▶ If $s \neq s'$.	

Rule system 2: System $\mathcal{S}_{member}$. Propagate or delete membership constraints.

finding a criterion (or a strategy) to limit such expansion while preserving both termination and completeness.

It seems that difficulty arises as soon as one collapsing rule of the form $f(\alpha_1, \alpha_2) \to \alpha_1$ is allowed and even in the case of positive constraints. For a better understanding of the problem, a first step could be the study of the case when all variables occur at the same height. Another way a research could be the study of strategies in the application of our inference rules in order to obtain termination while preserving completeness.

Connections with Context Unification Recent works on Context Unification and One Step Rewriting ([NPR97,NTT98]) show that both problems are closely related. But the transition from context unification to rewriting constraints uses collapsing rules which are not quasi-shallow. This connection suggests to study techniques applied for stratified unification in [SS94,SS97] in order to reformulate them in our framework.

Acknowledgments

The authors thank Joachim Niehren, Martin Müller and Ralf Treinen for interesting remarks and comments on preliminary versions of this paper.

A rule may be applied at the root position...	...or below the root position (propagation). Let us choose subterms where the rewriting applies and let say that other subterms are pairwise equal.

$$\frac{f(\overrightarrow{x}) \to^r g(\overrightarrow{y}) \wedge P}{f(\overrightarrow{x}) \to^r_\bullet g(\overrightarrow{y}) \wedge P} \quad (7)$$

$$\frac{f(\overrightarrow{x}) \to^r f(\overrightarrow{y}) \wedge P}{x_i \to^r y_i \bigwedge_{j \neq i} x_j = y_j \wedge P} \quad (8)$$

Now for non rewrite relations.

$$\frac{f(\overrightarrow{x}) \not\to^r g(\overrightarrow{y}) \wedge P}{f(\overrightarrow{x}) \not\to^r_\bullet g(\overrightarrow{y}) \wedge P} \quad (9)$$

▶ Where $f \neq g$.

There are three cases for non-rewrite relations below the root position. In all cases, the non rewriting relation has to be satisfied at the root position. First case: propagation where subterms differ.

$$\frac{f(\overrightarrow{x}) \not\to^r f(\overrightarrow{y}) \wedge P}{f(\overrightarrow{x}) \not\to^r_\bullet f(\overrightarrow{y}) \wedge x_i \not\to^r y_i \wedge x_i \neq y_i \bigwedge_{j \neq i} x_j = y_j \wedge P} \quad (10)$$

Second case: elimination. Two subterms differ.

$$\frac{f(\overrightarrow{x}) \not\to^r f(\overrightarrow{y}) \wedge P}{f(\overrightarrow{x}) \not\to^r_\bullet f(\overrightarrow{y}) \wedge x_l \neq y_l \wedge x_m \neq y_m \wedge P} \quad (11)$$

▶ Where $l \neq m$.

Third case: membership. When all subterms are pairwise equal, the non-rewrite relation has to be propagated to every subterm position. But we use membership constraints to say that a term cannot be rewritten in itself.

$$\frac{f(\overrightarrow{x}) \not\to^r f(\overrightarrow{y}) \wedge P}{f(\overrightarrow{x}) \not\to^r_\bullet f(\overrightarrow{y}) \bigwedge_i (x_i = y_i \wedge x_i \in L_{\mathcal{A}}(s)) \wedge P} \quad (12)$$

▶ Where s is a state such that $L_{\mathcal{A}}(s) \subseteq L_r$.

Rule system 3: The $\mathcal{S}_{gen}$ system expresses that a (non-)rewriting relation has to be satisfied either at the root position or below. It is correct for every term rewriting system.

Apply a rewrite relation. The first part concern equalities and the second part concerns ground terms.

$$\frac{f(\overrightarrow{x^1}) \to_\bullet^r g(\overrightarrow{x^2}) \wedge P}{\bigwedge_{(j,k,l,m)\in I} (x_l^j = x_m^k) \wedge \bigwedge_{(o,i)\in J} (x_o^i \in L_\mathcal{A}(s_{\alpha_o^i})) \wedge P} \tag{13}$$

▶ Where r is of the form $f(\widetilde{\alpha^1}) \to g(\widetilde{\alpha^2})$ and $I = \{(j,k,l,m) \mid \alpha_l^j = \alpha_m^k, \{j,k\} \subseteq \{1,2\}, (j,l) \neq (k,m)\}$ and $J = \{(o,i) \mid \alpha_o^i \in T(\mathcal{F})\}$ and $s_{\alpha_o^i}$ is such that $L_\mathcal{A}(s_{\alpha_o^i}) = \{\alpha_o^i\}$.

A rewrite relation clashes. $$\frac{f(\overrightarrow{x^1}) \to_\bullet^r g(\overrightarrow{x^2}) \wedge P}{\bot} \tag{14}$$ ▶ Where r is of the form $f'(\widetilde{\alpha^1}) \to g'(\widetilde{\alpha^2})$ and $f' \neq f$ or $g' \neq g$.	*For non-rewrite relations. First case: the rule has not the good shape.* $$\frac{f(\overrightarrow{x}) \not\to_\bullet^r g(\overrightarrow{y}) \wedge P}{P} \tag{15}$$ ▶ Where $r \equiv \alpha \to \beta$ and $\mathcal{H}ead(\alpha) \neq f$ or $\mathcal{H}ead(\beta) \neq g$.
Second case: An equality is not satisfied. An equality constraint concerns subterms in the left hand side, or the right hand side or subterms in both sides. $$\frac{f(\overrightarrow{x^1}) \not\to_\bullet^r g(\overrightarrow{x^2}) \wedge P}{x_l^j \neq x_m^k \wedge P} \tag{16}$$ ▶ Where r is of the form $f(\widetilde{\alpha^1}) \to g(\widetilde{\alpha^2})$ where $\alpha_l^j = \alpha_m^k$ and $\{j,k\} \subseteq \{1,2\}$ and $(j,l) \neq (k,m)$.	*Third case: The rule is of the good shape and there is no equality constraint in the rule.* $$\frac{f(\overrightarrow{x}) \not\to_\bullet^r g(\overrightarrow{y}) \wedge P}{\bot} \tag{17}$$ ▶ Where $r \equiv \alpha \to \beta$ and $\mathsf{Var}(\alpha) \cap \mathsf{Var}(\beta) = \emptyset$ and $\mathcal{H}ead(\alpha) = f$ and $\mathcal{H}ead(\beta) = g$.

Fourth case: Ground parts in the rule assure the non rewriting relation.

$$\frac{f(\overrightarrow{x}) \not\to_\bullet^r g(\overrightarrow{y}) \wedge P}{x_o^i \in L_\mathcal{A}(s) \wedge P} \tag{18}$$

▶ Where r is of the form $f(\widetilde{\alpha^1}) \to g(\widetilde{\alpha^2})$ and s is such that $L_\mathcal{A}(s) \cap \{\alpha_o^i\} = \emptyset$ and $(i,o) \in J = \{(o,i) \mid \alpha_o^i \in T(\mathcal{F})\}$.

Rule system 4: The $\mathcal{S}_{apply}$ system consists in rules that apply (non-)rewritings. It is correct only for quasi-shallow rewrite systems.

Clash. $$\frac{x \neq x \wedge P}{\bot} \tag{19}$$	*Substitution.* $$\frac{x = y \wedge P}{P[x/y]} \tag{20}$$

Rule system 5: System $\mathcal{S}_{equal}$. Treatment of equalities and differences.

References

[BT92]   B. Bogaert and S. Tison. Equality and disequality constraints on direct subterms in tree automata. In Patrice Enjalbert, Alain Finkel, and Klaus W. Wagner, editors, *9th Annual Symposium on Theoretical Aspects of Computer Science*, volume 577 of *Lecture Notes in Computer Science*, pages 161–171, 1992.

[CDG⁺97]   H. Comon, M. Dauchet, R. Gilleron, , F. Jacquemard, D. Lugiez, S. Tison, and M. Tommasi. Tree automata techniques and applications. Available on: http://www.grappa.univ-lille3.fr/tata, 1997.

[Com91]   Hubert Comon. Completion of rewrite systems with membership constraints. Rapport de Recherche 699, L.R.I., Université de Paris-Sud, September 1991.

[Com97]   H. Comon, editor. *Proceedings. Eighth International Conference on Rewriting Techniques and Applications*, volume 1232 of *Lecture Notes in Computer Science*, Sitges, Spain, 1997.

[CSTT99]   A.-C. Caron, F. Seynhaeve, S. Tison, and M. Tommasi. Deciding the satisfiability of quantifier free formulae on one-step rewriting. Technical Report TR-99-08, Lab. D'info. Fond. de Lille, 1999. www.grappa.univ-lille3.fr/~tommasi.

[Gan96]   H. Ganzinger, editor. *Proceedings. Seventh International Conference on Rewriting Techniques and Applications*, volume 1103 of *Lecture Notes in Computer Science*, 1996.

[Jac96]   F. Jacquemard. *Automates d'arbres et réécriture de termes*. PhD thesis, Université de Paris XI, 1996.

[Lev96]   Jordi Levy. Linear second-order unification. In Ganzinger [Gan96], pages 332–346.

[Mar97]   Jerzy Marcinkowski. Undecidability of the first order theory of one-step right ground rewriting. In Comon [Com97], pages 241–253.

[NPR97]   J. Niehren, M. Pinkal, and P. Ruhrberg. On equality up-to constraints over finite trees, context unification and one-step rewriting. In *Procedings of the International Conference on Automated Deduction*, volume 1249 of *Lecture Notes in Computer Science*, pages 34–48, Townsville, Australia, 14-17July 1997. Springer-Verlag.

[NTT98]   J. Niehren, S. Tison, and R. Treinen. Context unification and rewriting constraints. Presented at CCL2 Workshop, 1998.

[RTA]   The RTA List of Open Problems. http://www.lri.fr/~rtaloop/. Maintained by Nachum Dershowitz and Ralf Treinen.

[SS94]   Manfred Schmidt-Schauß. Unification of stratified second-order terms. Internal Report 12/94, Johann-Wolfgang-Goethe-Universität, Frankfurt, Germany, 1994.

[SS96]   Manfred Schmidt-Schauß. An algorithm for distributive unification. In Ganzinger [Gan96], pages 287–301.

[SS97]   Manfred Schmidt-Schauß. A unification algorithm for distributivity and a multiplicative unit. *Journal of Symbolic Computation*, 22(3):315–344, 1997.

[STT97]   F. Seynhaeve, M. Tommasi, and R. Treinen. Grid structures and undecidable constraint theories. In *Proceedings of the 7ᵗʰ International Joint Conference on Theory and Practice of Software Development*, volume 1214 of *Lecture Notes in Computer Science*, pages 357–368, 1997.

[STTT98]   F. Seynhaeve, S. Tison, M. Tommasi, and R. Treinen. Grid structures and undecidable constraint theories. To appear in Theoretical Computer Science, 1998.

[Tre96]   Ralf Treinen. The first-order theory of one-step rewriting is undecidable. In Ganzinger [Gan96], pages 276–286.

[Vor97]   Sergey Vorobyov. The first-order theory of one-step rewriting in linear noetherian systems is undecidable. In Comon [Com97], pages 254–268.

A New Result about the Decidability of the Existential One-Step Rewriting Theory[*]

Sébastien Limet and Pierre Réty

LIFO, Université d'Orléans (France), {limet|rety}@lifo.univ-orleans.fr

Abstract. We give a decision procedure for the whole existential fragment of one-step rewriting first-order theory, in the case where rewrite systems are linear, non left-left-overlapping (i.e. without critical pairs), and non ϵ-left-right-overlapping (i.e. no left-hand-side overlaps on top with the right-hand-side of the same rewrite rule[2]). The procedure is defined by means of tree-tuple synchronized grammars.

1 Introduction

Given a signature Σ, the theory of one-step rewriting for a finite rewrite system is the first order theory over the universe of ground Σ-terms that uses the only predicate symbol $\rightarrow$, where $x \rightarrow y$ means x rewrites into y by one step.

It has been shown undecidable in [11]. Sharper undecidability results have been obtained for some subclasses of rewrite systems, about the $\exists^*\forall^*$-fragment [10, 8] and the $\exists^*\forall^*\exists^*$-fragment [12].

It has been shown decidable for the positive existential fragment [9], in the case of unary signatures [3], in the case of linear rewrite systems whose left and right members do not share any variables [2][3], and for the whole existential fragment in the case of shallow rewrite systems without collapsing rules [1].

In comparison with [1], our result also excludes collapsing rules because of the stronger restriction: no ϵ-left-right-overlaps. On the other hand, it includes non shallow rewrite rules. In comparison with [2], linearity of rewrite rules is also assumed, but it includes the rewrite rules whose members share variables. To get this possibility using a technique based on tree automata or tree grammars, we must ensure that the two occurrences of each shared variable are replaced by identical terms. This cannot be performed by ground tree transducers. Automata with equality and disequality constraints can, but they cannot perform the closure of (one-step) rewriting by context application.

On the other hand tree-tuple synchronized grammars (*TTSG* for short) can correctly perform both variable instantiations and context applications, thanks to synchronizations. TTSG's have been introduced by the authors to solve some equational unification [4] and disunification [6] problems.

---

[*] A full version of this paper is available in [5].

[2] Formally, for each rewrite rule $l- > r$, the two terms l and r' are not unifiable, where r' is a renaming of r such that $Var(l) \cap Var(r') = \emptyset$.

[3] Even the theory of several-step rewriting is decidable.

Given an existential formula in the prenex form, our decision procedure consists in the following steps:

First: because of the restrictions, every predicate of the form $x \to x$ has no solutions (see Lemma 3 in Section 3.3): they are replaced by the predicate without solutions $\perp$. Next, the formula is transformed into a disjunction of conjunctions of items of the form $x \to y$ or $\neg(x \to y)$[4]. The solutions of $x \to y$ and of $\neg(x \to y)$ are tree-pair languages, which can both be generated by TTSG's (see Section 3).

Second: the solutions of a conjunctive factor are obtained by making natural joins (called intersection in this paper). Let $x \to y \wedge C$ (or $\neg(x \to y) \wedge C$) be a conjunctive factor and assume that the solutions of C have been computed already, as a tree-tuple language. The size of tuples is the number of distinct variables appearing in C. Three kinds of intersections are needed: over n components, $n \in \{0, 1, 2\}$, according to the number of variables shared by $x \to y$ and C. Of course, if $n = 0$ this is the cartesian product.

Third: the solutions of the entire formula (without quantifiers) are obtained by making unions of languages of tree-tuples of different sizes. By making cartesian products with the language of all ground terms, we can lengthen tuples so that they all have the same length[5].

Fourth: the validity of the existentially quantified formula is tested thanks to an emptiness test performed on the solutions of the formula without quantifiers.

2 Overview of TTSG's

This section briefly recalls notions and existing results about TTSG's. For a more formal presentation and proofs see [4]. We use the classic notions of terms and term rewriting systems. Let us just precise a few notations. For any occurrences u, v, $u < v$ means that u is a strict prefix of v, i.e. $v = u.w$ for some $w \neq \epsilon$. For any term t, $O(t)$ is the set of occurrences of t and $t(u)$ is the symbol of t that appears at occurrence u. Moreover we extend the notion of occurrences to tuples in a natural way i.e. $(t_1, \ldots, t_n)|_{i.u} = t_i|_u$ if $i \in [1, n]$. The arity of any symbol f is denoted by $ar(f)$. $\Sigma_{>i}$ contains the symbols of Σ whose arity is greater than i. *lhs* means left-hand-side.

A *TTSG* is a tree-tuple grammar whose terminals are the symbols of Σ, and that contains:

- *free productions*, i.e. productions like those of a regular tree language: $X \Rightarrow f(X_1, \ldots, X_n)$.
- *synchronized productions*: $\{X_1 \Rightarrow Y_1, \ldots, X_n \Rightarrow Y_n\}$, which means that X_1 is replaced by $Y_1, \ldots, X_n$ by Y_n at the same time (and only at the same

---

[4] The symbols of Σ are not allowed in formulas.

[5] Lengthening tuples can be necessary even if the lengths are equal. Consider for example two tree-pair languages. If the one provides the valuations of the variables x, y while the other provides those of x, z, we have to extend both into triple languages providing the valuations of x, y, z, before making their union.

time). They are always empty, i.e. they do not generate any terminals. A set of productions synchronized together, like $\{X_1 \Rightarrow Y_1, \ldots, X_n \Rightarrow Y_n\}$, is called *a pack of synchronized productions.*

A TTSG is denoted by (Σ, NT, P, I) where NT is the set of non-terminals, P is the set of productions, and I is the axiom. Languages generated by TTSG's are called *synchronized languages*. Note that regular tree-tuple languages are a particular case of synchronized languages.

Example 1. Let $\Sigma = \{f, g, a, b\}$ where f, g are monadic symbols and a, b are constants. Consider the TTSG whose productions are:

$$\{X \Rightarrow F, X' \Rightarrow G\}, \{X \Rightarrow A, X' \Rightarrow B\}$$
$$F \Rightarrow f(X), G \Rightarrow g(X'), A \Rightarrow a, B \Rightarrow b$$

Starting from the axiom (X, X'), we can derive

$$(X, X') \Rightarrow (F, G) \Rightarrow^* (f(X), g(X')) \Rightarrow (f(F), g(G)) \Rightarrow^* (f(f(X)), g(g(X')))$$
$$\Rightarrow^* (f^n(X), g^n(X')) \Rightarrow (f^n(A), g^n(B)) \Rightarrow^* (f^n(a), g^n(b))$$

Thus we get the language of pairs $\{(f^n(a), g^n(b)) \mid n \in I\!N\}$.

However when dealing with non monadic signatures, there might be confusions when some non-terminals occur several times in a tree. For example if X occurred twice and X' also did in a tree-tuple derived from the axiom, there would be two possible combinations of synchronizations. To avoid this ambiguity, a *control integer* is associated to each non-terminal occurrence. To make understanding easier, the way it works is explained in the sequel by means of Example 3.

A TTSG is *without internal synchronizations* if for each tree-tuple tt derived from the axiom and for each pack of productions PP that can be applied to tt, PP does not apply to several occurrences of the same component of tt.

The TTSG defined in the above example is without internal synchronizations, whereas the below one is with.

Example 2. Consider the 1-tuple language defined by the TTSG whose productions are: $\{X \Rightarrow F, Y \Rightarrow A\}, \{X \Rightarrow A, Y \Rightarrow A\}, F \Rightarrow f(X, Y), A \Rightarrow a$

Starting from the axiom F we have the derivation: $F \Rightarrow f(X, Y)$

Now, the first (and also the second) pack of synchronized productions applies to the single component $f(X, Y)$, and derives X and Y at the same time. Since X and Y belong to the same component there are internal synchronizations.

The *cartesian product* of two TTSG's is a TTSG, obtained by making the union of their productions and the concatenation of their axioms.

The *union* of two TTSG's that generate tuples of the same length is still a TTSG. Let $G = (\Sigma, NT, P, (A_1, \ldots, A_n))$ and $G' = (\Sigma', NT', P', (A'_1, \ldots, A'_n))$ s.t. $NT \cap NT' = \emptyset$. Then $G \cup G' = (\Sigma \cup \Sigma', NT \cup NT' \cup \{B_1 \ldots, B_n\}, P \cup P' \cup \{\{B_1 \Rightarrow A_1, \ldots, B_n \Rightarrow A_n\}, \{B_1 \Rightarrow A'_1, \ldots, B_n \Rightarrow A'_n\}\}, (B_1, \ldots, B_n))$.

The *intersection* over p components consists in making the cartesian product, except that p fixed components in each grammar are merged (by making intersections) together. It is the same as the natural join in relational data-bases. Performing the intersection of two TTSG's is much more difficult than making their union. It has been shown in [4] that the intersection over one component of two TTSG's is still a TTSG if at least one of the two components to be merged has no internal synchronizations. In this case, the resulting component has none either. The resulting TTSG needs a more precise control, carried out thanks to integer pairs[6] instead of single integers. Fortunately, the TTSG's defined in the sequel have no internal synchronizations.

The *emptiness* of TTSG's is decidable [4], assuming that no non-terminal appears more than once with the same control value. Some TTSG's defined in the sequel do not satisfy this property, but they will if modifying them slightly, as explained in [5].

3 Languages for positive and negative predicates

3.1 Basic languages

Let us define a few basic regular or synchronized languages we need in the following. The sets of non-terminals used by the below grammars are pairwise disjoint in order to avoid confusions when considering the union of their productions.

1. The language of ground terms.
   This regular language is generated by the grammar GrG whose axiom is G (G for ground) and whose productions are $G \Rightarrow f(G, \ldots, G)$ for each $f \in \Sigma$.
2. The language of pairs of arbitrary ground terms.
   This regular language is generated by the grammar $GrTT'$ whose axiom is (T, T') (T for term) and whose productions are:
$$\left. \begin{array}{l} T \Rightarrow f(T, \ldots, T) \\ T' \Rightarrow f(T', \ldots, T') \end{array} \right\} \text{ for each } f \in \Sigma$$
3. The language of pairs of identical ground terms.
   To get two identical terms, synchronizations are obviously necessary. This synchronized language is generated by the grammar $GrTT$ whose axiom is (T_1, T_1'), whose synchronized productions are:
$$\{T_i \Rightarrow T_f,\ T_i' \Rightarrow T_f'\} \text{ for each } i \in [1, MaxArity] \text{ and each } f \in \Sigma$$
   and whose free productions are:
$$T_f \Rightarrow f(T_1, \ldots, T_n),\ T_f' \Rightarrow f(T_1', \ldots, T_n') \quad \text{for each } f \in \Sigma$$
   where $MaxArity$ is the maximal arity of symbols of Σ, and $T_1, T_2, \ldots, T_f, \ldots$ are non-terminals (respectively $T_1', T_2', \ldots, T_f', \ldots$).

Example 3. Let $\Sigma = \{a, b, f\}$ where a, b are constants and f is a binary symbol. The productions are:
$$\{T_1 \Rightarrow T_a,\ T_1' \Rightarrow T_a'\},\ \{T_1 \Rightarrow T_b,\ T_1' \Rightarrow T_b'\},\ \{T_1 \Rightarrow T_f,\ T_1' \Rightarrow T_f'\},$$
$$\{T_2 \Rightarrow T_a,\ T_2' \Rightarrow T_a'\},\ \{T_2 \Rightarrow T_b,\ T_2' \Rightarrow T_b'\},\ \{T_2 \Rightarrow T_f,\ T_2' \Rightarrow T_f'\},$$
$$T_a \Rightarrow a,\ T_a' \Rightarrow a,\ T_b \Rightarrow b,\ T_b' \Rightarrow b,\ T_f \Rightarrow f(T_1, T_2),\ T_f' \Rightarrow f(T_1', T_2')$$

---

[6] And integer tuples when making several intersections incrementally.

From the axiom (T_1, T_1') we can derive (the control is written just above the non-terminals):

$$(\overset{0}{T_1},\ \overset{0}{T_1'}) \Rightarrow (\overset{1}{T_f},\ \overset{1}{T_f'}) \Rightarrow^* (f(\overset{1}{T_1},\overset{1}{T_2}),\ f(\overset{1}{T_1'},\overset{1}{T_2'}))$$

$$\Rightarrow (f(\overset{2}{T_f},\overset{1}{T_2}),\ f(\overset{2}{T_f'},\overset{1}{T_2'})) \Rightarrow^* (f(f(\overset{2}{T_1},\overset{2}{T_2}),\overset{1}{T_2}),\ f(f(\overset{2}{T_1'},\overset{2}{T_2'}),\overset{1}{T_2'}))$$

$$\Rightarrow (f(f(\overset{3}{T_a},\overset{2}{T_2}),\overset{1}{T_2}),\ f(f(\overset{3}{T_a'},\overset{2}{T_2'}),\overset{1}{T_2'})) \Rightarrow^* (f(f(a,\overset{2}{T_2}),\overset{1}{T_2}),\ f(f(a,\overset{2}{T_2'}),\overset{1}{T_2'}))$$

Free productions leave control unchanged, whereas synchronized productions increase it into a value not yet used. At this stage, T_2 appears twice in the first component, while T_2' appears twice in the second one. Of course the leftmost T_2 must not be derived together with the rightmost T_2' into identical terms. This is the role of control integers: only non-terminals that have the same control integer can be derived (synchronized) together. So:

$$\Rightarrow (f(f(a,\overset{4}{T_b}),\overset{1}{T_2}),\ f(f(a,\overset{4}{T_b'}),\overset{1}{T_2'})) \Rightarrow^* (f(f(a,b),\overset{1}{T_2}),\ f(f(a,b),\overset{1}{T_2'}))$$

$$\Rightarrow (f(f(a,b),\overset{5}{T_a}),\ f(f(a,b),\overset{5}{T_a'})) \Rightarrow^* (f(f(a,b),a),\ f(f(a,b),a))$$

4. **The language of contexts.**
A *context* is a ground term that contains one hole denoted by $\perp$ ($\perp \notin \Sigma$). This regular language is generated by the grammar GrC whose axiom is C (C for context) and whose productions are those of GrT plus:

$$C \Rightarrow \perp$$
$$C \Rightarrow f(T,\ldots,T,C,T,\ldots,T) \left\{ \begin{array}{l} \text{for each } f \in \Sigma_{>0} \\ \text{and each position of } C \text{ as an argument of } f \end{array} \right.$$

5. **The language of pairs of identical contexts.**
The hole of the second component is denoted by $\perp'$ instead of $\perp$. This synchronized language is generated by the grammar $GrCC$ whose axiom is (C_1, C_1'), whose productions are those of $GrTT$ plus the synchronized productions:

$$\{C_1 \Rightarrow C_\perp,\ C_1' \Rightarrow C_\perp'\}$$
$$\{C_1 \Rightarrow C_{f,i},\ C_1' \Rightarrow C_{f,i}'\} \text{ for each } f \in \Sigma_{>0} \text{ and each } i \in \{1,\ldots,ar(f)\}$$

and the free productions:

$$C_\perp \Rightarrow \perp,\ C_\perp' \Rightarrow \perp'$$
$$C_{f,i} \Rightarrow f(T_1,\ldots,T_{i-1},C_1,T_{i+1},\ldots,T_n) \Big| \text{ for each } f \in \Sigma_{>0}$$
$$C_{f,i}' \Rightarrow f(T_1',\ldots,T_{i-1}',C_1',T_{i+1}',\ldots,T_n') \Big| \text{ and each } i \in \{1,\ldots,ar(f)\}$$

where $C_1, C_\perp, C_{f,i}, \ldots$ are non-terminals (respectively $C_1', C_\perp', C_{f,i}', \ldots$).

6. **The language of pairs of different ground terms.**
The idea consists in generating a branch that contains at least one clash (only the first one is forced), and anything elsewhere. This synchronized language is generated by the grammar $GrT\overline{T}$ whose axiom is (D, D') (D for different), whose productions are those of $GrTT'$ plus the synchronized productions (G_f is for generating every term whose root symbol is f):

$$\{D \Rightarrow D_{f,i},\ D' \Rightarrow D_{f,i}'\} \text{ for each } f \in \Sigma_{>0} \text{ and each } i \in \{1,\ldots,ar(f)\}$$
$$\{D \Rightarrow G_f,\ D' \Rightarrow G_g'\} \text{ for each } f,g \in \Sigma \text{ s.t. } f \neq g$$

and the free productions:

$$D_{f,i} \Rightarrow f(T,\ldots,T,D,T,\ldots,T) \,\Big|\, \text{for each } f \in \Sigma_{>0},\, i \in \{1,\ldots,ar(f)\},$$
$$D'_{f,i} \Rightarrow f(T',\ldots,T',D',T',\ldots,T') \,\Big|\, \text{s.t. } D\ (D') \text{ is the } i^{th} \text{ argument of } f$$
$$\left.\begin{array}{l} G_f \Rightarrow f(T,\ldots,T) \\ G'_f \Rightarrow f(T',\ldots,T') \end{array}\right\} \text{ for each } f \in \Sigma$$

where $D_{f,i},\ldots,G_f,\ldots$ are non-terminals (respectively $D'_{f,i},\ldots,G'_f,\ldots$).

We can also deal with formulas that contain the equality predicate since the solutions of $x \doteq y$ are generated by $GrTT$, and those of $\neg(x \doteq y)$ by $GrT\overline{T}$.

3.2 The language $\mathcal{L}$ of solutions of $x \to y$

Now, the grammar Gr that generates $\mathcal{L}$ can be easily defined. Roughly speaking, we have to generate all $(C[\sigma l], C[\sigma r])$.

1. The language of rewrite rules.
   This is the set of pairs of terms (l,r) where $l \to r$ is a rewrite rule whose variables are replaced in the following way: each variable x of l (resp. x of r) is replaced by the new symbol X (resp. X'). Note that $X, X' \notin \Sigma$. This language is finite, then regular. Let $GrLR$ be a grammar that generates this language, and let (L, R) be its axiom.
2. The language $\mathcal{L}$.
   $\mathcal{L}$ is generated by the grammar Gr whose axiom is (C_1, C'_1), whose productions are those of $GrCC$ (for generating two identical contexts), plus those of $GrLR$ (for generating left and corresponding right-hand-sides), plus those of GrG and $GrTT$ (for generating ground instances), plus the synchronized productions (for making links between the four used grammars):

$$\{\perp \Rightarrow L,\ \perp' \Rightarrow R\}$$
$$\left.\begin{array}{r} \{X \Rightarrow T_1,\ X' \Rightarrow T'_1\} \text{ for each } x \in Var(l) \cap Var(r) \\ \{X \Rightarrow G\} \text{ for each } x \in Var(l) \setminus Var(r) \\ \{X' \Rightarrow G\} \text{ for each } x \in Var(r) \setminus Var(l) \end{array}\right\} \text{ for each rule } l \to r$$

Note that $\perp, \perp', X, X',\ldots$ are now considered as non-terminals.

3.3 The language $\neg\mathcal{L}$ of solutions of $\neg(x \to y)$

We guess that the languages generated by TTSG's are not closed by complement. Fortunately, we need only to complement the language $\mathcal{L}$, which is a particular case. Instead of computing the complement of Gr, we directly represent $\neg\mathcal{L}$ by a TTSG (denoted $\overline{Gr}$), by studying the solutions of $\neg(x \to y)$ carefully.

Definition 1. Let t_1, t_2 be two different ground terms. Let $u_{c_1},\ldots,u_{c_n}$ (exhaustive list) be the outer clash occurrences between t_1 and t_2, i.e.

$$\forall i,\ u_{c_i} \in O(t_1) \cap O(t_2) \wedge t_1(u_{c_i}) \neq t_2(u_{c_i}) \wedge \forall u < u_{c_i},\ t_1(u) = t_2(u)$$

A *father* of clashes is any occurrence u_{fa} located outer than or equal to all outer clash occurrences (i.e. $\forall i,\ u_{fa} \leq u_{c_i}$). Note that all fathers are located on the same branch, going from occurrence ϵ to the *innermost father* u_{if}.

The notions of father and innermost father are essential with respect to one-step rewriting. Indeed, if $t_1 \rightarrow_{[u, l \rightarrow r, \sigma]} t_2$, all clashes between t_1 and t_2 must disappear when replacing σl by σr. Then necessarily u is a father occurrence (otherwise at least one clash located outer than u or in another branch would not disappear when rewriting t_1). In other words, if u is not a father occurrence, then $t_1 \not\rightarrow_{[u]} t_2$.

Assume now that some left-hand-side l overlaps in t_1 at some father occurrence u, i.e. $t_1 = t_1[u \leftarrow \sigma l]$ and $u \leq u_{if}$.

Lemma 2. *Within $t_1 = t_1[u \leftarrow \sigma l]$, if l does not cover the innermost father occurrence u_{if}, i.e. $u_{if} = u.v.w$ where v is a variable occurrence of l, then $t_1 \not\rightarrow_{[u, l \rightarrow r]} t_2$.*

Lemma 3. *No term can rewrite by one step into itself.*

Consequently, $t_1 \not\rightarrow t_2$ iff one of the three following cases holds.

1. Some left-hand-sides $l_1, \ldots, l_k$ (exhaustive list) overlap in t_1 at father occurrences, and cover the innermost father occurrence u_{if}, but the terms reached by one-step rewriting thanks to these redexes are all different from t_2.
2. No left-hand-side overlaps in t_1 at father occurrences.
3. t_1 and t_2 are identical.

Let $\mathcal{L}_{case-1}$, $\mathcal{L}_{case-2}$, $\mathcal{L}_{case-3}$ be the languages of pairs of ground terms (t_1, t_2) defined by respectively case 1, case 2, case 3. Thus $\neg \mathcal{L} = \mathcal{L}_{case-1} \cup \mathcal{L}_{case-2} \cup \mathcal{L}_{case-3}$.

The language $\mathcal{L}_{case-3}$ is generated by the grammar $GrTT$. For $\mathcal{L}_{case-1}$ (resp. $\mathcal{L}_{case-2}$), we split the problem into two simpler ones. We define:

- the language $\mathcal{L}_0$ of pairs of different ground terms (t_1, t_2), where the innermost father occurrence u_{if} is marked by replacing in t_1 the symbol (say g) appearing at u_{if} by a new symbol g_{if}. This amounts to deal with the extended set of symbols $\Sigma' = \Sigma \cup \Sigma_{if}$ that contains an additional symbol g_{if} for each symbol $g \in \Sigma$.
- and the language $\mathcal{L}_1$ (resp. $\mathcal{L}_2$) of pairs of ground terms (t_1, t_2), such that t_1 contains one symbol of Σ_{if} at an arbitrary position u_{if} and such that the condition of case 1 (resp. case 2) is satisfied, u_{if} being considered as the innermost father.

Note that $\mathcal{L}_1$ (resp. $\mathcal{L}_2$) does not ensure that u_{if} is the actual innermost father position with respect to the existing clashes between t_1 and t_2. This is the role of $\mathcal{L}_0$. Therefore $\mathcal{L}_{case-1} = \mathcal{L}_0 \cap \mathcal{L}_1$ (resp. $\mathcal{L}_{case-2} = \mathcal{L}_0 \cap \mathcal{L}_2$) after replacing each symbol of Σ_{if} by the corresponding symbol of Σ. Note that this symbol replacement does not change anything with respect to emptiness.

Generating $\mathcal{L}_0$: $\mathcal{L}_0$ contains all pairs of the form (C is any context and '...' means any ground term(s)):

1. $(C[f_{if}(\ldots)], C[g(\ldots)])$ where $f \neq g$. This is the particular case where there is only one outer clash. And then the innermost father and the outer clash coincide.

2. $(C[g_{if}(\ldots, s_1, \ldots, s'_1, \ldots)], (C[g(\ldots, s_2, \ldots, s'_2, \ldots)])$ where $s_1 \neq s_2$, $s'_1 \neq s'_2$.

The grammar Gr_0 that generates $\mathcal{L}_0$ can be easily defined thanks to the previously defined grammars. Its axiom is (C_1, C'_1) and its productions are those of $GrCC$ (for generating two identical contexts), plus those of $GrTT'$ (for generating pairs of arbitrary ground terms), plus those of $GrT\overline{T}$ (for generating pairs of different ground terms), plus the productions for case 1 (O as Outer clash, NO as Non Outer clash) which are:

$$\{\bot \Rightarrow O_f, \bot' \Rightarrow O'_g\} \text{ for each } f, g \in \Sigma \text{ s.t. } f \neq g$$

$$\left.\begin{array}{l} O_f \Rightarrow f_{if}(T, \ldots, T) \\ O'_f \Rightarrow f(T', \ldots, T') \end{array}\right\} \text{ for each } f \in \Sigma$$

and the productions for case 2 (**T** means $T, \ldots, T$):

$$\{\bot \Rightarrow NO_{g,i,j}, \bot' \Rightarrow NO'_{g,i,j}\} \left\{\begin{array}{l} \text{for each } g \in \Sigma_{\geq 2} \\ \text{and each } i, j \in \{1, \ldots, ar(g)\} \text{ s.t. } i < j \end{array}\right.$$

$$\begin{array}{l} NO_{g,i,j} \Rightarrow g_{if}(\mathbf{T}, D_1, \mathbf{T}, D_2, \mathbf{T}) \\ NO'_{g,i,j} \Rightarrow g(\mathbf{T'}, D'_1, \mathbf{T'}, D'_2, \mathbf{T'}) \end{array}\left|\begin{array}{l} \text{for each } g \in \Sigma_{\geq 2}, \ i, j \in \{1, \ldots, ar(g)\} \text{ s.t. } i < j \\ \text{where } D_1 \ (D'_1) \text{ is the } i^{th} \text{ argument} \\ D_2 \ (D'_2) \text{ is the } j^{th} \text{ argument} \end{array}\right.$$

$$\{D_1 \Rightarrow D, D'_1 \Rightarrow D'\}$$
$$\{D_2 \Rightarrow D, D'_2 \Rightarrow D'\}$$

where $O_f, \ldots, NO_{g,i,j}, \ldots$ are non-terminals (respectively $O'_g, \ldots, NO'_{g,i,j}, \ldots$).

Generating $\mathcal{L}_1$: Assume some left-hand-side l overlaps in t_1 at some father occurrence u, s.t. l covers u_{if} within t_1. We have $u \leq u_{if}$ and $t_1 = t_1[u \leftarrow \sigma l]$. Let t' be the term obtained by rewriting t_1 at occurrence u: $t' = t_1[u \leftarrow \sigma r]$.

Let us first show that the following assumption is impossible: Suppose there is another (or the same) left-hand-side l' that overlaps in t_1 at some position $u' \leq u_{if}$, s.t. l' also covers u_{if}. Since l and l' are both linear, and moreover they share the same occurrence u_{if} of t_1, they overlap necessarily, which is in contradiction with the non left-left-overlapping restriction.

Therefore, to ensure that $t_1 \not\rightarrow t_2$, checking that t_2 is different from t' is enough. In other words, $\mathcal{L}_1$ contains all pairs (t_1, t_2) of the form $(C[\sigma l_{if}], \overline{C[\sigma r]})$ where C is any context, σ is any substitution, $l \rightarrow r$ is any rewrite rule, l_{if} is obtained from l by replacing one arbitrary symbol (say f) by the corresponding symbol of Σ_{if} (i.e. f_{if}), and $\overline{C[\sigma r]}$ denotes any term different from $C[\sigma r]$.

However, the above explanation does not take into account the possibility of having several t' for fixed t_1, u, $l \rightarrow r$; in other words several $C[\sigma r]$ for one $C[\sigma l_{if}]$. This case happens when r contains some variables that do not appear in l, since these variables can be instantiated by any ground term. So, forcing a clash with respect to some t' within the instances of such variables does not ensure that we get a term different from all potential t'. To process this case correctly, we just have to force the clash elsewhere.

Besides, the clashes between t_1 and t_2 occur only below u_{if}[7]. As u_{if} belongs to the part brought by l_{if}, there is no clash in the part brought by C. So computing $\overline{C[\sigma r]}$ as $C[\overline{\sigma} r] \cup C[\overline{\theta r}]$ where $\overline{\sigma} r$ denotes an instance of r different from σr and $\overline{\theta r}$ denotes any term non instance of r, is enough. Or again, $\mathcal{L}_1$ can be computed as the union of the languages $\mathcal{L}'_1$ and $\mathcal{L}''_1$ defined below.

- $\mathcal{L}'_1 = (C[\sigma l_{if}], C[\overline{\sigma} r])$.
  The TTSG that generates $\mathcal{L}'_1$ looks like Gr (language $\mathcal{L}$ in Section 3.2). The main difference is that an arbitrary variable of $Var(l) \cap Var(r)$ (don't know choice) is instantiated in l and r by two different terms, thanks to $GrT\overline{T}$, while other variable occurrences are replaced by any ground term. Thus there is at least one clash between the substitution applied to l_{if} and the one applied to r.
- $\mathcal{L}''_1 = (C[\sigma l_{if}], C[\overline{\theta r}])$.
  The TTSG that generates $\mathcal{L}''_1$ looks like Gr as well, except that to generate $\overline{\theta r}$, we generate an arbitrary branch of r that reaches to a clash with a non variable occurrence of r, and any ground terms on the other branches and below the clash.

Generating $\mathcal{L}_2$: $\mathcal{L}_2$ contains all pairs (t_1, t_2) such that t_1 is any ground term that does not belong to the projection on the first component of $\mathcal{L}_1$, and t_2 is any ground term. The projection on the first component of $\mathcal{L}_1$ contains the terms of the form $C[\sigma l_{if}]$, which is obviously a regular language. Its complement is also a regular language. $\mathcal{L}_2$ is obtained by computing the cartesian product between the complement and the language of all ground terms.

The complement is computed with respect to $\Sigma \cup \Sigma_{if}$. So it owns, among others, terms that contain no or several symbols of Σ_{if}. These terms are wrong because they do not make sense[8]. But it does not matter since these wrong terms will be deleted when making intersection with $\mathcal{L}_0$, all terms of which just contain one symbol of Σ_{if}.

4 Intersection over several components

The intersection over several components is needed because of conjunctions in formulas and also for computing $\mathcal{L}_{case-1}$ and $\mathcal{L}_{case-2}$. Lengthening the tuples by making cartesian products with the language of all ground terms[9] (to get tuples of the same size), and nesting tree-tuples below a new symbol (to get single trees), allows to transform the intersection over several components into a classical intersection between tree languages. Thus in the following, we deal with tree synchronized grammars (*TSG* for short) instead of TTSG's. However these TSG's contain internal synchronizations even if the initial TTSG's do not. As

---

[7] Actually, this is not ensured by $\mathcal{L}_1$, but this will hold when computing $\mathcal{L}_0 \cap \mathcal{L}_1$.

[8] There is only one innermost father.

[9] In the same way as done when making unions. See footnote 5, or [7] for more details.

guessed in [4], we think that in general, the resulting language cannot be generated by a TSG. Fortunately, the TTSG's to be processed satisfy the following particular property: the difference of depth between any two synchronized points is bounded. Of course it still holds in the corresponding TSG's. However this difference is not necessarily equal to 0, which causes a major difficulty, as explained in Section 4.1. This difficulty is solved in Section 4.2 by first transforming the grammars to make the depth difference equal to 0.

4.1 Intersection of TSG's

Recall that in the most general case, control numbers are actually integer-tuples. So in the following, the synchronized grammars are denoted by 5-tuples instead of 4-tuples because a new component (the first one) has been added to mention the size of these integer-tuples. Packs of synchronized productions are denoted by $\{\ldots\}_k$ where k is an integer (called *the level of the pack*) meaning that only the kth component of the control tuple is incremented when applying the pack (see [4] for more details). $Rec(G)$ denotes the language recognized (generated) by the grammar G.

The TSG intersection is mainly based on the regular tree grammar intersection. The difference is that we have to take care of control numbers. In order to make the intersection of G_1 and G_2, consider G_3 as defined below.

Definition 4. Let $G_1 = (S_1, C_1, NT_1, P_1, I_1)$ and $G_2 = (S_2, C_2, NT_2, P_2, I_2)$ be two TSG's.
Let $G_3 = (S_1 + S_2, C_1 \cup C_2, NT_1 \times NT_2, Fr \cup Sn_1 \cup Sn_2, I_1 I_2)$ where

- $Fr = \{X_1 X_2 \Rightarrow c(Y_{1,1} Y_{2,1}, \ldots, Y_{1,n} Y_{2,n})$ such that
  $X_1 \Rightarrow c(Y_{1,1}, \ldots, Y_{1,n}) \in P_1$ and $X_2 \Rightarrow c(Y_{2,1}, \ldots, Y_{2,n}) \in P_2\}$
- $Sn_1 = \{\{X_1 Y_1 \Rightarrow X_1' Y_1, \ldots, X_n Y_n \Rightarrow X_n' Y_n\}_k$ such that
  $\{X_1 \Rightarrow X_1', \ldots, X_n \Rightarrow X_n'\}_k \in P_1$ and $\forall i \in [1, n] Y_i \in NT_2\}$
- $Sn_2 = \{X_1 Y_1 \Rightarrow X_1 Y_1', \ldots, X_n Y_n \Rightarrow X_n Y_n'\}_{S_1 + k}$ such that
  $\{\{Y_1 \Rightarrow Y_1', \ldots, Y_n \Rightarrow Y_n'\}_k \in P_2$ and $\forall i \in [1, n] X_i \in NT_1\}$

The control tuples of G_3 are obtained by concatenating those of G_1 with those of G_2. This allows to avoid confusions between the control coming from G_1 and that coming from G_2 when deriving G_3.

It is easy to verify that $Rec(G_3) \subseteq Rec(G_1) \cap Rec(G2)$ because of the way G_3 is built (The proof is done with a classical induction on the length of the derivations of G_3). But unfortunately $Rec(G_1) \cap Rec(G_2) \not\subseteq Rec(G_3)$ as illustrated by the following example.

Example 4. Let $C = \{c, s, 0\}$, $NT_1 = \{X_i | i \in [1, 6]\}$, $NT_2 = \{Y_i | i \in [1, 6]\}$,
$P_1 = \{X_1 \Rightarrow c(X_2, X_3), X_2 \Rightarrow s(X_4), X_5 \Rightarrow s(X_6), X_6 \Rightarrow 0, \{X_3 \Rightarrow X_5, X_4 \Rightarrow X_6\}_1\}$,
$P_2 = \{Y_1 \Rightarrow c(Y_3, Y_2), Y_2 \Rightarrow s(Y_4), Y_5 \Rightarrow s(Y_6), Y_6 \Rightarrow 0, \{Y_3 \Rightarrow Y_5, Y_4 \Rightarrow Y_6\}_1\}$
Let $G_1 = (1, C, NT_1, P_1, X_1)$ and $G_2 = (1, C, NT_2, P_2, Y_1)$
$G_3 = (2, C, NT_1 \times NT_2, Fr \cup Sn_1 \cup Sn_2, X_1 Y_1)$ where
$Fr = \{X_1 Y_1 \Rightarrow c(X_2 Y_3, X_3 Y_2), X_2 Y_2 \Rightarrow s(X_4 Y_4), X_5 Y_5 \Rightarrow s(X_6 Y_6),$

$X_2Y_5 \Rightarrow s(X_4Y_6), X_5Y_2 \Rightarrow s(X_6Y_4), X_6Y_6 \Rightarrow 0\}$
$Sn_1 = \{\{X_3Y_i \Rightarrow X_5Y_i, X_4Y_j \Rightarrow X_6Y_j\}_1$ such that $i, j \in [1,6]\}$ and
$Sn_2 = \{\{X_iY_3 \Rightarrow X_iY_5, X_jY_4 \Rightarrow X_jY_6\}_2\}$ such that $i, j \in [1,6]\}$

Obviously $Rec(G_1) = Rec(G_2) = \{c(s(0), s(0))\}$. Unfortunately $Rec(G_3) =$
$\emptyset$. Indeed $\overset{0,0}{X_1Y_1} \Rightarrow c(\overset{0,0}{X_2Y_3}, \overset{0,0}{X_3Y_2})$ and no more productions can be applied because X_2Y_3 requires a non terminal of the form X_iY_4 to perform a synchro-
nization and X_3Y_2 requires a non terminal of the form X_4Y_j. So each branch
of the term is waiting for the other, which is a kind of deadlock. This happens
because G_1 and G_2 have leaning synchronizations, i.e. in their derivation appear
some synchronization points that are at different depths. So a sufficient condition to get $Rec(G_3) \subseteq Rec(G_1) \cap Rec(G_2)$ is that G_1 and G_2 have no leaning
synchronizations. This condition clearly still holds for G_3.

In fact, a leaning synchronization is a set of non-terminals such that their reduction may cause depth differences between further synchronization points.
In the following definition, $|u|$ is the depth of the occurrence u, $\Rightarrow_i$ is a step
that applies a pack of productions of level i, and $\Rightarrow^*_{\neq i} tt$ denotes a derivation
that does not apply any packs of level i s.t. tt is either irreducible or reducible
only by packs of level i.

Definition 5. Let $G = (S, C, NT, P, I)$ be a TSG.
A *synchronization of level* i is a set of non-terminals which is either $\{I\}$ or
$\{X_1, \ldots, X_n\}$ such that $\exists\{X_1 \Rightarrow X'_1, \ldots, X_n \Rightarrow X'_n\}_i \in P$.
For $\{X_1, \ldots, X_n\}$ a synchronization of level i, $Next(\{X_1, \ldots, X_n\})$ denotes the
set defined by $\{\{t_1, \ldots, t_n\}$ s.t. $(\overset{ct_1}{X_1}, \ldots, \overset{ct_n}{X_n}) \Rightarrow_i (\overset{ct'_1}{X'_1}, \ldots, \overset{ct'_n}{X'_n}) \Rightarrow^*_{\neq i} (t_1, \ldots, t_n)\}$.
For the synchronization $\{I\}$ and for each level i, $Next(\{I\}) = \{\{t\}|I \Rightarrow^*_{\neq i} t\}$.
A synchronization $\{X_1, \ldots, X_n\}$ of level i is said *leaning* if $\exists tt \in Next(\{X_1, \ldots,
X_n\})$, $\exists u, u' \in O(tt)$ such that both $tt(u), tt(u')$ are non-terminals and $|u| \neq |u'|$.
A synchronization $\{X_1, \ldots, X_n\}$ of level i is said *finite* if $Next(\{X_1, \ldots, X_n\})$ is
a finite set.
A TSG without leaning synchronizations is called a *balanced TSG*.

Lemma 6. *Let G be a balanced TSG, in any derivation of G, if*
$\{X_1 \Rightarrow Y_1, \ldots, X_n \Rightarrow Y_n\}_l$ *applies at occ. $u_1 \ldots u_n$ then $|u_1| = \ldots = |u_n|$.*

This property prevents from deadlocks when making intersection because all synchronized non terminals are always at the same depth. Unfortunately, the TSG's
built from one-step rewriting formulas are not necessarily balanced because variables may appear at different depths in the rewrite rules. The only non-balanced
languages are $\mathcal{L}$ and $\mathcal{L}'_1$, due to the leaning synchronization $\{\bot, \bot'\}$ and the
pack $\{\bot \Rightarrow L, \bot' \Rightarrow R\}$. This leaning synchronization cannot be applied infinitely many times (actually only once), which prevents from increasing the
depth difference between synchronization points in an unbounded way. So the
object of the next subsection is to give an algorithm to transform such grammars
(called quasi-balanced) into balanced ones.

4.2 From quasi-balanced to balanced TSG's

Definition 7. Let G be a TSG and $\{X_1, ..., X_n\}$ a synchronization of level i. $\{X_1, ..., X_n\}$ is said *recursive* if there exists a tuple tt s.t. $(X_1^{ct_1}, ..., X_n^{ct_n}) \Rightarrow^+ tt$ where $\forall j \in [1, n]$ $ct_j|_i = ct_1|_i$ (the other components of the controls being any) and there exist $u_1, ..., u_n \in O(tt)$ such that $\forall j$, $tt(u_j) = X_j^{ct_j''}$ and $ct_j''|_i = ct_1''|_i$. Otherwise $\{X_1, ..., X_n\}$ is said *non-recursive*.

A TSG G is said *quasi-balanced* if any leaning synchronization is finite and non-recursive.

Now, we have to prove that the class of languages recognized by quasi-balanced TSG's is the same as the class of languages recognized by balanced TSG's. For that we give an algorithm that transforms any quasi-balanced TSG into a balanced one.

The main idea of this algorithm is to re-organize synchronizations so that the synchronized points appear at the same depth. To do that, the algorithm "cuts" the slopes at the depth of the higher synchronized non-terminal. This cutting operation needs the algorithm to look in advance at the tree-parts that will be generated later. For that it creates new 13¡

non-terminals whose names encode the tree-parts to be produced. For sake of simplicity we present the algorithm for TSG's whose control numbers are single integers (which is enough for the one-step rewriting problem).

First, let us define the new non-terminals and the way to decode their names.

Definition 8. Let $G = (1, C, NT, PP, I)$ be a TSG. $AheadNT(G) = \{\overset{i}{X^t} \mid X \in NT, i \in \mathbb{N}, t \in T(C, NT \times \mathbb{N}) \cup \{\epsilon\}\}$ where ϵ is a new symbol. Let $\overset{i}{X^t} \in AheadNT(G)$, $term(\overset{i}{X^t}) = \overset{i}{X}$ when $t = \epsilon$ and t otherwise.

Roughly speaking, X is the actual non-terminal, t is the term to be generated from X (if $t = \epsilon$ there is no restriction on the term generated from X). So the function $term$ allows to look ahead in the derivation. Note that $AheadNT(G)$ is infinite, but thanks to the non-recursivity of leaning synchronizations only a finite part will be used by the algorithm. In the following $term(\{A_1, ..., A_n\})$ where $\forall i$, $A_i \in AheadNT(G)$, will denote $\{term(A_1), ..., term(A_n)\}$. Now let us define some notations needed in the algorithm.

Definition 9. Let $G = (1, C, NT, P, I)$ be a TSG and s be a finite subset of $AheadNT(G)$. s is said

- *free* if $\forall \overset{i}{X^t} \in s$ there exists a free production $X \Rightarrow c(X_1, ..., X_n) \in P$ and either $t = \epsilon$ or $t = c(t_1, ..., t_n)$. Moreover we denote,

$$NF(s) = \cup_{\overset{i}{X^t} \in s} \{\overset{i}{X_1^{t_1}}, ..., \overset{i}{X_n^{t_n}}\} \text{ and } NFP(s) = \cup_{\overset{i}{X^t} \in s} \{\overset{i}{X^t} \Rightarrow c(\overset{i}{X_1^{t_1}}, ..., \overset{i}{X_n^{t_n}})\}.$$

- *minimally synchronized* if $\cup_{X^t \in s}\ term(X^t)$ contains only synchronized non-terminals and all these non-terminals can be derived (i.e. none of them are waiting for another non-terminal of its synchronization) and $\forall\ X^t \in s,\ s \setminus \{X^t\}$ is not minimally synchronized. In that case, $NS(s) = \{\{X_1^{t_1 i_1}, \ldots, X_n^{t_n i_n}\}$ such that $term(s) \Rightarrow^* term(\{X_1^{t_1 i_1}, \ldots, X_n^{t_n i_n}\})$ where one and only one synchronized production has been applied to each non-terminal of $term(s)$. $NSP(s) = \{Y_1^{s_1 j_1} \Rightarrow X_1^{t_1 i_1}, \ldots, Y_n^{s_n j_n} \Rightarrow X_n^{t_n i_n}\}$ such that $s = \{Y_1^{s_1 j_1}, \ldots, Y_n^{s_n j_n}\}$, $\{X_1^{t_1 i_1}, \ldots, X_n^{t_n i_n}\} \in NS(s)$ and $term(s) \Rightarrow^* term(\{X_1^{t_1 i_1}, \ldots, X_n^{t_n i_n}\})$.

- *completely synchronized* if $\forall\ X^t \in s,\ term(X^t)$ contains only synchronized non-terminals. In that case, $Pa(s) = \{\{X_1^{t_1 i_1}, \ldots, X_n^{t_n i_n}\}$ such that this set is minimally synchronized $\}$.

Now the algorithm is given thanks to these 4 inference rules.

$$\frac{D \quad TD \cup \{s\} \quad R}{D \cup \{s\}\ TD \cup \{s'\} \setminus D\ R \cup \{p\}}(1) \qquad \frac{D \quad TD \cup \{s\} \quad R}{D \cup \{s\}\ TD \cup Pa(s) \setminus D\ R}(2)$$

$$\frac{D \quad TD \cup \{s\} \quad R}{D \cup \{s\}\ TD \cup NF(s) \setminus D\ R \cup NFP(s)}(3)$$

$$\frac{D \quad TD \cup \{s\} \quad R}{D \cup \{s\}\ TD \cup NS(s) \setminus D\ R \cup NSP(s)}(4)$$

(1) Look Ahead if s is not free and $Y^t \in s$ such that $term(t) \Rightarrow term(t')$ with the free production $X \Rightarrow c(X_1, \ldots, X_n)$.

p denotes the production $Y^t \Rightarrow Y^{t'}$ and s' the set $(s \setminus \{Y^t\}) \cup \{Y^{t'}\}$

(2) Partition if s is completely synchronized but not minimally synchronized.

(3) Free if s is free.

(4) Sync if s is minimally synchronized.

Comments: D stands for Done, TD for To Do and R for Result. To transform a quasi-balanced TSG G into a balanced one, one has to initialize the sets D and R to $\emptyset$, and TD to $\{\overset{0}{I}\}$ (i.e. the axiom of G). Then the inference rules are applied in a don't care way as long as possible. At the end, R contains the productions of the new TSG (the non-terminals being deduced from the productions). For each step, s is a subset of AheadNT representing non-terminals appearing at the same depth in a derivation of G, associated with their ahead terms when leaning synchronizations are detected. The role of each inference rule is the following:

Look Ahead is applied when s is neither free nor totally synchronized, which means that at the same depth in a synchronization of G there are some non-terminals waiting for synchronization and some others that can be derived with a free production. This case happens while visiting leaning synchronizations. So **Look Ahead** will apply the free productions on the Ahead terms of the non-terminals instead of the actual non-terminal to find the next synchronization points. It generates empty free productions (i.e. production that do not produce any terminal symbols). These production can be easily eliminated afterwards but they simplify all the proofs.

Partition is applied when s is completely synchronized, which means that $term(s)$ contains only synchronized non terminals, so the ahead terms are big enough to know what will happen when applying a synchronized pack of productions on non-terminals appearing at different depths. So partition divides s in subsets which are minimally synchronized, i.e. the non-terminals of each subset are actually bound by synchronization constraints.

Free is applied when s is free i.e. when all the non-terminals of s can be derived without synchronization constraints. So the free productions are applied on the Ahead non-terminals at the same time in order to keep the new non-terminals at the same depth. The free productions are stored in the result.

Sync is applied when s is minimally synchronized (i.e. all the non-terminals of $term(s)$ are actually to be synchronized) and the ahead terms are big enough to know what happens next, all the possibilities are explored and all the corresponding new synchronized productions are created. Note that thanks to **Free** we know that the ahead non-terminals of s always appear at the same depth in a derivation of the new TSG.

The application of the inference rules terminates because the depth between two synchronization points is bounded, so the depth of the Ahead terms is bounded, so the number of AheadNT's used in the algorithm is bounded, so the number of different sets of AheadNT's is bounded.

So from a quasi-balanced TSG $G = (1, \mathcal{C}, NT, P, I)$ we get $G' = (1, \mathcal{C}, NT, R, \overset{0}{I^\epsilon})$. G' is balanced from its construction. The last step is to prove that $Rec(G) = Rec(G')$. For that we make a correspondence between derivations of G' and derivations of G using the terms of G' where AheadNT's have been expanded with the function $term$.

5 Further work : weakening the restrictions

A referee has pointed out that our method can be easily extended to deal with rewrite systems that have critical pairs between different rules. If the rewrite system is $R = \{r_1, \ldots, r_n\}$, just replace each $x \rightarrow y$ in the formula by $x \rightarrow_{[r_1]} y \vee \ldots \vee x \rightarrow_{[r_n]} y$, bring the formula into a disjunctive form, and adjust the method (which is easy) to take into account the new predicate symbols.

Besides, we conjecture that the method could be extended to deal with rewrite systems that have non linear right-hand-sides, and collapsing rules[10].

---

[10] I.e. non ϵ-left-right-overlapping is still assumed, except for collapsing rules.

The non-linearity of rhs's creates internal synchronizations in languages. However the difference of depth between synchronization points is bounded. The intersection problem in this case has been solved already in Section 4, and the proof of lemma 2 seems to still hold. On the other hand, it seems nearly impossible to weaken the linearity of left-hand-sides because $\mathcal{L}_2$ is obtained by complementing a regular language, which would not be regular any more. Of course, a reduction automaton could be used, but it would not be deterministic, which prevents from computing the complement.

Within $\mathcal{L}_1$, the verification that the rewrite step is impossible is ensured by the failure of one rewrite rule applied at one occurrence. This is enough thanks to the restrictions. But we think that the number of failures to be ensured becomes unbounded if there are ϵ-left-right-overlapping rules that are non-collapsing.

So the question still arises : is the existential fragment of one-step rewriting theory without any restrictions decidable ?

References

1. A.C. Caron, F. Seynhaeve, S. Tison, and M. Tommasi. Solving One-step Rewriting Formulae. In *12th Workshop on Unification, Roma (Italy)*. U. of Roma, 1998.
2. H. Comon, M. Dauchet, R. Gilleron, D. Lugiez, S. Tison, and M. Tommasi. *Tree Automata Techniques and Applications (TATA)*. http://l3ux02.univ-lille3.fr/tata.
3. F. Jacquemard. *Automates d'Arbres et Réécriture de Termes*. Thèse de Doctorat d'Université, Université de Paris-sud, 1996. In french.
4. S. Limet and P. Réty. E-Unification by Means of Tree Tuple Synchronized Grammars. *DMTCS (http://dmtcs.loria.fr/)*, 1:69–98, 1997.
5. S. Limet and P. Rety. A new result about the decidability of the existential one-step rewriting theory. Research Report RR-LIFO-98-12, LIFO, 1998.
6. S. Limet and P. Réty. Solving Disequations modulo some Class of Rewrite Systems. In *Proc. of RTA '98, Tsukuba (Japon)*, vol 1379 of *LNCS*, pages 121–135.1998.
7. S. Limet and F. Saubion. A general framework for r-unification. In C. Palamidessi, H. Glaser, and K. Meinke, editors, *proc of PLILP-ALP'98*, volume 1490 of *LNCS*, pages 266–281. Springer Verlag, 1998.
8. J. Marcinkowski. Undecidability of the First-order Theory of One-step Right Ground Rewriting. In *Proceedings 8th Conference RTA, Sitges (Spain)*, volume 1232 of *LNCS*, pages 241–253. Springer-Verlag, 1997.
9. J. Niehren, M. Pinkal, and P. Ruhrberg. On Equality up-to Constraints over Finite Trees ,Context Unification and One-step Rewriting. In W. Mc Cune, editor, *Proc. of CADE'97, Townsville (Australia)*, volume 1249 of *LNCS*, pages 34–48, 1997.
10. F. Seynhaeve, M. Tommasi, and R. Treinen. Grid Structures and Undecidable Constraint Theories. In *Proceedings of 6th Colloquium on Trees in Algebra and Programming*, volume 1214 of *LNCS*, pages 357–368. Springer-Verlag, 1997.
11. R. Treinen. The First-order Theory of One-step Rewriting is Undecidable. In H. Ganzinger, editor, *Proceedings 7th Conference RTA, New Brunswick (USA)*, volume 1103 of *LNCS*. Springer-Verlag, 1996.
12. S. Vorobyov. The First-order Theory of One-step Rewriting in Linear Noetherian Systems is Undecidable. In *Proceedings 8th Conference RTA, Sitges (Spain)*, volume 1232 of *LNCS*, pages 241–253. Springer-Verlag, 1997.

A Fully Syntactic AC-RPO

Albert Rubio*

Universitat Politècnica de Catalunya, Dept. L.S.I.
Modul C6, C/ Jordi Girona, 1-3, 08034 Barcelona, Spain.
E-mail: rubio@lsi.upc.es
phone/fax: +34 93 4017988 / 4017014

Abstract. We present the first fully syntactic (i.e., non-interpretation-based) AC-compatible *recursive path ordering* (RPO). It is simple, and hence easy to implement, and its behaviour is intuitive as in the standard RPO. The ordering is AC-total, and defined uniformly for both ground and non-ground terms, as well as for partial precedences. More importantly, it is the first one that can deal *incrementally* with partial precedences, an aspect that is essential, together with its intuitive behaviour, for interactive applications like Knuth-Bendix completion.

1 Introduction

Rewrite-based methods with built-in associativity and commutativity (AC) properties for some of the operators are well-known to be crucial in theorem proving and programming. Therefore a lot of work has been done on the development of suitable *AC-compatible* reduction or simplification orderings, like [DHJP83,BP85,GL86,BCL87,KSZ90,NR91,Bac92,DP93,RN95,KS97]. An essential additional property of the ordering that is needed in order to preserve the completeness of most rewrite-based theorem proving techniques (modulo AC) is *AC-totality*, i.e. the totality on (AC-different) ground terms.

Since the initial attempts, it has always been an aim to obtain AC-compatible versions of Dershowitz' *recursive path ordering* [Der82], as it is simple, easy to automate and use, and normally orients the rules in an adequate direction. In [RN95] we gave the first RPO-based AC-total and AC-compatible reduction ordering without any restriction on the number of AC-symbols or on the precedence over the signature. Unfortunately, although being defined in terms of RPO, it does not behave like RPO; e.g. it does not orient the distributivity rule in the "right" (i.e. distributing) way, since a transformation on the terms is applied before using RPO (this approach, with different transformations, is also used in [BP85] among others). Therefore, a better approach seems to be to directly apply an RPO-like scheme, treating as the only special case the *AC-equal-top case*, that is, when both terms to be compared are headed by the same AC-symbol. In this direction the first AC-compatible simplification ordering with an RPO

* Partially supported by the ESPRIT working group CCL-II, ref. WG # 22457. and the CICYT project HEMOSS ref. TIC98-0949-C02-01

scheme was defined in [KSZ90] and the first one AC-total on ground terms in [KS97]. Other simpler proposals for AC-orderings with RPO scheme were given in [Rub97] and in [KS98].

However, all these AC-orderings need to interpret terms (apart from *flattening*) in some way, which makes their behaviour less intuitive, unlike it happens with the standard RPO, whose simple fully syntactic definition has been an important reason for its success.

In this paper we propose the first fully syntactic AC-RPO, i.e., no interpretation is needed apart from flattening. It is simple, and hence easy to implement, and its behaviour is intuitive as for the standard RPO. The ordering is AC-total, and defined uniformly for both ground and non-ground terms, as well as for partial precedences.

Moreover, precisely due to the fact that it is not interpretation-based, it is the first AC-RPO that can deal *incrementally* with partial precedences, i.e. if $s \succ t$, then $s \succ t$ under any extension of the precedence. This aspect is essential, together with its intuitive behaviour, for interactive applications like Knuth-Bendix completion. Of course, previously existing orderings could work with partial precedences, but in a useless way, simply by considering an arbitrarily chosen total extension of the partial precedence, and hence loosing incrementality.

In order to introduce the concepts smoothly we give the ordering in three steps, first for ground terms and total precedences, then for terms with variables and total precedences and finally for terms with variables and partial precedences, each definition strictly extending the previous one. For this reason we prove all properties only for the last one, showing that it is indeed an AC-compatible simplification ordering.

The paper is organized as follows. In the following section we give some basic notions and definitions. In section 3 we introduce the ordering for ground terms and total precedences. Section 4 is devoted to terms with variables and total precedences. In section 5 we generalize the previous ordering for dealing with partial precedences and in section 6 we prove that it is an AC-compatible simplification ordering. Conclusions are given in section 7.

2 Preliminaries

In the following we consider that $\mathcal{F}$ is a finite set of function symbols that is (partially) ordered by a precedence $\succ_{\mathcal{F}}$, where $\mathcal{F}_{AC}$ is the subset containing all AC-symbols of $\mathcal{F}$.

The arity of a function symbol f is a natural number that indicates the number of arguments that f may take. If $f \in \mathcal{F}_{AC}$ then its arity is greater than or equal to 2. $\mathcal{T}(\mathcal{F})$ and $\mathcal{T}(\mathcal{F}, \mathcal{X})$ are defined as usual according to these arities, if $\mathcal{X}$ is a set of variables, whose elements will be denoted by $x, y, z, \ldots$, possibly with subscripts. The size of a term t, i.e. the number of symbols of t, is denoted by $|t|$.

We denote by $=_{AC}$ the congruence generated on $\mathcal{T}(\mathcal{F}, \mathcal{X})$ by the associativity and commutativity axioms for the symbols in $\mathcal{F}_{AC}$. In what follows we will ambiguously use $=_{AC}$ to also denote the standard extension of AC-equality to multisets (and in fact to any other structure).

A term rewriting system (TRS) is a (possibly infinite) set of rules $l \to r$ where l and r are terms. Given a TRS R, s rewrites to t with R, denoted by $s \to_R t$, if there is some rule $l \to r$ in R, $s|_p = l\sigma$ for some position p and substitution σ and $t = s[r\sigma]_p$.

In the following terms are flattened wrt. the AC-symbols. The flattening of t, denoted by $\bar{t}$, is the normal for of t wrt. the infinite TRS containing the rules

$$f(x_1, \ldots, x_n, f(y_1, \ldots, y_m), z_1, \ldots, z_r) \to f(x_1, \ldots, x_n, y_1, \ldots, y_m, z_1, \ldots, z_r)$$

for every $f \in \mathcal{F}_{AC}$ and $n, m, r \geq 0$. Due to flattening, the AC-symbols have a variable arity. We assume that all other symbols have a fixed arity.

Let s and t be two terms such that $\bar{s} = f(s_1, \ldots, s_m)$ and $\bar{t} = g(t_1, \ldots, t_n)$. If $s =_{AC} t$ then $f = g$, $m = n$ and $\bar{s}$ is equal to $\bar{t}$ up to permutation of arguments for the AC-symbols. We will denote this equality up to permutation of arguments also by $=_{AC}$. The top-flattening of a term s wrt. an AC-symbol f, denoted by $tf_f(s)$, is a string of terms defined as $tf_f(f(s_1, \ldots, s_n)) = s_1, \ldots, s_n$ and $tf_f(s) = s$ if $top(s) \neq f$.

Let s, t, s' and t' be arbitrary terms in $\mathcal{T}(\mathcal{F}, \mathcal{X})$, let u be a non-empty context in $\mathcal{T}(\mathcal{F}, \mathcal{X})$ and let σ be a substitution. Then an ordering on $\mathcal{T}(\mathcal{F}, \mathcal{X})$ (a transitive irreflexive relation) $\succ$ is *monotonic* if $s \succ t$ implies $u[s] \succ u[t]$, and *stable under substitution* if $s \succ t$ implies $s\sigma \succ t\sigma$. Monotonic orderings that are stable under substitution are called *rewrite orderings*. An ordering $\succ$ fulfills the *subterm property* if $u[t] \succ t$ and the *deletion property* if $f(\ldots s \ldots) \succ f(\ldots \ldots)$ for every variadic symbol f. A rewrite ordering that fulfills the subterm property and the deletion property is called a *simplification ordering* and is *well-founded*: there are no infinite sequences $t_1 \succ t_2 \succ \ldots$ An ordering is AC-total on ground terms if when s and t are ground terms, either $s \succ t$ or $t \succ s$ or $s =_{AC} t$. Finally an ordering $\succ$ is *AC-compatible* if $s' =_{AC} s \succ t =_{AC} t'$ implies $s' \succ t'$.

Given a TRS R, a terms s rewrites to t with R modulo AC, denoted by $s \to_{R/AC} t$, if $s =_{AC} s'$, $s'|_p = l\sigma$ for some term s', position p and substitution σ, and $t =_{AC} s'[r\sigma]_p$. A TRS R is terminating for rewriting modulo AC if there is some AC-compatible simplification ordering $\succ$, such that $l \succ r$ for all rules $l \to r$ in R.

Given a relation $\succ$, the (AC-)lexicographic extension of $\succ$ on sequences, denoted by $\succ_{lex}$, is defined by: $\langle s_1, s_2, \ldots \rangle \succ_{lex} \langle t_1, t_2, \ldots \rangle$ if there is some s_j s.t. $s_j \succ t_j$ and we have $s_i =_{AC} t_i$ for all $i < j$.

Given a relation $\succ$, the (AC-)multiset extension of $\succ$ on finite multisets, denoted by $\gg$, is defined as $M = \{s_1, \ldots, s_m\} \gg \{t_1, \ldots, t_n\} = N$ if (i) $M \neq \emptyset$ and $N = \emptyset$; or (ii) $s_i =_{AC} t_j$ and $M \setminus \{s_i\} \gg N \setminus \{t_j\}$, for some i in $1 \ldots m$ and j in $1 \ldots n$; or (iii) $s_i \succ t_{j_1} \wedge \ldots \wedge s_i \succ t_{j_k}$ and $(M \setminus \{s_i\} \gtrsim N \setminus \{t_{j_1}, \ldots, t_{j_k}\}$ for some i in $1 \ldots m$ and $1 \leq j_1 < \ldots < j_k \leq n$ ($k \geq 0$), where $\gtrsim$ is the union of $\gg$ and $=_{AC}$. Alternatively (and equivalently if $\succ$ is AC-compatible), it can be defined as the smallest transitive relation containing

$$X \cup \{s\} \not\succ Y \cup \{t_1, \ldots, t_n\} \qquad \text{if } X =_{AC} Y \text{ and } s \succ t_i \text{ for all } i \in \{1 \ldots n\}$$

In general we will consider that $\succeq$ is the union of a given ordering $\succ$ and $=_{AC}$.

If $\succ$ is an AC-compatible ordering on a set S then $\not\succ$ and $\succ_{lex}$ are respectively an AC-compatible ordering on multisets of elements in S and an AC-compatible ordering on sequences of elements in S. Being more precise, in order to fulfil transitivity we need $\succ$ to be both transitive and AC-compatible.

3 The ordering for ground terms

In this section we consider only ground terms, and assume that the precedence is total on the set of function symbols. First we introduce two different sets of terms obtained from a term headed by an AC-symbol.

Definition 1. *Let s be a term of the form $f(s_1, \ldots, s_n)$ with $f \in \mathcal{F}_{AC}$.*

- *The set of terms embedded in s through an argument headed by a small symbol, denoted by $EmbSmall(s)$, is defined as*

$$\{f(s_1, \ldots, tf_f(v_j), \ldots, s_n) \mid s_i = h(v_1, \ldots, v_r) \wedge f \succ_{\mathcal{F}} h \wedge j \in \{1 \ldots r\}\}$$

- *The set of arguments of s headed by a big symbol, denoted by $BigHead(s)$, is defined as $\{s_i \mid 1 \leq i \leq n \wedge top(s_i) \succ_{\mathcal{F}} f\}$*

We will now give the definition of the ordering $\succ$ on ground terms.

Definition 2. *Let s and t be terms in $\mathcal{T}(\mathcal{F})$. Then $s = f(s_1, \ldots, s_n) \succ g(t_1, \ldots, t_m) = t$ if and only if*

1. *$s_i \succeq t$ for some $i \in \{1 \ldots n\}$, or*
2. *$f \succ_{\mathcal{F}} g$ and $s \succ t_i$ for all $i \in \{1 \ldots n\}$, or*
3. *$f = g \notin \mathcal{F}_{AC}$ and $\langle s_1, \ldots, s_n \rangle \succ_{lex} \langle t_1, \ldots, t_n \rangle$ and $s \succ t_i$ for all $i \in \{1 \ldots n\}$, or*
4. *$f = g \in \mathcal{F}_{AC}$ and $s' \succeq t$ for some $s' \in EmbSmall(s)$, or*
5. *$f = g \in \mathcal{F}_{AC}$ and $s \succ t'$ for all $t' \in EmbSmall(t)$ and $BigHead(s) \not\succ BigHead(t)$ and either*
   - *(a) $BigHead(s) \not\succ BigHead(t)$ or*
   - *(b) $n > m$ or*
   - *(c) $n = m$ and $\{s_1, \ldots, s_n\} \not\succ \{t_1, \ldots, t_m\}$.*

The first three cases of this definition of course correspond to the standard RPO. Cases 4 and 5 apply when both terms are headed by the same AC-symbol f. The intuition behind them is very simple. On the one hand, in order to obtain AC-compatibility, terms are considered in flattened form. On the other, the symbols that disappear under flattening must still be taken into account in order to obtain monotonicity. Let us consider an example.

Assume $f \succ_{\mathcal{F}} g$. Then, as in the standard RPO, we have of course $f(a,a) \succ g(a)$. By monotonicity, if we add the context $f(a,[])$ and flatten, we must have $f(a,a,a) \succ f(a,g(a))$, that is, the symbol f that has been removed under flattening is important in order to "take care" of the g. The number of such *implicit* f's depends of course on the number of arguments.

But, similarly, if $g \succ_{\mathcal{F}} f$, then $g(a) \succ f(a,a)$ and by monotonicity we should have $f(a,g(a)) \succ f(a,a,a)$. Clearly, in this kind of situations where the comparison of arguments headed by big symbols is conclusive, the number of such implicit f's is not important.

This motivates the three stage hierarchy in case 5: (a) first consider the multisets of arguments headed by symbols bigger than f; (b) if these sets coincide, then compare the number of arguments (i.e., the number of implicit f's); (c) finally, if both terms are equal under the previous two measures, then we can safely compare the multisets of all (or only the small-headed ones) arguments in the usual (multiset) way.

Of course, since any simplification ordering must contain the *embedding* relation, we must have $s[g(\ldots t \ldots)]_p \succ s[t]_p$ for all s, t, g and p. This indicates that the use of $EmbSmall(s)$ and $EmbSmall(t)$ in cases 4 and 5 are no real restriction.

But the ideas of the three stage approach of case 5 can be safely applied precisely due to the precondition stating that $s \succ t'$ for all $t' \in EmbSmall(t)$, which prevents situations where t is a term like $f(a, h(\ldots u \ldots))$, and where by removing h (with $f \succ_{\mathcal{F}} h$) we get $f(a, u)$, where u can be headed by a big symbol, or, if u is headed by f, the number of arguments increases.

The following examples show the behaviour of the ordering when comparing terms headed by the same AC-symbol.

Example 1. Let $h \succ_{\mathcal{F}} f \succ_{\mathcal{F}} g \succ_{\mathcal{F}} a \succ_{\mathcal{F}} b$ be the precedence. Then we have

1. $f(g(f(h(a),a)),a) \succ f(h(a),a,a)$ by case 4.
2. $s = f(h(a),g(a)) \succ f(g(h(a)),a) = t$ by case 5a, since $s \succ f(h(a),a) \in EmbSmall(t)$ by case 4, and $BigHead(s) = \{h(a)\} \nsucc \emptyset = BigHead(t)$.
3. $s = f(g(h(a)),b,b,b) \succ f(g(f(h(a),a)),a) = t$ by case 5b, since $n = 4 > 2 = m$ and $BigHead(s) = \emptyset = BigHead(t)$ and $s \succ f(h(a),a,a) = t' \in EmbSmall(t)$ by applying first case 4 and then $s' = f(h(a),b,b,b) \succ t'$ by case 5b, since $BigHead(s') = \{h(a)\} = BigHead(t')$, $EmbSmall(t') = \emptyset$ and $n = 4 > 3 = m$.
4. $s = f(h(a),a) \succ f(h(a),b) = t$, by case 5c, since we have $EmbSmall(t) = \emptyset$, $BigHead(s) = \{h(a)\} = BigHead(t)$, $n = m = 2$ and $\{h(a),a\} \nsucc \{h(a),b\}$.

Lemma 1. *If $\succ_{\mathcal{F}}$ is total on the set of function symbols then $\succ$ is AC-total on ground terms.*

Proof. Let $s = f(s_1, \ldots, s_n)$ and $t = g(t_1, \ldots, t_m)$ be ground terms. Then either $s \succ t$ or $t \succ s$ or $s =_{AC} t$. We proceed by induction on $|s| + |t|$.

By induction hypothesis for every s_i we have either $s_i \succeq t$ or $t \succ s_i$, and for every t_j we have either $s \succ t_j$ or $t_j \succeq s$. On the other hand by totality of the precedence, either $f \succ_{\mathcal{F}} g$ or $g \succ_{\mathcal{F}} f$ or $f = g$. Therefore, either we conclude $s \succ t$ or $t \succ s$ by cases 1 or 2 or $s \succ t_j$ for all t_j and $t \succ s_i$ for all s_i and $f = g$.

If $f \notin \mathcal{F}_{AC}$ then, by induction hypothesis either $s =_{AC} t$ or we can conclude $s \succ t$ or $t \succ s$ by case 3.

Finally if $f \in \mathcal{F}_{AC}$ then by induction hypothesis, either $s \succ t'$ or $t' \succeq s$ for all $t' \in EmbSmall(t)$; and either $t \succ s'$ or $s' \succeq t$ for all $s' \in EmbSmall(s)$. Therefore either $s \succ t$ or $t \succ s$ by case 4 or $s \succ t'$ for all $t' \in EmbSmall(t)$ and $t \succ s'$ for all $s' \in EmbSmall(s)$. By induction hypothesis, either $BigHead(s) \succ\!\!\succ BigHead(t)$ or $BigHead(t) \succ\!\!\succ BigHead(s)$ or $BigHead(s) =_{AC} BigHead(t)$. Therefore either $s \succ t$ or $t \succ s$ by case 5a or $BigHead(s) =_{AC} BigHead(t)$ and then either $s \succ t$ or $t \succ s$ by case 5b, or $m = n$, and by induction hypothesis, we have either $s \succ t$ or $t \succ s$ by case 5c, or $s =_{AC} t$.   □

The following theorem follows from corollary 1 and theorem 3.

Theorem 1. $\succ$ *is an AC-total AC-compatible simplification ordering on* $\mathcal{T}(\mathcal{F})$.

4   Terms with variables

In this section we consider terms with variables, but we still assume that the precedence is total on the set of function symbols. First, due to the presence of variables, the counting of arguments has to be adapted, since one cannot know how many arguments a variable will include when instantiated and flattened.

Therefore in cases 5b and 5c, instead of n and m we will use the following notion of $\#(s)$ and $\#(t)$, and $n > m$ and $n = m$ become diophantine inequations over the positive integers.

Definition 3. *Let s be a term. Then $\#(s)$ is an expression with variables on the positive integers, defined as $\#(f(s_1, \ldots, s_n)) = \#_v(s_1) + \ldots + \#_v(s_n)$, where $\#_v(x) = x$ and $\#_v(t) = 1$ if t is not a variable.*

For example, we have $\#(f(x, y, g(x))) = x + y + 1 > x + y = \#(f(x, y))$, which is necessary to achieve stability under substitution.

In addition we have to replace the set $BigHead(s)$ by $NoSmallHead(s)$, which may include variables, in one of its uses.

Definition 4. *Let s be of the form $f(s_1, \ldots, s_n)$ with $f \in \mathcal{F}_{AC}$. The set of arguments of s headed by a symbol not smaller than f, denoted by $NoSmallHead(s)$, is defined as $\{s_i \mid 1 \leq i \leq n \wedge f \not\succ_{\mathcal{F}} top(s_i)\}$*

Definition 5. *Let s and t be terms in $\mathcal{T}(\mathcal{F}, \mathcal{X})$. Then $s = f(s_1, \ldots, s_n) \succ_v g(t_1, \ldots, t_m) = t$ if and only if*

1. $s_i \succeq_v t$ *for some $i \in \{1 \ldots n\}$, or*
2. $f \succ_{\mathcal{F}} g$ *and $s \succ_v t_i$ for all $i \in \{1 \ldots n\}$, or*
3. $f = g \notin \mathcal{F}_{AC}$ *and $\langle s_1, \ldots, s_n \rangle (\succ_v)_{lex} \langle t_1, \ldots, t_n \rangle$ and $s \succ_v t_i$ for all $i \in \{1 \ldots n\}$, or*

4. $f = g \in \mathcal{F}_{AC}$ and there is some $s' \in EmbSmall(s)$ s.t. $s' \succeq_v t$, or
5. $f = g \in \mathcal{F}_{AC}$ and $s \succ_v t'$ for all $t' \in EmbSmall(t)$ and
   $NoSmallHead(s) \succcurlyeq_v NoSmallHead(t)$ and either

   (a) $BigHead(s) \succ\!\!\succ_v BigHead(t)$ or
   (b) $\#(s) > \#(t)$ or
   (c) $\#(s) \geq \#(t)$ and $\{s_1, \ldots, s_n\} \succ\!\!\succ_v \{t_1, \ldots, t_m\}$.

Note that the difference between $NoSmallHead(s)$ and $BigHead(s)$ is that the latter does not include the variables. Then, on the one hand, the condition $NoSmallHead(s) \succcurlyeq_v NoSmallHead(t)$ ensures that every variable in t is taken care of by a variable in s or by a argument of s headed by a big symbol. Then if, by instantiation, some variable becomes a term headed by a big symbol, we know that some argument of the (instantiation) of s headed by a big symbol takes care of it. On the other, the condition $BigHead(s) \succ\!\!\succ_v BigHead(t)$, prevents us from using variables that can become small terms by instantiation. The combination of both conditions is crucial to prove stability under substitutions.

Example 2. Let $h \succ_{\mathcal{F}} f \succ_{\mathcal{F}} g$ be the precedence. Then we have

1. $f(g(f(h(x), x)), x) \succ_v f(h(x), x, x)$ by case 4.
2. $s = f(h(x), g(x)) \succ_v f(g(h(x)), x) = t$ by case 5a, since we have $s \succ_v$
   $f(h(x), x) \in EmbSmall(t)$ by case 4, and $NoSmallHead(s) = \{h(x)\} \succ\!\!\succ_v \{x\} = $
   $NoSmallHead(t)$ and $BigHead(s) = \{h(x)\} \succ\!\!\succ_v \emptyset = BigHead(t)$.
3. $s = f(g(h(x)), x, x, y) \succ_v f(g(f(h(x), y)), x) = t$ by case 5b, since we have
   $\#(s) = 2x + y + 1 > x + 1 = \#(t)$ and $NoSmallHead(s) = \{x, x, y\} \succ\!\!\succ_v \{x\} = $
   $NoSmallHead(t)$ and $s \succ f(h(x), y, x) = t' \in EmbSmall(t)$ by applying first
   case 4 and then $s' = f(h(x), x, x, y) \succ t'$ by case 5b, since $NoSmallHead(s') = $
   $\{h(x), x, x, y\} \succ\!\!\succ_v \{h(x), y, x\} = NoSmallHead(t')$, $EmbSmall(t') = \emptyset$ and
   $\#(s') = 2x + y + 1 > x + y + 1 = \#(t')$ (since x is a positive integer).
4. $s = f(g(g(x)), x) \succ f(g(x), g(x)) = t$, by case 5c, since $s \succ_v f(g(x), x) \in $
   $EmbSmall(t)$ by case 4 (note that the symmetric follows in the same way),
   $NoSmallHead(s) = \{x\} \succ\!\!\succ_v \emptyset = NoSmallHead(t)$ and $\#(s) = x + 1 \geq 2 = $
   $\#(t)$ and $\{g(g(x)), x\} \succ\!\!\succ_v \{g(x), g(x)\}$.

Lemma 2. *Let s and t be ground terms. Then $s \succ t$ if and only if $s \succ_v t$.*

Proof. The result is trivial since both definitions coincide when applied to ground terms. Note that if $s = f(s_1, \ldots, s_n)$ is ground then we have $\#(s) = n$ and $NoSmallHead(s) = BigHead(s)$. □

The following theorem follows from lemma 3 and theorem 3.

Theorem 2. $\succ_v$ *is an AC-compatible simplification ordering on $\mathcal{T}(\mathcal{F}, \mathcal{X})$, that is AC-total on ground terms.*

Example 3. Rings. With $+, * \in \mathcal{F}_{AC}$ and $* \succ_{\mathcal{F}} I \succ_{\mathcal{F}} + \succ_{\mathcal{F}} 0$, the ordering orients (and hence proves termination of) the following TRS:

$$
\begin{aligned}
x + 0 &\to x \\
x + I(x) &\to 0 \\
I(0) &\to 0 \\
I(I(x)) &\to x \\
I(x + y) &\to I(x) + I(y) \\
x * (y + z) &\to (x * y) + (x * z) \\
x * 0 &\to 0 \\
x * I(y) &\to I(x * y)
\end{aligned}
$$

Example 4. With $+, * \in \mathcal{F}_{AC}$ and $* \succ_{\mathcal{F}} + \succ_{\mathcal{F}} s \succ_{\mathcal{F}} 0$, the ordering orients (and hence proves termination of) the following TRS:

$$
\begin{aligned}
x + 0 &\to x \\
x + s(y) &\to s(x + y) \\
x * 0 &\to 0 \\
x * s(y) &\to x * y + x \\
x * (y + z) &\to (x * y) + (x * z)
\end{aligned}
$$

5 Partial precedences

First, in order to deal with partial precedences, we now weaken the multiset extension of the ordering when applied to the sets *NoSmallHead(s)* and *NoSmallHead(t)*, since otherwise we cannot ensure incrementality.

Definition 6. *Let $\succ$ be an ordering on terms and let $\succ_{\mathcal{F}}$ be a (partial) precedence. The multiset extension of $\succ$ wrt. an AC-symbol f in the precedence $\succ_{\mathcal{F}}$, denoted by $\gg_f$, is defined as the smallest transitive relation containing*

$$
X \cup \{s\} \gg_f Y \cup \{t_1, \ldots, t_n\} \text{ if } \begin{cases} X =_{AC} Y \text{ and } s \succ t_i \text{ and} \\ \text{if } top(s) \not\succ_{\mathcal{F}} f \text{ then } top(s) \succeq_{\mathcal{F}} top(t_i) \\ \text{for all } i \in \{1 \ldots n\} \end{cases}
$$

Now we adapt the set *EmbSmall(s)* to allow embeddings through symbols not bigger than the head.

Definition 7. *Let s be a term of the form $f(s_1, \ldots, s_n)$ with $f \in \mathcal{F}_{AC}$. The set of terms embedded in s through an argument headed by a non-big symbol, denoted by EmbNoBig(s), is defined as*

$$
\{f(s_1, \ldots, tf_f(v_j), \ldots, s_n) \mid s_i = h(v_1, \ldots, v_r) \wedge h \not\succ_{\mathcal{F}} f \wedge j \in \{1 \ldots r\}\}
$$

Definition 8. *Let s and t be terms in $\mathcal{T}(\mathcal{F}, \mathcal{X})$. Then $s = f(s_1, \ldots, s_n) \succ_p g(t_1, \ldots, t_m) = t$ if and only if*

1. *$s_i \succeq_p t$ for some $i \in \{1 \ldots n\}$, or*

2. $f \succ_{\mathcal{F}} g$ and $s \succ_p t_i$ for all $i \in \{1 \ldots n\}$, or
3. $f = g \notin \mathcal{F}_{AC}$ and $\langle s_1, \ldots, s_n \rangle (\succ_p)_{lex} \langle t_1, \ldots, t_n \rangle$ and $s \succ_p t_i$ for all $i \in \{1 \ldots n\}$, or
4. $f = g \in \mathcal{F}_{AC}$ and there is some $s' \in EmbNoBig(s)$ s.t. $s' \succeq_p t$, or
5. $f = g \in \mathcal{F}_{AC}$ and $s \succ_p t'$ for all $t' \in EmbNoBig(t)$ and $NoSmallHead(s) \succeq_{pf} NoSmallHead(t)$ and either

   (a) $BigHead(s) \gg_p BigHead(t)$ or
   (b) $\#(s) > \#(t)$ or
   (c) $\#(s) \geq \#(t)$ and $\{s_1, \ldots, s_n\} \gg_p \{t_1, \ldots, t_m\}$.

The reason to ask for $NoSmallHead(s) \succeq_{pf} NoSmallHead(t)$, instead of using simply $\succeq_p$ is that if, by extending the precedence, an argument t' of t headed by a symbol incomparable with f becomes a term headed by a big symbol then the argument in s that takes care of t' becomes a term headed by a big symbol as well; and if, by extending the precedence, an argument s' of s headed by a symbol incomparable with f becomes a term headed by a small symbol, then all arguments in t taken care of by s' become terms headed by a small symbol as well.

Lemma 3. *Let s and t be terms. If the precedence is total then $s \succ_v t$ if and only if $s \succ_p t$.*

Proof. The result is trivial since both definitions coincide when applied with a total precedence. Note that if the precedence is total then we have $EmbNoBig(s) = EmbSmall(s)$ and that for all $s' \in NoSmallHead(f(s_1, \ldots, s_n))$ either $top(s') \succ_{\mathcal{F}} f$ or s' is a variable, which implies that $\gg_{pf}$ and $\gg_p$ coincide. $\square$

Corollary 1. *Let s and t be ground terms. If the precedence is total then $s \succ t$ if and only if $s \succ_p t$.*

The proof of the following theorem is given in the next section.

Theorem 3. *$\succ_p$ is an AC-compatible simplification ordering on $\mathcal{T}(\mathcal{F}, \mathcal{X})$, AC-total on ground terms and incremental wrt. the precedence.*

Example 5. Let f be an AC-symbol.

1. With any precedence we have $s = f(g(g(x)), x) \succ_p f(g(x), g(x)) = t$ by case 5c, since $s \succ_p t' = f(g(x), x) \in EmbNoBig(t)$, by case 4, and $NoSmallHead(s) = \{g(g(x)), x\} \succeq_{pf} \{g(x), g(x)\} = NoSmallHead(t)$, and $\#(s) = 1 + x \geq 2 = \#(t)$ and $\{g(g(x)), x\} \succeq \{g(x), g(x)\}$.
2. With precedence $g \succ_{\mathcal{F}} h$ we have $s = f(x, x, g(x)) \succ_p f(x, h(x)) = t$ by case 5b, since $s \succ_p t' = f(x, x) \in EmbNoBig(t)$, by case 5b, and $NoSmallHead(s) = \{x, x, g(x)\} \succeq_{pf} \{x, h(x)\} = NoSmallHead(t)$, and $\#(s) = 2x + 1 > x + 1 = \#(t)$ (note that x is a positive integer).

Example 6. Milners's nondeterministic machines. With $+ \in \mathcal{F}_{AC}$ and $T \succ_{\mathcal{F}} +$ and $L \succ_{\mathcal{F}} +$, the ordering orients (and hence proves termination of) the following TRS. Note that the decision about the precedence relation between T and $+$ is not needed until the last rule.

$$
\begin{aligned}
0 + x &\rightarrow x \\
x + x &\rightarrow x \\
L(T(x)) &\rightarrow L(x) \\
L(T(y) + x) &\rightarrow L(x + y) + L(y) \\
T(T(x)) &\rightarrow T(x) \\
T(x) + x &\rightarrow T(x) \\
T(x + y) + x &\rightarrow T(x + y) \\
T(T(y) + x) &\rightarrow T(x + y) + T(y)
\end{aligned}
$$

6 The Properties of the Ordering

Here we include the properties of the ordering $\succ_p$. Most proofs (not in full detail) are provided for the parts that are different from the ones for normal RPO. All proofs are quite simple, except the one for stability under substitutions, which involves some technical problems caused by the arguments of t that are variables instantiated by terms headed by a non-big symbol. Detailed proofs can be found in [Rub98][1].

Property 1. If $s =_{AC} t$ then $NoSmallHead(s) =_{AC} NoSmallHead(t)$, $BigHead(s) =_{AC} BigHead(t)$ and $\#(s) = \#(t)$.

Lemma 4. $\succ_p$ *is AC-compatible.*

Proof. Since after flattening AC-equal terms are equal up to permutation of arguments of AC-symbols, we have to prove the compatibility of $\succ_p$ wrt. this permutative equality, which we also call $=_{AC}$. Then $s' =_{AC} s \succ_p t =_{AC} t'$ implies $s' \succ_p t'$. We proceed by induction on $|s| + |t|$ and case analysis on the proof of $s \succ_p t$. Here we only include the case in which $s = f(s_1, \ldots, s_n) \succ_p f(t_1, \ldots, t_m) = t$ by case 5.

Then $f \in \mathcal{F}_{AC}$ and $s \succ_p u$ for all $u \in EmbNoBig(t)$ and $NoSmallHead(s) \succcurlyeq_{pf} NoSmallHead(t)$. Since $EmbNoBig(t) =_{AC} EmbNoBig(t')$, $NoSmallHead(s) =_{AC} NoSmallHead(s')$ and $NoSmallHead(t) =_{AC} NoSmallHead(t')$, by induction hypothesis we have $s' \succ_p u'$ for all $u' \in EmbNoBig(t')$ and $NoSmallHead(s') \succcurlyeq_{pf} NoSmallHead(t')$.

Finally, if $s \succ_p t$ by case 5a then, since $BigHead(s') =_{AC} BigHead(s) \gg_p BigHead(t) =_{AC} BigHead(t')$, by induction hypothesis we have $BigHead(s') \gg_p BigHead(t')$, and hence $s' \succ_p t'$ by case 5a. If $s \succ_p t$ by case 5b, then, since $\#(s') = \#(s) > \#(t) = \#(t')$, we have $s' \succ_p t'$ by case 5b. Otherwise, $s \succ_p t$ by case 5c, and $\#(s') = \#(s) \geq \#(t) = \#(t')$ and $\{s_1', \ldots, s_n'\} =_{AC} \{s_1, \ldots, s_n\} \gg_p \{t_1, \ldots, t_n\} =_{AC} \{t_1', \ldots, t_n'\}$, which implies by induction hypothesis that $\{s_1', \ldots, s_n'\} \gg_p \{t_1', \ldots, t_n'\}$, and hence $s' \succ_p t'$ by case 5c. □

---

[1] Available in http://www-lsi.upc.es/~albert/papers/aclong.ps.gz

Lemma 5. *Let f be an AC-symbol. If $n > m$ and $1 \leq i_1 < \ldots < i_m \leq m$ then $f(s_1, \ldots, s_n) \succ_p f(s_{i_1}, \ldots, s_{i_m})$.*

Lemma 6. *Let f be an AC-symbol. If $n > m$ and $1 \leq i_1 < \ldots < i_m \leq m$ then $s \succeq_p f(t_1, \ldots, t_n)$ implies $s \succ_p f(t_{i_1}, \ldots, t_{i_m})$.*

Lemma 7. *If $s \succeq_p t$ then $s \succ_p t_i$, for all t_i argument of t.*

Proof. By induction on the $|s| + |t|$. Let s be $f(s_1, \ldots, s_n)$ and t be $g(t_1, \ldots, t_m)$. If $s =_{AC} t$ then for every t_i there is some s_j s.t. $s_j =_{AC} t_i$ and therefore by case 1 we have $s \succ_p t_i$. Otherwise $s \succ_p t$ and we distinguish several cases according to the definition. We again only consider the case in which $s \succ_p t$ by case 5. Then there are three cases:

1. If $f \succ_{\mathcal{F}} top(t_i)$ and t_i is a constant, it holds trivially, by case 2.
2. If $f \succ_{\mathcal{F}} top(t_i)$ and $t_i = h(v_1, \ldots, v_r)$ with $r > 0$, then for all $j \in \{1 \ldots r\}$ there is some $t' \in EmbNoBig(t)$ s.t. $t' = f(t_1, \ldots, tf_f(v_j), \ldots, t_m)$ and $s \succ_p t'$. Now, for all v_j either $top(v_j) \neq f$ and $tf_f(v_j) = v_j$ and hence by induction hypothesis $s \succ_p v_j$, or $top(v_j) = f$, and then by lemma 6 we have $s \succ_p f(tf_f(v_j)) = v_j$. Therefore, since $f \succ_{\mathcal{F}} h$, by case 2, we have $s \succ_p t_i$.
3. If $f \not\succ_{\mathcal{F}} top(t_i)$ then, since $NoSmallHead(s) \succeq\!\!\!\succ_{pf} NoSmallHead(t)$ there is some s_j (with $f \not\succ_{\mathcal{F}} top(s_j)$), s.t. $s_j \succeq t_i$ and hence $s \succ_p t_i$ by case 1. □

Corollary 2. $\succ_p$ *fulfils the subterm property.*

Property 2. Let s and t be terms. Then $NoSmallHead(s) \succeq\!\!\!\succ_{pf} NoSmallHead(t)$ implies $BigHead(s) \succeq\!\!\!\succ_p BigHead(t)$.

Lemma 8. $\succ_p$ *is transitive.*

Proof. We prove $s \succ_p t$ and $t \succ_p u$ implies $s \succ_p u$ by induction on $|s| + |t| + |u|$ and case analysis on the definition. Let s be $f_1(s_1, \ldots, s_n)$, t be $f_2(t_1, \ldots, t_m)$ and u be $f_3(u_1, \ldots, u_p)$. We just consider here some of the cases.

- If $s \succ_p t$ by case 4 and $t \succ_p u$ by case 4 or 5, then there is some $s' \in EmbNoBig(s)$ s.t. $s' \succeq_p t$, and by induction hypothesis and AC-compatibility $s' \succ_p u$, which implies $s \succ_p u$ by case 4.
- If $s \succ_p t$ by case 5 and $t \succ_p u$ by case 4 then then there is some $t' \in EmbNoBig(t)$ s.t. $t' \succeq_p u$, and since $s \succ_p t'$ for all $t' \in EmbNoBig(t)$, by induction hypothesis and AC-compatibility we have $s \succ_p u$.
- If $s \succ_p t$ by case 5 and $t \succ_p u$ by case 5 then $f_1 = f_2 = f_3 = f$ and by induction hypothesis and AC-compatibility we have $s \succ_p u'$ for all $u' \in EmbNoBig(u)$ and $NoSmallHead(s) \succeq\!\!\!\succ_{pf} NoSmallHead(u)$.
  Now if either $s \succ_p t$ or $t \succ_p u$ by case 5a then, by property 2, induction hypothesis and AC-compatibility we have $s \succ_p u$ by case 5a. Otherwise, if either $s \succ_p t$ or $t \succ_p u$ by case 5b then we have $s \succ_p u$ by case 5b. Otherwise, $s \succ_p t$ and $t \succ_p u$ by case 5c, and then $\{s_1, \ldots, s_n\} \succeq\!\!\!\succ_p \{t_1, \ldots, t_m\}$ and $\{t_1, \ldots, t_m\} \succeq\!\!\!\succ_p \{u_1, \ldots, u_p\}$, implies, by induction hypothesis and AC-compatibility $\{s_1, \ldots, s_n\} \succeq\!\!\!\succ_p \{u_1, \ldots, u_p\}$, and hence $s \succ_p u$ by case 5c. □

Lemma 9. $\succ_p$ *is irreflexive.*

Lemma 10. $\succ_p$ *is monotonic.*

Proof. If $s \succ_p t$ then $\overline{f(\ldots s \ldots)} \succ_p \overline{f(\ldots t \ldots)}$ for every flattened context $f(\ldots [] \ldots)$, by induction on $|\overline{f(\ldots s \ldots)}| + |\overline{f(\ldots t \ldots)}|$. Note that if the context is not flattened, we can flatten it and then apply the result.

We only consider the case in which f is in $\mathcal{F}_{AC}$. Then $cs = \overline{f(\ldots s \ldots)} = f(\ldots tf_f(s) \ldots)$ and $ct = \overline{f(\ldots t \ldots)} = f(\ldots tf_f(t) \ldots)$. By induction hypothesis, it is easy to show that $cs \succ_p t'$ for every $t' \in EmbNoBig(ct)$ obtained from some $u_i = h(v_1, \ldots, v_r)$ with $h \neq_{\mathcal{F}} f$ in the context. Now, let $s = g_1(s_1, \ldots, s_n)$ and $t = g_2(t_1, \ldots, t_m)$, there are several cases to be considered according to the proof of $s \succ_p t$ and the head symbols of s and t.

We analyze the case in which $g_1 = f$. Then $tf_f(s) = s_1 \ldots s_n$. We consider three cases.

1. $s \succ_p t$ by case 1. Then $s_i \succeq t$ for some s_i, and by induction hypothesis $f(\ldots tf_f(s_i) \ldots) \succeq f(\ldots tf_f(t) \ldots)$. By lemma 5 we have $f(\ldots s_1 \ldots s_n \ldots) \succeq f(\ldots s_i \ldots)$ and, since $tf_f(s_i) = s_i$, by induction hypothesis $f(\ldots s_i \ldots) \succeq f(\ldots tf_f(t) \ldots)$, and therefore, by transitivity and AC-compatibility, we obtain $f(\ldots s_1 \ldots s_n \ldots) \succ_p f(\ldots tf_f(t) \ldots)$.
2. $g_1 \succ_{\mathcal{F}} g_2$. Then $f \succ_p g_2$ and $tf_f(t) = t$. Since $s \succ_p t_i$ for all t_i, by induction hypothesis, we have $f(\ldots s_1 \ldots s_n \ldots) \succ_p f(\ldots tf_f(t_i) \ldots)$ for all t_i, and hence $f(\ldots s \ldots) \succ_p t'$ for all $t' \in EmbNoBig(f(\ldots, t, \ldots))$. Then $NoSmallHead(f(\ldots s_1 \ldots s_n \ldots)) \supseteq NoSmallHead(f(\ldots t \ldots))$ and, since $n \geq 2$, $\#(f(\ldots s_1 \ldots s_n \ldots)) > \#(f(\ldots t \ldots))$, which implies $f(\ldots s_1 \ldots s_n \ldots) \succ_p f(\ldots t \ldots)$ by case 5b.
3. $g_1 = g_2$ and $s \succ_p t$ by case 4 or by case 5. Then $tf_f(t) = t_1 \ldots t_m$ and it is easy to prove that $f(\ldots s_1 \ldots s_n \ldots) \succ_p f(\ldots t_1 \ldots t_m \ldots)$ holds by the same case as $s \succ_p t$.

The other two cases are $g_1 \not\succ_{\mathcal{F}} f$ and $g_1 \succ_{\mathcal{F}} f$. The first one is solved again by case 4 if $s \succ_p t$ by case 1, and by case 5c if $g_1 \succeq_{\mathcal{F}} g_2$. The second one is always solved by case 5a considering the cases $g_2 \not\succ_{\mathcal{F}} f$, $g_2 \succ_{\mathcal{F}} f$ and $g_2 = f$. $\quad\square$

Property 3. Let x be a variable and let s and t be terms.
If $NoSmallHead(s) \succeq_{pf} NoSmallHead(t)$ and $BigHead(s) \succ_p BigHead(t)$ then $BigHead(s) \cup X_s \succ_p BigHead(t) \cup X_t$, where X_s and X_t are respectively the multisets containing all x in $NoSmallHead(s)$ and $NoSmallHead(t)$.

Lemma 11. *Let f be in $\mathcal{F}_{AC}$. If $f(s_1, \ldots, x, \ldots, s_n) \succ_p f(t_1, \ldots, x, \ldots, t_m)$ then $f(s_1, \ldots\ldots, s_n) \succ_p f(t_1, \ldots\ldots, t_m)$.*

In fact if either n or m is equal to 1 then (due to the arity of f) we do not include the f on top for $f(s_1, \ldots\ldots, s_n)$ or $f(t_1, \ldots\ldots, t_m)$ and we just have s_1 or t_1.

The following lemma is used to prove stability under substitution, i.e. $s \succ_p t$ implies $s\sigma \succ_p t\sigma$. The proof of this lemma becomes rather technical due to the new embedded terms in $EmbNoBig(t\sigma)$ which come from an embedding trough an instantiated variable. Due to this problem we have to generalize the stability property in order to prove that if $s \succ_p t$ then $s\sigma \succ_p t'$ for every t' (equal to or) embedded in $t\sigma$ through the instantiated variables. On the other hand, we have first considered the simplest substitution σ.

Lemma 12. *Let σ be $\{x \mapsto q(y_1, \ldots, y_k)\}$ for some symbol q. If $s \succ_p t$ then $\overline{s\sigma} \succ_p \overline{t'}$ for all $t' \in M(t)$, where $M(t)$ is defined as*

- *if $top(t) = f \notin \mathcal{F}_{AC}$ or $q \succeq_{\mathcal{F}} f$ then $M(t) = \{t\sigma\}$, and*
- *if $t = f(t_1, \ldots, t_m)$ and $f \in \mathcal{F}_{AC}$ and $q \not\succeq_{\mathcal{F}} f$ then*
  $M(t) = \{f(w_1, \ldots w_m) \mid w_i \in \{q(y_1, \ldots, y_k), y_1, \ldots, y_k\} \text{ if } t_i = x, \text{ and}$
  $w_i = t_i\sigma \text{ otherwise, and } 1 \leq i \leq m\}$

Proof. Note that $t\sigma \in M(t)$, and therefore, this property implies $\overline{s\sigma} \succ_p \overline{t\sigma}$.

We prove that $s \succ_p t$ implies $\overline{s\sigma} \succ_p \overline{t'}$ for every s, t and $t' \in M(t)$, by induction on the triple $(|s|, |t|, |t'|)$ ordered lexicographically and case analysis on the proof of $s \succ_p t$. We only consider the case in which $s \succ_p t$ by case 5. First it is proved that for all $w' \in EmbNoBig(\overline{t'})$ we have $\overline{s} \succ_p w'$. If the embedding is made through an instantiated variable then since $w' \in M(t)$, it holds by induction hypothesis. Otherwise, there is some $t'' \in EmbNoBig(t)$ s.t. $w' \in M(t'')$, and since $s \succ_p t''$, by induction hypothesis, we have $s \succ_p w'$.

Now, using what we have proved above, we can prove that $\overline{s\sigma} \succ_p \overline{t'}$. There are three cases depending on the occurrences of the variable x as argument in s and t. If some $s_i = x$ and some $t_j = x$ then we can solve it using lemma 11 and induction hypothesis, and then monotonicity and either subterm, if $q \neq f$, or deletion property, otherwise. If all $t_i \neq x$ then by induction hypothesis, it is quite easy to see that $\overline{s\sigma} \succ_p \overline{t'}$ holds by the same case as $s \succ_p t$. If all $s_i \neq x$ for all $i \in \{1 \ldots n\}$ then we use the fact that in this case $NoSmallHead(s) \not\succeq_{pf} NoSmallHead(t)$ implies $BigHead(s) \not\succeq_p BigHead(t)$, since some term should take care of the x's in t. Finally, by induction hypothesis, and using property 3 we can show that $\overline{s\sigma} \succ_p \overline{t'}$ holds by case 5a. □

Lemma 13. $\succ_p$ *is stable under substitution.*

Proof. The proof of $s \succ_p t$ implies $\overline{s\sigma} \succ_p \overline{t\sigma}$ for every substitution σ, is done by induction on the $|\sigma|$ defined as the multiset $\{|w| \mid (x \mapsto w) \in \sigma\}$ and compared by the multiset extension of $>$, and using lemma 12. □

Lemma 14. $\succ_p$ *is incremental.*

Proof. We prove that if $s \succ_p t$ with the precedence $\succ_{\mathcal{F}}$ then $s \succ_p t$ with the precedence $\succ_{\mathcal{F}} \cup \{f' \succ g'\}$ where f' and g' are symbols in $\mathcal{F}$ not related by $\succ_{\mathcal{F}}$. We proceed by induction on $|s| + |t|$ and case analysis on the proof of $s \succ_p t$ with the precedence $\succ_{\mathcal{F}}$.

We only consider the case in which $s = f(s_1, \ldots, s_n) \succ_p f(t_1, \ldots, t_m) = t$ by case 5. Then we have $s \succ_p t'$ for all $t' \in EmbNoBig(t)$ and $NoSmallHead(s) \not\succeq_{pf}$

$NoSmallHead(t)$. First, since $EmbNoBig(t)$ wrt. $\succ_{\mathcal{F}}$ is included in $EmbNoBig(t)$ wrt. $\succ_{\mathcal{F}} \cup \{f' \succ g'\}$, by induction hypothesis $s \succ_p t'$ for all $t' \in EmbNoBig(t)$ wrt. $\succ_{\mathcal{F}} \cup \{f' \succ g'\}$. Second, by definition of $\gg_{pf}$ all terms s' in $NoSmallHead(s)$ s.t. $top(s') \not\succ_{\mathcal{F}} f$ only takes care of terms t' in $NoSmallHead(t)$ s.t. $top(s') \succeq_{\mathcal{F}} top(t')$, then if s' is not in $NoSmallHead(s)$ wrt. $\succ_{\mathcal{F}} \cup \{f' \succ g'\}$ then none of the t' is in $NoSmallHead(t)$ wrt. $\succ_{\mathcal{F}} \cup \{f' \succ g'\}$. Therefore, by induction hypothesis, $NoSmallHead(s) \gg_{pf} NoSmallHead(t)$ wrt. $\succ_{\mathcal{F}} \cup \{f' \succ g'\}$.

Now if we have applied case 5b it holds trivially and if we have applied case 5c then it holds by by induction hypothesis. If we have applied case 5a then we have $BigHead(s) \gg_p BigHead(t)$ wrt. $\succ_{\mathcal{F}}$. Since $NoSmallHead(s) \gg_{pf} NoSmallHead(t)$, wrt. $\succ_{\mathcal{F}}$, by definition of $\gg_{pf}$, for every term t' in $NoSmallHead(t)$ wrt. $\succ_{\mathcal{F}}$ s.t. $t' \notin BigHead(t)$ wrt. $\succ_{\mathcal{F}}$ and $t' \in BigHead(t)$ wrt. $\succ_{\mathcal{F}} \cup \{f' \succ g'\}$ there is some s' in $NoSmallHead(s)$ wrt. $\succ_{\mathcal{F}}$ s.t. $s' \in BigHead(s)$ wrt. $\succ_{\mathcal{F}} \cup \{f' \succ g'\}$ that takes care of t' (note that for every t' in $NoSmallHead(t)$ wrt. $\succ_{\mathcal{F}}$ there is some s' which takes care of it and either $top(s') \succ_{\mathcal{F}} f$ or $top(s') \succeq_{\mathcal{F}} top(t')$). Therefore it holds by case 5a, since $BigHead(s) \gg_p BigHead(t)$ wrt. $\succ_{\mathcal{F}} \cup \{f' \succ g'\}$. □

7 Conclusions

We have presented the first fully syntactic AC-compatible *recursive path ordering* (RPO). The ordering is AC-total on ground terms, and defined uniformly for both ground and non-ground terms, as well as for partial precedences, being the first incremental one (note that due to this, we can allow as well signature extensions without any restriction).

Our ordering does not coincide (even for ground terms and total precedences) with any of the ones given in [KS97], [Rub97] and [KS98] (with *fcount* abstraction). With the precedence $h \succ_{\mathcal{F}} f \succ_{\mathcal{F}} g \succ_{\mathcal{F}} a$, and $f \in \mathcal{F}_{AC}$ the terms $f(h(a), g(a))$ and $f(g(h(a)), a)$ are compared in a different way (only in our case the first one is greater than the second one). The reason is that in our approach the arguments headed by big symbols are more important than in the others. However, it could be the case that by taking another abstraction function for [KS98], the orderings coincide. On the other hand we have also found another syntactic definition in which the number of arguments of an AC-symbol is more important than its arguments headed by big symbols, which we believe to coincide with the orderings in [Rub97] and [KS98] (with *fcount* abstraction). A weakness of this new definition is that it is only monotonic for ground terms, although, in fact, this is not a problem for practical applications.

As simple improvements to the presented ordering, we can allow the user to use multiset status for non-AC symbols as well as defining equivalences between the non-AC symbols in the precedence.

Regarding efficiency of implementation, it is easy to show that some of the recursive comparisons in case 5 can be avoided. Being more precise, when comparing terms s and t, some of the recursive comparisons $s \succ_p t'$ for every $t' \in EmbNoBig(t)$ are not necessary since they follow from other ones, by monotonicity and transitivity. Currently, we are looking for properties, in the ground case and total precedences, which, by comparing the arguments of t, allow us to choose a single term u in $EmbNoBig(t)$ such that $s \succ u$ ensures $s \succ t'$ for every $t' \in EmbSmall(t)$. Then, by extending these properties to the general case, we would further reduce the amount of recursive comparisons to be performed.

As a future development, due to its simplicity and, mainly, the fact that it is not interpretation-based, it opens the door to finding practically feasible ordering constraint solvers for the AC-case [CNR95].

References

[Bac92]   L. Bachmair. Associative-commutative reduction orderings. *Information Processing Letters*, 43:21–27, 1992.

[BCL87]   A. Ben-Cherifa and P. Lescanne. Termination of rewriting systems by polynomial interpretations and its implementation. *Science of Computer Programming*, 9:137–160, 1987.

[BP85]   L. Bachmair and D. A. Plaisted. Termination orderings for associative-commutative rewriting systems. *J. of Symbolic Comput.*, 1:329–349, 1985.

[CNR95]   H. Comon, R. Nieuwenhuis, and A. Rubio. Orderings, AC-Theories and Symbolic Constraint Solving. In *10th Annual IEEE Symp. on Logic in Computer Science*, pp. 375–385, San Diego, 1995. IEEE Comput. Society Press.

[Der82]   N. Dershowitz. Orderings for term-rewriting systems. *Theoretical Computer Science*, 17(3):279–301, 1982.

[DHJP83]   N. Dershowitz, J. Hsiang, A. Josephson, and D. Plaisted. Associative-commutative rewriting. In *Int. Joint Conf. on Artificial Inteligence*, pp. 940–944, 1983.

[DP93]   C. Delor and L. Puel. Extension of the associative path ordering to a chain of associative commutative symbols. In *5th Int. Conf. on Rewriting Techniques and Applications*, LNCS 690, pp. 389–404, Montreal, 1993. Springer-Verlag.

[GL86]   I. Gnaedig and P. Lescanne. Proving termination of associative-commutative rewrite systems by rewriting. In *8th International Conference on Automated Deduction*, LNCS 230, pp. 52–61, Oxford, England, 1986. Springer-Verlag.

[KS97]   D. Kapur and G. Sivakumar. A total, ground path ordering for proving termination of ac-rewrite systems. In *8th International Conference on Rewriting Techniques and Applications*, LNCS 1232, Sitges, Spain, 1997.

[KS98]   D. Kapur and G. Sivakumar. A recursive path ordering for proving associative-commutative termination, 1998.

[KSZ90]   D. Kapur, G. Sivakumar and H. Zhang. A new method for proving termination of ac-rewrite systems. In *Conf. Found. of Soft. Technology and Theor. Comput. Science*, LNCS 472, pp. 134–148, New Delhi, 1990. Springer-Verlag.

[NR91]   P. Narendran and M. Rusinowitch. Any ground associative commutative theory has a finite canonical system. In *4th Int. Conf. on Rewriting Techniques and Applications*, LNCS 488, pp. 423–434, Como, 1991. Springer-Verlag.

[RN95]   A. Rubio and R. Nieuwenhuis. A total AC-compatible ordering based on RPO. *Theoretical Computer Science*, 142(2):209–227, 1995.

[Rub97]   A. Rubio. A total AC-compatible ordering with RPO scheme. Technical Report UPC-LSI-97, Univ. Polit. Catalunya, September 1997.

[Rub98]   A. Rubio. A fully syntactic AC-RPO. Technical Report UPC-LSI-98, Univ. Polit. Catalunya, December 1998.

Theory Path Orderings[*]

Jürgen Stuber

Max-Planck-Institut für Informatik
Im Stadtwald, D-66123 Saarbrücken, Germany
Tel: +49-681-9325-228, fax: +49-681-9325-299
juergen@mpi-sb.mpg.de
http://www.mpi-sb.mpg.de/~juergen/

Abstract. We introduce the notion of a theory path ordering (TPO), which simplifies the construction of term orderings for superposition theorem proving in algebraic theories. To achieve refutational completeness of such calculi we need total, E-compatible and E-antisymmetric simplification quasi-orderings. The construction of a TPO takes as its ingredients a status function for interpreted function symbols and a precedence that makes the interpreted function symbols minimal. The properties of the ordering then follow from related properties of the status function. Theory path orderings generalize associative path orderings.

1 Introduction

Refutationally complete superposition calculi for algebraic theories require simplification orderings which are total on ground terms, and which obey additional restrictions imposed by the algebraic theories. In particular, theorem proving modulo some equational theory E requires that the term ordering is E-compatible. The most important theory in practice is AC (Bachmair and Plaisted 1985, Delor and Puel 1993, Rubio and Nieuwenhuis 1995, Kapur and Sivakumar 1997, Baader 1997). Additionally, presenting the theory by a term rewriting system modulo E (Bachmair, Ganzinger and Stuber 1995, Marché 1996, Stuber 1998a, Stuber 1998b) requires that the ordering orients the rules in this system in the right direction. The same applies to rules in symmetrizations which arise from the interaction of nontheory and theory equations. In some cases, for instance for modules and algebras over a fixed ring, these requirements cannot be met by combining orderings known from the literature and necessitate the construction of new orderings (Stuber 1998a).

We present a general construction of such an ordering. We distinguish between *free* and *interpreted* function symbols, where interpreted function symbols have a special meaning in the given theory. In particular, the function symbols in the set of equations E are contained in the interpreted function symbols. For free function symbols the ordering is defined analogously to the lexicographic path

---

[*] Part of this work has been supported by Deutsche Forschungsgemeinschaft under grant GA-261/7-1.

ordering. That is, we assume as given a precedence on free function symbols, and use lexicographic status to compare terms with the same free root symbol. Contexts consisting of interpreted function symbols will be handled as a whole by a status function, which is the main ingredient of our ordering. Such a status function can be presented as a function that extends an ordering on constants to an ordering on terms over constants and interpreted function symbols. Finally, the interaction between free and interpreted function symbols is handled by extending the precedence to a quasi-precedence where the interpreted function symbols form the minimal equivalence class.

The main problem in the construction of E-compatible reduction orderings is to obtain compatibility with contexts (monotonicity). Considering contexts of depth greater than one in a single application of the status function essentially amounts to flattening the context. To obtain compatibility with contexts we let free function symbols have a greater influence on the ordering than interpreted function symbols. Formally this is reflected by interpreted function symbols being minimal in the precedence, and by requiring the status to have certain *multiset properties*, which ensure that the status function is compatible with the path ordering also within a context of interpreted function symbols.

Our construction can be viewed as a generalization of the associative path ordering (Bachmair and Plaisted 1985). In view of this we call our ordering *theory path ordering* (TPO). On the other hand our construction is related to the general path ordering (Dershowitz and Hoot 1995, Geser 1996). We use Geser's approach to prove that the ordering has all the properties of a simplification quasi-ordering except compatibility with contexts. There are also some overlaps of the properties we require of an extension function with those required by Baader (1997). He combines an E_1- and an E_2-compatible reduction ordering into an $E_1 \cup E_2$-compatible one. However, his combination ordering compares terms first by the maximal number of alternations between E_1 and E_2. This makes it unsuitable for our purposes, since we need that free function symbols dominate the ordering.

We use quasi-orderings throughout our presentation, since these integrate an ordering and an equivalence relation in a natural way. We split the notion of a quasi-ordering being total up to E into being E-antisymmetric and total. E-antisymmetry has the advantage that it is also meaningful for partial orderings, which is useful in the nonground case. However, here we consider only the ground case.

In Section 2 we state some preliminaries, in particular about quasi-orderings. Section 3 presents our adaption of the general path ordering of Geser (1996). In Section 4 we define the TPO and the notion of a TPO-status, and we show that the TPO has the desired properties. In Section 5 we show that a TPO-status can be given in a natural way as a function that extends an ordering on constants to an ordering on terms. In Section 6 we give examples of theory path orderings.

2 Preliminaries

We assume the reader is familiar with term rewriting and the properties of orderings arising in that context (Dershowitz and Jouannaud 1990, Baader and Nipkow 1998). Following the latter we call a binary relation R on terms *compatible with contexts* if $s \, R \, t$ implies $u[s] \, R \, u[t]$ for any context u. A binary relation R is *compatible* with an equivalence relation $\sim$ if $x \sim x' \, R \, y' \sim y$ implies $x \, R \, y$. An equation is called *collapse-free* if neither side is a variable. A set of equations is called *collapse-free* if all its equations are collapse-free. We write

$$\mathrm{args}_{\mathrm{AC}(f)}(t) = \begin{cases} \mathrm{args}_{\mathrm{AC}(f)}(t_1) \cup \mathrm{args}_{\mathrm{AC}(f)}(t_2) & \text{for } t = f(t_1, t_2), \\ \{t\} & \text{otherwise,} \end{cases}$$

which is the multiset of arguments of f after flattening.

A *quasi-ordering* is a binary relation that is reflexive and transitive. Each quasi-ordering $\succsim$ can be split into its *strict part* $(\succ) = (\succsim) \setminus (\precsim)$ and its *equivalence kernel* $(\sim) = (\succsim) \cap (\precsim) = (\succsim) \setminus (\succ)$. In the context of some quasi-ordering $\succsim$ we always use $\succ$ for its strict part and $\sim$ for its equivalence kernel. On the other hand, if $\sim$ is an equivalence relation and $\succ$ is a strict partial ordering that is compatible with $\sim$ then $(\succsim) = (\sim) \cup (\succ)$ is a quasi-ordering. A quasi-ordering $\succsim$ is *total* on a set S if $x \succsim y$ or $y \succsim x$ for any two elements of S. It is well-founded if $\succ$ is well-founded. We say a quasi-ordering $\succsim$ on terms is *E-compatible* if $\succsim$ is compatible with $=_E$. It is *E-antisymmetric* if $s \sim t$ implies $s =_E t$ for all terms s and t. A quasi-ordering may also be viewed as a partial ordering on $\sim$-equivalence classes. Via this construction the multiset extension of partial orderings induces the multiset extension $\succsim^{mul}(\succsim)$ of a quasi-ordering $\succsim$. Similarly, quasi-orderings can be combined lexicographically. The lexicographic extension of $\succsim$ to tuples of fixed length is denoted by $\succsim^{lex}(\succsim)$. Multiset extension and the lexicographic combination preserve well-foundedness and totality. For a given quasi-ordering $\succsim$ we will also use subscripts to denote their multiset or lexicographic extension, e.g., $\succsim_{mul}$. Note that in this case $\succ_{mul}$ denotes $\succ^{mul}(\succsim)$, that is, the strict part of the multiset extension of the (non-strict) quasi-ordering. The *subterm ordering* $\rhd$ is defined by $s \rhd t$ if and only if t is a proper subterm of s. Since $\rhd$ is well-founded, $\rhd_{mul} = \succ^{mul}(\unrhd)$ is also well-founded.

In general we will use $\succsim$ both to denote quasi-orderings and quasi-ordering functionals, because then $\succ$ and $\sim$ allow easy access to the strict part and equivalence kernel of the result of the functional applied to some quasi-ordering. We will use superscripts on quasi-ordering functionals to distinguish them from quasi-orderings, where we use no superscripts.

We say that a quasi-ordering is *strictly compatible with contexts* if both $\succsim$ and $\succ$ are compatible with contexts. A *reduction quasi-ordering* is a quasi-ordering on terms that is well-founded, strictly compatible with contexts and strictly closed under substitutions. If in addition it has the subterm property, it is called a *simplification quasi-ordering*. The strict part of a simplification quasi-ordering is a simplification ordering. If $R \subseteq (\succ)$ and $E \subseteq (\sim)$ for a term rewriting system R modulo E and a reduction quasi-ordering $\succsim$ then R is terminating modulo E.

3 General path orderings

We use the general path ordering (GPO) of Geser (1996) as a starting point for constructing quasi-orderings modulo E on ground terms. However, Geser's method of proving compatibility with contexts from a status being "prepared for contexts" cannot cope with flattening, hence it is not applicable in this setting. Consequently, we weaken the notion of a status to that of a prestatus, which need not be prepared for contexts. By inspection of Geser's proofs one sees that his proofs use only the properties of a prestatus, with the exception of the proofs of reflexivity and of compatibility with contexts. Fortunately reflexivity can easily be proved from the properties of a prestatus alone. Compatibility with contexts will be the main topic of Section 4. Beyond the work of Geser we show that natural conditions on a prestatus imply that the induced GPO is total and E-antisymmetric.

Given two terms s and t we let $\mathrm{fin}(s,t)$ be the set $\{\langle s',t'\rangle \mid \{s,t\} \rhd_{mul} \{s',t'\}\}$. That is, a pair $\langle s',t'\rangle$ is in $\mathrm{fin}(s,t)$ if either both terms are proper subterms of s or t, or if one term is equal to s or t and the other is a proper subterm of the other. For instance,

$$\mathrm{fin}(f(a),g(b)) = \{\langle f(a),b\rangle, \langle b,f(a)\rangle, \langle a,g(b)\rangle, \langle g(b),a\rangle, \langle a,a\rangle, \langle a,b\rangle, \langle b,a\rangle, \langle b,b\rangle\}.$$

A *quasi-ordering functional* is a function $\succsim^{st}$ which maps any quasi-ordering $\succsim$ on ground terms to a quasi-ordering $\succsim^{st}(\succsim)$ on ground terms. A quasi-ordering functional $\succsim^{st}$ is *subterm founded* on a set of pairs of terms S if $s \succsim^{st}(\succsim) t$ is equivalent to $s \succsim^{st}(\succsim \cap \mathrm{fin}(s,t)) t$ for any quasi-ordering $\succsim$ and any pair $\langle s,t\rangle$ in S. We say that $\succsim^{st}$ is *subterm founded* if it is subterm founded on the set of all pairs of ground terms. Subterm foundedness ensures that the GPO is well-defined and allows to use induction to prove its properties. A quasi-ordering functional $\succsim^{st}$ *decreases infinite derivations* if for every infinite derivation

$$s_1 \succ^{st}(\succsim) s_2 \succ^{st}(\succsim) \ldots$$

there exists an infinite derivation $t_1 \succ t_2 \succ \ldots$ such that $s_i \rhd t_1$ for some $i \geq 1$. This property ensures well-foundedness of the resulting GPO. A quasi-ordering functional $\succsim^{st}$ is called a *prestatus* if (i) $\succsim^{st}$ is subterm founded, and (ii) $\succsim^{st}$ decreases infinite derivations. The *general path ordering* $\succsim_{gpo} = \succsim^{gpo}(\succsim^{st})$ induced by a prestatus $\succsim^{st}$ is the smallest quasi-ordering such that $s = f(s_1,\ldots,s_m) \succsim_{gpo} g(t_1,\ldots,t_n) = t$ if

1. $s_i \succsim_{gpo} t$ for some $i = 1,\ldots,m$, or
2. $s \succ_{gpo} t_j$ for each $j = 1,\ldots,n$ and $s \succ^{st}(\succsim_{gpo}) t$.

As an example consider a language consisting of a binary function symbol f and a constant a, and define $\succsim^{st}$ by $s \succsim^{st}(\succsim) t$ if and only if either

1. $s = a$, or
2. $s = f(s_1,s_2)$, $t = f(t_1,t_2)$ and $\mathrm{args}_{\mathrm{AC}(f)}(s) \succsim_{mul} \mathrm{args}_{\mathrm{AC}(f)}(t)$.

Then $\succsim^{st}$ preserves reflexivity and transitivity, since the multiset extension does. It is subterm founded, since it invokes its argument ordering only on multisets of its subterms, and it decreases infinite derivations, since the multiset extension does. By inspecting the definitions one sees that $\succsim^{gpo}(\succsim^{st})$ is the recursive path ordering with respect to the precedence $a \succ_p f$ and with multiset status for f.

Lemma 1 *(Geser 1996) Let $\succsim^{st}$ be a prestatus.*

1. *If $s \succsim_{gpo} t$ and $t \rhd t'$ then $s \succ_{gpo} t'$.*
2. *If $s \rhd s'$ and $s' \succsim_{gpo} t$ then $s \succ_{gpo} t$.*
3. *$\succsim_{gpo}$ is transitive and well-founded.*

Geser (1996) proof of transitivity and the proofs of the lemmas below use the following scheme for proving that $\succsim_{gpo}$ has some property P, based on subterm foundedness and preservation of P by $\succsim^{st}$:

- Consider some instance $P[t_1, \ldots, t_n]$ of P. First prove $P[t_1, \ldots, t_n]$ for the case where some atom $t_i \succsim t_j$ in P becomes true by case 1 of the definition of $\succsim_{gpo}$. Lemma 1(1+2) can often be used in this case. It remains to consider only case 2, where the prestatus is used.
- Let fin be the union of all sets $\text{fin}(t_i, t_j)$ where $t_i \succsim t_j$ is an atom in P.
- Restrict $\succsim_{gpo}$ to fin, that is, consider $(\succsim_{gpo}) \cap \text{fin}$.
- Extend $(\succsim_{gpo}) \cap \text{fin}$ to some $\succsim_P$ that satisfies P, and that coincides with $\succsim_{gpo}$ on fin. That is, $(\succsim_{gpo}) \cap \text{fin} = (\succsim_P) \cap \text{fin}$.
- Then by subterm foundedness

$$t_i \succsim^{st}(\succsim_{gpo}) \, t_j \quad \text{if and only if} \quad t_i \succsim^{st}(\succsim_{gpo} \cap \text{fin}) \, t_j$$
$$\text{if and only if} \quad t_i \succsim^{st}(\succsim_P \cap \text{fin}) \, t_j$$
$$\text{if and only if} \quad t_i \succsim^{st}(\succsim_P) \, t_j.$$

- Use that $\succsim^{st}$ preserves P to conclude that $P[t_1, \ldots, t_n]$ holds for $\succsim^{st}(\succsim_{gpo})$ and hence for $\succsim_{gpo}$.

Lemma 2 *Let $\succsim^{st}$ be a prestatus. Then $\succsim_{gpo}$ is reflexive.*

Lemma 3 *Let $\succsim^{st}$ be a prestatus that preserves totality. Then $\succsim_{gpo}$ is total.*

Our example prestatus preserves totality. Either one of the terms is a, or both have f at their root and we may use that the multiset extension preserves totality.

Lemma 4 *Let $\succsim^{st}$ be a prestatus that preserves E-antisymmetry. Then $\succsim_{gpo}$ is E-antisymmetric.*

For our example let $E = \text{AC}(f) = \text{AC}$. We claim that $\succsim^{st}$ preserves AC-antisymmetry. If $s \sim_{st} t$ then either $s = t = a$ or $\text{args}_{\text{AC}(f)}(s) = \{s_1, \ldots, s_k\} \sim_{mul} \{t_1, \ldots, t_k\} = \text{args}_{\text{AC}(f)}(t)$. Then there exists a permutation π of $\{1, \ldots, k\}$ such that $s_i \sim t_{\pi(i)}$ for $i = 1, \ldots, k$, and $s_i =_{\text{AC}} t_{\pi(i)}$ by AC-antisymmetry of $\succsim$. By combining the equational proofs for $s_i =_{\text{AC}} t_{\pi(i)}$ with a proof derived from the permutation π we can then obtain $s =_{\text{AC}} t$.

We say that a quasi-ordering functional $\succsim^{st}$ is *prepared for E-compatibility* if $s\sigma \sim^{st}(\succsim) t\sigma$ for any ground instance $s\sigma \approx t\sigma$ of an equation $s \approx t$ in E and for any quasi-ordering $\succsim$.

Lemma 5 *Let E be a set of equations, let $\succsim^{st}$ be a prestatus that is prepared for E-compatibility, and suppose that $\succsim_{gpo}$ is compatible with contexts. Then $\succsim_{gpo}$ is E-compatible.*

Proof: By induction on $\langle s,t \rangle$ with respect to $\rhd_{mul}$. Let $s = f(s_1,\ldots,s_m)$ and $t = g(t_1,\ldots,t_n)$. We have to show $s \succsim_{gpo} t$ for any E-step $s \Leftrightarrow_E t$. Symmetry of $\Leftrightarrow_E$ and reflexivity and transitivity of $\succsim_{gpo}$ then imply $(=_E) \subseteq (\succsim_{gpo})$.

(1) Suppose the E-step is not at the root of s and t. Then $s_i \Leftrightarrow_E t_i$ for some $i = 1,\ldots,n$ and $s_j = t_j$ for all $j \neq i$ in $1,\ldots,n$. By using the induction hypothesis we get $s_i \succsim_{gpo} t_i$ and $s \succsim_{gpo} t$ by compatibility with contexts.

(2) It remains to consider an E-step at the root position. We can write s as $s'\sigma$ and t as $t'\sigma$ where $s' \approx t'$ or $t' \approx s'$ is the equation in E that is used. Since $\succsim^{st}$ is prepared for E-compatibility, $s \succsim^{st}(\succsim_{gpo}) t$ follows. $\quad\square$

By construction our example prestatus is prepared for E-compatibility. We conclude for the moment that our example GPO is a well-founded E-compatible and E-antisymmetric quasi-ordering on terms that has the subterm property. It remains to consider compatibility with contexts.

4 Theory path orderings

Theory path orderings generalize the idea underlying the APO that compatibility with contexts can be achieved for path orderings if interpreted function symbols are minimal in the precedence. It combines a lexicographic path ordering on nontheory function symbols with a special treatment of symbols in the theory, which is formalized by a status function.

We let F denote the set of all function symbols, F_E the function symbols in E and F_T the function symbols in the theory T. Since in general not all function symbols in F_T need to be treated specially by the ordering, we select a set F_I of *interpreted function symbols* such that $F_E \subseteq F_I \subseteq F_T$. Function symbols not in F_I are called *free*. We let $\mathcal{I}$ be the set of terms over F with an interpreted function symbol at the root, and $\mathcal{A}$ the set of terms with a free function symbol at the root. Terms in $\mathcal{A}$ are called *atomic*. A *precedence* $\succsim_p$ is a quasi-ordering on function symbols whose strict part is well-founded. A precedence $\succsim_p$ is called *TPO-admissible* for F_I if $f \sim_p g$ for any pair of function symbols from F_I, $f \succ_p g$ whenever $f \notin F_I$ and $g \in F_I$, and $f = g$ whenever $f \sim_p g$ for f and g not in F_I. A well-founded partial ordering $\succeq$ on $F \setminus F_I$ can be extended to a TPO-admissible quasi-ordering on F by letting $f \succsim g$ whenever either (i) f and g not in F_I and $f \succeq g$, or (ii) g in F_I. If $\succeq$ is total this is the only TPO-admissible extension. E.g., for $F_I = \{+,0\}$ and free function symbols $\{a,f\}$ with a given precedence $f \succ_p a$ the TPO-admissible extension is $f \succ_p a \succ_p + \sim_p 0$.

We now come to the properties of a quasi-ordering functional which ensure compatibility with contexts. A quasi-ordering functional $\succsim^{st}$ is *strictly internally prepared for contexts* with respect to F_I if

$$s \succsim^{st}(\succsim) t \text{ implies } f(\ldots, s, \ldots) \succsim^{st}(\succsim) f(\ldots, t, \ldots) \qquad \text{and}$$
$$s \succ^{st}(\succsim) t \text{ implies } f(\ldots, s, \ldots) \succ^{st}(\succsim) f(\ldots, t, \ldots)$$

for any f in F_I. We say that a quasi-ordering functional $\succsim^{st}$ has the *multiset properties* for F_I if it satisfies

$$s \succsim t \iff s \succsim_{st} t \qquad \text{for } s \in A \text{ and } t \in A \qquad \text{(M0)}$$
$$s_1 \succsim_{st} t \vee \ldots \vee s_m \succsim_{st} t \iff f(s_1, \ldots, s_m) \succsim_{st} t \quad \text{for } t \in A \qquad \text{(M1)}$$
$$s_1 \succsim_{st} t \vee \ldots \vee s_m \succsim_{st} t \implies f(s_1, \ldots, s_m) \succsim_{st} t \qquad \text{(M2)}$$
$$s \succ_{st} t_1 \wedge \ldots \wedge s \succ_{st} t_n \implies s \succsim_{st} f(t_1, \ldots, t_n) \quad \text{for } s \in A \qquad \text{(M3)}$$
$$s \succ_{st} t_1 \wedge \ldots \wedge s \succ_{st} t_n \iff s \succsim_{st} f(t_1, \ldots, t_n) \qquad \text{(M4)}$$

for any f in F_I, where $\succsim_{st}$ denotes $\succsim^{st}(\succsim)$. A quasi-ordering functional $\succsim^{st}$ is called a *TPO-status* for F_I if it is subterm founded on $\mathcal{I}^2$, decreases infinite derivations in $\mathcal{I}$, is strictly internally prepared for contexts with respect to F_I, and has the multiset properties for F_I.

The theory path ordering $\succsim_{tpo} = \succsim^{tpo}(\succsim_p, \succsim^{st})$ induced by a TPO-admissible precedence $\succsim_p$ and a TPO-status $\succsim^{st}$ is defined as the smallest binary relation such that $s = g(s_1, \ldots, s_m) \succsim_{tpo} h(t_1, \ldots, t_n) = t$ if

1. $s_i \succsim_{tpo} t$ for some $i = 1, \ldots, m$, or
2. $s \succ_{tpo} t_j$ for each $j = 1, \ldots, n$ and either
   (a) $g \succ_p h$,
   (b) $g \sim_p h \notin F_I$ and $\langle s_1, \ldots, s_m \rangle \succsim^{lex}(\succsim_{tpo}) \langle t_1, \ldots, t_n \rangle$, or
   (c) $g \sim_p h \in F_I$ and $s \succsim^{st}(\succsim_{tpo}) t$.

We assume that each function symbol has a fixed arity, hence $m = n$ in case 2b. To view $\succsim_{tpo}$ as a general path ordering we have to define a suitable quasi-ordering functional. We define $\succsim^{st}_{tpo} = \succsim^{tpost}(\succsim_p, \succsim^{st})$ by

$$s = g(s_1, \ldots, s_m) \succsim^{st}_{tpo}(\succsim) h(t_1, \ldots, t_n) = t$$

if and only if

(a) $g \succ_p h$,
(b) $g \sim_p h \notin F_I$ and $\langle s_1, \ldots, s_m \rangle \succsim^{lex}(\succsim) \langle t_1, \ldots, t_n \rangle$, or
(c) $g \sim_p h \in F_I$ and $s \succsim^{st}(\succsim) t$.

Then clearly $\succsim^{tpo}(\succsim_p, \succsim^{st}) = \succsim^{gpo}(\succsim^{tpost}(\succsim_p, \succsim^{st}))$ for any precedence $\succsim_p$ and TPO-status $\succsim^{st}$.

For our running example we let $F_I = \{f\}$ and $a \succ_p f$. To satisfy the multiset properties we have to modify $\succsim^{st}$ such that it uses $\succsim$ on the (only) atomic term a. We let $s \succsim^{st}(\succsim) t$ if and only if $\text{args}_{AC(f)}(s) \succsim_{mul} \text{args}_{AC(f)}(t)$. Note that a

becomes a singleton multiset, hence (M0) is satisfied. The other properties of a TPO-status are also easily verified. Moreover, the resulting $\succsim_{tpo}^{st}$ is the prestatus of Section 3.

To obtain E-compatibility in the context of a simplification ordering we require E to be collapse-free. A collapsing equation $t \approx x$ would lead to a conflict with the subterm property, since x has to occur in t for any nontrivial theory.

Lemma 6 *Let E be a set of collapse-free equations, let $F_I \supseteq F_E$, let $\succsim_p$ be a precedence that is TPO-admissible for F_I, and let $\succsim^{st}$ be a TPO-status.*

1. *Then $\succsim_{tpo}^{st}$ is a prestatus.*
2. *If $\succsim_p$ is total and if $\succsim^{st}$ preserves totality on $\mathcal{I}^2$ then $\succsim_{tpo}^{st}$ preserves totality.*
3. *If $\succsim^{st}$ preserves E-antisymmetry on $\mathcal{I}^2$ then $\succsim_{tpo}^{st}$ preserves E-antisymmetry.*
4. *If $\succsim^{st}$ is prepared for E-compatibility then so is $\succsim_{tpo}^{st}$.*

Proof: The proof is straightforward. We consider only the last item in order to show where collapse-freeness is needed.

(*Prepared for E-compatibility*) Let $s = \hat{s}\sigma \approx \hat{t}\sigma = t$ be a ground instance of an equation $\hat{s} \approx \hat{t}$ in E, where $s = g(s_1, \ldots, s_m)$ and $t = h(t_1, \ldots, t_n)$. Since E is collapse-free, g and h are both in F_I. Then $g \sim_p h$ and $s \succsim^{st}(\succsim) t$ for any quasi-ordering $\succsim$, since $\succsim^{st}$ is prepared for E-compatibility. This implies $s \succsim_{tpo}^{st}(\succsim) t$. $\qquad\square$

Next we show that due to the multiset properties a TPO-status and the resulting TPO are identical within contexts of interpreted function symbols.

Lemma 7 *Let $\succsim_p$ be a precedence that is TPO-admissible for F_I, and let $\succsim^{st}$ be a quasi-ordering functional that is subterm founded on $\mathcal{I}^2$ and has the multiset properties for F_I. Then $s \succsim_{tpo} t$ if and only if $s \succsim^{st}(\succsim_{tpo}) t$.*

Proof: We use induction on the pairs $\langle s, t \rangle$ with respect to $\rhd_{mul}$. Let $s = g(s_1, \ldots, s_m)$ and $t = h(t_1, \ldots, t_n)$.

(1) Suppose g and h are not in F_I. Then $s \succsim_{tpo} t$ if and only if $s \succsim^{st}(\succsim_{tpo}) t$ by (M0).

(2) Suppose g is in F_I and h is not, which implies $h \succ_p g$. Then $s \succsim_{tpo} t$ if and only if there exists some $i = 1, \ldots, m$ such that $s_i \succsim_{tpo} t$. This is equivalent to $s_i \succsim^{st}(\succsim_{tpo}) t$ by the induction hypothesis, and to $s \succsim^{st}(\succsim_{tpo}) t$ by (M1) and (M2).

(3) Suppose g is not in F_I and h is, which implies $g \succ_p h$. Then $s \succsim_{tpo} t$ is equivalent to $s \succ_{tpo} t_j$ for all $j = 1, \ldots, n$ by case 2 of the definition of $\succsim_{tpo}$. Note that case 1 implies case 2 in this context. By induction hypothesis this is equivalent to $s \succ^{st}(\succsim_{tpo}) t_j$ for all $j = 1, \ldots, n$, and to $s \succsim^{st}(\succsim_{tpo}) t$ by (M3) and (M4).

(4) Otherwise g and h are in F_I.

For the only-if-direction suppose $s \succsim_{tpo} t$. If $s \succsim_{tpo} t$ by case 1 of the definition of $\succsim_{tpo}$ then there exists some $i = 1, \ldots, m$ such that $s_i \succsim_{tpo} t$. This implies $s_i \succsim^{st}(\succsim_{tpo}) t$ by induction hypothesis, and $s \succsim^{st}(\succsim_{tpo}) t$ by (M2). Otherwise case 2 of the definition of $\succsim_{tpo}$ holds, which explicitly includes $s \succsim^{st}(\succsim_{tpo}) t$.

For the if-direction assume $s \succsim^{st}(\succsim_{tpo}) t$. Then $s \succ^{st}(\succsim_{tpo}) t_j$ for all $j = 1, \ldots, n$ by (M4), and $s \succsim_{tpo} t_j$ for all $j = 1, \ldots, n$ by induction hypothesis. Together with $s \succsim^{st}(\succsim_{tpo}) t$ this satisfies case 2 of the definition of $\succsim_{tpo}$, hence $s \succsim_{tpo} t$. □

Theorem 8 *Let E be a set of collapse-free equations, let $F_I \supseteq F_E$, let $\succsim_p$ be a precedence that is TPO-admissible for F_I, and let $\succsim^{st}$ be a TPO-status for F_I.*

1. *Then $\succsim_{tpo}$ is a simplification quasi-ordering.*
2. *If $\succsim_p$ is total and $\succsim^{st}$ preserves totality on $\mathcal{I}^2$ then $\succsim_{tpo}$ is total.*
3. *If $\succsim^{st}$ preserves E-antisymmetry on $\mathcal{I}^2$ then $\succsim_{tpo}$ is E-antisymmetric.*
4. *If $\succsim^{st}$ is prepared for E-compatibility then $\succsim_{tpo}$ is E-compatible.*

Proof: (*Simplification quasi-ordering*) By Lemma 6 $\succsim^{st}_{tpo}$ is a prestatus. Thus $\succsim_{tpo}$ is a well-founded quasi-ordering on ground terms that has the subterm property. It remains to show compatibility with contexts and strict compatibility with contexts.

Let $s = g(s_1, \ldots, s_m)$, $t = h(t_1, \ldots, t_n)$, $s' = f(u_1, \ldots, u_i, s, u_{i+1}, \ldots, u_k)$, $t' = f(u_1, \ldots, u_i, t, u_{i+1}, \ldots, u_k)$, and suppose $s \succsim_{tpo} t$. By the subterm property $s' \succ_{gpo} u_j$ for $j = 1, \ldots, k$, and $s' \succ_{tpo} t$ by Lemma 1.
(*Compatibility with contexts*) We have to show $s' \succsim_{tpo} t'$.
(1) Suppose $f \notin F_I$. Then $s' \succsim^{st}_{tpo}(\succsim_{tpo}) t'$ because

$$\langle u_1, \ldots, u_i, s, u_{i+1}, \ldots, u_k \rangle \succsim^{lex}(\succsim_{tpo}) \langle u_1, \ldots, u_i, t, u_{i+1}, \ldots, u_k \rangle$$

by definition of $\succsim_{lex}$.
(2) Otherwise $f \in F_I$. From $s \succsim_{tpo} t$ we get $s \succsim^{st}(\succsim_{tpo}) t$ by Lemma 7, and by internal preparedness for contexts $s' \succsim^{st}(\succsim_{tpo}) t'$, which is equivalent to $s' \succsim^{st}_{tpo}(\succsim_{tpo}) t'$ for f in F_I.
(*Strict compatibility with contexts*) We have to show $s' \succ_{tpo} t'$ under the assumption $s \succ_{tpo} t$. Since there cannot exist a $j = 1, \ldots, n$ such that $t_j \succsim_{tpo} s$, case 1 of the definition of $\succsim_{tpo}$ cannot be used to obtain $t' \succsim_{tpo} s'$. Hence it suffices to show $s' \succ^{st}_{tpo}(\succsim_{tpo}) t'$ in order to conclude $s' \succ_{tpo} t'$.
(1) Suppose $f \notin F_I$. Then $s' \succ^{st}_{tpo}(\succsim_{tpo}) t'$ if and only if

$$\langle u_1, \ldots, u_i, s, u_{i+1}, \ldots, u_k \rangle \succ^{lex}(\succsim_{tpo}) \langle u_1, \ldots, u_i, t, u_{i+1}, \ldots, u_k \rangle,$$

which follows from the definition of $\succsim_{lex}$.
(2) Otherwise $f \in F_I$. From $s \succ_{tpo} t$ we get $s \succ^{st}(\succsim_{tpo}) t$ by Lemma 7, and by strict internal preparedness for contexts $s' \succ^{st}(\succsim_{tpo}) t'$, which implies $s' \succ^{st}_{tpo}(\succsim_{tpo}) t'$ and in turn $s' \succ_{tpo} t'$.
(*Totality*) Since $\succsim_p$ is total and $\succsim_{st}$ preserves totality on $\mathcal{I}^2$, we get that $\succsim^{st}_{tpo}$ preserves totality by Lemma 6(2). Hence $\succsim_{tpo}$ is total by Lemma 3.
(*E-antisymmetry*) $\succsim_{st}$ preserves E-antisymmetry on $\mathcal{I}^2$, hence $\succsim^{st}_{tpo}$ preserves E-antisymmetry by Lemma 6(3), and $\succsim_{tpo}$ is E-antisymmetric by Lemma 4.
(*E-compatibility*) Since $\succsim_{st}$ is prepared for E-compatibility we get that $\succsim^{st}_{tpo}$ is prepared for E-compatibility by Lemma 6(4). Since $\succsim_{tpo}$ is compatible with contexts it is E-compatible by Lemma 5. □

5 From extension function to TPO-status

When defining a TPO-status it is often useful to represent atomic subterms by constants. This ensures that the ordering obtained from the status for atomic subterms is determined only by its argument quasi-ordering. For instance, the definition of a TPO-status often involves normalizing with respect to some distributivity rules. By hiding atomic subterms in constants no rewriting can take place in atomic subterms. Also, extending an ordering on constants to terms is more natural and allows to reuse known simplification quasi-orderings in a status function.

We let F_C be the set of new constants $\{c_t \mid t \in \mathcal{A}\}$. That is, we assume that F_C and F are disjoint, where $\mathcal{A}$ contains only terms over F. Then for a given ordering $\succsim$ on terms over F we define the ordering $\succsim^c(\succsim)$ on constants in F_C by $c_s \succsim^c(\succsim) c_t$ if and only if $s \succsim t$. We will pack atomic subterms into constants from F_C and compare them according to $\succsim^c(\succsim)$. Technically, we let U be the convergent term rewriting system $\{c_t \Rightarrow t \mid t \in \mathcal{A}\}$, and write $U(t)$ for the normal form of t with respect to U. We use the term rewriting system

$$P = \{t \Rightarrow c_{t'} \mid t \text{ ground term over } F \cup F_C,\ t \notin F_C,\ t' = U(t) \text{ and } t' \in \mathcal{A}\}$$

for packing atomic subterms into constants. The unpacking of t in the definition of P is needed to remove nested constants in t. For termination of P observe that the number of symbols from F decreases in each step. For confluence observe that P contains a rule $u[c_{s'}] \Rightarrow c_{t'}$ for each critical pair

$$c_{t'} \Leftarrow_P u[s] \Rightarrow_P u[c_{s'}]$$

where $t' = U(u[s])$, $s' = U(s)$ and u is a nonempty F-context, since $U(u[c_{s'}]) = U(u[s]) = t' \in \mathcal{A}$.

It remains to obtain a quasi-ordering on the packed terms. Let $\succsim^t$ be a function that maps any quasi-ordering $\succsim_c$ on constants F_C to a quasi-ordering on terms over $F_I \cup F_C$, with the following properties:

1. $\succsim^t(\succsim_c)$ extends $\succsim_c$, is strictly compatible with contexts and has the subterm property.
2. Whenever there is an infinite descending chain $t_1 \succ^t(\succsim_c) t_2 \succ^t(\succsim_c) \ldots$ of terms over $F_I \cup F_C$ then there exists an infinite descending chain $c_1 \succ_c c_2 \succ_c \ldots$ of constants in F_C such that c_1 occurs in some t_j for $j \geq 1$.
3. Let c be a constant in F_C. If $c \succ_c c'$ for all constants c' occurring in a term t then $c \succ^t(\succsim_c) t$.
4. If $t \succsim^t(\succsim_c) c$ then $c' \succsim_c c$ for some constant c' in t.

Then we will call $\succsim^t$ an *extension function*. Property 3 is the constant dominance condition of Baader (1997). Note that it implies property 4 for total $\succsim_c$. We can now define a TPO-status $\succsim_t^{st}$ by $s \succsim_t^{st}(\succsim) t$ if and only if $P(s) \succsim^t(\succsim^c(\succsim)) P(t)$.

Lemma 9 *Let $\succsim^t$ be an extension function. Then $\succsim_t^{st}$ is a TPO-status for F_I.*

Proof: The proof is straightforward. □

Proposition 10 *Let $\succsim^t$ be an extension function such that $\sim^t(\succsim_c)$ is total for any total quasi-ordering $\succsim_c$ on constants. Then $\succsim_t^{st}$ preserves totality.*

Lemma 11 *Let $\succsim^t$ be an extension function such that $\succsim^t(\succsim_c)$ is $(E \cup \sim_c)$-antisymmetric for any quasi-ordering $\succsim_c$ on constants. Then $\succsim_t^{st}$ preserves E-antisymmetry.*

Proof: Suppose $\succsim$ is E-antisymmetric and $s \sim_t^{st}(\succsim) t$. Then $P(s) \sim^t(\succsim_c) P(t)$ and hence $P(s) =_{E \cup \sim_c} P(t)$. Since $\succsim$ is E-antisymmetric, $c_{s'} \sim_c c_{t'}$ implies $s' =_E t'$ for any atomic subterms s' and t' of s and t. Hence $s =_E t$. □

Lemma 12 *Suppose $F_I \supseteq F_E$, and let $\succsim^t$ be an extension function such that $\succsim^t(\succsim_c)$ is E-compatible for any quasi-ordering $\succsim_c$ on constants. Then $\succsim_t^{st}$ is prepared for E-compatibility.*

Proof: Observe that due to $F_I \supseteq F_E$ contexts at the root consisting of function symbols in E are left intact by packing. Hence for any instance of an equation in E packing both sides results again in an instance of the same equation. Thus for an equation $s \approx t$ in E we have $P(s\sigma) = s\sigma' \succsim^t(\succsim_c) t\sigma' = P(t\sigma)$ by E-compatibility of $\succsim_t$, where we define σ' by $x\sigma' = P(x\sigma)$ for all variables x in E. We conclude that $s\sigma \succsim_t^{st}(\succsim) t\sigma$ for any ground instance $s\sigma \approx t\sigma$ of an equation $s \approx t$ in E. □

Theorem 13 *Let E be a set of collapse-free equations, let $\succsim^t$ be an extension function such that $\succsim^t(\succsim_c)$ is total, $(E \cup \sim_c)$-antisymmetric and E-compatible for any total quasi-ordering $\succsim_c$ on F_C, and let $\succsim_p$ be a precedence that is TPO-admissible for F_I.*
  Then $\succsim^{tpo}(\succsim_p, \succsim_t^{st})$ is a total E-antisymmetric and E-compatible simplification quasi-ordering.

6 Examples of theory path orderings

In a trivial way any simplification quasi-ordering can be constructed as a TPO, by taking $F_I = F$ and letting $\succsim_t$ be the original ordering. Then $\mathcal{A}$ and F_C are empty, and properties (3) and (4) of an extension function become void. Being a simplification quasi-ordering, $\succsim_t$ satisfies properties (1) and (2).

On the other end of the spectrum is the lexicographic path ordering, which is obtained for $F_I = \emptyset$.

The simplest nontrivial example is the associative path ordering for a single associative and commutative symbol f which we already have used as an example in Sections 3 and 4. That is, we have $E = \mathrm{AC}(f)$ and $F_I = F_E = \{f\}$. Terms over $F_I \cup F_C$ are ordered according to the multiset of constants from F_C they contain. Formally, we associate the *complexity* $\kappa(t) = \mathrm{args}_{\mathrm{AC}(f)}(t)$ to each term t over $F_I \cup F_C$, and define $\succsim^t(\succsim_c)$ by $s \succsim^t(\succsim_c) t$ if and only if $\kappa(s) \succsim^{mul}(\succsim_c) \kappa(t)$. We now show that this extension function satisfies the requirements of Theorem 13. Associativity and commutativity are collapse-free. Clearly $\succsim^t(\succsim_c)$ extends $\succsim_c$ and satisfies properties 3 and 4. The multiset extension of a quasi-ordering has the following properties:

$$M_1 \succsim_{mul} M_2 \text{ implies } N \cup M_1 \succsim_{mul} N \cup M_2 \tag{1}$$

$$M_1 \succ_{mul} M_2 \text{ implies } N \cup M_1 \succ_{mul} N \cup M_2 \tag{2}$$

$$M_1 \subsetneq M_2 \text{ implies } M_1 \succ_{mul} M_2. \tag{3}$$

Strict compatibility with F_I-contexts is a consequence of (1) and (2). The subterm property follows by (3). The multiset extension preserves well-foundedness, hence an infinite descending chain in $\succsim^t(\succsim_c)$ can only arise from an infinite descending chain in $\succsim_c$. Since we can restrict the ordering to the constants occurring in the infinite descending chain of multisets, there is an infinite descending chain of constants occurring in these multisets. Thus $\succsim^t$ is an extension function. If $\succsim_c$ is total then $\succsim^t(\succsim_c)$ is total, since the multiset extension preserves totality. AC-compatibility is obvious from the construction. Finally we show that the quasi-ordering $\succsim^t(\succsim_c)$ is $(AC \cup \sim_c)$-antisymmetric. Suppose $s \sim^t(\succsim_c) t$. To take care of $\sim_c$ we select a representative $rep(c)$ for each $\sim_c$-equivalence class in F_C. That is, $c \sim_c d$ if and only if $rep(c) = rep(d)$ for any two constants c and d in F_C. We replace each constant c in s and t by its representative and obtain terms s' and t', respectively. Then $s' \sim^t(\succsim_c) t'$,

$$\kappa(s') = M_1 = \{c_1, \ldots, c_k\} \text{ and}$$
$$\kappa(t') = M_2 = \{d_1, \ldots, d_l\},$$

and $M_1 \sim^{mul}(\succsim_c) M_2$. Since $\sim_c$-equivalent constants are equal in s' and t', we even have $M_1 = M_2$ and hence $s' =_{AC} t'$. Combining this with $s =_{\sim_c} s'$ and $t =_{\sim_c} t'$ we get $s =_{AC \cup \sim_c} t$. We conclude that $\succsim^t(\succsim_c)$ is $AC \cup \sim_c$-antisymmetric. Thus $\succsim^{tpo}(\succsim_p, \succsim_t^{st})$ is a total, E-antisymmetric and E-compatible simplification quasi-ordering.

To express a general associative path ordering as a TPO one would put the AC-function symbols and the symbols below them in the precedence into F_I and keep the precedence on the other symbols in $\succsim_p$. The TPO-status would consist of the APO obtained by reusing the precedence on F_I and extending it above by the ordering on the constants in F_C. One easily sees that this is an extension function, and that the resulting TPO is equal to the original APO.

As a larger example we present a quasi-simplification ordering for modules over some fixed ring R. We assume as given a well-ordering $\succ_R$ on R such that $a \succ_R 1 \succ_R 0$ for any $a \in R \setminus \{0, 1\}$. The module is described by the following term rewriting system modulo AC, where the v_i are variables for scalars and $*$ is scalar multiplication:

$$x + 0 \approx x \tag{M.1}$$

$$0 * x \approx 0 \tag{M.2}$$

$$1 * x \approx x \tag{M.3}$$

$$v * 0 \approx 0 \tag{M.4}$$

$$v * (x + y) \approx v * x + v * y \tag{M.5}$$

$$v_1 * (v_2 * x) \approx v * x \, [v = v_1 \cdot v_2] \tag{M.6}$$

$$x + x \approx v * x \, [v = 1 + 1] \tag{M.7}$$

$$v_1 * x + x \approx v * x \, [v = v_1 + 1] \tag{M.8}$$

$$v_1 * x + v_2 * x \approx v * x \, [v = v_1 + v_2] \tag{M.9}$$

We denote the set of ground instances of these rules that satisfy the constraints by M. The ordering should orient rules in M from left to right. Furthermore, the symmetrization requires that $a * t \succ_M a'' * t + a' * r$ for any terms t and r and $a, a', a'' \in R$ such that t is atomic, $t \succ_M r$ and $a \succ_R a''$. The particular problem that rules out standard orderings is that (M.9) combines arbitrary coefficients to a possibly greater coefficient, while the reduction in the coefficient from a to a'' must suffice to make room for any term r smaller than t.

We let $F_I = F_M = \{+, *, 0, 1\} \cup R$, assume a total precedence $\succeq_p$ on $F \setminus F_I$, and let $\succsim_p$ denote its TPO-admissible extension to F. Let D be the convergent term rewriting system consisting of the ground instances of the distributivity rule (M.5). We denote the normal form of a term t with respect to D by $D(t)$. Let t be a ground term over $F_I \cup F_C$ in D-normal form. Then t is of the form

$$t = a_{11} * \cdots * a_{1k_1} * c_1 + \cdots + a_{n1} * \cdots * a_{nk_n} * c_n$$

where $n \geq 1$, $k_i \geq 0$, $c_i \in F_C \cup \{0, 1\}$, and $a_{ij} \in R$ for $i = 1, \ldots, n$ and $j = 1, \ldots, k_i$. We assign a complexity κ to any such t as follows. Again we assume a function rep : $F_C \to F_C$ such that $\text{rep}(c) = \text{rep}(d)$ if and only if $c \sim_c d$. We extend rep to $F_C \cup \{0, 1\}$ by $\text{rep}(0) = 0$ and $\text{rep}(1) = 1$. The ordering $\succsim_c$ is extended to $F_C \cup \{0, 1\}$ such that $c \succ_c 1 \succ_c 0$ for any constant c in F_C. We let $\text{occ}(t, c_i) = \{j \mid c_j \sim_c c_i\}$, $\#(t, c_i) = |\text{occ}(t, c_i)|$, and $\text{cs}(t, c_i) = \bigcup_{j \in \text{occ}(t, c_i)} \{\langle a_{j1}, \ldots, a_{jk_1} \rangle\}$. That is, $\text{occ}(t, c_i)$ is the set of indices of the occurrences of constants in the same $\sim_c$-equivalence class as c_i, $\#(t, c_i)$ is the number of these occurrences, and $\text{cs}(t, c_i)$ is the multiset of the tuples of coefficients associated with these occurrences. To each equivalence class we associate the tuple $\langle \text{rep}(c_i), \#(t, c_i), \text{cs}(t, c_i) \rangle$. Finally, we let $\kappa(t)$ be the set of tuples for the constants from F_C that occur in t. We order these complexities according to the multiset extension of the lexicographic combination of $\succsim_c$, $>$ and the multiset extension of the length-lexicographic extension of $\succeq_R$. We denote the ordering on complexities by $\succsim_\kappa$. Then we define the ordering $\succsim_t$ on terms over $F_I \cup F_C$ by $s \succsim_t t$ if and only if $\kappa(D(s)) \succsim_\kappa \kappa(D(t))$ where s and t are terms over $F_I \cup F_C$. Finally we get the TPO-status $\succsim_1^{st}$ as the status derived from $\succsim_t$, and let $\succsim_1(\succeq_p) = \succsim^{tpo}(\succsim_p, \succsim_1^{st})$.

Theorem 14 $\succsim_1(\succeq_p)$ *is a total* $AC \cup D$*-compatible and* $AC \cup D$*-antisymmetric simplification quasi-ordering on ground terms that contains* $M \setminus D$.

Proof: We do not have enough space for the complete proof that $\succsim^t$ satisfies the requirements of Theorem 13. As an example, we show how strict compatibility with contexts can be obtained:

Let $f_{u[]}$ be the function that maps any complexity $\kappa(t)$ to $\kappa(D(u[t]))$, where

$$t = a_{11} * \cdots * a_{1k_1} * c_1 + \cdots + a_{n1} * \cdots * a_{nk_n} * c_n$$

is a ground term in D-normal form. To see that $f_{u[]}$ is well-defined, observe that t can be reconstructed up to $AC \cup \sim_c$-equivalence from $\kappa(t)$, and that κ maps terms in an $AC \cup \sim_c$-equivalence class to the same complexity. We show that $f_{u[]}$ is strictly monotonic for any context $u[]$ by considering contexts of depth one.

(1) Consider $u = s + []$.

(1.1) Suppose $s = b_1 * \cdots * b_l * d$.

(1.1.1) Suppose no c_i is $\sim_c$-equivalent to d. Then

$$f_{u[]}(\kappa(t)) = \kappa(t) \cup \{\langle \text{rep}(d), 1, \{\langle b_1, \ldots, b_l \rangle\}\rangle\}$$

and $f_{u[]}$ is strictly monotonic by properties (1) and (1) of the multiset extension.

(1.1.2) Otherwise $d \sim_c c_i$ for some $i = 1, \ldots, n$. Suppose without loss of generality $d \sim_c c_1$. Then

$$f_{u[]}(\kappa(t)) = \{\langle c_1, \#(t, c_1) + 1, \text{cs}(t, c_1) \cup \{\langle b_1, \ldots, b_l \rangle\}\rangle\}$$
$$\cup \{\langle c_i, \#(t, c_i), \text{cs}(t, c_i)\rangle \mid c_i \not\sim_c c_1\}.$$

Consider the function that maps a tuple $\langle c, n, M \rangle$ to $\langle c, n + 1, M \cup \{\langle b_1, \ldots, b_l \rangle\}\rangle$ if $c \sim_c d$ and to $\langle c, n, M \rangle$ otherwise. This function is strictly monotonic. Hence its multiset extension $f_{u[]}$ is strictly monotonic.

(1.2) Otherwise s is a proper sum, and $f_{u[]}$ can be obtained as a finite composition of the strictly monotonic functions of case (1.1). Hence $f_{u[]}$ is strictly monotonic.

(2) Consider $u = a * []$. Then $f_{u[]}$ maps any tuple $\langle a_1, \ldots, a_k \rangle$ of coefficients in the multiset in the third component to $\langle a, a_1, \ldots, a_k \rangle$. This mapping is strictly monotonic. By multiset extension, lexicographic product with identity functions for the first two components and again multiset extension we obtain $f_{u[]}$, which thus is strictly monotonic. □

From $\succsim_1$ we obtain an ordering that orients D from left to right and that is AC-antisymmetric by combining $\succsim_1$ lexicographically with a polynomial ordering and the ACRPO of Rubio and Nieuwenhuis (1995).

7 Conclusion and Further Work

We have presented a general path ordering for the purpose of superposition theorem proving in algebraic theories. We have used this construction successfully for the cases of abelian groups, commutative rings, and modules and algebras over a fixed ring.

In practice it will be necessary to consider also nonground terms. Clearly lifting depends on the particular TPO-status chosen. An approach that should be applicable in many cases is to compare two nonground terms by considering a finite set of their instances (Bachmair and Plaisted 1985).

Waldmann (1998) shows how to extend an arbitrary reduction ordering on terms over free function symbols to an ACU-compatible reduction ordering by semantic labeling, using essentially an APO for a single AC-symbol. It seems feasible to extend this approach to other theories via TPOs.

Acknowledgments

I thank Uwe Waldmann and the anonymous referees for their detailed and helpful comments and Harald Ganzinger for his support.

References

BAADER, F. (1997). Combination of compatible reduction orderings that are total on ground terms. In *12th Ann. IEEE Symp. on Logic in Computer Science*, Warsaw, pp. 2–13. IEEE Computer Society Press.

BAADER, F. AND NIPKOW, T. (1998). *Term rewriting and all that*. Cambridge University Press, Cambridge, UK.

BACHMAIR, L. AND PLAISTED, D. (1985). Termination orderings for associative-commutative rewriting systems. *Journal of Symbolic Computation* 1: 329–349.

BACHMAIR, L., GANZINGER, H. AND STUBER, J. (1995). Combining algebra and universal algebra in first-order theorem proving: The case of commutative rings. In *Proc. 10th Workshop on Specification of Abstract Data Types*, Santa Margherita, Italy, LNCS 906, pp. 1–29. Springer.

DELOR, C. AND PUEL, L. (1993). Extension of the associative path ordering to a chain of associative commutative symbols. In *Proc. 5th Int. Conf. on Rewriting Techniques and Applications*, Montreal, LNCS 690, pp. 389–404. Springer.

DERSHOWITZ, N. AND HOOT, C. (1995). Natural termination. *Theoretical Computer Science* **142**: 179–207.

DERSHOWITZ, N. AND JOUANNAUD, J.-P. (1990). Rewrite systems. In J. van Leeuwen (ed.), *Handbook of Theoretical Computer Science: Formal Models and Semantics*, Vol. B, chapter 6, pp. 243–320. Elsevier/MIT Press.

GESER, A. (1996). An improved general path order. *Applicable Algebra in Engineering, Communication and Computing* **7**: 469–511.

KAPUR, D. AND SIVAKUMAR, G. (1997). A total ground path ordering for proving termination of AC-rewrite systems. In *Proc. 8th Int. Conf. on Rewriting Techniques and Applications*, Sitges, Spain, LNCS 1103, pp. 142–156. Springer.

MARCHÉ, C. (1996). Normalised rewriting: an alternative to rewriting modulo a set of equations. *Journal of Symbolic Computation* **21**: 253–288.

RUBIO, A. AND NIEUWENHUIS, R. (1995). A total AC-compatible ordering based on RPO. *Theoretical Computer Science* **142**: 209–227.

STUBER, J. (1998a). Superposition theorem proving for abelian groups represented as integer modules. *Theoretical Computer Science* **208**(1–2): 149–177.

STUBER, J. (1998b). Superposition theorem proving for commutative rings. In W. Bibel and P. H. Schmitt (eds), *Automated Deduction - A Basis for Applications. Volume III. Applications*, chapter 2, pp. 31–55. Kluwer, Dordrecht, The Netherlands.

WALDMANN, U. (1998). Extending reduction orderings to ACU-compatible reduction orderings. *Information Processing Letters* **67**(1): 43–49.

A Characterisation of Multiply Recursive Functions with Higman's Lemma

Hélène Touzet

LIFL – Université Lille 1
59 655 Villeneuve d'Ascq
France
touzet@lifl.fr

Abstract. We prove that string rewriting systems which reduce by Higman's lemma exhaust the multiply recursive functions. This result provides a full characterisation of the expressiveness of Higman's lemma when applied to rewriting theory. The underlying argument of our construction is to connect the order type and the derivation length via the Hardy hierarchy.

1 Introduction

Higman's lemma

Recall the statement of Higman's lemma for strings. Given an alphabet Σ, define the *division ordering* $\trianglelefteq$ as the least pre-order on the set of finite strings Σ^* satisfying the following properties

- *sub-term property*: $\forall a \in \Sigma$, $\forall u \in \Sigma^*$, $u \triangleleft au$,
- *monotonicity*: $\forall u, v \in \Sigma^*$, $\forall a \in \Sigma$, $u \trianglelefteq v \Rightarrow au \trianglelefteq av$.

Theorem 1 (Higman [7]).
For any finite alphabet Σ, $(\Sigma^*, \trianglelefteq)$ is a well-quasi-ordering.

A well-quasi-ordering is a well-founded ordering with no infinite anti-chain. In other words, every ordering extending $\trianglelefteq$ is still well-founded. So Higman's lemma provides a syntactic criterion for the definition of well-founded orderings on strings. Let's mention the Knuth-Bendix ordering, the recursive path ordering, the polynomial orderings. What concerns us is the expressiveness of string rewriting systems (SRS). Given a Noetherian finite SRS $\mathcal{R}$ on an alphabet Σ, define the *derivation length function* $Dl_{\mathcal{R}}$ by

$$
\begin{aligned}
dl_{\mathcal{R}} :\ & \Sigma^* \to \mathbb{N} \\
& w \mapsto \max\{dl_{\mathcal{R}}(u),\ w \to_{\mathcal{R}} u\} + 1 \\
Dl_{\mathcal{R}} :\ & \mathbb{N} \to \mathbb{N} \\
& m \mapsto \max\{n \in \mathbb{N},\ \exists\, w \in \Sigma^*, dl_{\mathcal{R}}(w) = n\ \wedge\ |w| \le m\}
\end{aligned}
$$

where $|w|$ is the size of the string w. The expressiveness of the main termination orderings was extensively studied and we know that most ensure primitive recursive derivation lengths on strings (see [8] for the Knuth-Bendix and polynomial orderings, [9] for the recursive path ordering). The purpose of this paper is to investigate the derivation length of the whole class of string rewriting systems reducing by Higman's lemma. Extending the result of [13], we establish that existing termination proof techniques do not reach the full strength of Higman's lemma: one can go far beyond primitive recursiveness and exhaust the class of multiply recursive functions.

Multiply recursive functions and the Hardy hierarchy

Multiply recursive functions are traditionally defined by closure under the schemes of k-recursion (see Peter [11] for instance). Grzegorzyck, Wainer and others teach us that classes of functions may also be described by hierarchies of functions indexed by ordinals. We adopt this alternative point of view here and introduce the class of multiply recursive functions by the means of the Hardy hierarchy. Let $\mathcal{CNF}(\varepsilon_0)$ be the set of notations in Cantor Normal Form for ordinals below ε_0. A canonical assignment of fundamental sequences for limit ordinals in $\mathcal{CNF}(\varepsilon_0)$ is defined recursively as follows:

$$\omega_n = n$$
$$(\alpha + \lambda)_n = \alpha + \lambda_n$$
$$(\omega^{\beta+1})_n = \omega^\beta n$$
$$(\omega^\lambda)_n = \omega^{\lambda_n}$$

where β is in $\mathcal{CNF}(\varepsilon_0)$ and λ is a limit ordinal in $\mathcal{CNF}(\varepsilon_0)$. With the fundamental sequences, one can define the family of *predecessor functions*[1]. For all $n \in \mathbb{N}$, $P_n : \mathcal{CNF}(\varepsilon_0) \to \mathcal{CNF}(\varepsilon_0)$ is

$$P_n(0) = 0$$
$$P_n(\alpha + 1) = \alpha$$
$$P_n(\lambda) = \lambda_n, \text{ if } \lambda \text{ is a limit ordinal.}$$

For each ordinal α of $\mathcal{CNF}(\varepsilon_0)$, the *Hardy function* $\mathcal{H}_\alpha$ is now defined as follows.

$$\mathcal{H}_0 : \mathbb{N} \to \mathbb{N}$$
$$n \mapsto 0$$

and for $\alpha > 0$

$$\mathcal{H}_\alpha : \mathbb{N} \to \mathbb{N}$$
$$n \mapsto \mathcal{H}_{P_n(\alpha)}(n+1) + 1$$

The class of *multiply recursive functions* is exactly described by the family of Hardy functions indexed by ordinals of ω^{ω^ω} (Robbin [12]).

---

[1] For the limit case, the predecessor function is sometimes defined as $P_n(\lambda) = P_n(\lambda_n)$. This does not affect the complexity of the Hardy hierarchy.

Why should we give a preponderant role to the Hardy hierarchy ? Consider the maximal order type of the division ordering $\trianglelefteq$. De Jongh and Parikh established in [3] that for all finite non-empty alphabet Σ, the maximal order type of $(\Sigma^*, \trianglelefteq)$ is $\omega^{\omega^{|\Sigma|-1}}$. Cichon and Tahhan Bittar took advantage of this result and produced a measure for sequences compatible with $\trianglelefteq$ using the Hardy hierarchy indexed by the maximal order type of the division ordering.

Theorem 2 (Cichon and Tahhan-Bittar [2]).
Let Σ be a finite alphabet and $k \in \mathbb{N}$. For each string u in Σ^*, $|u|$ denotes the size of u. There is a function $\phi : \mathbb{N} \to \mathbb{N}$ such that for all sequence $(u_i)_{i \in \mathbb{N}}$ of Σ^* satisfying

- $\forall i, j \in \mathbb{N}, \ i < j \Rightarrow \neg(u_i \trianglelefteq u_j)$,
- $\forall i \in \mathbb{N}, \ |u_i| \leq |u_0| + k \times i$,

the length of $(u_i)_{i \in \mathbb{N}}$ is bounded by $\phi(|u_0|)$. Moreover ϕ is an elementary function in $\mathcal{H}_{\omega^{\omega^{|\Sigma|}}}$.

This theorem provides an upper bound for rewrite derivations : each finite string rewriting system reducing by Higman's lemma has a multiply recursive derivation length.

We investigate here the intriguing role of Hardy hierarchy and show that it is possible to encode Hardy functions indexed by ordinals of $\omega^{\omega^{\omega}}$ by finite string rewriting systems which are compatible with the division ordering of Higman's lemma. This construction shows that Cichon and Tahhan's upper bound is essentially optimal. At a logical level it confirms the fact that the Hardy hierarchy is the right tool for connecting derivation length and order type. The proof goes as follows. The Hardy hierarchy enjoys an intuitive geometrical description : for each ordinal α and all integer n, consider the decreasing sequence of ordinals $(\alpha_i)_{i \in \mathbb{N}}$ given by

$$(\star) \qquad \begin{aligned} \alpha_0 &= \alpha, \\ \alpha_{i+1} &= P_{n+i}(\alpha_i). \end{aligned}$$

The sequence stops when it reaches 0. We call it a $(\star)$-*sequence*. $\mathcal{H}_\alpha(n)$ is simply the length of the $(\star)$-sequence generated by α and n. We use this representation and encode $(\star)$-sequences for ordinals below $\omega^{\omega^{\omega}}$ by rewrite systems. This idea is already present in [14], where it leads to a new lower bound of the complexity of simplifying *term* rewriting systems. Here we use strings instead of terms. First, we need to produce a specific notation system for ordinals below $\omega^{\omega^{\omega}}$ based on strings. For that we choose the recursive path ordering of Dershowitz (section 3). We are then able to construct the string rewriting systems and their proof of termination by Higman's lemma (section 4). We even establish *total termination*.

2 Rewriting theory and total termination

We do not recall fundamental notions on rewrite systems and termination (see [5] for instance). Let Σ be an alphabet and $\prec$ an ordering on Σ^*. We say that

$\prec$ is *strictly monotone* if $\forall u, v \in \Sigma^*$, $(u \prec v \Rightarrow \forall a \in \Sigma \ au \prec av)$.

$\prec$ is *monotone* if $\forall u, v \in \Sigma^*$, $(u \preceq v \Rightarrow \forall a \in \Sigma, \ au \preceq av)$, where $\preceq$ is the reflexive closure of $\prec$.

$\prec$ has the *sub-term property* if $\forall u \in \Sigma^*$, $\forall a \in \Sigma$, $u \prec au$.

We now come to the definition of total termination, due to Ferreira and Zantema [6].

Definition 1 (Total Termination). *Let Σ be an alphabet. A* total termination ordering *on Σ^* is a strictly monotone well-order. A rewrite system $\mathcal{R}$ of Σ^* is* totally terminating *if it is compatible with a total termination ordering.*

Total termination on a finite alphabet implies the sub-term property (see [4] or [6]). It follows that any totally terminating rewrite system is compatible with the division ordering of Higman's theorem. We now give another characterisation of total termination, which requires only monotonicity, instead of strict monotonicity. This result is useful in section 4.

Proposition 1. *Let Σ be a finite alphabet and let $\mathcal{R}$ be a rewrite system on Σ^*. $\mathcal{R}$ is totally terminating if and only if there exists an ordering $\prec$ on Σ^* such that*

(i) *for each $l \to r$ in $\mathcal{R}$, for each $u \in \Sigma^*$, $lu \succ ru$,*

(ii) *$\prec$ has the sub-term property,*

(iii) *$\prec$ is monotone.*

Proof. One direction of the proof is obvious. For the other direction, let $\mathrm{mul}(\Sigma^*)$ denote the set of finite multisets on Σ^*, $\mathrm{mul}(\prec)$ the multiset extension of $\prec$ on $\mathrm{mul}(\Sigma^*)$ and $\cup$ the union of multisets. For each string u in Σ^*, define $\mathcal{M}(u)$ as the multiset containing u and its suffixes:

$$\mathcal{M}(\varepsilon) = \emptyset,$$
$$\mathcal{M}(au) = \{au\} \cup \mathcal{M}(u),$$

and define $\prec'$ as

$$u \prec' v \ \Leftrightarrow \ \mathcal{M}(u)\mathrm{mul}(\prec)\mathcal{M}(v).$$

We claim that $\prec'$ is a total termination ordering for $\mathcal{R}$. Firstly, $\prec'$ is strictly monotone: let $u, v \in \Sigma^*$ such that $u \prec' v$ and let $a \in \Sigma$. We have

$$\mathcal{M}(au) = \{au\} \cup \mathcal{M}(u),$$
$$\mathcal{M}(av) = \{av\} \cup \mathcal{M}(v).$$

By hypothesis, we have $\mathcal{M}(u)\mathrm{mul}(\prec)\mathcal{M}(v)$. So it suffices to show that $au \preceq av$. Suppose $u \succ v$. By hypothesis (ii) on $\prec$, this would imply $\mathcal{M}(u)\mathrm{mul}(\succ)\mathcal{M}(v)$, which contradicts the hypothesis $\mathcal{M}(u)\mathrm{mul}(\prec)\mathcal{M}(v)$. So $u \preceq v$, which with (iii) ensures $au \preceq av$. Thus $au \prec' av$. Secondly, $\prec'$ is well-founded, since it is monotonic and enjoys the sub-term property. Finally, $\prec'$ is compatible with $\mathcal{R}$: let $l \to r$ in $\mathcal{R}$ and $u \in \Sigma^*$. By (i), $lu \succ ru$, which with (ii) implies $lu \ \mathrm{mul}(\succ) \ ru$. Hence $lu \succ' ru$. This completes the proof. $\square$

Remark 1. In definition 1, the notion of total termination on strings coincides with the usual definition on terms : it uses only (left-) monotonicity, and not stability (left- and right- monotonicity). So our total termination orderings are not total division orderings, such as studied in [10]. There is a slight difference between those two families of orderings: for instance the string rewriting system

$$\begin{cases} ff \to gf \\ gg \to fg \end{cases}$$

is totally terminating, but it does not reduce under any total division ordering.

3 Ordinal notations for ω^{ω^ω} with strings

The core of our construction is to simulate decreasing sequences of ordinals of ω^{ω^ω} by sequences of strings. For that, we are now going to introduce an ordinal notation system that is based on the recursive path ordering on strings.

Definition 2 (Recursive path ordering [4]). *Let Σ be an alphabet equipped with the precedence $\prec$. The recursive path ordering $\prec_{rpo}$ is the least stable ordering which satisfies*

- *if $u \lhd v$, then $u \prec_{rpo} v$,*
- *if $u \prec_{rpo} bv$ and $a \prec b$, then $au \prec_{rpo} bv$.*

Proposition 2 ([4]). *If $(\Sigma, \prec)$ is a well-order, then $(\Sigma^*, \prec_{rpo})$ is a total termination ordering.*

In the sequel of the paper, we choose $\Sigma = \{a_i \; ; \; i \in \mathbb{N}\}$ with the well-ordered precedence $a_i \prec a_{i+1}$. It is routine to prove that the order type of $\prec_{rpo}$ on Σ^* is ω^{ω^ω}, the maximal order type of the division ordering. It means that there exists an isomorphism $\mathcal{O}$ of $\omega^{\omega^\omega} \to (\Sigma^*, \prec_{rpo})$ such that each ordinal of ω^{ω^ω} may be denoted in an unique and non-ambiguous way by a string of Σ^*. The purpose of the remaining of the section is to make the construction of $\mathcal{O}$ explicit.

Proposition 3. *For each limit ordinal α in ω^{ω^ω}, there are unique $i \in \mathbb{N}$, $\beta, \gamma \in \omega^{\omega^\omega}$ such that*

$$\alpha = \gamma + \omega^{\omega^i}\beta$$

and satisfying

- (i) $0 < \beta < \omega^{\omega^{i+1}}$,
- (ii) $\forall\, 0 < \mu < \omega^{\omega^{i+1}}, \; \forall\, \delta \in \omega^{\omega^\omega} \; \gamma \neq \delta + \mu.$

Proof. Let $\omega^{\alpha_1} + \cdots + \omega^{\alpha_n}$ be the Cantor Normal Form of α. Since α is a limit ordinal of ω^{ω^ω}, we have $0 < \alpha_n \leq \ldots \leq \alpha_1 < \omega^\omega$. Let $i \in \mathbb{N}$ such that

$\omega^i \leq \alpha_n < \omega^{i+1}$ and let j be the smallest index such that $\omega^i \leq \alpha_j < \omega^{i+1}$. There are $\delta_j, \ldots, \delta_n$ in ω^{i+1} such that

$$\alpha_j = \omega^i + \delta_j,$$
$$\vdots$$
$$\alpha_n = \omega^i + \delta_n.$$

If we set $\gamma = \omega^{\alpha_1} + \cdots + \omega^{\alpha_{j-1}}$ (γ is possibly 0) and $\beta = \omega^{\delta_j} + \cdots + \omega^{\delta_n}$, then i, β, γ satisfy conditions (i), (ii) and $\alpha = \gamma + \omega^{\omega^i} \beta$.

We now prove that this decomposition is unique. Let β, γ, i and i' satisfying conditions (i),(ii) such that $\gamma + \omega^{\omega^i} \beta = \gamma' + \omega^{\omega^{i'}} \beta'$. Consider the Cantor Normal Form of β and β': $\beta = \omega^{\beta_1} + \cdots + \omega^{\beta_n}$ and $\beta' = \omega^{\beta'_1} + \cdots + \omega^{\beta'_m}$. If we require that γ and γ' are in Cantor Normal Form too, conditions (ii), (iii) guarantee that the notations $\gamma + \omega^{\omega^i + \beta_1} + \cdots + \omega^{\omega^i + \beta_n}$ and $\gamma' + \omega^{\omega^{i'} + \beta'_1} + \cdots + \omega^{\omega^{i'} + \beta'_m}$ are two Cantor Normal Forms of the same ordinal. They are identical. It implies $i = i'$. Suppose now that $\gamma < \gamma'$ (for instance). It would follow that $\gamma' = \gamma + \omega^{\omega^i + \beta_1} + \cdots + \omega^{\omega^i + \beta_k}$ for some $k \leq n$, which contradicts (ii). So $\gamma = \gamma'$ and then $n = m$, $\beta_1 = \beta'_1, \ldots, \beta_n = \beta'_n$. □

In proposition 3, we require that $\beta > 0$. In this case, we define $-1 + \beta$ as the unique ordinal such that $1 + (-1 + \beta) = \beta$. Recall that for infinite ordinals $1 + \beta = \beta$, and for finite ordinals $1 + \beta = \beta + 1$. So $-1 + \beta = \beta$ when β is infinite and $-1 + \beta = \beta - 1$ when β is finite and non-empty. This leads to the definition of the notation system $\mathcal{O}$.

Definition 3 (The notation system $\mathcal{O}$).

$$\mathcal{O}: \quad \omega^{\omega^\omega} \to \Sigma^*$$
$$0 \mapsto \varepsilon$$
$$\beta + 1 \mapsto a_0 \mathcal{O}(\beta)$$
$$\gamma + \omega^{\omega^i} \beta \mapsto a_{i+1} \mathcal{O}(-1 + \beta) \mathcal{O}(\gamma)$$

Example 1. $\mathcal{O}(n) = a_0^n$, $\mathcal{O}(\omega) = a_1$, $\mathcal{O}(\omega + n) = a_0^n a_1$, $\mathcal{O}(\omega + \omega) = a_1 a_0$, $\mathcal{O}(\omega^2) = a_1 a_1$, $\mathcal{O}(\omega^\omega) = a_2$.

Proposition 4. $\mathcal{O}$ *is an isomorphism of* $(\omega^{\omega^\omega}, <) \to (\Sigma^*, \prec_{rpo})$.

Proof. We first prove that for all $\alpha, \beta \in \omega^{\omega^\omega}$, if $\alpha < \beta$ then $\mathcal{O}(\alpha) \prec_{rpo} \mathcal{O}(\beta)$. The ordinal ordering $<$ is the transitive closure of the schemes

$$\beta < \beta + 1,$$
$$\forall n \in \mathbb{N} \quad \gamma + \omega^{\omega^i}(\beta + 1) < \gamma + \omega^{\omega^i} \beta + (\omega^{\omega^i})n,$$
$$\forall n \in \mathbb{N} \quad \gamma + \omega^{\omega^i} \beta < \gamma + \omega^{\omega^i} \beta_n \quad (\beta \text{ limit}).$$

On strings, it corresponds to the three following inequalities:

$$u \prec_{rpo} a_0 u,$$
$$\forall n \in \mathbb{N} \quad a_{i+1} a_0 \mathcal{O}(\beta) \mathcal{O}(\gamma) \prec_{rpo} a_i^n a_{i+1} \mathcal{O}(\beta) \mathcal{O}(\gamma),$$
$$\forall n \in \mathbb{N} \quad a_{i+1} \mathcal{O}(\beta) \mathcal{O}(\gamma) \prec_{rpo} a_{i+1} \mathcal{O}(\beta_n) \mathcal{O}(\gamma).$$

The proof is direct, using the definition of $\prec_{rpo}$ with an easy induction on β. As a consequence, $\mathcal{O}$ is an injective morphism. It remains to show that $\mathcal{O}$ is surjective. Let $u \in \Sigma^*$. We construct by induction on the length of u an ordinal α such that $\mathcal{O}(\alpha) = u$. If $u = a_0 v$ for some $v \in \Sigma^*$, then the induction hypothesis gives us an ordinal β such that $\mathcal{O}(\beta) = v$. Set $\alpha = \beta + 1$. The definition of $\mathcal{O}$ ensures $\mathcal{O}(\alpha) = a_0 v$. If $u = a_{i+1} v$ for some $v \in \Sigma^*$ and some $i \in \mathbb{N}$, we have to consider two sub-cases. When $v \in \{a_0, \ldots, a_{i+1}\}^*$, let β such that $\mathcal{O}(-1+\beta) = v$. In this case, $\beta < \omega^{\omega^{i+1}}$. So $\mathcal{O}(\omega^{\omega^i}(\beta)) = a_{i+1} v$. Otherwise, there exist $b \in \{a_{i+1}, \ldots\}, v_1 \in \{a_0, \ldots, a_{i+1}\}^*$ and $v_2 \in \Sigma^*$ such that $v = v_1 b v_2$. Let β and γ such that $\mathcal{O}(-1+\beta) = v_1$ and $\mathcal{O}(\gamma) = bv_2$. We have $\mathcal{O}(\gamma + \omega^{\omega^i} \beta) = a_{i+1} \, v_1 \, bv_2 = a_{i+1} v$. $\qquad \square$

From now on, we shall always consider that a string built up from the alphabet $\Sigma = \{a_0, \ldots, a_i, \ldots\}$ is a notation for an ordinal of ω^{ω^ω}. We express the predecessor functions in this notation system.

Proposition 5. *For all $u \in \Sigma^*$, for all $i, j, n \in \mathbb{N}$*

   (i) $P_n(a_0 u) = u$,

   (ii) $P_n(a_{i+1} a_0 u) = a_i^n a_{i+1} u$,

   (iii) $P_n(a_{i+1} a_j u) = a_{i+1} P_n(a_j u)$, *when $0 < j \leq i+1$*,

   (iv) $P_n(a_{i+1} u) = a_i^n u$, *otherwise.*

Proof. Given an ordinal α of ω^{ω^ω}, we distinguish four main cases for the computation of $P_n(\alpha)$:

Case 1: $\alpha = \delta + 1$. Then $P_n(\alpha) = \delta$.

In all remaining cases, α is a limit ordinal. Let i, β, γ satisfying conditions (i), (ii) of proposition 3 such that $\alpha = \gamma + \omega^{\omega^i} \beta$.

Case 2: $\alpha = \gamma + \omega^{\omega^i}(\beta' + 1)$, $i = 0, \beta' \neq 0$. Then $P_n(\alpha) = \gamma + \omega^{\omega^i}\beta' + (\omega^{\omega^i})_n$,

Case 3: $\alpha = \gamma + \omega^{\omega^i}\beta$, β limit. Then $P_n(\alpha) = \gamma + \omega^{\omega^i}\beta_n$,

Case 4: $\alpha = \gamma + \omega^{\omega^i}$. Then $P_n(\alpha) = \gamma + (\omega^{\omega^i})_n$.

We apply the morphism $\mathcal{O}$ to α and $P_n(\alpha)$. The proof is by induction on the length of u.

Case 1: $\mathcal{O}(\alpha) = a_0 \mathcal{O}(\delta)$ and $\mathcal{O}(P_n(\alpha)) = \mathcal{O}(\delta)$,

Case 2: $\mathcal{O}(\alpha) = a_{i+1} a_0 \mathcal{O}(\beta')\mathcal{O}(\gamma)$ and $\mathcal{O}(P_n(\alpha)) = a_i^n a_{i+1} \mathcal{O}(\beta')\mathcal{O}(\gamma)$,

Case 3: $\mathcal{O}(\alpha) = a_{i+1}\mathcal{O}(\beta)\mathcal{O}(\gamma)$ and $\mathcal{O}(P_n(\alpha)) = a_{i+1}\mathcal{O}(\beta_n)\mathcal{O}(\gamma)$. By induction hypothesis, note that for all $j \in \mathbb{N}$, $u \in \{a_0, \ldots, a_j\}^+$ and $v \in \Sigma^*$ $P_n(ua_{j+1}v) = P_n(u)a_{j+1}v$. So $\mathcal{O}(P_n(\alpha)) = a_{i+1}P_n(\mathcal{O}(\beta)\mathcal{O}(\gamma))$.

Case 4: $\mathcal{O}(\alpha) = a_{i+1}\mathcal{O}(\gamma)$ and $\mathcal{O}(P_n(\alpha)) = a_{i-1}^n \mathcal{O}(\gamma)$.

(i) comes from case *1*, (ii) from case *2*, (iii) from case *3* and (iv) from case *4*.

$\qquad \square$

4 Encoding multiply recursive functions by totally terminating SRS's

4.1 Construction of the SRS

In this subsection, we build a family of string rewriting systems that simulate the Hardy hierarchy using the notation system of the previous section. Our aim is to describe $(\star)$-sequences of the form

$$(u, n),\ (P_n(u), n+1),\ (P_{n+1}P_n(u), n+2) \ldots$$

In a couple (u, n), we represent the ordinal u by the corresponding notation of Σ^*. For n, we introduce a new symbol $|$ and denote the integer n in unary notation : $|^n$. To deal with technical details in the computation, we need two extra symbols, $\circ$ and $\bullet$. Each step (u, n) of the $(\star)$-sequence is finally encoded by the string $\bullet|^n u$.

It is hopeless to try to simulate the whole class of multiply recursive functions by a single finite string rewriting system. We define a family $\mathcal{R}_i$, $i \in \mathbb{N}$, such that $\mathcal{R}_i$ exhausts the Hardy hierarchy on ω^{ω^i}. For all $u \in \{a_0, \ldots, a_i\}^*$, the system $\mathcal{R}_i$ should then allow us to derive

$$\bullet|^n u \xrightarrow{+}_{\mathcal{R}_i} \bullet|^{n+1} P_n(u).$$

We set

$$\mathcal{R}_0 \left\{ \begin{array}{ll} a_0 \to \circ & (0,1) \\ \circ \to \bullet| & (0,2) \\ \bullet| \to |\bullet\bullet & (0,3) \\ |\circ \to \circ| & (0,4) \\ \bullet \to \varepsilon & (0,5) \end{array} \right.$$

and

$$\mathcal{R}_{i+1} = \mathcal{R}_i \cup \left\{ \begin{array}{ll} \bullet a_{i+1} a_0 \to k_{i+1} a_{i+1} & (i+1,1) \\ \bullet a_{i+1} \to a_{i+1}\bullet & (i+1,2) \\ \bullet k_{i+1} \to k_{i+1} a_i & (i+1,3) \\ a_{i+1}\circ \to \circ a_{i+1} & (i+1,4) \\ k_{i+1} \to \circ & (i+1,5) \\ \bullet a_{i+1} \to k_{i+1} & (i+1,6) \end{array} \right.$$

Proposition 6. *Let $u \in \{a_0, \ldots, a_{i+1}\}^*$. For all $n \geq 1$, $\bullet|^n u \xrightarrow{+}_{\mathcal{R}_{i+1}} \bullet|^{n+1} P_n(u)$.*

Proof. Define $(\spadesuit)$ $\quad \bullet^{n+1} u \xrightarrow{+} \circ P_n(u)$. $(\spadesuit)$ implies the desired result.

$$\begin{array}{ll} \bullet|^n u \xrightarrow{+} |^n \bullet^{2^n} u & (0,3)^+ \\ \xrightarrow{*} |^n \bullet^{n+1} u & (0,5)^* \\ \xrightarrow{+} |^n \circ P_n(u) & (\spadesuit) \\ \xrightarrow{+} \circ|^n P_n(u) & (0,4)^+ \\ \xrightarrow{+} \bullet|^{n+1} P_n(u) & (0,2) \end{array}$$

We now establish ($\spadesuit$) by induction on u. We consider the four cases introduced in proposition 5.

Case 1:

$$\bullet^{n+1} a_0 v \to a_0 v \qquad (0,5)^*$$
$$\to \circ v \qquad (0,1)$$

Case 2:

$$\bullet^{n+1} a_{i+1} a_0 v \to \bullet^n k_{i+1} a_{i+1} v \quad (i{+}1,1)$$
$$\overset{\cdot}{\to} k_{i+1} a_i^n a_{i+1} v \quad (i{+}1,3)^*$$
$$\to \circ a_i^n a_{i+1} v \quad (i{+}1,5)$$

Case 3:

$$\bullet^{n+1} a_{i+1} v \overset{+}{\to} a_{i+1} \bullet^{n+1} v \quad (i{+}1,2)^*$$
$$\overset{\cdot}{\to} a_{i+1} \circ P_n(v) \quad \text{(induction hypothesis)}$$
$$\to \circ a_{i+1} P_n(v) \quad (i{+}1,4)$$

Case 4:

$$\bullet^{n+1} a_{i+1} v \to \bullet^n k_{i+1} v \qquad (i{+}1,6)$$
$$\to k_{i+1} a_i^n v \qquad (i{+}1,3)^*$$
$$\to \circ a_i^n v \qquad (i{+}1,5)$$

$\square$

Corollary 1. *For each multiply recursive function f, there exists i in $\mathbb{N}$ such that f is eventually dominated by $Dl_{\mathcal{R}_i}$, the derivation length of $\mathcal{R}_i$.*

Proof. The proposition 6 implies that for all i in $\mathbb{N}$, $Dl_{\mathcal{R}_i}$ eventually dominates the Hardy function indexed by ω^{ω^i}. $\square$

4.2 For all $i \in \mathbb{N}$, $\mathcal{R}_i$ is totally terminating

We now come to the final argument of our construction and show that the string rewriting systems $\mathcal{R}_i$ are totally terminating. The proof relies on proposition 1: we define for $\mathcal{R}_i$ a monotone ordering which enjoys the sub-term property. Our starting point is the intentional meaning of the symbols of Σ: each string u built up from $a_0, \ldots, a_i, \ldots$ may simply be interpreted by the underlying ordinal or equivalently by the notation based on strings of Σ^*. For the symbol k_i, define the function ψ_i by

$$\psi_{i+1} : u \mapsto \sup\{a_i^n u; \ n \in \mathbb{N}\}.$$

(The supremum is wrt $\prec_{rpo}$.) We have the following properties.

Lemma 1. *For all u in Σ^*, for all i in $\mathbb{N}$*

(i) $\psi_{i+1}(u) = \psi_{i+1}(a_i u)$,

(ii) $a_{i+1} a_0 u \succeq_{rpo} \psi_{i+1}(a_{i+1} u)$,

(iii) $a_{i+1} u \succeq_{rpo} \psi_{i+1}(u)$,

(iv) $\psi_i(u) \succ_{rpo} u$,

(v) ψ_i *is an increasing function.*

Proof. (i) is by definition of ψ_i and (ii), (iii), (iv), (v) are easy consequences of the definition of $\preceq_{rpo}$. □

For the symbols •, ∘ and |, consider the sub-system

$$\mathcal{S} \left\{ \begin{array}{c} \circ \to \bullet| \\ \bullet| \to |\bullet\bullet \\ |\circ \to \circ| \\ \bullet \to \varepsilon \\ \bullet a_{i+1} \to a_{i+1}\bullet \\ \bullet \to a_i \\ a_{i+1}\circ \to \circ a_{i+1} \end{array} \right.$$

Lemma 2. *There exists a total termination ordering $\prec_{\mathcal{S}}$ for $\mathcal{S}$.*

Proof. We give an interpretation $\mathcal{I}$ on $\mathbb{N}^3$ for the rules of $\mathcal{S}$.

$$\mathcal{I}(\circ)(n, m, p) = (2n + 4, m, p)$$
$$\mathcal{I}(|)(n, m, p) = (2n + 1, m, p)$$
$$\mathcal{I}(\bullet)(n, m, p) = (n, n + m, 2p + 2)$$
$$\mathcal{I}(a_i)(n, m, p) = (n, n + m, p + 1)$$

Define $u \prec_{\mathcal{S}} v$ by $\mathcal{I}(u) < \mathcal{I}(v)$. □

Combining the ordering $\prec_{rpo}$ for a_i and k_i, and the ordering $\prec_{\mathcal{S}}$ for $|$, ∘ and •, we define the interpretation $[\,]$ on $(\Sigma \cup \{k_i, \ i > 0\})^* \times (\Sigma \cup \{\circ, \bullet, |\})^*$ as follows:

$$[a_i] = (u, v) \mapsto (a_i u, a_i v)$$
$$[k_{i+1}] = (u, v) \mapsto (\psi_{i+1}(u), v)$$
$$[\bullet] = (u, v) \mapsto (u, \bullet v)$$
$$[\circ] = (u, v) \mapsto (u, \circ v)$$
$$[|] = (u, v) \mapsto (u, |v)$$

$(\Sigma \cup \{k_i, \ i > 0\})^* \times (\Sigma \cup \{\circ, \bullet, |\})^*$ is ordered by the left-to-right lexicographic combination of $\prec_{rpo}$ and $\prec_{\mathcal{S}}$. We finally define $\prec$ by

$$u \prec v \ \Leftrightarrow \ [u] \ \mathsf{lex}(\prec_{rpo}, \prec_{\mathcal{S}}) \ [v].$$

Lemma 3.

   (i) $\prec$ *has the sub-term property,*

   (ii) $\prec$ *is monotone,*

   (iii) *for all $i \in \mathbb{N}$, for all $l \to r \in \mathcal{R}_i$, for all $w \in \Sigma^*$, $lw \succ rw$.*

Proof. (i) and (ii) are consequences of lemma 1-(iv), (v). We establish (iii) : we examine each rule of $\mathcal{R}_0$ and $\mathcal{R}_{i+1}$ and verify that it reduces under the interpretation $[\,]$ with the ordering $\prec$. Let $w \in \Sigma^*$ and $(u, v) = [w]$.

For $\mathcal{R}_0$:

$(0,1) : (a_0 u, a_0 v) \succ (u, \circ v),$
$(0,2) : (u, \circ v) \succ (u, \bullet | v),$
$(0,3) : (u, \bullet | v) \succ (u, | \bullet \bullet v),$
$(0,4) : (u, | \circ v) \succ (u, \circ | v),$
$(0,5) : (u, \bullet v) \succ (u, v).$

For $\mathcal{R}_{i+1}$:

$(i+1,1) : (a_{i+1} a_0 u, \bullet a_{i+1} a_0 v) \succ (\psi_{i+1}(a_{i+1} u), a_{i+1} v),$
$(i+1,2) : (a_{i+1} u, \bullet a_{i+1} v) \succ (a_{i+1} u, a_{i+1} \bullet v),$
$(i+1,3) : (\psi_{i+1}(u), \bullet v) \succ (\psi_{i+1}(a_i u), a_i v),$
$(i+1,4) : (a_{i+1} u, a_{i+1} \circ v) \succ (a_{i+1} u, \circ a_{i+1} v),$
$(i+1,5) : (\psi_{i+1}(u), v) \succ (u, \circ v),$
$(i+1,6) : (a_{i+1} u, \bullet a_{i+1} v) \succ (\psi_{i+1}(u), v).$

(See lemma 1-(i), (ii), (iii)) □

Proposition 7. *For all $i \in \mathbb{N}$, $\mathcal{R}_i$ is totally terminating.*

Proof. Consequence of lemma 3 and proposition 1. □

5 Conclusion : what about *term* rewriting systems?

We have encoded the maximal order type of the division ordering $\trianglelefteq$ by strings equipped with the recursive path ordering. Here is the key point of the construction. Then it is easy to simulate the Hardy hierarchy for this notation system. We believe that this approach would apply to term rewriting systems, using the lexicographic path ordering on terms instead of the recursive path ordering on strings: the order type of the lexicographic path ordering reaches the maximal order type of the homeomorphic embedding of Kruskal's theorem.

Acknowledgements. The author thanks Adam Cichon and the anonymous referee.

References

1. E.A. Cichon, *A short proof of two recently discovered independence results using recursion theoretic methods*. Proceedings of the American Mathematical Society, vol 97 (1983), p.704-706.
2. E.A. Cichon and E. Tahhan Bittar, *Ordinal recursive bounds for Higman's theorem*. To appear in Theoretical Computer Science.
3. D.H.J. De Jongh and R. Parikh, *Well-partial orderings and hierarchies*. Indagationes Mathematicae 14 (1977), p. 195-207.
4. N. Dershowitz, *Orderings for term rewriting systems*. Theoretical Computer Science 17-3 (1982), p. 279–301.

5. N. Dershowitz et J.P. Jouannaud, *Rewrite systems.* Handbook of Theoretical Computer Science vol.B, North-Holland.

6. M.C.F. Ferreira and H. Zantema, *Total termination of term rewriting.* Proceedings of RTA-93, Lecture notes in Computer Science 690, p. 213-227.

7. G. Higman, *Ordering by divisibility in abstract algebras.* Bull. London Mathematical Society 3-2 (1952), p. 326-336.

8. D. Hofbauer and C. Lautemann, *Termination proofs and the length of derivations.* Proceedings of RTA-88, Lecture Notes in Computer Science 355.

9. D. Hofbauer, *Termination proofs with multiset path orderings imply primitive recursive derivation lengths.* Theoretical Computer Science 105-1 (1992), p.129-140.

10. U. Martin et E.A. Scott, *The order types of termination orderings on monadic terms, strings and multisets.* 8th Annual Symposium on Logic in Computer Science, IEEE (1993), p. 356-363.

11. R. Péter, *Recursive Functions.* Academic Press (1967)

12. J.W. Robbin, *Subrecursive Hierarchies.* Ph.D. Princeton

13. H. Touzet, *A complex example of a simplifying rewrite system.* Proceedings of ICALP'98, Lecture Notes in Computer Science 1442 (1998).

14. H. Touzet, *Encoding the Hydra Battle as a rewrite system.* Proceedings of MFCS'98, Lecture Notes in Computer Science 1450 (1998).

Deciding the Word Problem in the Union of Equational Theories Sharing Constructors

Franz Baader[1]* and Cesare Tinelli[2]

[1] LuFg Theoretical Computer Science, RWTH Aachen,
Ahornstraße 55, 52074 Aachen, Germany
baader@informatik.rwth-aachen.de
[2] Department of Computer Science, University of Illinois at Urbana-Champaign,
1304 W. Springfield Ave, Urbana, IL 61801 – USA
tinelli@cs.uiuc.edu

Abstract. The main contribution of this paper is a new method for combining decision procedures for the word problem in equational theories sharing "constructors." The notion of constructors adopted in this paper has a nice algebraic definition and is more general than a related notion introduced in previous work on the combination problem.

1 Introduction

The integration of constraint solvers (that is, specialized decision procedures for restricted classes of problems) into general purpose deductive systems (such as Knuth-Bendix completion procedures, resolution-based theorem provers, or Logic Programming systems) aims at combining the efficiency of the specialized method with the universality of the general one. Many applications of the constraint-based systems obtained by such an integration require a combination of more than one constraint language, and thus a solver for the resulting mixed constraints. The development of general combination methods for constraint solvers tries to avoid the necessity of designing a new specialized decision procedure for each new combination of constraint languages.

For equational theories, one is usually interested in solvers for the following decision problems: the word problem, the matching problem, and the unification problem. In this setting, the research on combination of constraint solvers is mainly concerned with finding conditions under which the following question can be answered affirmatively: given two equational theories E_1 and E_2 with decidable word/matching/unification problems, is the word/matching/unification problem for $E_1 \cup E_2$ also decidable?

A very effective (but also rather strong) restriction is to require that E_1 and E_2 be equational theories over disjoint signatures. Under this restriction, decision procedures for the word problems in E_1 and E_2 can always be combined into a decision procedure for the word problem in $E_1 \cup E_2$ [10, 14, 13, 8, 7]. For the matching and the unification problem, there also exist very general combination

---

* Partially supported by the EC Working Group CCL II.

results under the disjointness restriction (see [12] for matching, and, e.g., [13, 4, 1] for unification). It is not hard to extend these results to theories sharing constant symbols [11, 7, 2]. The only work we are aware of that presents a general combination approach for the union of equational theories having more than constant symbols in common is [6], where the problem of combining algorithms for the unification, matching, and word problem is investigated for theories sharing so-called "constructors."

In this paper, we restrict our attention to the word problem. The combination result we obtain improves on the corresponding result in [6] in the following respects. Firstly, we introduce a notion of constructors, modeled after the one introduced in [15], which is strictly more general than the one in [6]. Whereas [6] does not allow for nontrivial identities between constructor terms, we only require the constructor theory to be collapse-free. Secondly, the definition of constructors in [6] depends strongly on technical details such as the choice of an appropriate well-founded and monotonic ordering. In contrast, our definition uses only abstract algebraic properties. Finally, the combination procedure described in [6], like the ones for the disjoint case [10, 13, 8, 7], directly transforms the terms for which the word problem is to be decided, by applying collapse equations[1] and abstracting alien subterms. This transformation process must be carried on with a rather strict strategy (in principle, going from the leaves of the terms to their roots) and it is not easy to describe. In contrast, our procedure extends the rule-based combination procedure for the word problem introduced in [2] for the case of shared constants. It works on a set of equations rather than terms, and its transformation rules can be applied in arbitrary order, that is, no strategy is needed. We claim that this difference makes the method more flexible and easier to describe and comprehend.

The next section introduces the word problem and describes a reduction of the word problem in the union of equational theories to satisfiability of a conjunction of two pure formulae. Before we can describe our combination procedure, we must introduce our notion of constructors (Section 3). Section 3 also contains some results concerning the union of theories sharing constructors. In Section 4 we describe the new combination procedure for theories sharing constructors, and prove its correctness. Section 5 investigates the connection between our notion of constructors and the one introduced in [6], and includes some remarks on how this work relates to the research on modularity properties of term rewriting systems. Because of the page limit, we cannot give detailed proofs of our results. They can be found in [3].

2 Word Problems and Satisfiability Problems

We will use V to denote a countably infinite set of variables, and $T(\Omega, V)$ to denote the set of all Ω-terms, that is, terms over the signature Ω with variables in V. An equational theory E over the signature Ω is a set of (implicitly universally

---

[1] i.e., equations of the form $x \equiv t$, where x is a variable occurring in the non-variable term t.

quantified) equations between Ω-terms. We use $s \equiv t$ to denote an equation between the terms s, t. For an equational theory E, the *word problem* is concerned with the validity in E of quantifier-free formulae of the form $s \equiv t$. Equivalently, the word problem asks for the (un)satisfiability of the *disequation* $s \not\equiv t$ in E— where $s \not\equiv t$ is an abbreviation for the formula $\neg(s \equiv t)$. As usual, we often write "$s =_E t$" to express that the formula $s \equiv t$ is valid in E. An equational theory E is *collapse-free* iff $x \neq_E t$ for all variables x and non-variable terms t.

Given an Ω-term s, an Ω-algebra $\mathcal{A}$, and a valuation α (of the variables in s by elements of $\mathcal{A}$), we denote by $[\![s]\!]_\alpha^{\mathcal{A}}$ the interpretation of the term s in $\mathcal{A}$ under the valuation α. Also, if Σ is a subsignature of Ω, we denote by $\mathcal{A}^\Sigma$ the reduct of $\mathcal{A}$ to the subsignature Σ. An Ω-algebra $\mathcal{A}$ is a model of E iff every equation in E is valid in $\mathcal{A}$. The equational theory E over the signature Ω defines an Ω-*variety*, i.e., the class of all models of E. When E is *non-trivial* i.e., has models of cardinality greater than 1, this variety contains free algebras for any set of generators. We will call these algebras E-*free algebras*. Given a set of generators (or variables) X, an E-free algebra with generators X can be obtained as the quotient term algebra $\mathcal{T}(\Omega, X)/=_E$. It is well-known that two E-free algebras with sets of generators of the same cardinality are isomorphic.

In this paper, we are interested in *combined* equational theories, that is, equational theories E of the form $E := E_1 \cup E_2$, where E_1 and E_2 are equational theories over two (not necessarily disjoint) signatures Σ_1 and Σ_2. The elements of $\Sigma_1 \cap \Sigma_2$ are called *shared* symbols. We call 1-*symbols* the elements of Σ_1 and 2-*symbols* the elements of Σ_2. A term $t \in T(\Sigma_1 \cup \Sigma_2, V)$ is an i-*term* iff its *top symbol* $t(\epsilon) \in V \cup \Sigma_i$, i.e., if t is a variable or has the form $t = f(t_1, ..., t_n)$ for some i-symbol f $(i = 1, 2)$. Note that variables and terms t with $t(\epsilon) \in \Sigma_1 \cap \Sigma_2$ are both 1- and 2-terms. A subterm s of a 1-term t is an *alien subterm* of t iff it is not a 1-term and every proper superterm of s in t is a 1-term. Alien subterms of 2-terms are defined analogously. For $i = 1, 2$, an i-term s is *pure* iff it contains only i-symbols and variables. A (dis)equation $s \equiv t$ $(s \not\equiv t)$ is i-*pure* iff s and t are pure i-terms. It is called pure iff it is i-pure for some $i \in \{1, 2\}$.

A given disequation $s \not\equiv t$ between $(\Sigma_1 \cup \Sigma_2)$-terms s, t can be transformed into an equisatisfiable formula $\varphi_1 \wedge \varphi_2$, where φ_i is a conjunction of i-pure equations and disequations $(i = 1, 2)$. This can be achieved by the usual *variable abstraction* process in which alien subterms are replaced by new variables (see, e.g., [1,3] for a detailed description of the process). Obviously, if we know that $\varphi_1 \wedge \varphi_2$ is satisfiable in a model $\mathcal{A}$ of $E_1 \cup E_2$, then φ_i is satisfiable in the reduct $\mathcal{A}^{\Sigma_i}$, which is a model of E_i $(i = 1, 2)$. However, the converse need not be true, that is, if φ_i is satisfiable in a model $\mathcal{A}_i$ of E_i $(i = 1, 2)$, then we cannot necessarily deduce that the conjunction $\varphi_1 \wedge \varphi_2$ is satisfiable in some model $\mathcal{A}$ of $E_1 \cup E_2$. One case in which we can is described by the proposition below.

Proposition 1. *Let $\mathcal{A}_i$ be a model of E_i $(i = 1, 2)$, and $\Sigma := \Sigma_1 \cap \Sigma_2$. Assume that the reducts $\mathcal{A}_1{}^\Sigma$ and $\mathcal{A}_2{}^\Sigma$ are both free in the same Σ-variety and their respective sets of generators Y_1 and Y_2 have the same cardinality. If φ_i is satisfiable in $\mathcal{A}_i$ with the variables in $Var(\varphi_1) \cap Var(\varphi_2)$ taking distinct values over Y_i for $i = 1, 2$, then there is a model of $E_1 \cup E_2$ in which $\varphi_1 \wedge \varphi_2$ is satisfiable.*

This proposition is a special case of more general results in [15]. A simpler direct proof in the special case can also be found in [3].

In the following, we will consider the case where the algebras $\mathcal{A}_i$ are E_i-free. Unfortunately, the property of being a free algebra is not preserved under signature reduction. The problem is that the reduct of an algebra may need more generators than the algebra itself. For example, consider the signature $\Omega := \{\mathsf{p}, \mathsf{s}\}$ and the equational theory E axiomatized by the equations

$$E := \{x \equiv \mathsf{p}(\mathsf{s}(x)),\ x \equiv \mathsf{s}(\mathsf{p}(x))\}.$$

The integers $\mathcal{Z}$ are a free model of E over a set of generators of cardinality 1 when s and p are interpreted as the successor and the predecessor function, respectively. Now, if $\Sigma := \{\mathsf{s}\}$, then $\mathcal{Z}^\Sigma$ is definitely not free because it does not even admit a non-redundant set of generators, which is a necessary condition for an algebra to be free.

Nonetheless, there are free algebras admitting reducts that are also free, although over a possibly larger set of generators. These algebras are models of equational theories that admit *constructors* in the sense explained in the next section.

3 Theories Admitting Constructors

In the following, Ω will be an at most countably infinite functional signature, and Σ a subset of Ω. For a given equational theory E over Ω we define the Σ-*restriction of* E as $E^\Sigma := \{s \equiv t \mid s, t \in T(\Sigma, V) \text{ and } s =_E t\}$.

Definition 2 (Constructors). *The subsignature Σ of Ω is a set of constructors for E if the following two properties hold:*

1. *The Σ-reduct of the countably infinitely generated E-free Ω-algebra is an E^Σ-free algebra.*
2. *E^Σ is collapse-free.*

This definition is a rather abstract formulation of our requirements on the theory E. In the following, we develop a more concrete characterization[2] of theories admitting constructors, which will make it easier to show that a given theory admits constructors. But first, we must introduce some more notation.

Given a subset G of $T(\Omega, V)$, we denote by $T(\Sigma, G)$ the set of terms over the "variables" G. To express this construction we will denote any such term by $s(\bar{r})$ where $\bar{r}$ is the tuple made of the terms of G that replace the variables of s. Notice that this notation is consistent with the fact that $G \subseteq T(\Sigma, G)$. In fact, every $r \in G$ can be represented as $s(r)$ where s is a variable of V. Also notice that $T(\Sigma, V) \subseteq T(\Sigma, G)$ whenever $V \subseteq G$. In this case, every $s \in T(\Sigma, V)$ can be trivially represented as $s(\bar{v})$ where $\bar{v}$ are the variables of s.

---

[2] This *characterization* of constructors is a special case of the *definition* of constructors in [15].

For every equational theory E over the signature Ω and every subset Σ of Ω, we define the following subset of $T(\Omega, V)$:

$$G_E(\Sigma, V) := \{r \in T(\Omega, V) \mid r \neq_E f(\bar{t}) \text{ for all } f \in \Sigma \text{ and } \bar{t} \text{ in } T(\Omega, V)\}.$$

We will show that, if Σ is a set of constructors for E, then $G_E(\Sigma, V)$ determines a set of free generators for the Σ-reduct of the countably infinitely generated E-free algebra. But first, let us point out the following properties of $G_E(\Sigma, V)$:

Lemma 3. *Let E be an equational theory over Ω and $\Sigma \subseteq \Omega$.*

1. *$G_E(\Sigma, V)$ is nonempty iff $V \subseteq G_E(\Sigma, V)$;*
2. *If $V \subseteq G_E(\Sigma, V)$, then E^Σ is collapse-free.*

Theorem 4 (Characterization of constructors). *Let $\Sigma \subseteq \Omega$, E a nontrivial equational theory over Ω, and $G := G_E(\Sigma, V)$. Then Σ is a set of constructors for E iff the following holds:*

1. *$V \subseteq G$.*
2. *For all $t \in T(\Omega, V)$, there is an $s(\bar{r}) \in T(\Sigma, G)$ such that $t =_E s(\bar{r})$.*
3. *For all $s_1(\bar{r}_1), s_2(\bar{r}_2) \in T(\Sigma, G)$,*

$$s_1(\bar{r}_1) =_E s_2(\bar{r}_2) \text{ iff } s_1(\bar{v}_1) =_E s_2(\bar{v}_2),$$

where $\bar{v}_1, \bar{v}_2$ are fresh variables abstracting $\bar{r}_1, \bar{r}_2$ so that two terms in $\bar{r}_1, \bar{r}_2$ are abstracted by the same variable iff they are equivalent in E.

Actually, the proof of the theorem—which can be found in [3]—provides a little more information than stated in the formulation of the theorem.

Corollary 5. *Let Σ be a set of constructors for E, $\mathcal{A}$ an E-free Ω-algebra with the countably infinite set of generators X, and α a bijective valuation of V onto X. Then, the reduct $\mathcal{A}^\Sigma$ is an E^Σ-free algebra with generators $Y := \{[\![r]\!]_\alpha^{\mathcal{A}} \mid r \in G_E(\Sigma, V)\}$, and $X \subseteq Y$.*

Condition 2 of Theorem 4 says that, when Σ is a set of constructors for E, every Ω-term t is equivalent in E to a term $s(\bar{r}) \in T(\Sigma, G)$ where $G := G_E(\Sigma, V)$. We will call $s(\bar{r})$ a *normal form of t in E*—in general, a term may have more than one normal form. We will say that a term t is *in normal form* if it is already of the form $t = s(\bar{r}) \in T(\Sigma, G)$. Because $V \subseteq G$, it is immediate that Σ-terms are in normal form, as are terms in G. We will say that a term t is *E-reducible* if it is not in normal form. Otherwise, it is *E-irreducible*.

We will make use of normal forms in our combination procedure. In particular, we will consider normal forms that are computable in the following sense.

Definition 6 (Computable Normal Forms). *Let Σ be a set of constructors for the equational theory E over the signature Ω. We say that normal forms are computable for Σ and E if there is a computable function*

$$\mathrm{NF}_E^\Sigma : T(\Omega, V) \longrightarrow T(\Sigma, G)$$

such that $\mathrm{NF}_E^\Sigma(t)$ is a normal form of t, i.e., $\mathrm{NF}_E^\Sigma(t) =_E t$.

Notice that Definition 6 does not entail that the variables of $\mathrm{NF}_E^\Sigma(t)$ are included in the variables of t. However, if $V_0 := Var(\mathrm{NF}_E^\Sigma(t)) \setminus Var(t)$ is nonempty, then $\pi(\mathrm{NF}_E^\Sigma(t))$ is also a normal form of t for any injective renaming π of the variables in V_0. Consequently, if V_1 is a given finite subset of V, we can always assume without loss of generality that $Var(\mathrm{NF}_E^\Sigma(t)) \setminus Var(t)$ and V_1 are disjoint.[3] As a rule then we will *always* assume that the variables occurring in a normal form $\mathrm{NF}_E^\Sigma(t)$ but not in t, if any, are *fresh* variables.

An important consequence of Definition 6 is that, when normal forms are computable for Σ and E, it is always possible to tell whether a term is in normal form or not.

Proposition 7. *Let Σ be a set of constructors for the equational theory E over the signature Ω and assume that normal forms are computable for Σ and E. Then, the E-reducibility of terms in $T(\Omega, V)$ is decidable.*

We provide below two examples of equational theories admitting constructors in the sense of Definition 2. But first, let us consider some counter-examples:

- The signature $\Sigma := \Omega := \{f\}$ is not a set of constructors for the theory E axiomatized by $\{x \equiv f(x)\}$ because Definition 2(2) is not satisfied.
- The signature $\Sigma := \{f\} \subseteq \{f, g\} =: \Omega$ is not a set of constructors for the theory E axiomatized by $\{g(x) \equiv f(g(x))\}$ because Theorem 4(2) is not satisfied. In fact, the term $g(x)$ does not have a normal form. (The signature $\{f, g\}$, however, is a set of constructors for the same theory.)
- Finally, take $\Omega := \{f, g\}$ and $\Sigma := \{f\}$ and consider the theory $E := \{f(g(x)) \equiv f(f(g(x)))\}$. Then we have $G_E(\Sigma, V) = V \cup \{g(t) \mid t \in T(\Omega, V)\}$. It is easy to see that conditions (1) and (2) of Theorem 4 hold. However, condition (3) does not hold since $f(g(x)) =_E f(f(g(x)))$, although $f(y) \neq_E f(f(y))$.

Example 8. The theory of the natural numbers with addition is the most immediate example of a theory with constructors. Consider the signature $\Sigma_1 := \{0, \mathsf{s}, +\}$ and the equational theory E_1 axiomatized by the equations below:

$$x + (y + z) \equiv (x + y) + z, \quad x + y \equiv y + x, \quad x + \mathsf{s}(y) \equiv \mathsf{s}(x + y), \quad x + 0 \equiv x.$$

It can be shown that the signature $\Sigma := \{0, \mathsf{s}\}$ is a set of constructors for E_1 in the sense of Definition 2. The proof in [3] uses the fact that orienting the third and fourth equation from left to right yields a canonical term rewrite system modulo the first two equations. Note that the restriction of E_1 to Σ (i.e., the theory $E_1{}^\Sigma$) is the syntactic equality of Σ-terms.

Example 9. Consider the signature $\Sigma_2 := \{0, 1, \mathsf{rev}, \cdot\}$ and the equational theory E_2 axiomatized by the equations below:

$$\begin{aligned} x \cdot (y \cdot z) &\equiv (x \cdot y) \cdot z, & \mathsf{rev}(0) &\equiv 0, & \mathsf{rev}(1) &\equiv 1, \\ \mathsf{rev}(x \cdot y) &\equiv \mathsf{rev}(y) \cdot \mathsf{rev}(x), & \mathsf{rev}(\mathsf{rev}(x)) &\equiv x. \end{aligned}$$

---

[3] Otherwise, we apply an appropriate renaming that produces a normal form of t satisfying such disjointness condition.

The signature $\Sigma' := \{0, 1, \cdot\}$ is a set of constructors for E_2 in the sense of Definition 2. The proof in [3] depends on the fact that orienting the equations from left to right yields a canonical term rewriting system. This example differs from the previous one in that the restriction of the theory to the constructor signature is no longer syntactic equality: $E_2{}^\Sigma$ expresses associativity of ".".

Combination of Theories Sharing Constructors

For the next results, in which we go back to the problem of combining equational theories, we will consider two non-trivial equational theories E_1, E_2 with respective countable signatures Σ_1, Σ_2 such that $\Sigma := \Sigma_1 \cap \Sigma_2$ is a set of constructors for E_1 and for E_2, and $E_1{}^\Sigma = E_2{}^\Sigma$.

The proposition below—which is important in the proof of correctness of our combination procedure—is an easy consequence of Proposition 1 and Corollary 5.

Proposition 10. *For $i = 1, 2$, let $\mathcal{A}_i$ be an E_i-free Σ_i-algebra with a countably infinite set X_i of generators, and let $Y_i := \{[\![r]\!]_{\alpha_i}^{\mathcal{A}_i} \mid r \in G_E(\Sigma_i, V)\}$, where α_i is any bijective valuation of V onto X_i. Let φ_1, φ_2 be conjunctions of equations and disequations of respective signature Σ_1, Σ_2. If φ_i is satisfiable in $\mathcal{A}_i$ with $Var(\varphi_1) \cap Var(\varphi_2)$ taking distinct values over Y_i for $i = 1, 2$, then $\varphi_1 \wedge \varphi_2$ is satisfiable in $E_1 \cup E_2$.*

The following theorem shows that being a set of constructors is a modular property. Thus, the application of the combination procedure described in the next section can be iterated.

Theorem 11. *Let E_1, E_2 be two non-trivial equational theories with respective signatures Σ_1, Σ_2 such that $\Sigma := \Sigma_1 \cap \Sigma_2$ is a set of constructors for E_1 and for E_2, $E_1{}^\Sigma = E_2{}^\Sigma$, the word problem for E_i is decidable, and normal forms are computable for Σ and E_i for $i = 1, 2$. Then, the following holds:*

1. *Σ is a set of constructors for $E := E_1 \cup E_2$.*
2. *$E^\Sigma = E_1{}^\Sigma = E_2{}^\Sigma$.*
3. *Normal forms are computable for Σ and E.*

The (quite involved) proof in [3] shows that the three conditions in Theorem 4 are satisfied. It depends on an appropriate characterization of $G_E(\Sigma, V)$. Modulo E, this set is identical to the set G' defined below.

Definition 12. *For $i = 1, 2$, let $G_i := G_{E_i}(\Sigma, V)$. The set G' is inductively defined as follows:*

1. *Every variable is an element of G', that is, $V \subseteq G'$.*
2. *Assume that $r(\bar{v}) \in G_i$ for $i \in \{1, 2\}$ and $\bar{r}$ is a tuple of elements of G' such that the following conditions are satisfied:*
   (a) *$r(\bar{v}) \neq_E v$ for all variables $v \in V$;*
   (b) *$r_k(\epsilon) \notin \Sigma_i$ for all components r_k of $\bar{r}$;*
   (c) *the tuple $\bar{v}$ consists of all variables of r without repetitions;*

(d) the tuples $\bar{v}$ and $\bar{r}$ have the same length;
(e) $r_k \neq_E r_\ell$ if r_k, r_ℓ occur at different positions in the tuple $\bar{r}$.
Then $r(\bar{r}) \in G'$.

Notice that $G_i \subseteq G'$ for $i = 1, 2$ because the components of $\bar{r}$ above can also be variables. Also notice that no element r of G' can have a shared symbol as top symbol since r is either a variable or a term "starting" with an element of G_i.

4   A Combination Procedure for the Word Problem

In this section, we will present a combination procedure that allows us to derive the following decidability result for the word problem in the union of equational theories sharing constructors:

Theorem 13. *Let E_1, E_2 be two non-trivial equational theories of signature Σ_1, Σ_2, respectively, such that $\Sigma := \Sigma_1 \cap \Sigma_2$ is a set of constructors for both E_1 and E_2, and $E_1^\Sigma = E_2^\Sigma$. If for $i = 1, 2$,*

- *normal forms are computable for Σ and E_i, and*
- *the word problem in E_i is decidable,*

then the word problem in $E_1 \cup E_2$ is also decidable.

From Theorem 11 it follows that, given the right conditions, the combination procedure applies immediately by recursion to more than two theories:

Corollary 14. *Let Σ be a signature and $E_1, \dots, E_n$ be n equational theories of signature $\Sigma_1, \dots, \Sigma_n$, respectively, such that $\Sigma = \Sigma_i \cap \Sigma_j$ and $E_i^\Sigma = E_j^\Sigma$ for all distinct $i, j \in \{1, \dots, n\}$. Also, assume that Σ is a set of constructors for every E_i. If for all $i \in \{1, \dots, n\}$,*

- *normal forms are computable for Σ and E_i, and*
- *the word problem in E_i is decidable,*

then the word problem in $E_1 \cup \cdots \cup E_n$ is decidable and normal forms are computable for Σ and $E_1 \cup \cdots \cup E_n$.

As shown in Section 2, the word problem for $E := E_1 \cup E_2$ can be reduced to the satisfiability problem for disequations of the form $s_0 \not\approx t_0$, where s_0 and t_0 are $(\Sigma_1 \cup \Sigma_2)$-terms. By variable abstraction, this disequation can be transformed into an equisatisfiable formula $\varphi_1 \wedge \varphi_2$, where φ_i is a conjunction of i-pure equations and disequations $(i = 1, 2)$. We will use finite sets of (dis)equations in place of conjunctions of such formulae, and say that a set of (dis)equations is satisfiable in a theory iff the conjunction of its elements is satisfiable in that theory. It turns out that the finite set of (dis)equations obtained by applying variable abstraction is what we call an *abstraction system*. Before we can define this notion, we must introduce some notation.

Let $x, y \in V$ and T be a set of equations of the form $v \equiv t$ where $v \in V$ and $t \in T(\Sigma_1 \cup \Sigma_2, V) \setminus V$. The relation $\prec$ is the smallest binary relation on $\{x \not\equiv y\} \cup T$ such that, for all $u \equiv s, v \equiv t \in T$,

$$(x \not\equiv y) \prec (v \equiv t) \text{ iff } v \in \{x, y\},$$
$$(u \equiv s) \prec (v \equiv t) \text{ iff } v \in Var(s).$$

By $\prec^+$ we denote the transitive and by $\prec^*$ the reflexive-transitive closure of $\prec$. The relation $\prec$ is *acyclic* if there is no equation $v \equiv t$ in T such that $(v \equiv t) \prec^+ (v \equiv t)$.

Definition 15 (Abstraction System). *The set $S := \{x \not\equiv y\} \cup T$ is an abstraction system with initial formula $x \not\equiv y$ iff $x, y \in V$ and the following holds:*

1. *T is a finite set of equations of the form $v \equiv t$ where $v \in V$ and $t \in (T(\Sigma_1, V) \cup T(\Sigma_2, V)) \setminus V$;*
2. *the relation $\prec$ on S is acyclic;*
3. *for all $(u \equiv s), (v \equiv t) \in T$,*
   (a) if $u = v$ then $s = t$;
   (b) if $(u \equiv s) \prec (v \equiv t)$ and $s \in T(\Sigma_i, V)$ with $i \in \{1, 2\}$ then $t(\epsilon) \notin \Sigma_i$.

Condition (1) above states that T consists of equations between variables and pure non-variable terms; Condition (2) implies that for all $(u \equiv s), (v \equiv t) \in T$, if $(u \equiv s) \prec^* (v \equiv t)$ then $u \notin Var(t)$; Condition (3a) implies that a variable cannot occur as the left-hand side of more than one equation of T; Condition (3b) implies, together with Condition (1), that the elements of every $\prec$-chain of T have *strictly* alternating signatures $(\ldots, \Sigma_1, \Sigma_2, \Sigma_1, \Sigma_2, \ldots)$.

Every abstraction system S induces a finite graph $\mathcal{G}_S := (S, \prec)$ whose set of *nodes* is S and whose set of *edges* consists of all pairs $(n_1, n_2) \in S \times S$ such that $n_1 \prec n_2$. According to Definition 15, $\mathcal{G}_S$ is in fact a directed acyclic graph (or *dag*). Assuming the standard definition of path between two nodes and of length of a path in a dag, the *height* $h(n)$ of the node n is the maximum of the lengths of all the paths in the dag that end with n.[4]

We say that an equation of an abstraction system S is *reducible* iff its right-hand side is E_i-reducible (i.e., not in normal form) for $i = 1$ or $i = 2$. The disequation in S is always irreducible. In the previous section, we would have represented the normal form of a term in $T(\Sigma_i, V)$ $(i = 1, 2)$ as $s(\bar{q})$ where s was a term in $T(\Sigma, V)$ and $\bar{q}$ a tuple of terms in $G_{E_i}(\Sigma, V)$. Considering that $G_{E_i}(\Sigma, V)$ contains V because of the assumption that Σ is a set of constructors, we will now use a more descriptive notation. We will distinguish the variables in $\bar{q}$ from the non-variables terms and write $s(\bar{y}, \bar{r})$ instead, where $\bar{y}$ collects the elements of $\bar{q}$ that are in V and $\bar{r}$ those that are in $G_{E_i}(\Sigma, V) \setminus V$.

The combination procedure described in Fig. 1 decides the word problem for the theory $E := E_1 \cup E_2$ by deciding the satisfiability in E of disequations of the form $s_0 \not\equiv t_0$ where s_0, t_0 are $(\Sigma_1 \cup \Sigma_2)$-terms. During the execution of the

---

[4] Since $\mathcal{G}_S$ is acyclic and finite, this maximum exists.

Input: $(s_0, t_0) \in T(\Sigma_1 \cup \Sigma_2, V) \times T(\Sigma_1 \cup \Sigma_2, V)$.

1. Let S be the abstraction system obtained by applying variable abstraction to $s_0 \not\equiv t_0$.
2. Repeatedly apply (in any order) **Coll1**, **Coll2**, **Ident**, **Simpl**, **Shar1**, **Shar2** to S until none of them is applicable.
3. Succeed if S has the form $\{v \not\equiv v\} \cup T$ and fail otherwise.

Fig. 1. The Combination Procedure.

procedure, the set S of formulae on which the procedure works is repeatedly modified by the application of one of the derivation rules defined in Fig. 2. We describe these rules in the style of a sequent calculus. The premise of each rule lists all the formulae in S before the application of the rule, where T stands for all the formulae not explicitly listed. The conclusion of the rule lists all the formulae in S after the application of the rule. It is understood that any two formulae explicitly listed in the premise of a rule are distinct.

In essence, **Coll1** and **Coll2** remove from S collapse equations that are valid in E_1 or E_2, while **Ident** identifies any two variables equated to equivalent Σ_i-terms and then discards one of the corresponding equations. The restriction that the height of $y \equiv t$ be not smaller than the height of $x \equiv s$ is there to preserve the acyclicity of $\prec$. In these rules we have used the notation $t[y]$ to express that the variable y occurs in the term t, and the notation $T[x/t]$ to denote the set of formulae obtained by substituting every occurrence of the variable x by the term t in the set T.

Simpl eliminates those equations that have become unreachable along a $\prec$-path from the initial disequation because of the application of previous rules. This rule is not essential but it reduces clutter in S by eliminating equations that do not contribute to the solution of the problem anymore. It can be used to obtain optimized, complete implementations of the combination procedure.

The main idea of **Shar1** and **Shar2** is to push shared symbols towards lower positions of the $\prec$-chains they belong to so that they can be processed by other rules. To do that the rules replace the reducible right-hand side t of an equation $x \equiv t$ by its normal form, and then plug the "shared part" of the normal form into all equations whose right-hand sides contain x. The exact formulation of the rules is somewhat more complex since we must ensure that the resulting system is again an abstraction system. In particular, the "alternating signature" condition (3b) of Definition 15 must be respected.

In the description of the rules, an expression like $\bar{z} \equiv \bar{r}$ denotes the set $\{z_1 \equiv r_1, \ldots, z_n \equiv r_n\}$ where $\bar{z} = (z_1, \ldots, z_n)$ and $\bar{r} = (r_1, \ldots, r_n)$, and $s(\bar{y}, \bar{z})$ denotes the term obtained from $s(\bar{y}, \bar{r})$ by replacing the subterm r_j with z_j for each $j \in \{1, \ldots, n\}$. Observe that this notation also accounts for the possibility that t reduces to a non-variable term of $G_{E_i}(\Sigma, V)$. In that case, s will be a variable, $\bar{y}$ will be empty, and $\bar{r}$ will be a tuple of length 1. Substitution expres-

$$\textbf{Coll1} \quad \frac{T \qquad u \not\equiv v \qquad x \equiv t[y] \quad y \equiv r}{T[x/r] \quad (u \not\equiv v)[x/y] \qquad y \equiv r}$$

if t is an i-term and $y =_{E_i} t$ for $i = 1$ or $i = 2$.

$$\textbf{Coll2} \quad \frac{T \qquad x \equiv t[y]}{T[x/y]}$$

if t is an i-term and $y =_{E_i} t$ for $i = 1$ or $i = 2$
and there is no $(y \equiv r) \in T$.

$$\textbf{Ident} \quad \frac{T \qquad x \equiv s \quad y \equiv t}{T[x/y] \qquad y \equiv t}$$

if s, t are i-terms and $s =_{E_i} t$ for $i = 1$ or $i = 2$
and $x \neq y$ and $\mathsf{h}(x \equiv s) \leq \mathsf{h}(y \equiv t)$.

$$\textbf{Simpl} \quad \frac{T \quad x \equiv t}{T}$$

if $x \notin \mathcal{V}ar(T)$.

$$\textbf{Shar1} \quad \frac{T \qquad\qquad\qquad u \not\equiv v \quad x \equiv t \qquad \bar{y}_1 \equiv \bar{r}_1}{T[x/s(\bar{y}, \bar{z})[\bar{y}_1/\bar{r}_1]] \quad \bar{z} \equiv \bar{r} \quad u \not\equiv v \quad x \equiv s(\bar{y}, \bar{r}) \quad \bar{y}_1 \equiv \bar{r}_1}$$

if (a) t is an E_i-reducible i-term for $i = 1$ or $i = 2$,
(b) $\mathrm{NF}_{E_i}^{\Sigma}(t) = s(\bar{y}, \bar{r}) \notin V$,
(c) $\bar{r}$ non-empty,
(d) $\bar{z}$ fresh variables with no repetitions,
(e) $\bar{r}_1$ irreducible (for both theories),
(f) $\bar{y}_1 \subseteq \mathcal{V}ar(s(\bar{y}, \bar{r}))$ and $(x \equiv s(\bar{y}, \bar{r})) \prec (y \equiv r)$ for no $(y \equiv r) \in T$.

$$\textbf{Shar2} \quad \frac{T \qquad\qquad u \not\equiv v \quad x \equiv t \qquad \bar{y}_1 \equiv \bar{r}_1}{T[x/s[\bar{y}_1/\bar{r}_1]] \quad u \not\equiv v \quad x \equiv s[\bar{y}_1/\bar{r}_1] \quad \bar{y}_1 \equiv \bar{r}_1}$$

if (a) t is an E_i-reducible i-term for $i = 1$ or $i = 2$,
(b) $\mathrm{NF}_{E_i}^{\Sigma}(t) = s \in T(\Sigma, V) \setminus V$,
(c) $\bar{r}_1$ irreducible (for both theories),
(d) $\bar{y}_1 \subseteq \mathcal{V}ar(s)$ and $(x \equiv s) \prec (y \equiv r)$ for no $(y \equiv r) \in T$.

Fig. 2. The Derivation Rules.

sions containing tuples are to be interpreted accordingly; e.g., $[\bar{z}/\bar{r}]$ replaces the variable z_j by r_j for each $j \in \{1, \ldots, n\}$.

In both **Shar** rules it is assumed that the normal form is not a variable. The reason for this restriction is that the case where an i-term is equal modulo E_i to a variable is already taken care of by the rules **Coll1** and **Coll2**. By requiring that $\bar{r}$ be non-empty, **Shar1** excludes the possibility that the normal form of the term t is a shared term. It is **Shar2** that deals with this case. The reason for a separate case is that we want to preserve the property that every $\prec$-chain is made of equations with alternating signatures (cf. Definition 15(3b)). When the equation $x \equiv t$ has immediate $\prec$-successors, the replacement of t by the Σ-term

s may destroy the alternating signatures property because $x \equiv s$, which is both a Σ_1- and a Σ_2-equation, may inherit some of these successors from $x \equiv t$.[5] **Shar2** restores this property by merging into s all the immediate successors of $x \equiv s$— which are collected, if any, in the set $\bar{y}_1 \equiv \bar{r}_1$. Condition (d) in **Shar2** makes sure that the tuple $\bar{y}_1 \equiv \bar{r}_1$ collects all these successors. The replacement of $\bar{y}_1$ by $\bar{r}_1$ in **Shar1** is done for similar reasons. In both **Shar** rules, the restriction that all the terms in $\bar{r}_1$ be in normal form is necessary to ensure termination.

A sketch of the Correctness Proof

As a first step to proving the correctness of the combination procedure, we can show that an application of one of the rules of Fig. 2 transforms abstraction systems into abstraction systems, preserves satisfiability, and leads to a decrease w.r.t. a certain well-founded ordering. This ordering can be obtained as follows: every node in the dag corresponding to the abstraction system S is associated with a pair (h, r), where h is the height of the node, and r is 1 if the corresponding (dis)equation is reducible, and 0 otherwise. The abstraction system S is associated with the multiset $M(S)$ consisting of all these pairs. Let $\sqsupset$ be the multiset ordering [5] induced by the lexicographic ordering on pairs.

Lemma 16. *Assume that S' is obtained from S by an application of one of the rules of Fig. 2.*

1. *If S is an abstraction system, then so is S'.*
2. *S is satisfiable in $E_1 \cup E_2$ iff S' is satisfiable in $E_1 \cup E_2$.*
3. *$M(S) \sqsupset M(S')$.*

The second point of the lemma implies soundness of our combination procedure, that is, if the combination procedure succeeds on an input (s_0, t_0), then $s_0 =_{E_1 \cup E_2} t_0$. Since the multiset ordering $\sqsupset$ is well-founded, the third point implies that the procedure always terminates. The first point implies that the final system obtained after the termination of the procedure is an abstraction system. This fact plays an important rôle in the proof of completeness of the procedure. The completeness of the combination procedure, meaning that the procedure succeeds on an input (s_0, t_0) whenever $s_0 =_{E_1 \cup E_2} t_0$, can be proved by showing that Proposition 10 can be applied (see [3] for details).

5 Related work

In this section, we investigate the connection between our notion of constructors and the one introduced in [6]. Before we can define the notion of constructors according to [6], called DKR-constructors in the following, we need to introduce

---

[5] Recall that we assume, without loss of generality, that the variables in $Var(s) \setminus Var(t)$ do not occur in the abstraction system (cf. the remark after Definition 6). Thus, the equations in $\bar{y} \equiv \bar{r}$ are in fact successors of $x \equiv t$.

the notion of a monotonic ordering. An ordering on $T(\Omega, V)$ is called monotonic if $s > t$ implies $f(\ldots, s, \ldots) > f(\ldots, t, \ldots)$ for all $s, t \in T(\Omega, V)$ and all function symbols $f \in \Omega$. In the rest of the section, we will consider a non-trivial equational theory E of signature Ω and a subsignature Σ of Ω.

Definition 17. *Let $>$ be a well-founded and monotonic ordering on $T(\Omega, V)$. The signature Σ is a set of* DKR-*constructors for E w.r.t. $>$ if*

1. *the $=_E$ congruence class of any term $t \in T(\Omega, V)$ contains a least element w.r.t. $>$, which we denote by $t{\downarrow}^>_E$, and*
2. *$f(t_1, \ldots, t_n){\downarrow}^>_E = f(t_1{\downarrow}^>_E, \ldots, t_n{\downarrow}^>_E)$ for all $f \in \Sigma$ and Ω-terms $t_1, \ldots, t_n$.*

We will call $t{\downarrow}^>_E$ the DKR-normal form of t, and then say that t is in DKR-normal form whenever $t = t{\downarrow}^>_E$. For the theory E_1 in Example 8, it is not hard to show that the signature Σ is set of DKR-constructors for E_1 w.r.t. an appropriate well-founded and monotonic ordering.

Example 9 shows that a set of constructors in the sense of Definition 2 need not be a set of DKR-constructors. In fact, as shown in [6], the definition of DKR-constructors implies that, if Σ is a set of DKR-constructors for E, then E^Σ is the theory of syntactic equality on Σ-terms. This implies that, in Example 9, the signature Σ' is not a set of DKR-constructors for E_2.

To show that the notion of DKR-constructors is a special case of our notion of constructors, we need a representation of the set $G_E(\Sigma, V)$.

Lemma 18. *Let Σ be a set of* DKR-*constructors for E w.r.t. $>$. Then $G_E(\Sigma, V) = \{r \in T(\Omega, V) \mid r{\downarrow}^>_E(\epsilon) \notin \Sigma\}$.*

Using this lemma, it is not hard to show the next proposition.

Proposition 19. *If Σ is a set of* DKR-*constructors for E w.r.t. $>$, then Σ is a set of constructors for E according to Definition 2.*

The definition of DKR-constructors does not assume that DKR-normal forms are computable. In [6], this is achieved by additionally assuming that the so-called symbol matching problem is decidable.

Definition 20. *We say that the symbol matching problem on Σ modulo E is decidable in $T(\Omega, V)$ if there exists an algorithm that decides, for all $t \in T(\Omega, V)$, whether there exists a function symbol $f \in \Sigma$ and a tuple of Ω-terms $\bar{t}$ such that $t =_E f(\bar{t})$. We say that t matches onto Σ modulo E if $t =_E f(\bar{t})$ for some $f \in \Sigma$ and some tuple $\bar{t}$ of Ω-terms.*

As pointed out in [6], if the symbol matching problem and the word problem are decidable for E, then a symbol $f \in \Sigma$ and a tuple of terms $\bar{t}$ satisfying $t =_E f(\bar{t})$ can be effectively computed, whenever it exists. In fact, once we know that an appropriate function symbol in Σ and a tuple of Ω-terms exists, we can simply enumerate all pairs consisting of a symbol $f \in \Sigma$ and a tuple $\bar{t}$ of Ω-terms, and test whether $t =_E f(\bar{t})$. We call an algorithm that realizes such a computation a *symbol matching algorithm on Σ modulo E*. Using such a symbol matching algorithm, we can define a function NF^Σ_E for E and Σ with the following recursive definition.

Definition 21. *Assume that Σ is set of* DKR-*constructors for E w.r.t. $>$, the word problem for E and the symbol matching problem on Σ modulo E are decidable, and let M be any symbol matching algorithm on Σ modulo E. Then, let* NF_E^Σ *be the function defined as follows: For every $t \in T(\Omega, V)$,*

1. $\mathrm{NF}_E^\Sigma(t) := f(\mathrm{NF}_E^\Sigma(t_1), \dots, \mathrm{NF}_E^\Sigma(t_n))$ *if t matches onto Σ modulo E and f is the Σ-symbol and $(t_1, \dots, t_n)$ the tuple of Ω-terms returned by M on input t.*
2. $\mathrm{NF}_E^\Sigma(t) := t$, *otherwise.*

Lemma 22. *Under the assumptions of Definition 21 the function NF_E^Σ is well-defined and satisfies the requirements of Definition 6.*

This lemma, together with Proposition 19, entails that Theorem 14 in [6] can be obtained as a corollary of our Theorem 13.

Corollary 23. *Let E_1, E_2 be non-trivial equational theories of signature Σ_1, Σ_2, respectively, such that $\Sigma := \Sigma_1 \cap \Sigma_2$ is a set of* DKR-*constructors for both E_1 and E_2. If for $i = 1, 2$, the symbol matching problem on Σ modulo E_i is decidable, and the word problem in E_i is decidable, then the word problem in $E_1 \cup E_2$ is also decidable.*

A third notion of constructors has been introduced in term rewriting in the context of modularity properties for term rewriting systems: a constructor is a function symbol that does not occur at the top of a left-hand side of a rule. It is easy to see that, for complete (i.e., confluent and strongly normalizing) term rewriting systems, this notion of constructors is a special case of the notion of DKR-constructors. A finite complete term rewriting system provides a decision procedure for the word problem. Although the union of two complete term rewriting systems sharing constructors need not be complete, this union is at least semi-complete (i.e., confluent and weakly normalizing), which is sufficient to obtain a decision procedure for the word problem (see, e.g., [9] for details). The main difference between this combination result and ours, in addition to the greater generality of our constructors, is that we do not assume that the word problem in the component theories can be decided by a complete or semi-complete term rewriting system, that is, our approach also applies in cases where the decision procedure is not based on term rewriting.

6 Future Work

As mentioned in the introduction, [6] also contains combination results for unification and matching, whereas the present paper is concerned only with the word problem. Thus, one direction for future research would be to extend our approach to the combination of decision procedures for the matching and the unification problem as well.

Another direction would be to extend the class of theories even further by relaxing the restriction that the equational theory over the constructors be collapse-free. A crucial artifact to our completeness proof is the set $G_E(\Sigma, V)$, which is used to obtain the (countably infinite) set of generators of a certain free algebra. When the equational theory over the constructors is not collapse-free, $G_E(\Sigma, V)$ is empty, and thus cannot be used to describe this set of generators. An appropriate alternative characterization of the set of generators might allow us to remove altogether the restriction that the equational theory over the constructors be collapse-free.

References

1. F. Baader and K.U. Schulz. Unification in the union of disjoint equational theories: Combining decision procedures. *J. Symbolic Computation* 21, 1996.
2. F. Baader and C. Tinelli. A new approach for combining decision procedures for the word problem, and its connection to the Nelson-Oppen combination method. In *Proc. CADE-14*, Springer LNAI 1249, 1997.
3. F. Baader and C. Tinelli. Deciding the word problem in the union of equational theories. UIUCDCS-Report 98-2073, Department of Computer Science, University of Illinois at Urbana-Champaign, 1998. Available at http://www-lti.informatik.rwth-aachen.de/Forschung/Papers.html.
4. A. Boudet. Combining unification algorithms. *J. Symbolic Computation* 16, 1993.
5. N. Dershowitz and Z. Manna. Proving termination with multiset orderings. *Communications of the ACM* 22, 1979.
6. E. Domenjoud, F. Klay, and C. Ringeissen. Combination techniques for non-disjoint equational theories. In *Proc. CADE-12*, Springer LNAI 814, 1994.
7. H. Kirchner and Ch. Ringeissen. Combining symbolic constraint solvers on algebraic domains. *J. Symbolic Computation* 18, 1994.
8. T. Nipkow. Combining matching algorithms: The regular case. In *Proc. RTA'89*, Springer LNCS 335, 1989.
9. E. Ohlebusch. Modular properties of composable term rewriting systems. *J. Symbolic Computation* 20, 1995.
10. D. Pigozzi. The join of equational theories. *Colloquium Mathematicum* 30, 1974.
11. Ch. Ringeissen. Unification in a combination of equational theories with shared constants and its application to primal algebras. In *Proc. LPAR'92*, Springer LNAI 624, 1992.
12. Ch. Ringeissen. Combination of matching algorithms. In *Proc. STACS-11*, Springer LNCS 775, 1994.
13. M. Schmidt-Schauß. Combination of unification algorithms. *J. Symbolic Computation* 8, 1989.
14. E. Tidén. *First-Order Unification in Combinations of Equational Theories*. Phd thesis, The Royal Institute of Technology, Stockholm, 1986.
15. C. Tinelli and Ch. Ringeissen. Non-disjoint unions of theories and combinations of satisfiability procedures: First results. Technical Report UIUCDCS-R-98-2044, Department of Computer Science, University of Illinois at Urbana-Champaign, April 1998. (Also available as INRIA research report no. RR-3402.)

Normalization via Rewrite Closures*

L. Bachmair, C. R. Ramakrishnan, I. V. Ramakrishnan and A. Tiwari

Department of Computer Science, SUNY at Stony Brook,
Stony Brook, NY 11794, U.S.A
{leo,cram,ram,astiwari}@cs.sunysb.edu

Abstract. We present an abstract completion-based method for finding normal forms of terms with respect to given rewrite systems. The method uses the concept of a rewrite closure, which is a generalization of the idea of a congruence closure. Our results generalize previous results on congruence closure-based normalization methods. The description of known methods within our formalism also allows a better understanding of these procedures.

1 Introduction

Efficient procedures for normalization of expressions are crucial for the practical performance of rewrite-based systems. A straightforward approach to normalization, by a "straight-line" sequence of individual reduction steps, each consisting of matching and subterm replacement, may be expensive if it requires many steps. For instance, suppose a term $f(a)$ can be rewritten to a normal form a in n reduction steps. Then the normalization of the term $f(f(a))$ may require twice as many steps, as $f(f(a))$ is first rewritten to $f(a)$ (in n steps) and then to a (in n more steps). Note that the two subsequences consists essentially of the same reduction steps, though applied to different terms. There have been attempts to store the "history" of reductions in a suitable way so as to avoid repetition of such "equivalent" reduction sequences. The aim is to generate, once the subterm in $f(f(a))$ has been reduced to a, the normal form a from the intermediate term $f(a)$ in a *single* additional step. Such a "non-oblivious" normalization method requires $n + 1$ steps in this case, though the individual steps are usually different from standard matching and subterm replacement. The key question is how to store the application of rewrite rules in a way that can be efficiently exploited for performing future reductions.

Chew [3, 4] had the fundamental insight to adapt techniques developed for congruence closure algorithms (cf., Nelson and Oppen [8]) to normalization with history. Congruence closure algorithms apply to finite sets of variable-free equations and yield a compact representation of the underlying equational theory, in that unique representatives are assigned to equivalent terms. Normalization usually needs to be done for rewrite systems (i.e., sets of directed equations)

---

* The research described in this paper was supported in part by the National Science Foundation under grants CCR-9510072, CCR-9705998 and CCR-9711386.

with variables, which may represent infinitely many variable-free equations over the given term domain. Chew's method therefore combines a "dynamic" version of congruence closure with a method for selecting rewrite rule instances needed to normalize a given input term. Whenever additional rule instances are selected, the congruence closure algorithm is applied incrementally to update the representation of term equivalences. Once a term has been rewritten, the congruence closure represents it by its current normal form and thus effectively stores the history of previous reductions. If no further useful rule instances can be selected, one either obtains a normal form for the input term or else detects non-termination of the rewriting process. (Non-terminating rewrite systems may also cause the selection process to continue indefinitely.)

Chew's work applies to orthogonal rewrite systems, but was extended by Verma [10] to priority rewrite systems. The description of these methods is technically involved. We develop a different, more abstract view of this approach to normalization by formulating it in terms of standard techniques from term rewriting, such as completion and narrowing. More specifically, the basic method, and various optimizations, are described in terms of transformation rules in the style of Bachmair and Dershowitz [2].

We briefly describe congruence closure by transformation rules in Section 2, and explain its application to normalization in Section 3. Optimizations of the basic method by using a modified congruence closure, called a *rewrite closure*, are discussed in Section 4. Finally we outline further optimizations for the special cases of orthogonal and convergent rewrite systems in Section 5.

2 Abstract Congruence Closure

First we briefly introduce the concept of an "abstract congruence closure" which forms the basis of our approach to nonoblivious normalization. We assume that the reader is familiar with the basic notions and terminology of term rewriting; for details see [5].

Let Σ be a signature consisting of constants and function symbols, and $\mathcal{V}$ be a set of variables. We mostly deal with variable-free, or *ground*, terms; and denote by $\mathcal{T}(\Sigma)$ the set of all ground terms over Σ. The symbols $s, t, u, \ldots$ are used to denote terms; $f, g, \ldots$, function symbols; and $x, y, z, \ldots$, variables. We write $E[t]$ to indicate that an expression E contains t as a subterm and (ambiguously) denote by $E[u]$ the result of replacing a particular occurrence of t by u. (The same notation will be employed if E is a set of expressions.) By $E\sigma$ we denote the result of applying a substitution σ to E.

An *equation* is a pair of terms, written $s \approx t$. We usually identify the two equations $s \approx t$ and $t \approx s$; if the distinction is important, we call the equation a *rewrite rule* and write $s \to t$. A *rewrite system* is a set of rewrite rules. The *rewrite relation* $\to_R$ induced by a set of equations R is defined by: $u \to_R t$ if, and only if, u contains a subterm $l\sigma$ and $t = u[r\sigma]$, for some rewrite rule $l \to r$ in R and some substitution σ. The *equational theory* induced by R is the reflexive, symmetric, and transitive closure of this rewrite relation.

A term t is said to be in *normal form* with respect to a rewrite system R, or in *R-normal form*, if there is no term u, such that $t \to_R u$. We write $s \to^!_R t$ to indicate that t is a R-normal form of s. A rewrite system is said to be (ground) *confluent* if all (ground) terms have a unique normal form. Rewrite systems that are (ground) confluent and terminating are called (ground) *convergent*.

Congruence closure algorithms may be viewed as methods for constructing ground convergent rewrite systems for given equations, but over an extended signature. The extension of the given signature is limited to the introduction of new constants and can be combined with techniques similar to (ground) completion, as pointed out by Kapur [6].

For example, let E_0 be the set of equations $\{fa \approx fb, ffb \approx a, fb \approx a\}$ over the signature $\Sigma = \{a, b, f\}$. New constants $c_0, \ldots, c_4$ are introduced to represent the different subterms in E_0, as specified by the following rewrite rules:

$$D_0 = \{a \to c_0, \ b \to c_1, \ fc_0 \to c_2, \ fc_1 \to c_3, \ fc_3 \to c_4\}.$$

Rewrite rules of the form

$$f(c_1, \ldots, c_k) \to c_0$$

where $f \in \Sigma$ and $c_0, c_1, \ldots, c_k$ are constants in a set K disjoint from Σ, are called *D-rules* (with respect to Σ and K).

The D-rules represent the structure of the given terms, whereas equations between these terms can be represented by equations between constants from K, which we call *C-equations* or *C-rules*. For example, E_0 is represented by three C-equations, $c_2 \approx c_3$, $c_4 \approx c_0$ and $c_3 \approx c_0$.

Let R be a set of C-rules and D-rules (with respect to Σ and K). We say that a constant c in K *represents* a term t in $\mathcal{T}(\Sigma \cup K)$ (via R) if $t \to^*_R c$. A term t is said to be *represented* by R if it is represented by some constant via R. For example, the constant c_2 represents the term fa via D_0.

Definition 1. *Let Σ be a signature and K be a set of constants disjoint from Σ. In addition, let E be a set of ground equations over $\mathcal{T}(\Sigma \cup K)$. A ground rewrite system $R = C \cup D$ of C-rules and D-rules is called an (abstract) congruence closure for E (with respect to Σ and K) if*

(i) each constant $c \in K$ that is in normal form with respect to R represents a term $t \in \mathcal{T}(\Sigma)$ via R,

(ii) R is ground convergent (over $\mathcal{T}(\Sigma \cup K)$), and

(iii) for all terms s and t in $\mathcal{T}(\Sigma)$, we have $s \leftrightarrow^*_E t$ if, and only if, $s \to^!_R \circ \leftarrow^!_R t$.

The rewrite system $R_0 = D_0 \cup C_0$, where $C_0 = \{c_2 \to c_3, c_0 \to c_4, c_0 \to c_3\}$, is not a congruence closure for E_0, as it is not ground convergent. But we can obtain a congruence closure from R_0 by a completion-like process described next.

Our description is fairly abstract, in terms of *transformation rules* in the style of Bachmair and Dershowitz, see [1, 2]. The transformation rules operate on triples (K, E, R), where K is the set of newly introduced constants (the original signature Σ is fixed); E is a set of ground equations over Σ yet to be processed; and R is the set of C-rules and D-rules that have been derived so far. Triples may be viewed as possible *states* in the process of constructing a closure.

A key transformation rule is the introduction of new constants.

Extension:
$$\frac{(K, E[f(c_1, \ldots, c_n)], R)}{(K \cup \{c\}, E[c], R \cup \{f(c_1, \cdots, c_k) \to c\})}$$

where $f \in \Sigma$, $c_1, \ldots, c_k$ are constants in K, and $c \notin \Sigma \cup K$.

Once a D-rule $f(c_1, \ldots, c_k) \to c$ has been introduced, it can be used to eliminate other occurrences of $f(c_1, \ldots, c_k)$.

Simplification:
$$\frac{(K, E[s], R \cup \{s \to t\})}{(K, E[t], R \cup \{s \to t\})}$$

where s occurs in some equation in E.

Evidently, any equation in E can be transformed to a C-equation by suitable extension and simplification steps. We orient C-equations into rewrite rules.

Orientation:
$$\frac{(K, E \cup \{c \approx d\}, R)}{(K, E, R \cup \{c \to d\})}$$

if c and d are constants in K with $c \succ d$.[1]

Trivial equations can be deleted.

Deletion:
$$\frac{(K, E \cup \{c \approx c\}, R)}{(K, E, R)}$$

Construction of convergent rewrite systems requires primarily the following transformation rule.

Superposition:
$$\frac{(K, E, R \cup \{t \to c, t \to d\})}{(K, E \cup \{c \approx d\}, R \cup \{t \to d\})}$$

if $t \to c$ and $t \to d$ are D-rules.

The efficiency of completion also depends on additional transformation rules for simplification of rewrite rules, usually called "collapse" and "composition" rules; see [1, 2] for details.

We write $\xi \vdash_{CC} \xi'$ to indicate that a state ξ can be transformed to ξ' in one step by one of the above rules. By a *derivation* we mean a sequence of such transformation steps, $\xi_0 \vdash_{CC} \xi_1 \vdash_{CC} \cdots \xi_n$ from an *initial state* $(\emptyset, E, \emptyset)$. (It can be shown that derivations based on the above transformation rules are always finite.)

---

[1] We assume that a total reduction ordering $\succ$ on ground terms is supplied initially and extended appropriately whenever a new constant is introduced. For our purposes, it is sufficient to use a lexicographic path ordering based on a total precedence for $\Sigma \cup K$, such that $f \succ c$, whenever $f \in \Sigma$ and $c \in K$.

The following table shows some of the intermediate states of a derivation from $(\emptyset, E_0, \emptyset)$, where $E_0 = \{fa \approx fb,\ ffb \approx a,\ fb \approx a\}^2$.

i	Constants K_i	Equations E_i	Rules R_i
0	$\emptyset$	E_0	$\emptyset$
1	$\{c_0, \ldots, c_3\}$	$\{ffb \approx a, fb \approx a\}$	$\{a \to c_0,\ b \to c_1,\ fc_0 \to c_3,$ $fc_1 \to c_3,\ c_2 \to c_3\}$
2	$K_1 \cup \{c_4\}$	$\{fb \approx a\}$	$R_1 \cup \{fc_3 \to c_4,\ c_0 \to c_4\}$
3	K_2	$\emptyset$	$R_2 \cup \{c_0 \to c_3\}$
4	K_2	$\emptyset$	$\{a \to c_4,\ b \to c_1,\ fc_4 \to c_4,\ fc_1 \to c_4$ $c_2 \to c_4,\ c_0 \to c_4,\ c_3 \to c_4\}$

The final rewrite system R_4 is a congruence closure.

In general, exhaustive application of the transformation rules will result in a *final state* of the form $(K, \emptyset, R)$, where R is a congruence closure.

Theorem 1. *If $(K, \emptyset, R)$ is the final state of a derivation from $(\emptyset, E, \emptyset)$, then R is a congruence closure for E.*

3   Normalization Using Congruence Closure

We next outline how to apply congruence closures to the problem of finding, given a rewrite system $\mathcal{R}$ and a ground term t, a normal form of t with respect to $\mathcal{R}$.[3] Let us first consider the simple case when $\mathcal{R}$ is a ground rewrite system.

For example, suppose we want to normalize the term $f^5 a$ with respect to the rewrite system $\mathcal{R}_0 = \{fa \to fb, ffb \to a, fb \to a\}$. We already know that that R_4 is a congruence closure for $\mathcal{R}_0$. First we compute the normal form of $f^5 a$ by R_4, which is c_4. Then we identify an irreducible term over the signature Σ that is represented by c_4. The definition of a congruence closure guarantees that such a term exists, in this example, we get a. Thus, we have $f^5 a \to^*_{R_4} c_4 \leftarrow^*_{\mathcal{R}_0} a$ and conclude that a is a normal form of $f^5 a$.

This approach is simple, but needs to be generalized to rewrite systems with variables. The basic idea is to select one or more instances $l\sigma \to r\sigma$ of rules in $\mathcal{R}$ that can be used to reduce (a subterm of) t and to apply congruence closure to them, so that a normal form t' of t (with respect to the selected instances) can be identified. If t' is also in normal form with respect to $\mathcal{R}$, we are done; otherwise, further rule instances need to be selected to reduce t'. This yields a method that incrementally applies congruence closure to selected instances of given rewrite rules. Two key issues to be addressed are selection– how to select instances, and termination– how to efficiently identify that a term is in $\mathcal{R}$-normal form. Selection, it turns out, can be done by a simple narrowing process.

---

[2] In all examples, new constants will be ordered as follows : $c_i \succ c_j$ if $i < j$.

[3] It is sufficient to consider *ground* terms t, The normal form of a general term u can be easily obtained from a normal form of the ground term $\hat{u}$, where u is obtained from u by replacing each variable by a new constant.

3.1 Narrowing

We say that a term t *narrows* to a term t' (with respect to a rewrite system R) if there exist a non-variable subterm s of t and a rewrite rule $l \to R \in R$, such that (i) s is unifiable with l and (ii) t' is obtained from $t\sigma$ by replacing the subterm $s\sigma$ by $r\sigma$, where σ is a most general unifier of s and l.

In our context, the term t to be normalized, and all its subterms, are represented by constants via a congruence closure R. We will present a simple narrowing procedure to determine whether some left-hand side l of a rule in R can be narrowed to a constant via R, i.e., whether there exists a substitution σ, such that $l\sigma \to_R^* c$. Any rule instance $l\sigma \to r\sigma$ selected by narrowing in this way will then be used for incremental extension of the congruence closure R.

We use transformation rules to describe the narrowing process for selection of rule instances from a rewrite system $R = \{l_1 \to r_1, \ldots, l_n \to r_n\}$. The transformation rules operate on states (K, R, S), where K is a set of constants disjoint from Σ, R is a congruence closure, and S is sequence of n sets $\langle S_1, \ldots, S_n \rangle$. Each set S_i consists of pairs (l'_i, σ_i), where $l_i\sigma_i \to_R^* l'_i$. The pairs (l'_i, σ_i) indicate candidates to be selected among rule instances. Selection of $l_i\sigma \to r_i\sigma$ is possible if a term l'_i is a constant. If a term l'_i is not a constant, but can not be reduced further, then the corresponding candidate pair can be deleted, as selection will be impossible.

The formal transformation rule is as follows.

Narrowing:
$$\frac{(K, R, \langle \ldots, L_j \cup \{(s[t], \sigma)\}, \ldots \rangle)}{(K, R, \langle \ldots, L_j \cup L'_j, \ldots \rangle)}$$

where t is either a constant or a non-variable non-constant innermost subterm of s and either (i) t can be narrowed by R, in which case L'_j is the set of all pairs $(s[c]\sigma_1, \sigma\sigma_1)$ such that $t\sigma_1 = u$ for some rule $u \to c$ in R, or (ii) t is not a constant and cannot be narrowed, in which case $L'_j = \emptyset$.

We write $\eta \vdash_S \eta'$ to indicate that a state η can be obtained from η' by application of this narrowing rule. It can easily be shown that derivations by narrowing are always finite. The *final state* of such a derivation is a triple $(K, R, \langle L_1, \ldots, L_n \rangle)$ where each set L_i contains only pairs $(c, \sigma), c \in K$.

3.2 Normalization

We now have the two main components of a non-oblivious normalization method–the congruence closure transformation relation $\vdash_{CC}$ and the narrowing transformation relation $\vdash_S$. We describe normalization by rules operating on tuples (R, t, K, E, R, S), where R is a rewrite system, t is the term to be normalized, (K, E, R) represents the current state of a congruence closure computation, and S indicates current candidates for selection. We define:

$$(R, t, K, E, R, \Lambda) \vdash_N (R, t, K', E', R', \Lambda)$$

if $(K, E, R) \vdash_{CC} (K', E', R')$, where Λ denotes a sequence of empty sets; and

$$(R, t, K, \emptyset, R, S) \vdash_N (R, t, K, \emptyset, R, S')$$

if $(K, R, S) \vdash_S (K, R, S')$ and $(K, \emptyset, R)$ is a congruence closure final state.

In short, we distinguish between two phases during normalization: congruence closure rules are applied when no candidates for selection are available, whereas narrowing is only performed in the presence of a (completed) congruence closure.

An *initial state* for a normalization derivation is a tuple $(\mathcal{R}, t, \{c\}, \{c \approx t\}, \emptyset, \Lambda)$, where Λ is a sequence of empty sets and c is a constant not contained in Σ. The first stage will consist of a congruence closure computation, which has the effect of representing all subterms of the term t to be normalized.

The following transformation rules are used to connect congruence closure and narrowing stages and to determine when the normalization process is done.

Narrowing is initiated as follows.

Initialization:
$$\frac{(\mathcal{R}, t, K, \emptyset, R, \Lambda)}{(\mathcal{R}, t, K, \emptyset, R, \langle \{(l_1, \mathrm{id})\}, \ldots, \{(l_n, \mathrm{id})\} \rangle)}$$

if the state $(K, \emptyset, R)$ is a congruence closure final state. (The symbol id denotes the identity mapping.)

If the narrowing phase is successful, further rule instances can be selected.

Selection:
$$\frac{(\mathcal{R}, t, K, \emptyset, R, \langle L_1, \ldots, L_j \cup \{(c, \sigma)\}, \ldots, L_n \rangle)}{(\mathcal{R}, t, K, \{l_j \sigma \to r_j \sigma\}, R, \Lambda)}$$

if $c \in K$. The rule $l_j \sigma \to r_j \sigma$ which is moved to the set E in this rule, will be called a *selected*, or, *processed* rule. In general, we can move more than one rule instance to the E component without affecting any results.

Computation of a congruence closure may change the representation of equivalences. Instead of initiating another narrowing phase, we may check whether a normal form term is already represented.

Detection:
$$\frac{(\mathcal{R}, t, K, \emptyset, R, \Lambda)}{t^*}$$

if (i) the state $(K, \emptyset, R)$ is a congruence closure final state, (ii) there is a $t^* \in \mathcal{T}(\Sigma)$ such that $t \to_R^! c \leftarrow_R^* t^*$, and (iii) t^* is not further reducible by $\mathcal{R}$.[4]

Terminating in a state t^* means that we output t^* as the normal form of t.

For example, consider the problem of normalizing the term fa with respect to $\mathcal{R} = \{a \to b, fx \to gfx, gfb \to c\}$. The initial state obtained by the initialization rule is $\xi_0 = (\mathcal{R}, fa, \{c_0\}, \{c_0 \approx fa\}, \emptyset, \Lambda)$. We first obtain a congruence closure,

$$\xi_1 = (\mathcal{R}, fa, \{c_0, c_1, c_2\}, \emptyset, \{a \to c_1, fc_1 \to c_2, c_0 \to c_2\}, \Lambda),$$

and then initialize a narrowing phase, with initial candidates $\langle \{(a, \mathrm{id})\}, \{(fx, \mathrm{id})\}, \{(gfb, \mathrm{id})\} \rangle$. Observe that a narrows to c_0 by the identity mapping. We select

---

[4] This inference rule can be effectively applied. We simply non-deterministically guess for each constant $d \in K$, a D rule $f(\cdots) \to d$. Then the required t^* is one which reduces to c using these guessed rules, and which satisfies condition (iii).

the rule $a \rightarrow b$ and continue with a congruence closure computation, obtaining $R_2 = \{b \rightarrow c_3, a \rightarrow c_3, fc_3 \rightarrow c_2, c_1 \rightarrow c_3, c_0 \rightarrow c_2\}$. Now fx can be narrowed to a constant with the substitution $x \mapsto c_3$. Thus we select the rule instance $fc_3 \rightarrow gfc_3$ next. Eventually, we identify c as a normal form-term of fa and terminate using the detection rule.

We emphasize that the above transformation rules are only correct when the rewrite system $\mathcal{R}$ is confluent. For example, if we attempt to normalize the term gfa with respect to the non-confluent rewrite system $\mathcal{R} = \{a \rightarrow b, fa \rightarrow gfa, gfb \rightarrow c\}$, then we could terminate with fb as the normal form of gfa.

3.3 Soundness and Completeness

Let $i_K(\mathcal{R})$ denote the set of all ground instances of the rules in $\mathcal{R}$ over the extended signature $\Sigma \cup K$. Corresponding to every derivation we have a set $F \subset i_K(\mathcal{R})$ of rules processed during that derivation. The R-*extension* of F is defined to be the set F_K^R of all equations $s\rho \rightarrow t\rho$, where $s \rightarrow t$ is an equation in F and ρ is a mapping from K to $\mathcal{T}(\Sigma)$, such that $c \leftrightarrow_R^* c\rho$, for all constants c in K. This notion of F_K^R is crucial in establishing completeness of the procedure.

Theorem 2. *(Soundness) If a rewrite system $\mathcal{R}$ is confluent and a derivation from $(\mathcal{R}, t, \{c\}, \{c \approx t\}, \emptyset, \Lambda)$ terminates with t^*, then t^* is an $\mathcal{R}$-normal form of t.*

In order to establish completeness of the procedure, we need to make sure that some normal form of t is eventually represented. For this, we require fairness conditions.

Definition 2. *A rule instance $l_i\sigma \rightarrow r_i\sigma$ is said to be selectible with respect to K and R if l_i can be narrowed to a constant $c \in K$ via R, i.e., $l_i\sigma \rightarrow_R^* c$.*

Definition 3. *A derivation is said to be* fair *if (i) the Detection rule is (eventually) applied if it is ever applicable, or, (ii) every selectible rule instance with respect to any intermediate sets K_i and R_i of the derivation, is eventually selected.*

A fair reduction strategy ensures that enough rule instances are processed so as to guarantee the representation of normal form term.

Theorem 3. *(Completeness) If $t \in \mathcal{T}(\Sigma)$ has a $\mathcal{R}$-normal form then a fair derivation starting from state $(\mathcal{R}, t, \{c\}, \{c \approx t\}, \emptyset, \Lambda)$ terminates in state t^*, where t^* is in $\mathcal{R}$-normal form.*

One difficulty with the inference rules for normalization is that the termination checks essentially involve a non-deterministic step. To make this more efficient, we can store some information about which rules to use to find normal form terms in the congruence closure itself. This will be discussed next.

4 Rewrite Closure

In order to make the termination checks more efficient and useful, the basic congruence closure procedure requires additional refinements, so that one can determine whether a given represented term is in normal form or not. We will achieve this by *marking* certain rules, or, in other words, we will partition the set D into two sets: marked rules X, and unmarked rules N. The idea would be that terms represented by the left-hand sides of N rules will be F_K^D-irreducible. Hence while searching for normal forms in the termination rules, instead of guessing, we will use the N-rules directly.

Definition 4. *An abstract congruence closure $R = D \cup C$ for F is called a rewrite closure if, the set D can be partitioned into $N \cup X$ such that for all terms t in $\mathcal{T}(\Sigma)$ represented by D, t is in normal form with respect to F_K^R if, and only if, it is represented by N.*

Equivalently, we can also say that a congruence closure $D = N \cup X$ is a rewrite closure for F if for every $t \in \mathcal{T}(\Sigma)$, t is F_K^R-irreducible, iff, any reduction sequence starting with t contains only N steps, and no X steps. For example, let $F = \{ffa \to fffa, fffa \to fa\}$. If we let N_0 be the set of two rules, $a \to c_0$ and $fc_0 \to c_1$; and X_0 be a set of one rule, $fc_1 \to c_1$, then $N_0 \cup X_0$ is a rewrite closure for F. Rewrite closures need not always exist, though. Consider the set of equations $F' = \{fffa \to ffa, fffa \to fa\}$. We cannot get a rewrite closure from the abstract congruence closure $D_1 = \{a \to c_0, fc_0 \to c_1, fc_1 \to c_1\}$ for F'. Since a, fa and ffa are all in normal forms, we are forced to have all the D-rules in R_1 in the set N_0.

Note that there are several ways in which the set of D-rules can be partitioned. If $s \to t$ is a rewrite rule in F, then its left-hand side s is called an *F-redex* (or simply a *redex*). One method is to put all D-rules, whose left-hand sides represent F-redexes, into the set X. Rules in X are therefore also called *redex rules*. We write $s \xrightarrow{n} t$ to indicate that $s \to t$ is a rule in N, and $s \xrightarrow{x} t$ to indicate that $s \to t$ is a rule in X. If $f(c_1, \ldots, c_k) \to c_0$ is a rule in X, then the term $f(c_1, \ldots, c_k)$ is also called a *redex template*. However, using this scheme for marking rules we may not still get a rewrite closure. We need the additional property of *persistence*.

Let $t[l\sigma] \to_R t[r\sigma]$ be a (one step) reduction using the rewrite rule $l \to r \in R$. This reduction will be called a *non-root reduction*, denoted by $\to_R^{nr}$ if $l\sigma$ is a proper subterm of t; otherwise this is called a *root reduction*, denoted by $\to_R^{root}$.

Definition 5. *Let R be a abstract congruence closure for a set of ground rules F over $\mathcal{T}(\Sigma \cup K)$. The set F is said to have the persistence property with respect to R if whenever, there exist terms $f(t_1, \cdots, t_n), f(t'_1, \cdots, t'_n) \in \mathcal{T}(\Sigma)$ such that, $f(t_1, \cdots, t_n)$ is F_K^R-reducible at the top (root) position, and $f(t_1, \cdots, t_n) \leftrightarrow_{F_K^R}^{*,nr} f(t'_1, \cdots, t'_n)$, it is always the case that, $f(t'_1, \cdots, t'_n)$ is F_K^R-reducible.*

The idea behind the persistence property is simple. Since we put every redex-template in the set X, this simply means that we assume that all the terms represented by that template are reducible. The persistence property is true whenever this is actually the case.

Lemma 1. *Let F be a finite set of equations over $\mathcal{T}(\Sigma \cup K)$. A congruence closure R of F can be extended to a rewrite closure if F has the* persistence *property with respect to R.*

The converse of this theorem is however false, as the set $F = \{a \rightarrow b,\ fa \rightarrow c,\ c \rightarrow fb\}$ is *not* persistent (with respect to its abstract congruence closure), but the congruence closure can be extended to a rewrite closure.

4.1 Construction of Rewrite Closures

We give a set of transition rules (similar to the ones for congruence closure), that would compute the rewrite closure for a given F,assuming that the persistence property holds.

The *extension* inference rule, which introduces new constants as names for subterms, is the same as before except that now it creates N-rules. We have to be a little careful in simplification rules, as we cannot simplify at the top of the left hand side.

$$\textbf{Simplification:} \quad \frac{(K, F[s], R \cup \{s \rightarrow t\})}{(K, F[t], R \cup \{s \rightarrow t\})}$$

where s is either a subterm of a right-hand side of a rule in F, or else a *proper* subterm of a left-hand side. Note that only proper subterms of left-hand sides of rules in F can be replaced.

It is fairly easy to see that any rewrite rule in F can be transformed to a D-rule by suitable extension and simplification steps. The final D-rules are eliminated from F as follows.

$$\textbf{Orientation:} \quad \frac{(K, F \cup \{f(c_1, \cdots, c_k) \rightarrow c\}, R)}{(K, F, R \cup \{f(c_1, \cdots, c_k) \xrightarrow{x} c\})}$$

if $f(c_1, \ldots, c_k) \rightarrow c$ is a D-rule with respect to Σ and K. Note that orientation generates a *redex* rule.

In the superposition rule too, we have to be careful with the markings.

$$\textbf{Superposition:} \quad \frac{(K, F, R \cup \{t \xrightarrow{\alpha_1} c, t \xrightarrow{\alpha_2} d\})}{(K, F, R \cup \{t \xrightarrow{\alpha_3} d, c \rightarrow d\})}$$

if (i) $t \rightarrow c$ and $t \rightarrow d$ are D-rules, (ii) $c \succ d$, and (iii) α_3 is n only in the case when both α_1 and α_2 are n; in all other cases, α_3 is x.[5]

The other rules like deletion and collapse can be similarly formulated.

---

[5] In other words, $t \rightarrow d$ is a redex rule in the new state if, and only if, at least one of the two superposed rules is a redex rule.

The following table shows some states of a derivation from the initial state $(\emptyset, E_0, \emptyset)$, where $E_0 = \{fa \approx fb, \; ffb \approx a, \; fb \approx a\}$.

i	Constants K_i	Equations E_i	Rules R_i
0	$\emptyset$	E_0	$\emptyset$
1	c_0, c_1, c_2	$ffb \approx a, fb \approx a$	$a \xrightarrow{n} c_0, \; b \xrightarrow{n} c_1, \; fc_1 \xrightarrow{n} c_2, \; fc_0 \xrightarrow{x} c_2$
2	K_1	$fb \approx a$	$R_1, \; fc_2 \xrightarrow{x} c_0$
3	K_2	$\emptyset$	$R_2, \; fc_1 \xrightarrow{x} c_0$
4	K_2	$\emptyset$	$a \xrightarrow{n} c_2, \; b \xrightarrow{n} c_1, \; fc_1 \xrightarrow{x} c_2 \; fc_2 \xrightarrow{x} c_2 \; c_0 \to c_2$

The rewrite system R_4 is a rewrite closure for E_0.

Using lemma 1 we know that any derivation constructs a rewrite closure whenever F satisfies the persistence property. We use the symbol $\vdash_{RC}$ to denote the one-step transformation relation on states induced by the above transformation rules. *Final states* are states of the form $(K, \emptyset, R)$ such that no further transition rules can be applied to them.

4.2 Normalization via Rewrite Closure

The normalization procedure described earlier, can now be optimized by replacing the use of *congruence-closure* by *rewrite-closure*. The additional marking information in the rewrite closure can be used to optimize the *Detection* inference rule: Essentially all we are saying is that in order to find a normal form in the equivalence class c, we need to check for only those term t' such that $t' \to_N^* c$.

We just mention those inference rules of the normalization procedure which now look different. The *congruence-closure* phase now is replaced by the *rewrite-closure* phase. The *initialization*, and *selection* rules are the same as in section 3. The new *Marked-Detection* rule uses the unmarked N rules to find a normal form term.

Marked-Detection: $\dfrac{(\mathcal{R}, t, K, \emptyset, R, \Lambda)}{t^*}$

if (i) $(K, \emptyset, R)$ is a rewrite closure final state, (ii) there is a $t^* \in \mathcal{T}(\Sigma)$ such that $t \to_R^! c \leftarrow_N^* t^*$, where $R = N \cup X \cup C$ and (iii) t^* is in $\mathcal{R}$-normal form.

To establish soundness and completeness, all we need is (a) the confluence of $\mathcal{R}$, (b) persistence of the processed set of rules, and (c) a strategy that ensures that the normal form term is eventually represented. The soundness theorem 2 holds under these new rules. To ensure (c) (for completeness), we still have to use all the D rules for narrowing (and not just N rules). Fairness then would guarantee (c).

Theorem 4. *(Completeness) If $t \in \mathcal{T}(\Sigma)$ has a $\mathcal{R}$-normal form t^* then a fair derivation starting from state $(\mathcal{R}, t, \{c\}, \{c \approx t\}, \emptyset, \Lambda)$ in which the processed rule instances F is always (eventually) persistent, terminates in state t^*.*

The conditions of *fairness* and *persistence* are complementary. In order to satisfy *persistence*, we should process fewer and only particular rules. On the other hand, to satisfy fairness we are required to process as many rules as possible. For example, informally, an innermost strategy in choosing instances to process shall always process sets of instances that are persistent. But, unfortunately, such a strategy may violate fairness. In the next section, we consider two special cases of rewrite systems $\mathcal{R}$ where we can effectively satisfy both conditions together and use the normalization transition rules to find normal forms.

5 Special Cases

Next we further improve the method for normalization with orthogonal and convergent rewrite systems. It appears wasteful to use all of the D-rules in the narrowing process. To compute normal forms we only need to select rule instances that reduce current *irreducible* terms. Intuitively, only N-rules need to be employed by narrowing, as they represent irreducible terms. In fact, soundness is preserved by this restriction. Completeness, however, is only preserved in certain special cases, which we discuss next.

We redefine the narrowing phase of normalization as follows:

$$(\mathcal{R}, t, K, \emptyset, N \cup X \cup C, S) \vdash_N (\mathcal{R}, t, K, \emptyset, N \cup X \cup C, S')$$

if $(K, N, S) \vdash_S (K, N, S')$ and $(K, \emptyset, N \cup X \cup C)$ is a rewrite closure final state. Now the detection rule can be refined also.

Refined-Detection: $\dfrac{(\mathcal{R}, t, K, \emptyset, R, S)}{t^*}$

if assuming $R = N \cup X \cup C$, we have (i) (K, N, S) is a narrowing final state, (ii) there is a $t^* \in \mathcal{T}(\Sigma)$ such that $t \to^!_R c \leftarrow^*_N t^*$, and (iii) none of the constants that appear in the derivation $t^* \to_N c$ occur in (as the first component in) any element of any set in S.

Since we are narrowing with a restricted set N, we can also terminate if a narrowing phase produces no candidates for selection.

Termination: $\dfrac{(\mathcal{R}, t, K, \emptyset, R, \Lambda)}{\Omega}$

if assuming $R = N \cup X \cup C$ we have (i) the state $(K, \emptyset, R)$ is a rewrite closure final state, (ii) the state (K, N, Λ) is a narrowing final state, and (iii) $\Omega = t^*$ if there is a $t^* \in \mathcal{T}(\Sigma)$ such that $t \to^!_R c \leftarrow^*_N t^*$; and $\Omega = \bot$, otherwise.

Terminating in a state $\bot$ means that we output "no normal form of t exists".

5.1 Normalization in Orthogonal Systems

We first discuss normalization of terms with respect to so-called "orthogonal" rewrite systems. A rewrite system $\mathcal{R}$ is called *orthogonal* if (i) all its rewrite rules are *left-linear* (i.e., contain no multiple occurrences of the same variable in any left-hand side) and (ii) it is nonoverlapping (i.e., without critical pairs except trivial ones of the form $t \approx t$). We need the following well-known result [7].

Lemma 2. *Every orthogonal term rewriting system is confluent.*

In addition, for orthogonal rewrite systems root reducibility of terms is preserved in the following sense.

Lemma 3. *Let $\mathcal{R}$ be an orthogonal system. Let t be root reducible (by an instance of the rule $l \to r \in \mathcal{R}$) and also reducible to t' at a non-root position by some rule in $\mathcal{R}$. Then t' is root-reducible by an instance of the rule $l \to r$.*

Using the previous two lemmas, it can be shown that irrespective of the strategy chosen to select rewrite rule instances, the set of these instances is confluent, and persistent. Hence, we can conclude the following.

Theorem 5. *If $\mathcal{R}$ is an orthogonal rewriting system, then given any term and a fair strategy, the inference system outlined above finds its normal form with respect to $\mathcal{R}$, if one exists.*

5.2 Normalization in Convergent Systems

In order to perform normalization of terms with respect to convergent systems, we need a strategy to ensure that normal form term will be eventually represented and that the processed rule instances are persistent. Intuitively, for convergent systems, the normal form term will always eventually get represented under any strategy for choosing the next instance, as convergent systems are confluent and *terminating*. Using an innermost strategy allows us to satisfy persistence.

Definition 6. *Let $l\sigma \to r\sigma$ and $l'\sigma' \to r'\sigma'$ be two selectible instances (with respect to only the unmarked rules now). Say $l\sigma \to_N^* c$ and $l'\sigma' \to_N^* c'$. An innermost strategy is one that makes sure that if there exists a term $t[c]$ containing c such that $t[c] \to_N^* c'$, then the rule $l\sigma \to r\sigma$ is chosen first.*

Lemma 4. *Suppose that in a derivation, we choose the next instance to process using an innermost strategy. If we choose to process the instance $l\sigma^e \to r\sigma^e$ at some point, then for every proper subterm t of $l\sigma^e$, there exists a constant c such that $t \to_{N_l \cup C_l}^! c$ always eventually (assuming constants introduced earlier are smaller than constants introduced later).*

When $\mathcal{R}$ is convergent, by performing an induction on the number of applications of the *selection* rule, we can show that an (i) innermost strategy is unambiguous; (ii) the processed set of rules satisfy the persistence property whenever rules are chosen using an innermost strategy; (iii) the R-extension of the processed rules is convergent; and (iv) each new constant in normal form represents exactly one term in $\mathcal{T}(\Sigma)$ via N-rules. Once this result is proven, it is straightforward to establish the correctness result.

Theorem 6. *Let $\mathcal{R}$ be a convergent rewriting system. Then given any term, the inference system outlined above finds its normal form.*

6 Conclusion

Normalization of terms by a given set of rewrite rules is critical for the efficient implementation of rewrite-based systems. Simple straight-line reduction methods can be made more efficient by incorporating a history of reduction steps into the normalization process, so as to avoid repeating similar rewrite steps. Chew [4] adapted congruence closure techniques to obtain a practical technique such a non-oblivious normalization procedure. Chew's procedure applies to orthogonal systems, but was refined and generalized by Verma and Ramakrishnan [11] and Verma [10].

We have presented a general formalism, based on transformation rules, within which both Chew's original method and variants thereof, as well as more general normalization procedures, can be described. The most comprehensive previous results were obtained by Verma [10], who specified several postulates on a (priority) rewrite relation, which suffice to ensure completeness of a rewrite closure-based procedure for normalization. The postulates that are relevant for standard rewrite relations imply that the given rewrite system be confluent and non-overlapping, which means that our results cover a broader class of rewrite systems.

We believe that our approach sheds new light on the basic concepts underlying non-oblivious normalization in that it relates normalization methods to standard term rewriting techniques, such as completion and narrowing. (For instance, our results seem to indicate that the concept of a "strong closure," which plays a critical role in other completeness proofs, may not be intrinsic to non-oblivious normalization in general.) The transformation rules for the two basic components of non-oblivious normalization–congruence closure and narrowing–are essentially specialized versions of standard rules used to describe rewrite-based deduction and computation methods. Rules specific to non-oblivious normalization control the interface between the two components and the termination of the overall process.

There are some similarities between these transformation rules and the "calculus of rewriting with sharing" designed by Sherman [9]. This calculus operates on relations that can be described by D-rules and C-rules, and employs several transformations rules, most of which can be derived from transformations used in completion. The calculus contains a rule similar to superposition, though surprisingly, its application is not obligatory. This may reflect the fact that Sherman uses his calculus to provide an equational semantics for an implementation of a symbolic computation system, but does not address issues such as termination or completeness.

Acknowledgements. We would like to thank the anonymous reviewers for their helpful comments and for directing our attention to Sherman's work.

References

[1] L. Bachmair. *Canonical equational proofs.* Birkhäuser, Boston, 1991.

[2] L. Bachmair and N. Dershowitz. Equational inference, canonical proofs, and proof orderings. *JACM*, 41:236–276, 1994.

[3] L. P. Chew. An improved algorithm for computing with equations. In *21st Annual Symposium on Foundations of Computer Science*, 1980.

[4] L. P. Chew. *Normal forms in term rewriting systems.* PhD thesis, Purdue University, 1981.

[5] N. Dershowitz and J. P. Jouannaud. Rewrite systems. In J. van Leeuwen, editor, *Handbook of Theoretical Computer Science (Vol. B: Formal Models and Semantics)*, Amsterdam, 1990. North-Holland.

[6] D. Kapur. Shostak's congruence closure as completion. In H. Comon, editor, *Proc. 8th Intl. RTA*, pages 23–37, 1997. LNCS 1232, Springer, Berlin.

[7] J. W. Klop. Term rewriting systems. In S. Abramsky, D. M. Gabbay, and T. S. E. Maibaum, editors, *Handbook of Logic in Computer Science*, volume 1, chapter 6, pages 2–116. Oxford University Press, Oxford, 1992.

[8] G. Nelson and D. Oppen. Fast decision procedures based on congruence closure. *JACM*, 27(2):356–364, 1980.

[9] D. J. Sherman and N. Magnier. Factotum: Automatic and systematic sharing support for systems analyzers. In *Proc. TACAS, LNCS 1384*, 1998.

[10] R. M. Verma. A theory of using history for equational systems with applications. *JACM*, 42:984–1020, 1995.

[11] R. M. Verma and I. V. Ramakrishnan. Nonoblivious normalization algorithms for nonlinear systems. In *Proc. of the Int. Colloquium on Automata, Languages and Programming*, New York, 1990. Springer-Verlag.

Test Sets for the Universal and Existential Closure of Regular Tree Languages

Dieter Hofbauer and Maria Huber

Universität GH Kassel, Fachbereich 17 Mathematik/Informatik,
D–34109 Kassel, Germany
{dieter,maria}@theory.informatik.uni-kassel.de

Abstract. Finite test sets are a useful tool for deciding the membership problem for the universal closure of a given tree language, that is, for deciding whether a term has all its ground instances in the given language. A uniform test set for the universal closure must serve the following purpose: In order to decide membership of a term, it is sufficient to check whether all its test set instances belong to the underlying language.
A possible application, and our main motivation, is ground reducibility, an essential concept for many approaches to inductive reasoning. Ground reducibility modulo some rewrite system is membership in the universal closure of the set of reducible ground terms. Here, test sets always exist, and several algorithmic approaches are known. The resulting sets, however, are often unnecessarily large.
In this paper we consider regular languages and linear closure operators. We prove that universal as well as existential closure, defined analogously, preserve regularity. By relating test sets to tree automata and to appropriate congruence relations, we show how to characterize, how to compute, and how to minimize ground and non-ground test sets. In particular, optimal solutions now replace previous ad hoc approximations for the ground reducibility problem.

1 Introduction

Many approaches to inductive reasoning rely on the notion of *ground reducibility*, or *inductive reducibility*, in particular methods based on the so-called inductionless induction paradigm, see [32, 23, 24, 1] among many others. For a 'rational reconstruction' of the completion based versions as Noetherian induction on term orderings we refer to [22, 29, 19, 36, 34, 6]. A recent application can be found in [17].

A term is said to be ground reducible if all its ground instances are reducible. Ground reducibility was first shown decidable by Plaisted [35] and Kapur, Narendran and Zhang [26]; they compute a finite ground test set depending on the term to be checked. Kounalis first proved the effective existence of finite *uniform* (non-ground) test sets, that is, test sets not depending on the input term [28], cf. [20, 37], leading to the following ground reducibility check: A term is ground reducible if and only if all its test set instances are reducible. Several approaches for the construction of test sets for left-linear rewrite systems have been suggested, see [24, 25, 27, 8, 7] among others. A recent inductive theorem prover based on test sets is reported in [4, 3]. An alternative to the test set approach is the use of tree automata, in the non-linear case automata with constraints. For such (non-uniform) automata see [2, 11, 9].

The universal closure operator introduced in this paper generalizes ground reducibility, replacing the language of reducible ground terms by an arbitrary tree language: A term belongs to the universal closure of a language if all its ground instances belong to that language. Also for this more general problem we introduce the notion of a test set. Using classical results from formal language theory, we show how to compute and minimize finite ground or non-ground test sets in the case where the underlying language is regular. This is general enough to cover the computation of test sets for ground reducibility in the linear case. Our main motivation to write this paper was to give a precise characterization of test sets, rendering obsolete the race for smaller and smaller test sets. All previously proposed approaches rely on concepts like *tops*, *extensible positions*, *expandedness* etc. and therefore seem to be somewhat ad hoc in retrospective as they only approximate optimal test sets. Those criteria, however, may still be very useful when complexity issues matter, or when additional properties (like completeness) are needed. Our contribution leads to a better understanding of test sets and bridges the gap between test set based and automata based approaches to ground reducibility.

Restricting ourselves to linear terms is less limiting than it seems at first glance when we are interested in functional constructor based rewrite systems. Consider a ground convergent system which is sufficiently complete and constructor preserving relative to some constructor part of the signature, and where the set of ground normal forms is regular. These assumptions are very natural for many applications. Then a strongly equivalent left-linear system can be computed ([20] Sect. 7, cf. [30, 40]).

After briefly describing the algebraic and the formal language background in Section 2 and 3, we introduce the universal and the existential closure operator in Section 4, and prove that they preserve regularity. Ground test sets are treated in 5.1, test sets with variables in 5.2. The interested reader is also refered to [21], a more detailed version of this paper containing all proofs and some further examples.

2 Term Monoids

In order to describe the algebraic framework used in the sequel we begin with fixing notations for terms and instantiation operations. Let Σ be a one-sorted first order signature where each symbol has a fixed arity, and let $\mathcal{T}_\Sigma$ denote the set of ground terms over Σ. When dealing with *linear* terms and their instances, variable names are redundant. Only variable positions have to be indicated. For this purpose we use the constant symbol $\square$, assuming $\square \notin \Sigma$. The set of *linear terms* (or *contexts*) over Σ then consists of terms possibly containing the 'wildcard symbol' $\square$ and is denoted by $\mathcal{C}_\Sigma = \mathcal{T}_{\Sigma \cup \{\square\}}$. We further write $\mathcal{T}_\Sigma(n)$ and

$\mathcal{T}_\Sigma[n]$ to denote terms with at most n or exactly n occurrences of $\square$ respectively. Hence $\mathcal{C}_\Sigma = \bigcup_{n \geq 0} \mathcal{T}_\Sigma[n]$ and $\mathcal{T}_\Sigma[0] = \mathcal{T}_\Sigma$; the set $\mathcal{T}_\Sigma[1]$ corresponds to what is called 'special trees' in [39] and 'pointed trees' in [33]. The basic concatenation operation $\circ\colon \mathcal{T}_\Sigma(1) \times \mathcal{T}_\Sigma \to \mathcal{T}_\Sigma$ is recursively defined by

$$\square \circ t = t, \qquad f(s_1, \dots, s_n) \circ t = f(s_1 \circ t, \dots, s_n \circ t).$$

The same recursion equations naturally extend to an operation $\circ\colon \mathcal{C}_\Sigma \times \mathcal{C}_\Sigma \to \mathcal{C}_\Sigma$. Most important, however, is the extension $\circ\colon \mathcal{C}_\Sigma \times \mathcal{PC}_\Sigma \to \mathcal{PC}_\Sigma$ to sets of terms[1] via

$$\square \circ T = T, \qquad f(s_1, \dots, s_n) \circ T = f(s_1 \circ T, \dots, s_n \circ T)$$

using the operation $f(T_1, \dots, T_n) = \{f(t_1, \dots, t_n) \mid t_i \in T_i\}$ for sets of terms T_i from the power set term algebra; note that this gives $\{f\}$ for $n = 0$. Different occurrences of $\square$ may be instanciated by different terms, e.g., $f(\square, \square) \circ \{t, t'\} = \{f(t,t), f(t,t'), f(t',t), f(t',t')\}$, whereas $f(\square, \square) \circ t = f(t,t)$. Finally, we will use $\circ\colon \mathcal{PC}_\Sigma \times \mathcal{PC}_\Sigma \to \mathcal{PC}_\Sigma$ with $S \circ T = \bigcup_{s \in S} s \circ T$.

If $R \subseteq \mathcal{C}_\Sigma$ is the set of left-hand sides of a left-linear term rewrite system then $\mathrm{Red}(R) = \mathcal{T}_\Sigma[1] \circ R \circ \mathcal{T}_\Sigma$ is the set of *reducible ground terms*. (In different terminology, $\mathrm{Red}(R)$ is the set of ground terms *encompassing* some term from R [12, 11].) Its complement $\mathrm{Nf}(R) = \mathcal{T}_\Sigma \setminus \mathrm{Red}(R)$ contains the corresponding *irreducible* ground terms, or ground *normal forms*. The sets $\mathrm{RED}(R)$ and $\mathrm{NF}(R)$ consist of all R-reducible and R-irreducible terms from $\mathcal{C}_\Sigma$ respectively. The following two examples will be reconsidered later.

Example 1 *(1) Addition of naturals modulo $n > 0$ is specified by the rewrite system*

$$x + 0 \to x, \qquad x + s(y) \to s(x + y), \qquad s^n(x) \to x$$

over signature $\Sigma = \{0, s, +\}$. The set of left-hand side patterns in our setting is $R_1 = \{\square + 0, \square + s(\square), s^n(\square)\}$. Here $\mathrm{Nf}(R_1) = \{s^i(0) \mid i < n\}$, so $\mathrm{Red}(R_1)$ consists of all ground terms containing the symbol $+$ together with $\{s^i(0) \mid i \geq n\}$.
(2) Let $R_2 = \{g(h^i(a)) \mid 1 \leq i < n\} \cup \{g(h^n(\square)), h(g(\square))\}$ over $\Sigma = \{a, g, h\}$ for some $n \geq 1$. Then $\mathrm{Red}(R_2)$ consists of all ground terms containing both g and h, so $\mathrm{Nf}(R_2) = \{g^i(a), h^i(a) \mid i \geq 0\}$.

Monoid structures play an important role in this setting. The most prominent one is $(\mathcal{T}_\Sigma(1), \circ, \square)$; it has $(\mathcal{T}_\Sigma[1], \circ, \square)$ as a submonoid. A monoid for term sets is $(\mathcal{PC}_\Sigma, \circ, \{\square\})$; it has $(\{\{t\} \mid t \in \mathcal{T}_\Sigma(1)\}, \circ, \{\square\})$ as a submonoid, which is isomorphic to $(\mathcal{T}_\Sigma(1), \circ, \square)$. Slightly more general algebraic structures (*left-semi-modules* [13], or 'automata' [31, 14]) are needed when handling the heterogeneous operation $\circ\colon \mathcal{C}_\Sigma \times \mathcal{PC}_\Sigma \to \mathcal{PC}_\Sigma$. We will not go into details here, however.

---

[1] By $\mathcal{P}S$ we denote the power set of a set S.

3 Regular Tree Languages

Regularity of tree languages can be characterized in various ways; see [16, 10, 38, 18], among others. We will use tree automata, tree grammars, and congruence relations in the sequel.

A *(finite bottom-up) tree automaton* $\mathcal{A}$ is a finite rewrite system over $\Sigma \cup Q$ where Σ is the *terminal* signature and Q is a set of constants disjoint to Σ, called *states*. Rules have the form $f(X_1, \ldots, X_n) \to Y$ with $X_1, \ldots, X_n, Y \in Q$, $f \in \Sigma$ of arity n. The language *accepted by* $\mathcal{A}$ *in* $F \subseteq Q$ (the *accepting states*) is

$$L(\mathcal{A}) = \{t \in \mathcal{T}_\Sigma \mid \exists X \in F : t \xrightarrow{*}_{\mathcal{A}} X\}.$$

A *regular tree grammar* $\mathcal{G}$ is a finite rewrite system over $\Sigma \cup N$ where Σ is the *terminal* signature and N is a set of constants disjoint to Σ, called *non-terminals*. Rules have the form $X \to t$ with $X \in N$, $t \in \mathcal{T}_{\Sigma \cup N}$. The language *generated by* $\mathcal{G}$ *from* $S \subseteq N$ (the *starting symbols*) is

$$L(\mathcal{G}) = \{t \in \mathcal{T}_\Sigma \mid \exists X \in S : X \xrightarrow{*}_{\mathcal{G}} t\}.$$

Without changing the definition of $L(\mathcal{G})$ we can more generally allow arbitrary *ground rewrite rules*, i.e., rules of the form $s \to t$ with $s, t \in \mathcal{T}_{\Sigma \cup N}$, without leaving the class of regular tree languages [5].

Let $R \subseteq \mathcal{C}_\Sigma$ be a finite set of linear terms. Consider the regular tree grammar consisting of all rules according to one of the schemes

$$X \to f(\square, \ldots, \square, X, \square, \ldots, \square), \qquad X \to t, \qquad \square \to f(\square, \ldots, \square)$$

for $t \in R$, $f \in \Sigma$. Choosing $\{X, \square\}$ as non-terminals and X as the only starting symbol, the grammar generates $\mathrm{Red}(R)$. (Note that the grammar generates $\mathcal{T}_\Sigma$ from $\square$.) If instead we choose $\Sigma \cup \{\square\}$ as terminal signature and X as the only nonterminal then the same ground rewrite system[2] generates (from X) $\mathrm{RED}(R)$.

An algebraic characterization of regularity can be based on 'observational equivalence' of terms or contexts w.r.t. membership in a given tree language. For a set $L \subseteq \mathcal{T}_\Sigma$ define equivalence relations $\frown_{\in L}$ on $\mathcal{T}_\Sigma$ and $\smile_{\in L}$ on $\mathcal{T}_\Sigma(1)$ by

$$t \frown_{\in L} t' \qquad \text{iff} \qquad \forall s \in \mathcal{T}_\Sigma(1) : s \circ t \in L \leftrightarrow s \circ t' \in L,$$
$$s \smile_{\in L} s' \qquad \text{iff} \qquad \forall t \in \mathcal{T}_\Sigma : \quad s \circ t \in L \leftrightarrow s' \circ t \in L.$$

These 'syntactic' equivalence relations are left and right congruences respectively in the sense that $t \frown_{\in L} t'$ implies $s \circ t \frown_{\in L} s \circ t'$ for all $s \in \mathcal{T}_\Sigma(1)$ and that $s \smile_{\in L} s'$ implies $s \circ t \smile_{\in L} s' \circ t$ for all $t \in \mathcal{T}_\Sigma(1)$.

---

[2] Which is not a grammar since the rules $\square \to \ldots$ rewrite a terminal symbol.

Example 2 *(Example 1 cont'd) (1)*
For $L = \text{Red}(R_1)$ we get that $s^i(0)$
and $s^j(0)$, $i, j < n$, are left congru-
ent if and only if $i = j$. Indeed, for
$i < j$ the context $s^{n-j}(\Box)$ separates
these terms as $s^{n-j}(\Box) \circ s^i(0) \notin L$
and $s^{n-j}(\Box) \circ s^j(0) \in L$. In general,
all reducible ground terms form a
class themselves, separated from ir-
reducible terms by the 'empty' con-
text $\Box$. Thus the left congruence

	$\text{Red}(R_1)$	$s^{n-1}(0)$	$\ldots$	$s(0)$	0
A_1	+	+	+	+	+
$\text{Nf}(R_1)$	−	−	−	−	−
$\Box$	+	−	−	−	−
$s(\Box)$	+	+	−	−	−
$\ldots$	+	+	$\ldots$	−	−
$s^{n-1}(\Box)$	+	+	+	+	−

classes are $\text{Red}(R_1)$, $\{s^{n-1}(0)\}$, $\ldots$, $\{0\}$. Analogously, the right congruence
classes are the set A_1 of all terms in $\mathcal{T}_\Sigma(1)$ containing $+$ or being reducible, and
the sets $\text{Nf}(R_1)$, $\{\Box\}$, $\ldots$, $\{s^{n-1}(\Box)\}$. All tables in the rest of the paper are to
be understood in the following way: columns refer to left congruence classes (or
representatives) and rows to right congruence classes (or representatives) respec-
tively. A plus-entry in row s and column t means $s \circ t \in L$, a minus-entry means
$s \circ t \notin L$.

	$\text{Red}(R_2)$	a	$g(a)$	$h(a)$
A_2	+	+	+	+
$\text{Nf}(R_2)$	−	−	−	−
$\Box$	+	−	−	−
$g(\Box)$	+	−	−	+
$h(\Box)$	+	−	+	−

(2) It is not difficult to verify that $L = \text{Red}(R_2)$ yields the left congruence classes $\text{Red}(R_2)$, $\{a\}$, $\{g^i(a) \,|\, i \geq 1\}$, $\{h^i(a) \,|\, i \geq 1\}$, and the right congruence classes $A_2 = \text{RED}(R_2) \cup \{g^i(h^j(\Box)) \,|\, i \geq 1, 1 \leq j < n\}$, $\text{Nf}(R_2)$, $\{\Box\}$, $\{g^i(\Box) \,|\, i \geq 1\}$, $\{h^i(\Box) \,|\, i \geq 1\}$. Note that the index, i.e., the number of congruence classes, does neither depend on the size of the underlying term set nor on the depth of its terms in this example.

We conclude by summarizing the characterizations of regularity mentioned above.

Theorem 1 *For $L \subseteq \mathcal{T}_\Sigma$ the following properties are equivalent:*
(1) L is accepted by a tree automaton. (2) L is generated by a regular tree gram-
mar. (3) L is generated by a finite ground rewrite system. (4) $\frown_{\in L}$ has finite
index. (5) $\smile_{\in L}$ has finite index.

4 Universal and Existential Closure

In this section we introduce the universal and the existential closure operator and prove as a main result that they preserve regularity.

Definition 1 *The* universal closure *$L^\forall \subseteq \mathcal{C}_\Sigma$ of a tree language $L \subseteq \mathcal{T}_\Sigma$ consists of those terms that have all their ground instances in L, that is*

$$s \in L^\forall \qquad \textit{iff} \qquad s \circ \mathcal{T}_\Sigma \subseteq L.$$

Equivalently, $s \in L^\vee$ if and only if $\forall t \in s \circ T_\Sigma : t \in L$. Analogously we define the existential closure $L^\exists \subseteq C_\Sigma$ *of L as the set of terms that have some ground instance in L, that is $s \in L^\exists$ if and only if $\exists t \in s \circ T_\Sigma : t \in L$, or (with $\overline{L} = T_\Sigma \setminus L$)*

$$s \in L^\exists \quad \text{iff} \quad s \circ T_\Sigma \not\subseteq \overline{L}.$$

Example 3 *Linear propositional formulas are terms over the signature $\{\neg, \vee, \wedge,$ true, false, $\square\}$. When L is the set of all ground formulas whose value (under the natural interpretation) is* true *then $L^\vee$ is the set of* tautologies *whereas $L^\exists$ is the set of* satisfiable *formulas.*

If L is equal to $\text{Red}(R)$ for some set $R \subseteq C_\Sigma$ then $L^\vee$ is the set of (linear) *ground reducible* terms w.r.t. R, that is, it contains those terms that have R-reducible ground instances only.

Example 4 *(Example 1 cont'd) The set $\text{Red}(R_1)^\vee$ of ground reducible terms w.r.t. R_1 consists of $\text{RED}(R_1)$ together with all terms in C_Σ containing the symbol $+$, a 'defined' symbol in the given specification. And $\text{Red}(R_2)^\vee$ is the set of all terms that contain both g and h.*

Theorem 2 *The following properties are equivalent: (1) L is a regular subset of T_Σ. (2) $L^\vee$ is a regular subset of C_Σ. (3) $L^\exists$ is a regular subset of C_Σ.*

PROOF. Regularity of $L^\vee$ or $L^\exists$ directly implies regularity of L since $L^\vee \cap T_\Sigma = L^\exists \cap T_\Sigma = L$ and T_Σ is a regular subset of C_Σ.

To show that (1) implies (2) and (3) let $\mathcal{A}$ be an arbitrary, possibly non-deterministic, tree automaton over Σ accepting L. Let Q be its set of states and let $F \subseteq Q$ be the set of accepting states. In order to fix notations, let $\mathcal{A}(t) = \{X \in Q \,|\, t \xrightarrow{*}_\mathcal{A} X\}$ and $\mathcal{A}(T) = \bigcup_{t \in T} \mathcal{A}(t)$ for $t \in T_{\Sigma \cup Q}$ and $T \subseteq T_{\Sigma \cup Q}$. Without loss of generality we assume that $\mathcal{A}$ is complete and that all states are accessible, i.e., for any $X \in Q$ there is a ground term t with $X \in \mathcal{A}(t)$, or equivalently $\mathcal{A}(T_\Sigma) = Q$.

The basic observation is that a term $s \in C_\Sigma$ is in $L^\vee$ if and only if all terms in $s \circ Q$ are accepted by $\mathcal{A}$. Similarly, s is in $L^\exists$ if and only if some term in $s \circ Q$ is accepted by $\mathcal{A}$.

Using $\mathcal{A}(T_\Sigma) = Q$ we get $\mathcal{A}(s \circ T_\Sigma) = \mathcal{A}(s \circ \mathcal{A}(T_\Sigma)) = \mathcal{A}(s \circ Q)$, and therefore

$$s \in L^\vee \quad \text{iff} \quad s \circ T_\Sigma \subseteq L \quad \text{iff} \quad \mathcal{A}(s \circ T_\Sigma) = \mathcal{A}(s \circ Q) \subseteq F$$

in the case where $\mathcal{A}$ is deterministic[3], and

$$s \in L^\exists \quad \text{iff} \quad s \circ T_\Sigma \cap L \neq \emptyset \quad \text{iff} \quad \mathcal{A}(s \circ T_\Sigma) \cap F = \mathcal{A}(s \circ Q) \cap F \neq \emptyset$$

for arbitrary $\mathcal{A}$. Now, for $P \subseteq Q$, define the automaton

$$\mathcal{A}[P] = \mathcal{A} \cup \{\square \to X \,|\, X \in P\}$$

---

[3] Note that determinism of $\mathcal{A}$ is used for the implication from $s \circ T_\Sigma \subseteq L$ to $\mathcal{A}(s \circ T_\Sigma) \subseteq F$.

over $\Sigma \cup \{\Box\}$ and let $\mathcal{B}[P]$ be the standard power-set automaton for $\mathcal{A}[P]$. (We get the same automaton $\mathcal{B}[P]$ when we first construct the power-set automaton for $\mathcal{A}$ and then add the single rule $\Box \to P$.) By construction, $\mathcal{B}[P]$ is deterministic and complete; for $s \in \mathcal{C}_\Sigma$ we write $\mathcal{B}[P](s) = S$ for $S \subseteq Q$ in case $s \overset{*}{\to}_{\mathcal{B}[P]} S$. We conclude by proving that $L^\forall$ and $L^\exists$ are accepted by $\mathcal{B}[Q]$ and $\mathcal{A}[Q]$ respectively:

(i) $L(\mathcal{B}[Q]) = L^\forall$ with $\{S \subseteq Q \mid \forall X \in S : X \in F\} = \mathcal{P}F$ as set of accepting states if $\mathcal{A}$ is deterministic,

(ii) $L(\mathcal{A}[Q]) = L^\exists$ with F as set of accepting states, hence

$L(\mathcal{B}[Q]) = L^\exists$ with $\{S \subseteq Q \mid \exists X \in S : X \in F\}$ as set of accepting states.

Indeed, $\mathcal{A}(s \circ P) = \mathcal{A}[P](s) = \mathcal{B}[P](s)$ implies $s \in L^\forall$ iff $\mathcal{A}[Q](s) = \mathcal{B}[Q](s) \subseteq F$ for deterministic $\mathcal{A}$, which proves (i), and $s \in L^\exists$ iff $\mathcal{A}[Q](s) \cap F = \mathcal{B}[Q](s) \cap F \neq \emptyset$ for arbitrary $\mathcal{A}$, which proves (ii). □

For studying non-ground test sets in Section 5.2 we have to generalize the above closure operators to non-ground languages $L \subseteq \mathcal{C}_\Sigma$. Note, however, that already the ground part of L determines $L^\forall$ since $L^\forall = (L \cap \mathcal{T}_\Sigma)^\forall$. The above theorem then reads as follows:

$$L \cap \mathcal{T}_\Sigma \text{ is regular} \quad \text{iff} \quad L^\forall \text{ is regular} \quad \text{iff} \quad L^\exists \text{ is regular.}$$

Another reason to consider non-ground languages is to obtain *closure operators* in the proper sense. Now universal closure is *idempotent* (i.e., $(L^\forall)^\forall = L^\forall$), *monotone* (i.e., $L_1 \subseteq L_2$ implies $L_1^\forall \subseteq L_2^\forall$), and, under the additional assumption that L is closed under ground substitutions ($L \circ \mathcal{T}_\Sigma \subseteq L$), also *extensive* (i.e., $L \subseteq L^\forall$). For instance, languages of the form $\text{RED}(R)$ are closed under ground substitutions. The same remarks hold for existential closure as well.

5 Test Sets

By definition, a term is in the universal closure of a tree language L if all its ground instances are members of L. Instead of checking the typically infinite set of all ground instances it often suffices to check a smaller number of instances, preferably finitely many. Additionally we want the set of terms that is used for instantiations to be *uniform* for L, that is, independent of the term under consideration. Such a set is said to be a *test set* for the universal closure of L. Analogously we define test sets for the existential closure. The following definition is general enough to capture non-ground test sets.

Definition 2 *A set of terms $T \subseteq \mathcal{C}_\Sigma$ is a test set for the universal closure of $L \subseteq \mathcal{C}_\Sigma$ if, for all $s \in \mathcal{C}_\Sigma$,*

$$s \in L^\forall \quad \text{iff} \quad s \circ T \subseteq L,$$

and T is a test set for the existential closure of L if, for all $s \in \mathcal{C}_\Sigma$,

$$s \in L^\exists \quad \text{iff} \quad s \circ T \not\subseteq \overline{L}$$

with $\overline{L} = \mathcal{C}_\Sigma \setminus L$. In both cases, T is a ground test set if $T \subseteq \mathcal{T}_\Sigma$.

Example 5 *(Example 1(1) cont'd) Standard approaches from the literature yield* Nf(R_1) *of cardinality n as a test set. We will show in Example 6(1) that in fact a singleton test set is sufficient.*

Membership in $L^\forall$ and $L^\exists$ respectively becomes decidable if a finite test set effectively exists. In Section 5.1 we show how to characterize and how to compute (minimal) finite ground test sets in case L is regular. Perhaps surprising at first glance, finite ground test sets might exist even for non-regular L. In general, however, finite ground test sets do not exist in the non-regular case. This is possible, for instance, for non-regular languages of the form $L = \mathrm{Red}(R)$, whereas for languages $L = \mathrm{RED}(R)$ it is known that finite (non-ground) test sets always exist. Section 5.2 deals with non-ground test sets for regular languages. We explain why non-ground test sets can often be smaller than any ground test set, and again show how to obtain such sets.

5.1 Ground Test Sets

A set of ground terms T is a test set for the universal closure of a language $L \subseteq \mathcal{T}_\Sigma$ if $s \circ \mathcal{T}_\Sigma \subseteq L$ is equivalent to $s \circ T \subseteq L$ for any $s \in \mathcal{C}_\Sigma$. In particular, the set $\mathcal{T}_\Sigma$ of all ground terms is always a trivial test set. A general characterization of ground test sets can be given in terms of the equivalence relation $\frown_{\subseteq L}$ on $\mathcal{P}\mathcal{T}_\Sigma$ defined by

$$T \frown_{\subseteq L} T' \qquad \text{iff} \qquad \forall s \in \mathcal{C}_\Sigma : s \circ T \subseteq L \leftrightarrow s \circ T' \subseteq L.$$

(Note that this is a left congruence in the sense that $T \frown_{\subseteq L} T'$ implies $S \circ T \frown_{\subseteq L} S \circ T'$ for any set $S \subseteq \mathcal{C}_\Sigma$.) Then T is a ground test set for the universal closure of L if and only if

$$T \frown_{\subseteq L} \mathcal{T}_\Sigma.$$

It turns out that we can replace quantification over all terms in the definition of $\frown_{\subseteq L}$ by quantification over the much simpler set of terms with at most one variable. Defining $\frown^1_{\subseteq L}$ on $\mathcal{P}\mathcal{T}_\Sigma$ by

$$T \frown^1_{\subseteq L} T' \qquad \text{iff} \qquad \forall s \in \mathcal{T}_\Sigma(1) : s \circ T \subseteq L \leftrightarrow s \circ T' \subseteq L,$$

we get that both congruence relations are identical. Stated differently, test sets for $L^\forall(1) = L^\forall \cap \mathcal{T}_\Sigma(1)$ are already test sets for $L^\forall$. When looking for ground test sets it therefore suffices to consider contexts with one variable only (as proven in [21]).

Lemma 1 *The congruences $\frown_{\subseteq L}$ and $\frown^1_{\subseteq L}$ on $\mathcal{P}\mathcal{T}_\Sigma$ coincide.*

As a further characterization, we establish a bijection between ground test sets on the one hand and certain automata on the other. For regular L let the automata $\mathcal{A}$ and $\mathcal{B}$ be as in the proof of Theorem 2 and assume $\mathcal{A}$ to be complete and deterministic. Now there is direct connection between ground term sets T and state sets $P \subseteq Q$. Suppose $\mathcal{A}(T)$ equals P. Then the automaton $\mathcal{B}[P]$ accepts $L^{\forall}$ (in $\mathcal{P}F$ as before) if and only if T is a test set for the universal closure of L. This can be read in two directions: Given a test set T, the automaton $\mathcal{B}[\mathcal{A}(T)]$ accepts $L^{\forall}$. Conversely, given an automaton $\mathcal{B}[P]$ accepting $L^{\forall}$, any set T proving the accessibility of P in $\mathcal{A}$ is a test set. Minimal test sets, in particular, correspond to minimal such sets of states with exactly the same cardinality. The proof of the following theorem can be found in [21].

Theorem 3 (Characterizing ground test sets) *Let $L \subseteq \mathcal{T}_\Sigma$ be regular, let $\mathcal{A}$ be a deterministic automaton with states Q accepting L, and let $\mathcal{B}$ be defined as in the proof of Theorem 2. For $T \subseteq \mathcal{T}_\Sigma$ and $P \subseteq Q$ with $P = \mathcal{A}(T)$ the following properties are equivalent: (1) T is a test set for the universal closure of L. (2) $T \frown_{\subseteq L}^1 \mathcal{T}_\Sigma$. (3) $\mathcal{B}[P]$ accepts $L^{\forall}$ in $\mathcal{P}F$. Analogously, with $\overline{L} = \mathcal{T}_\Sigma \setminus L$ the following properties are equivalent: (1) T is a test set for the existential closure of L. (2) $T \frown_{\subseteq \overline{L}}^1 \mathcal{T}_\Sigma$. (3) $\mathcal{A}[P]$ accepts $L^{\exists}$ in F.*

The left congruence $\frown_{\in L}$ on $\mathcal{T}_\Sigma$ and the left congruence $\frown_{\subseteq L}^1$ on $\mathcal{P}\mathcal{T}_\Sigma$ are directly related by

$$t \frown_{\in L} t' \quad \text{iff} \quad L/t = L/t' \quad \text{and} \quad T \frown_{\subseteq L}^1 T' \quad \text{iff} \quad L/T = L/T'$$

for $t, t' \in \mathcal{T}_\Sigma$, $T, T' \subseteq \mathcal{T}_\Sigma$ using the quotients

$$L/t = \{s \in \mathcal{T}_\Sigma(1) \mid s \circ t \in L\} \quad \text{and} \quad L/T = \bigcap_{t \in T} L/t.$$

Therefore $\frown_{\in L}$ has finite index if and only if

$\frown_{\subseteq L}^1$ on $\mathcal{P}\mathcal{T}_\Sigma$ has finite index   if and only if $L \subseteq \mathcal{T}_\Sigma$ is regular.

Another consequence is $\mathcal{T}_\Sigma \frown_{\subseteq L}^1 T$ for any set T of representatives for $\frown_{\in L}$, hence those sets are test sets. Typically, however, much smaller test sets exist as shown by the examples below.

Theorem 4 *Each set of representatives for $\frown_{\in L}$ on $\mathcal{T}_\Sigma$ is a ground test set for the universal and the existential closure of L. Therefore, finite ground test sets always exist for the universal and the existential closure of regular languages.*

We can interpret these observations in terms of tables as used in Example 2. First note that computing this kind of tables amounts to figure out finite sets of representatives for $\frown_{\in L}$ on $\mathcal{T}_\Sigma$ and $\smile_{\in L}$ on $\mathcal{T}_\Sigma(1)$ respectively; this is possible just for regular L. Now, a term s is in $L^{\forall}(1)$ if and only if the corresponding row contains no minus-entry. (Thus $L^{\forall}(1)$ is always a single right class.) More generally, checking $T \frown_{\subseteq L}^1 T'$ amounts to collect rows that have a minus-entry below some term from T and T' respectively, and to check whether the two collections are equal. Hence T is a test set for the universal closure if and only if the corresponding columns 'cover' all existing minus-entries.

Theorem 5 *It is decidable whether a given finite (or more general, regular) set of ground terms is a test set for the universal or existential closure of a given regular language.*

Ground supersets of ground test sets are test sets as well. Therefore we are interested in *minimal* test sets w.r.t. set inclusion, or in *optimal* test sets, that is, test sets of least cardinality. Finding a minimal test set in this way amounts to solve a *minimum cover* problem which is NP-complete in general [15]. Note that we can also minimize arbitrary given ground test sets in the same manner.

Theorem 6 *Optimal ground test sets for the universal and the existential closure of regular tree languages effectively exist. Minimality of a given ground test set is decidable.*

Example 6 *(Example 2 cont'd) (1) For $L = \text{Red}(R_1)$ we obtain $L^\vee = A_1$, since the row for A_1 contains only plus-entries. Although the set of representatives has cardinality $n+1$, a singleton test set exists: The set $\{0\}$ is the only minimal (thus optimal) test set for the universal closure of L since the minus-entries of the last column cover the minus-entries of any other column, that is, $\{0\} \frown_{\subseteq L} T_\Sigma$.*
(2) For $L = \text{Red}(R_2)$ we get $L^\vee = A_2$. In this example, two different minimal test sets exist (relative to the representatives chosen):

$$\{a\} \frown_{\subseteq L} \{g(a), h(a)\} \frown_{\subseteq L} T_\Sigma$$

shows that $\{a\}$ as well as $\{g(a), h(a)\}$ is a possible ground test set and it is easy to see that both are minimal. The only optimal ground test set is $\{a\}$.

Example 7 *Consider the schematic examples given in Figures (a)–(c) where $T_0 = \{t_1, \ldots, t_n\}$ is a set of representatives for the left congruence on T_Σ. The singleton $\{t_n\}$ is the unique minimal test set (relative to T_0) in example (a). For (b) any two element subset of T_0 is a minimal test set. And T_0 is the unique minimal test set in example (c); note that here $T \frown_{\subseteq L} T'$ holds if and only if $T = T'$ for $T, T' \subseteq T_0$. All three schemes have instances of the form $L = \text{Red}(R)$ with finite linear R. For the sake of simplicity we left out the right class $L^\vee$ in (b) and (c).*

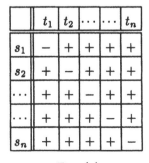

	t_1	t_2	$\cdots$	$\cdots$	t_n
s_1	+	−	−	−	−
s_2	+	+	−	−	−
$\cdots$	+	+	+	−	−
$\cdots$	+	+	+	+	−
s_n	+	+	+	+	+

Fig. (a)

	t_1	t_2	$\cdots$	$\cdots$	t_n
s_1	+	−	−	−	−
s_2	−	+	−	−	−
$\cdots$	−	−	+	−	−
$\cdots$	−	−	−	+	−
s_n	−	−	−	−	+

Fig. (b)

	t_1	t_2	$\cdots$	$\cdots$	t_n
s_1	−	+	+	+	+
s_2	+	−	+	+	+
$\cdots$	+	+	−	+	+
$\cdots$	+	+	+	−	+
s_n	+	+	+	+	−

Fig. (c)

We conclude this section with two remarks on ground test sets for non-regular languages.

(I) Finite ground test sets might exist also for the universal closure of non-regular languages, i.e., even in case the set of representatives for the left congruence is infinite (for an example see [21]). This shows that Theorem 4 cannot be extended to a characterization of regularity. Interestingly, our example is again a language of the form Red(R) with finite R. In this case, however, R is necessarily non-linear.

(II) One might conjecture that finite ground test sets exist for the universal closure of any language of the form Red(R) where R is an arbitrary (linear or non-linear) finite set. Indeed, finite non-ground test sets can always be effectively obtained for this class of languages [28, 20]. However, we cannot dispense with the generalization to non-ground test sets (again, see [21] for an example).

5.2 Non-Ground Test Sets

In connection with ground reducibility, usually test sets with variables are considered. They are guaranteed to exist (also in the non-regular case), and variables are indispensible when additional properties, e.g. completeness[4], are required. Furthermore, allowing variables often leads to smaller test sets.

The idea underlying the use of non-ground-test sets is to test membership of test set instances relative to a superset of L. Since we cannot extend L by ground terms without changing its universal closure, we add terms over an enlarged signature instead. Here we consider the case where the constant symbol $\square$ is added to Σ; this gives 'test sets with variables' as studied in the literature on ground reducibility. Let $L \subseteq T_\Sigma$ as before and choose any set $M \subseteq C_\Sigma$ with

$$L \subseteq M \subseteq L^\forall.$$

Thus $L = M \cap T_\Sigma$ and therefore the universal closure of L and M coincide since then $L^\forall = (M \cap T_\Sigma)^\forall = M^\forall$. In this setting, we call $T \subseteq C_\Sigma$ a *test set for the universal closure of M* if, for any $s \in C_\Sigma$,

$$s \in M^\forall \qquad \text{iff} \qquad s \circ T \subseteq M. \tag{1}$$

A good choice for M might be a set which is as large as possible, and at the same time simple enough to make membership for M easily decidable, i.e., a regular set. The two extrems, however, don't make much sense: For $M = L$ we are back to the ground case, and for $M = L^\forall$ we don't gain anything as with the singleton test set $T = \{\square\}$ condition (1) becomes a tautology. In all examples of this section we choose $L = \text{Red}(R)$ and $M = \text{RED}(R)$ for the respective set R.

First we restrict our attention to terms with at most one variable. Consider the schematic example below where the ground part of the table (cf. Example 7(c)) is enriched by three non-ground columns, i.e., we assume $s_i \in T_\Sigma(1)$, $t_i \in T_\Sigma$, $c_i \in C_\Sigma$. Now a plus-entry stands for membership in M.

---

[4] $T \subseteq C_\Sigma$ is said to be *complete* for R if $\text{Nf}(R) \subseteq T \circ T_\Sigma$.

Here the ground test set $\{t_1,\dots,t_n\}$ of cardinality n can be replaced by a test set with two elements, namely $\{c_1,c_2\}$, as these two terms cover all minus-entries in the ground part. It is important to note that c_2 cannot be replaced by c_3 in this test set, and c_3 is not even allowed in any test set; although $\{c_1,c_3\}$ covers all minus-entries, it also introduces new ones, thereby loosing the test set property: $s_0 \in M^{\vee}(1)$ but $s_0 \circ c_3 \notin M$.

	t_1	t_2	$\cdots$	$\cdots$	t_n	c_1	c_2	c_3
s_0	+	+	+	+	+	+	+	−
s_1	−	+	+	+	+	−	+	+
s_2	+	−	+	+	+	+	−	−
$\cdots$	+	+	−	+	+	−	+	+
$\cdots$	+	+	+	−	+	+	−	−
s_n	+	+	+	+	−	−	+	+

In particular, this example shows that in the non-ground case the property of being a test set is not preserved in general when we go from sets to supersets.

Example 8 (Singleton non-ground sets) *Let $R = \{g(a), h(b), g(h(\square)), h(g(\square))\}$ over $\Sigma = \{a, b, g, h\}$. For the corresponding left and right classes in the table we use the notations $g^*(t) = \{g^i(t) \mid 0 \le i\}$ and $g^+(t) = \{g^i(t) \mid 0 < i\}$. An optimal ground test set is $\{b, a\}$, whereas we can choose $\{\square\}$ as an optimal non-ground test set. The example is rather special as ground reducibility coincides with reducibility, that is, $\mathrm{RED}(R)^{\vee} = \mathrm{RED}(R)$. Note that this is the case if and only if the singleton $\{\square\}$ suffices as a test set.*

	$\mathrm{Red}(R)$	$g^*(b)$	$h^*(a)$	$\square$
$\mathrm{RED}(R)$	+	+	+	+
$\square$	+	−	−	−
$g^+(\square)$	+	−	+	−
$h^+(\square)$	+	+	−	−

As in the previous section we can characterize also non-ground test sets in terms of congruences. For this purpose we replace $\frown_{\subseteq L}$ by the left congruence $\frown_{\subseteq M}$ on $\mathcal{PC}_\Sigma$ where, for $T, T' \subseteq \mathcal{C}_\Sigma$, $T \frown_{\subseteq M} T'$ is defined to be equivalent to $\forall s \in \mathcal{C}_\Sigma : s \circ T \subseteq M \leftrightarrow s \circ T' \subseteq M$.

Theorem 7 (Characterizing non-ground test sets) *Let $M \subseteq \mathcal{C}_\Sigma$. A set $T \subseteq \mathcal{C}_\Sigma$ is a test set for the universal closure of M if and only if $T \frown_{\subseteq M} \mathcal{T}_\Sigma$.*

Unfortunately, in contrast to the ground case, test sets T for $M^{\vee}(1)$ are not necessarily test sets for $M^{\vee}$. They may cheat in every possible sense: In "$s \in M^{\vee}$ iff $s \circ T \subseteq M$" neither the 'if' nor the 'only if' implication holds in general.

Example 9 (Test sets for one-variable terms are not always general test sets) *(1) Consider $R = \{f(a, \square), f(\square, a), f(\square, f(\square, \square)), f(f(\square, \square), \square)\}$ over $\Sigma = \{a, f\}$. Then $\mathrm{NF}(R) = \mathcal{C}_\Sigma \setminus \mathrm{RED}(R) = \{a, \square, f(\square, \square)\}$ and $\mathrm{Nf}(R) = \{a\}$. Here, for terms in $\mathcal{T}_\Sigma(1)$ ground reducibility is equivalent to reducibility, and test sets for $\mathrm{RED}(R)^{\vee}(1)$ are $\{a\}$, $\{\square\}$, and $\{f(\square, \square)\}$, for instance. But $T = \{\square\}$ is not a test set for $\mathrm{RED}(R)^{\vee}$ since*

$$f(\square, \square) \in \mathrm{RED}(R)^{\vee} \quad but \quad f(\square, \square) \circ T = \{f(\square, \square)\} \nsubseteq \mathrm{RED}(R).$$

(2) Let $R = \{h(a), h(h(\square)), h(f(\square, \square)), f(h(\square), h(\square)), f(\square, f(\square, \square)), f(f(\square, \square), \square)\}$ over $\Sigma = \{a, h, f\}$. Then $\mathrm{Nf}(R) = \{a, f(a, a)\}$ and $\mathrm{NF}(R) = \{a, h(\square), f(a, a),$

$f(\square,\square), f(a, h(\square)), f(h(\square), a)\}$. *Among the test sets for* $\text{RED}(R)^{\vee}(1)$ *are* $\{a\}$ *and* $\{h(\square)\}$. *Nevertheless* $T = \{h(\square)\}$ *is not a test set for* $\text{RED}(R)^{\vee}$ *since*

$$f(\square,\square) \notin \text{RED}(R)^{\vee} \quad \text{but} \quad f(\square,\square) \circ T = \{f(h(\square), h(\square))\} \subseteq \text{RED}(R).$$

Hence an analogue to Lemma 1 does not hold. The question now is how to decide the relation $\frown_{\subseteq M}$ in order to apply Theorem 7. For this purpose we will consider several syntactic congruences for M over $\Sigma' = \Sigma \cup \{\square\}$ as intermediate steps. Since the variable symbol $\square$ now is already a constant in the signature Σ' we need another constant, $\blacksquare$ say, that will play the same role for Σ' as $\square$ in the 'ground case' for Σ. Denote by $\mathcal{C}_{\Sigma}(n)$ and $\mathcal{C}_{\Sigma}[n]$ the set of terms in $\mathcal{T}_{\Sigma \cup \{\square, \blacksquare\}}$ with at most n and exactly n occurrences of $\blacksquare$ respectively, and define the operation $\bullet : \mathcal{C}_{\Sigma}(1) \times \mathcal{C}_{\Sigma} \to \mathcal{C}_{\Sigma}$ by $\blacksquare \bullet t = t$ and $f(s_1, \ldots, s_n) \bullet t = f(s_1 \bullet t, \ldots, s_n \bullet t)$; thus $\text{RED}(R) = \mathcal{C}_{\Sigma}[1] \bullet R \circ \mathcal{C}_{\Sigma}$, e.g. Then left and right congruence relations $\approx_{\in M}$ on $\mathcal{C}_{\Sigma}$ and $\smile_{\in M}$ on $\mathcal{C}_{\Sigma}(1)$ are given by

$$t \approx_{\in M} t' \qquad \text{iff} \qquad \forall s \in \mathcal{C}_{\Sigma}(1): s \bullet t \in M \leftrightarrow s \bullet t' \in M,$$
$$s \smile_{\in M} s' \qquad \text{iff} \qquad \forall t \in \mathcal{C}_{\Sigma}: \quad s \bullet t \in M \leftrightarrow s' \bullet t \in M.$$

We conclude by sketching an algorithm for (minimal) non-ground test sets. Like in the ground case (Sect. 5.1) we first compute sets of representatives for $\approx_{\in M}$ and $\smile_{\in M}$ respectively. Now we can determine the relation $\approx^1_{\subseteq M}$ on $\mathcal{PC}_{\Sigma}$ – which coincides with $\approx_{\subseteq M}$ (Lemma 1) – where $\approx^1_{\subseteq M}$ and $\approx_{\subseteq M}$ are defined in analogy to $\frown^1_{\subseteq L}$ and $\frown_{\subseteq L}$. In the ground case the algorithm would stop here. In the non-ground case, we are still looking for the congruence $\frown_{\subseteq M}$. Since the left congruence $\frown_{\subseteq M}$ is refined by $\approx_{\subseteq M}$, we can use $\approx_{\subseteq M}$ to compute the corresponding right congruence $\smile_{\subseteq M}$ (defined in the obvious way), which in turn finally yields $\frown_{\subseteq M}$. Now, according to Theorem 7, test sets for the universal closure of M are those subsets of a set of representatives for $\approx_{\in M}$ that are congruent modulo $\frown_{\subseteq M}$ to $\mathcal{T}_{\Sigma}$.

Example 10 *(Exp. 9(1) cont'd) In order to determine a correct test set, we first compute sets of representatives for* $\approx_{\in M}$ *on* $\mathcal{C}_{\Sigma}$ *and for* $\smile_{\in M}$ *on* $\mathcal{C}_{\Sigma}(1)$ *using any standard algorithm; the result is shown in the left table below. The right table represents the congruences* $\frown_{\subseteq M}$ *on* $\mathcal{PC}_{\Sigma}$ *and* $\smile_{\subseteq M}$ *on* $\mathcal{C}_{\Sigma}$. *We get*

$$\mathcal{T}_{\Sigma} \frown_{\subseteq M} \text{Red}(R) \cup \{a\} \frown_{\subseteq M} \{f(a,a), a\} \frown_{\subseteq M} \{a\} \frown_{\subseteq M} \{f(\square,\square)\},$$

and $\{a\}$ *and* $\{f(\square,\square)\}$ *are the (only) minimal test sets. It becomes obvious now that* $\{\square\}$ *is incorrect as this set introduces a minus-entry in the last row where* $\mathcal{T}_{\Sigma}$ *has a plus-entry. (Note that* $\text{RED}(R)$ *is a left class in the first table whereas* Red(R) *and* $\mathcal{T}_{\Sigma}$ *are representatives for left classes in the second.)*

	$\text{RED}(R)$	$a,$ $f(\square,\square)$	$\square$
reducible $\mathcal{C}_{\Sigma}(1)$	+	+	+
$\text{NF}(R)$	−	−	−
$\blacksquare$	+	−	−
$f(\square, \blacksquare), f(\blacksquare, \square)$	+	+	−

	Red(R), $\emptyset$, $\{f(a,a)\}$	$\mathcal{T}_{\Sigma}$, $\{a\}$, $\{f(\square,\square)\}$	$\{\square\}$
$\text{RED}(R)$	+	+	+
Nf(R)	−	−	−
$\square$	+	−	−
$f(\square,\square)$	+	+	−

Theorem 8 *Optimal test sets for the universal and the existential closure of regular tree languages effectively exist. It is decidable whether a given finite (or regular) subset of C_Σ is a test set for the universal or existential closure of a given regular subset of C_Σ. Minimality of a given test set is decidable.*

References

1. L. Bachmair. Proof by consistency in equational theories. *Proc. LICS-88*, pp. 228–233. IEEE, 1988.
2. B. Bogaert and S. Tison. Equality and disequality constraints on direct subterms in tree automata. *Proc. STACS-92*, LNCS 577, pp. 161–172, 1992.
3. A. Bouhoula. Automated theorem proving by test set induction. *Journal of Symbolic Computation*, 23(1):47–77, 1997.
4. A. Bouhoula, E. Kounalis, and M. Rusinowitch. Automated mathematical induction. *Journal of Logic and Computation*, 5(5):631–668, 1995.
5. W. S. Brainerd. Tree generating regular systems. *Information and Control*, 14(2):217–231, 1969.
6. F. Bronsard, U. S. Reddy, and R. W. Hasker. Induction using term orderings. *Proc. CADE-94*, LNAI 814, pp. 102–117, 1994.
7. R. Bündgen. *Term completion versus algebraic completion*. Dissertation, Universität Tübingen, Germany, 1991.
8. R. Bündgen and W. Küchlin. Computing ground reducibility and inductively complete positions. *Proc. RTA-89*, LNCS 355, pp. 59–75, 1989.
9. H. Comon and F. Jacquemard. Ground reducibility is EXPTIME-complete. *Proc. LICS-97*, pp. 26–34, 1997.
10. B. Courcelle. On recognicable sets and tree automata. In *Resolution of Equations in Algebraic Structures*, vol. 1, pp. 351–367, Academic Press, 1989.
11. M. Dauchet, A.-C. Caron, and J.-L. Coquidé. Automata for reduction properties solving. *Journal of Symbolic Computation*, 20(2):215–233, 1995.
12. N. Dershowitz and J.-P. Jouannaud. Rewrite systems. In *Handbook of Theoretical Computer Science*, vol. B, pp. 243–320, 1990.
13. P. Deussen. *Halbgruppen und Automaten*. Springer, 1971.
14. S. Eilenberg. *Automata, languages, and machines*, vol. A. Academic Press, 1974.
15. M. R. Garey and D. S. Johnson. *Computers and intractability – A Guide to the Theory of NP-Completeness*. W.H. Freeman and Company, 1979.
16. F. Gécseg and M. Steinby. *Tree automata*. Akadémiai Kiadó, 1984.
17. A. Geser. Mechanized inductive proof of properties of a simple code optimizer. *Proc. TAPSOFT-95*, LNCS 915, pp. 605–619, 1995.
18. R. Gilleron and S. Tison. Regular tree languages and rewrite systems. *Fundamenta Informaticae*, 24:157–176, 1995.
19. B. Gramlich. Completion based inductive theorem proving: An abstract framework and its applications. *Proc. ECAI-90*, pp. 314–319, 1990.
20. D. Hofbauer and M. Huber. Linearizing term rewriting systems using test sets. *Journal of Symbolic Computation*, 17:91–129, 1994.
21. D. Hofbauer and M. Huber. Test sets for the universal and existential closure of regular tree languages. Mathematische Schriften 7/99, Universität–GH Kassel, Germany, 1999.
22. D. Hofbauer and R.-D. Kutsche. Proving inductive theorems based on term rewriting systems. *Proc. ALP-88*, LNCS 343, pp. 180–190, 1988.

23. G. Huet and J. M. Hullot. Proofs by induction in equational theories with constructors. *Journal of Computer and System Sciences*, 25:239–266, 1982.
24. J.-P. Jouannaud and E. Kounalis. Automatic proofs by induction in theories without constructors. *Information and Control*, 82(1):1–33, 1989.
25. D. Kapur, P. Narendran, and H. Zhang. Proof by induction using test sets. *Proc. CADE-86*, LNCS 230, pp. 99–117, 1986.
26. D. Kapur, P. Narendran, and H. Zhang. On sufficient completeness and related properties of term rewriting systems. *Acta Informatica*, 24(4):395–415, 1987.
27. D. Kapur, P. Narendran, and H. Zhang. Automating inductionless induction using test sets. *Journal of Symbolic Computation*, 11(1 and 2):83–112, 1991.
28. E. Kounalis. Testing for the ground (co-)reducibility property in term-rewriting systems. *Theoretical Computer Science*, 106(1):87–117, 1992.
29. E. Kounalis and M. Rusinowitch. Mechanizing inductive reasoning. *Bulletin of the EATCS*, 41:216–226, 1990.
30. G. Kucherov and M. Tajine. Decidability of regularity and related properties of ground normal form languages. *Information and Computation*, 118(1), pp. 91–100, 1995.
31. G. Lallement. *Semigroups and combinatorial applications*. Wiley & Sons, 1979.
32. D. R. Musser. On proving inductive properties of abstract data types. *Proc. POPL-80*, pp. 154–162, 1980.
33. M. Nivat and A. Podelski. Tree monoids and recognizable sets of trees. In *Resolution of equations in algebraic structures*, vol. 1, pp. 351–367, Academic Press, 1989.
34. P. Padawitz. *Deduction and declarative programming*. Cambridge Univ. Press, 1992.
35. D. Plaisted. Semantic confluence tests and completion methods. *Information and Control*, 65(2/3):182–215, 1985.
36. U. S. Reddy. Term rewriting induction. *Proc. CADE-90*, LNAI 449, pp. 162–177, 1990.
37. K. Schmid and R. Fettig. Towards an efficient construction of test sets for deciding ground reducibility. *Proc. RTA-95*, LNCS 914, pp. 86–100, 1995.
38. M. Steinby. Recognizable and rational subsets of algebras. *Fundamenta Informaticae*, 18:249–266, 1993.
39. W. Thomas. Logical aspects in the study of tree languages. *Proc. CAAP-84*, pp. 31–49, Cambridge Univ. Press., 1984.
40. S. Vágvölgyi and R. Gilleron. For a rewrite system it is decidable whether the set of irreducible, ground terms is decidable. *Bulletin of the EATCS*, 48:197–209, 1992.

Higher-Order Rewriting

Femke van Raamsdonk

Division of Mathematics and Computer Science
Faculty of Sciences, Vrije Universiteit
De Boelelaan 1081a, 1081 HV Amsterdam
The Netherlands
femke@cs.vu.nl

CWI
P.O. Box 94079, 1090 GB Amsterdam
The Netherlands

1   Introduction

The wish to consider rewriting systems with bound variables emerges naturally. The various equations with bound variables that are present in both logic and mathematics give rise to rewrite rules as soon as they are oriented. The β-axiom of lambda-calculus is oriented as $(\lambda x.M)N \rightarrow M[x := N]$. The so-obtained rewriting system was used to provide consistency proofs. Another well-known equation with bound variables is the axiom for μ-recursion. Its usual orientation gives rise to the rewrite rule $\mu x.M \rightarrow M[x := \mu x.M]$. Equivalences in logic may contain bound variables, like in $\neg \exists x.P(x) \leftrightarrow \forall x.\neg P(x)$. Also these equivalences can be turned into rewrite rules. Moreover, rules for proof normalisation may contain bound variables. Many equations occurring in mathematics contain bound variables, for instance if derivatives or integrals are present, like in $\int f(x) + g(x)dx = \int f(x)dx + \int g(x)dx$. Some aspects of the theory of a set of equations can be studied by considering the equations as rewrite rules, as is also done for first-order equations without bound variables. Further, functional programming languages may contain specifications of functions that take functions as arguments, like for instance the function map that takes as argument a function and a list, and that applies the function to every element of the list.

This explains the need for a unifying theory of higher-order rewriting, where the rewrite rules may contain bound variables.

The present paper aims at providing readers familiar with first-order term rewriting some intuitions of higher-order rewriting. The important feature of higher-order rewriting is that besides all first-order rewrite rules also rules with bound variables, like the ones for β and μ, can be expressed. I focus on two subjects in the theory of rewriting, namely confluence and termination, and discuss how some results concerning these subjects can be be generalised from the first-order to the higher-order case. It is certainly not the intention to present this generalisation in a completely formal way; I just try to explain what difficulties arise due to the presence of bound variables, and how they can be overcome.

Very generally speaking, the additional combinatorial complexity of higher-order rewriting compared to first-order rewriting is caused by the possibility of nesting. For instance in the higher-order term rewriting system $\{f(x.z(x)) \to z(z(x)), g(z) \to h(z)\}$ we have a rewrite step $f(x.g(x)) \to g(g(x))$ in which the redex $g(x)$ is duplicated and the two residuals are nested. One of the consequences of this phenomenon is that invariants used in proofs should be closed under substitution. This theme plays a rôle throughout the paper.

A systematic study of rewriting systems with bound variables starts with the work by Klop, who introduces in [16] the class of combinatory reduction systems. Another impulse for the study of rewriting with bound variables originates in the work by Nipkow, who defines in [22] the class of higher-order rewrite systems. Combinatory reduction systems form a generalisation of the class of contraction schemes defined in [1], and can more generally be seen as standing in a tradition where extensions of lambda-calculus are studied. Examples of such extensions are lambda-calculus extended with infinitely many δ-rules that test for equality of closed normal forms, defined by Church (see [3]), lambda-calculus with rules for surjective pairing, and the class of $\lambda(a)$-reductions which consists of lambda-calculus with constants and rewrite rules for these constants [10]. Higher-order rewrite systems were introduced with the aim to study the meta-theory of systems like Isabelle and λ-Prolog.

The presentation of higher-order rewriting systems in this paper is mainly based on [25, 32], which builds on [16, 22]. However, I would like to stress that the actual format of higher-order rewriting is not of the utmost importance here, since the paper is informal in nature. Moreover, the essence of concepts and proofs does not depend on the details of the chosen format.

For various purposes other classes of rewriting systems with bound variables are introduced in the literature. To mention a two of them: expression reduction systems [15] are similar to combinatory reduction systems, but have been introduced independently, and interaction systems [2] form a subclass of combinatory reduction systems that is introduced for the study of optimality. It is not hard to adapt the presentation of these two classes of systems, and of the one of combinatory reduction systems and higher-order rewrite systems, to the presentation of higher-order rewriting as chosen for this paper.

Although the theory of higher-order rewriting is not as widely developed as the one of lambda-calculus or first-order rewriting, it is nevertheless by no means possible to give a complete overview in the present paper. Readers interested in the theory of equational reasoning and narrowing for higher-order rewriting are referred to [23, 30] and the literature mentioned there. Further, results concerning confluence and termination can be found in detail in [16, 25, 19, 29, 32].

Acknowledgements. I am grateful to Jan Willem Klop and Vincent van Oostrom for discussions that in the course of the last years always have been, and still are, a source of inspiration. I wish to thank Jean-Pierre Jouannaud, Aart Middeldorp, Jaco van de Pol, and Roel de Vrijer for helpful discussions and comments. The diagrams are designed using the package Xy-pic of K.R. Rose.

2 Higher-Order Rewriting

We assume a set of *base types*. A *simple type*, usually shortly called a type, is either a base type or an expression of the form $A \to B$, with A and B types. For every type A a set consisting of infinitely many *variables* of type A is assumed. Variables are written as $x, y, z, \ldots$, if necessary decorated with their types.

A *signature* is a set of function symbols with a unique type. Function symbols are written as $f, g, h, a, b, \ldots$; sometimes a more suggestive notation is used.

A *preterm* of type A is a simply typed lambda-term of type A, inductively defined as follows, where $s : A$ denotes that s is a preterm of type A:

1. variables of type A and function symbols of type A are preterms of type A,
2. if x is a variable of type A, and $s : B$, then $x.s : A \to B$,
3. if $s : A \to B$ and $t : A$, then $st : B$.

Preterms are denoted by $s, t, r, \ldots$. Note that abstraction is written as $x.s$ instead of $\lambda x.s$. In the preterm $x.s$, occurrences of the variables x in s are bound. A variable occurrence that is not bound is said to be free. A preterm without bound variables is closed. Preterms are considered up to the equivalence relation generated by the renaming of bound variables, or α-conversion.

In the remainder of this paper, all preterms are supposed to be in long-η-normal form (also written as $\bar{\eta}$-normal form). That means that every subterm has the maximal number of arguments according to its type. Note that every type can be written as $A_1 \to \ldots \to A_n \to B$ with B a base type. In an $\bar{\eta}$-normal form, a function symbol of type $A_1 \to \ldots \to A_n \to B$ occurs always in a subterm of the form $f s_1 \ldots s_n$ with $s_1 : A_1, \ldots, s_n : A_n$, and similarly for variables. This permits to adopt the functional notation $f(s_1, \ldots, s_n)$ instead of the applicative notation $f s_1 \ldots s_n$, and similarly for variables; this is often done in the sequel. A sequence of expressions e is usually abbreviated as e.

We work modulo the equivalence generated by the β-reduction rule, which is given as $(x.s)t \to_\beta s[x := t]$, with s and t arbitrary preterms. Here $s[x := t]$ denotes the result of substituting t for every free occurrence of x in s, renaming variables if necessary in order to avoid unintended capture of variables.

We make use of the facts that β-reduction is confluent and terminating on simply typed lambda-calculus, and that the set of $\bar{\eta}$-normal forms is closed under β-reduction. Every β-equivalence class of preterms contains a unique β-normal form, which is used as a representative. Such a representative is called a *term*. Terms are the objects that are rewritten in a higher-order rewriting system.

For the definition of a rewrite rule we first need to introduce the notion of a rule-pattern, which is a slight adaptation of the notion of pattern introduced in [21]. A *rule-pattern* is a term of the form $x.f(s)$ such that, first, every $x_i \in x$ occurs free in $f(s)$, and second, every $x_i \in x$ occurs only in subterms of the form $x_i(y_1, \ldots, y_m)$ with $y_1, \ldots, y_m$ the $\bar{\eta}$-normal forms of different bound variables not among x. The terms $z.f(z)$ with z of some base type, and $z.f(x.z(x))$ with z of type $A \to B$ with B a base type, are examples of rule-patterns. The term $z.f(z(x))$ is not a rule-pattern because x is not bound, the term $z.f(x.z(x, x))$ is not a rule-pattern because the arguments of z are not different bound variables, and the term $z.z$ is not a rule-pattern because there is no occurrence of a function symbol after the outermost abstractions.

A *rewrite rule* of a higher-order rewriting system is defined as a pair of closed terms of the form $z.l \to z.r$, with $z.l$ a rule-pattern. In this paper, a higher-order rewriting system is usually described by giving the set of its rewrite rules.

The next thing to define is the rewrite relation of a higher-order rewriting system. To that end we need to introduce the notion of context. We use the symbol $\square^A$, or simply $\square$, to denote a variable of type A that is supposed to be free. A *context* of type A is a term with one occurrence of $\square^A$. The preterm obtained by replacing $\square^A$ by a term s of type A is denoted by $C[s]$.

Now the *rewrite relation* of a higher-order rewriting system, denoted by $\to$, is the relation on terms that is defined as follows: we have $s \to t$ if there is a context C of type A, and a rewrite rule $l \to r$ with l and r of type A, such that s is the β-normal form of $C[l]$ and t is the β-normal form of $C[r]$. That is, such a rewrite step can be decomposed as

$$s \ {}^!_\beta\!\!\leftarrow C[l] \to C[r] \twoheadrightarrow^!_\beta t$$

where $\twoheadrightarrow^!_\beta$ denotes β-reduction to β-normal form. The requirement that the left-hand side of a rewrite rule must be a pattern makes the rewrite relation decidable [21].

In first-order term rewriting, a rewrite step is usually defined as $C[l\sigma] \to C[r\sigma]$, with $l \to r$ a rewrite rule, C a context, and σ an assignment. In higher-order rewriting, simply typed lambda-calculus with β-reduction is used to assign values to free variables in rewrite rules. Therefore lambda-calculus is called the *substitution calculus* of higher-order rewriting as defined here. It is possible to consider other calculi as substitution calculus. As a matter of fact, the substitution calculus of combinatory reduction systems and of expression reduction systems is not simply typed lambda-calculus with reduction to β-normal form, but untyped lambda-calculus with complete developments.

The rewrite rule $f(z) \to g(z)$, in the usual format of first-order term rewriting, induces for instance a rewrite step $f(a) \to g(a)$. This rewrite rule is now written as $z.f(z) \to z.g(z)$, and the rewrite step $f(a) \to f(b)$ is obtained as follows: $f(a) \ {}^!_\beta\!\!\leftarrow (z.f(z)) a \to (z.f(z)) b \twoheadrightarrow^!_\beta f(b)$.

The signature of lambda-calculus consists of the two function symbols app : term $\to$ term $\to$ term and abs : (term $\to$ term) $\to$ term. Here term is the only base type; intuitively speaking it stands for the set of lambda-terms. The rewrite rules are given as follows:

$$z.z'. \operatorname{app}(\operatorname{abs}(x.z(x)), z') \to_b z.z'. z(z')$$

$$z. \operatorname{abs}(x.\operatorname{app}(z, x)) \to_e z.z$$

The rewrite step $\operatorname{app}(\operatorname{abs}(x.x), y) \to_{\mathsf{beta}} y$ is obtained as follows:

$$\operatorname{app}(\operatorname{abs}(x.x), y) \ {}^!_\beta\!\!\leftarrow (z.z'.\operatorname{app}(\operatorname{abs}(x.z(x)), z'))(x.x)y \to_b (z.z'.z(z'))(x.x)y \twoheadrightarrow^!_\beta y.$$

Usually we adopt the convention that the outermost abstractions of rewrite rules are not written, that is, we write $l \to r$ instead of $z.l \to z.r$.

3 Confluence

A rewriting system is *confluent* if every two divergent rewrite sequences can be joined; that is, whenever $t' \twoheadleftarrow s \twoheadrightarrow t$ there exists a term r with $t' \twoheadrightarrow r \twoheadleftarrow t$. Confluence is an important property because it guarantees that normal forms are unique. Moreover it provides a method to prove consistency. In this section we discuss some results concerning confluence of higher-order term rewriting.

3.1 Orthogonality

A first general confluence result for first-order term rewriting systems is the one stating that *orthogonal* term rewriting systems are confluent [33]. Orthogonality of a rewriting system roughly speaking expresses that rewrite steps are independent. In the setting of first-order term rewriting, it is usually formalised by two requirements imposed on the rewrite rules. These requirements arise from an analysis of the different ways in which contraction of a redex can destroy another redex that is present in the same term.

First, a redex u can be destroyed by contraction of a redex u' if two subterms in u must be identical and contraction of u' changes one of them. For example in the step $f(a, a) \to f(b, a)$ in the term rewriting system $\{a \to b, f(x, x) \to b\}$, the redex $f(a, a)$ is destroyed by contraction of the leftmost redex a. This kind of interference may cause a rewriting system to be non-confluent: in the term rewriting system $\{f(x, x) \to a, f(x, g(x)) \to b, c \to g(c)\}$ given in [11], we have both $f(c, c) \to a$ and $f(c, c) \twoheadrightarrow b$. Another example, given in [16], is the term rewriting system $\{d(x, x) \to e, c(x) \to d(x, c(x)), a \to c(a)\}$. We have $c(a) \twoheadrightarrow e$ and $c(a) \twoheadrightarrow c(e)$. The source of non-confluence as illustrated here is ruled out by requiring the term rewriting system to be *left-linear*, which means that variables are not allowed to occur more than once in the left-hand side of a rewrite rule.

Second, a redex u can be destroyed by contraction of a redex u' if that contraction changes one or more symbols in the pattern of u. Consider for instance the term rewriting system $\{a \to b, f(a) \to c\}$. In the step $f(a) \to f(b)$, the redex $f(a)$ is destroyed by the contraction of the redex a. The redexes a and $f(a)$ are said to be *overlapping* in the term $f(a)$. Clearly, the presence of overlapping redexes can cause a rewriting system to be non-confluent. Ruling out this source of non-confluence can be done by requiring that in all terms all redexes are non-overlapping. This requirement is guaranteed by a condition concerning the rewrite rules only; the formulation of this condition uses some additional terminology that we discuss now.

A *critical pair* arises from a most general way in which two redexes can be overlapping. Consider for example the term rewriting system $\{a \to b, f(a, x) \to g(x)\}$. The term $f(a, x)$ can be rewritten to $f(b, x)$ by an application of the first rewrite rule, and to $g(x)$ by an application of the second rewrite rule. The pair of terms $(f(b, x), g(x))$ is said to be a critical pair. The requirement of maximal generality concerns the surroundings of the two redexes that give rise to the critical pair. The intuition is that the surroundings should be taken to be minimal. For instance, the symbol g is not essential for the overlap between the redexes a and $f(a, x)$ in the term $g(f(a, x))$; hence $(g(f(b, x)), g(g(x)))$ is not a critical pair. The symbol b is not essential for the overlap between the redexes a and $f(a, b)$ in the term $f(a, b)$; hence $(f(b, b), g(b))$ is not a critical pair either.

The absence of critical pairs in a term rewriting system guarantees that in all terms all redexes are non-overlapping.

A first-order term rewriting system is defined to be *orthogonal* if all its rules are left-linear and it has no critical pairs. At some places in the literature the definition of orthogonality requires instead of the absence of critical pairs that in all terms all redexes are non-overlapping; the two formulations are equivalent.

Now the question arises whether the intuition of orthogonality, namely that rewrite steps are independent, is also for higher-order systems properly captured by requiring that rewrite rules are left-linear and that there are no critical pairs. As it turns out, the presence of bound variables does not yield another way in which contraction of a redex can destroy another redex. Note, however, that there is another way in which contraction of a redex can create another redex: by erasure of a bound variable a redex can be created for a rule with a so-called non-occur check. This happens for instance in the rewrite step $f(x.g(x)) \to f(x.a)$ in the rewriting system $\{f(x.z) \to a, g(x) \to a\}$. This phenomenon plays a rôle in the study of termination rather than in that of confluence.

So the definition of orthogonality for higher-order term rewriting systems is in essence the same as for first-order term rewriting systems. However, in particular the definition of critical pair is technically more complicated. Also unification of higher-order patterns is technically more complicated than that of first-order terms; it still decidable [21] and moreover has linear complexity [31]. Here we do not give a completely formal definition of critical pair, which can be found for instance in [22, 18].

Definition 1.

1. *A rewrite rule $x.l \to x.r$ of a higher-order rewriting system is* left-linear *if every variable $x \in x$ occurs exactly once in l.*
2. *Let $C[ls] = gt$ indicate a most general overlap between two redexes, with $l \to r$ and $g \to d$ rewrite rules. Then $(C[rs], dt)$ is a critical pair.*

Note that a critical pair is ordered. If the context C in the definition is the empty one, then the rewrite rules $l \to r$ and $g \to d$ must be different. The critical pair of lambda-calculus with $\beta\eta$-reduction is $(\mathsf{app}(z, z'), \mathsf{app}(z, z'))$; it arises from the most general overlap in $\mathsf{app}(\mathsf{abs}(x.\mathsf{app}(z, x)), z')$. The definition of orthogonality is now the same as the one for first-order term rewriting.

Definition 2. *A higher-order rewriting system is* orthogonal *if it has no critical pairs and all its rewrite rules are left-linear.*

There are two important methods to prove that orthogonal rewriting systems are confluent. The remainder of this subsection is devoted to a discussion of these two methods. The first one, that of simultaneous reduction, makes essential use of the structure of the expressions that are rewritten. The second one, that uses complete developments, is more abstract in nature. Both apply to higher-order rewriting systems, so we have the following result.

Theorem 1. *Orthogonal higher-order rewriting systems are confluent.*

Simultaneous Reduction. First we discuss a method to prove confluence that makes use of simultaneous reduction. It is introduced by Tait and Martin-Löf in their proof of confluence of lambda-calculus with β-reduction, see [3]. The outline of this method is as follows: we inductively define a relation $\Rightarrow$ satisfying the following two properties:

1. It has the diamond property, that is, $t' \Rightarrow r \Leftarrow t$ whenever $t' \Leftarrow s \Rightarrow t$.
2. Its reflexive-transitive closure equals the rewrite relation ($\twoheadrightarrow$).

It is not difficult to see that this indeed yields confluence of rewriting.

For first-order term rewriting systems, one can take for $\Rightarrow$ *parallel rewriting*. This is the relation that contracts a set of redexes that are in parallel in one step. For instance, in the rewriting system $\{a \to a', f(x) \to f'(x)\}$, where there is moreover a binary function symbol g, we have $g(a, a) \longmapsto\!\!\!+ g(a', a')$. The two redexes in the term $f(f(x))$ are not parallel but *nested*, and hence we do not have $f(f(x)) \longmapsto\!\!\!+ f'(f'(x))$. Note however that $f(f(x)) \longmapsto\!\!\!+ f'(f(x))$ and $f(f(x)) \longmapsto\!\!\!+ f(f'(x))$ both hold. The relation $\longmapsto\!\!\!+$ is defined inductively. The rule

$$l\sigma \longmapsto\!\!\!+ r\sigma$$

with $l \to r$ a rewrite rule and σ an assignment, expresses that redexes can be contracted, and the rule

$$\frac{s_1 \longmapsto\!\!\!+ s_1' \ldots s_n \longmapsto\!\!\!+ s_n'}{f(s_1, \ldots, s_n) \longmapsto\!\!\!+ f(s_1', \ldots, s_n')}$$

permits to do so in parallel. Finally, there is the rule $x \longmapsto\!\!\!+ x$, with x a variable. The relation $\longmapsto\!\!\!+$ is reflexive and compatible with the term structure. It can be shown that the reflexive-transitive closure of the parallel rewriting relation equals $\twoheadrightarrow$, and moreover that $\longmapsto\!\!\!+$ satisfies the diamond property.

It is possible adapt the definition of $\longmapsto\!\!\!+$ to the case of higher-order rewriting, and to show that its reflexive-transitive closure equals the rewrite relation. However, due to the possibility of nesting, $\longmapsto\!\!\!+$ doesn't have the diamond property. Consider for instance the higher-order rewriting system $\{f(x.z(x)) \to z(z(a)), g(z) \to h(z)\}$. We have $f(x.g(x)) \longmapsto\!\!\!+ g(g(a))$ and $f(x.g(x)) \longmapsto\!\!\!+ f(x.h(x))$ but not $g(g(a)) \longmapsto\!\!\!+ h(h(a))$ (note however that we do have $f(x.h(x)) \longmapsto\!\!\!+ h(h(a))$).

In lambda-calculus there is also the possibility of nesting of redexes. Indeed, the definition of $\longmapsto\!\!\!+$ adapted to lambda-calculus does not satisfy the diamond property either: we have $(\lambda x.(\lambda y.x)I))(II) \longmapsto\!\!\!+ (\lambda y.II)I$ and on the other hand $(\lambda x.(\lambda y.x)I))(II) \longmapsto\!\!\!+ (\lambda x.x)I$ but there is no lambda-term M such that $(\lambda y.II)I \longmapsto\!\!\!+ M$ and $(\lambda x.x)I \longmapsto\!\!\!+ M$.

The relation used by Tait and Martin-Löf in their proof of confluence is essentially different from parallel rewriting in that it permits to contract any set of redexes in a term in one step. The important clause in the definition of this relation, which we denote here by $\twoheadrightarrow$, is as follows:

$$\frac{M \twoheadrightarrow M' \qquad N \twoheadrightarrow N'}{(\lambda x.M)N \twoheadrightarrow M'[x := N']}$$

This suggests that for higher-order rewriting we should define instead of $\twoheadrightarrow\!\!\!\!\!+$ a relation, which we denote by $\twoheadrightarrow$ as the one for lambda-calculus, with

$$\frac{s_1 \twoheadrightarrow s_1' \ldots s_n \twoheadrightarrow s_n'}{l s_1 \ldots s_n \downarrow_\beta \twoheadrightarrow r s_1' \ldots s_n' \downarrow_\beta}$$

where $\downarrow_\beta$ denotes β-reduction to normal form. Further, $\twoheadrightarrow$ must be compatible with the term structure. Note that, reconsidering the example give above, we have $g(g(a)) \twoheadrightarrow h(h(a))$. We call the relation $\twoheadrightarrow$ here *simultaneous reduction*; it is also sometimes called parallel reduction.

The relation $\twoheadrightarrow$ can be defined for left-linear higher-order rewriting systems. For orthogonal higher-order rewriting systems, it can be shown that the reflexive-transitive closure of $\twoheadrightarrow$ equals $\twoheadrightarrow$, and further that $\twoheadrightarrow$ satisfies the diamond property. As a consequence, we have the result that orthogonal higher-order rewriting systems are confluent.

Finite Developments. The second method to prove confluence we discuss is the one using developments. It is more abstract in nature than the one using simultaneous reduction, and can therefore more easily be adapted to situations where the rewriting system is not quite orthogonal or where the structures that are rewritten are not quite terms. Developments are used to show confluence of lambda-calculus in [5] and to show confluence of combinatory reduction systems in [16]. We follow here basically the account presented in [25, 26].

The idea is to define a relation $\Rightarrow$ with $\rightarrow \subseteq \Rightarrow \subseteq \twoheadrightarrow$ that satisfies moreover the following diagram:

Confluence follows then by an easy induction, since the diagram above and the fact that $\rightarrow \subseteq \Rightarrow$ permit to construct the *projection* of a rewrite sequence d over a rewrite step, yielding a rewrite sequence e as follows:

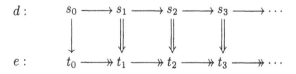

The second method to prove confluence now proceeds as follows. We define a *development* of a set of redexes U as a special kind of rewrite sequence, namely one in which only residuals of redexes in U are contracted. A development of U is said to be *complete* if it cannot be extended, that is, there are no residuals of redexes in U anymore. In general there can be different ways to perform a complete development, for instance $f(a, a) \rightarrow f(b, a) \rightarrow f(b, b)$ and $f(a, a) \rightarrow f(a, b) \rightarrow f(b, b)$

are both complete developments of the set consisting of the two redexes a in the term $f(a, a)$. Further, it is a priori not clear whether performing a complete development terminates at some point. The crucial step in the confluence proof is the following result, which is called the Finite Developments Theorem.

Theorem 2. *All complete developments of a set of redexes U are finite, end in the same term, and induce the same descendant relation.*

Hence all ways of sequentialising the contraction of a set of redexes are finite, and essentially the same. As a consequence, the complete developments relation satisfies the diamond property, so we can take it for $\Rightarrow$.

We denote the complete developments relation just as simultaneous reduction by $\twoheadrightarrow$. This is justified since there is the following relation between simultaneous reduction and complete developments: a proof of $s \twoheadrightarrow t$ using the rules for simultaneous reduction corresponds to a complete development in which the redexes are contracted from the inside to the outside.

The proof of the Finite Developments Theorem is essentially more complex for lambda-calculus and higher-order rewriting than for first-order term rewriting systems, due the possibility of nesting. Consider for instance the higher-order rewriting system $\{f(x.z(x)) \rightarrow z(z(a)), g(z) \rightarrow h(z, z)\}$. In the complete development $f(x.g(x)) \rightarrow g(g(a)) \rightarrow h(g(a), g(a)) \rightarrow h(h(a, a), g(a)) \rightarrow h(h(a, a), h(a, a))$, two residuals of the redex $g(x)$ in the initial term first get nested and then the innermost one is duplicated by the outermost one.

The Finite Developments Theorem is not only useful to derive confluence, but in any situation where a rewrite sequence is projected over a rewrite step.

3.2 Weak Orthogonality

The methods to prove confluence of orthogonal higher-order rewriting systems both can be adapted to the case where critical pairs are allowed, but only if they are of the form (s, s). Such a critical pair is said to be *trivial*. The notion of trivial critical pair is used to define the class of weakly orthogonal higher-order rewriting systems; the definition is analogous to the one for the first-order case.

Definition 3. *A higher-order rewriting system is weakly orthogonal if it is left-linear and all its critical pairs are trivial.*

Examples of weakly orthogonal rewriting systems that are not orthogonal are $\{a \rightarrow b, f(a) \rightarrow f(b)\}$ and $\{f(x) \rightarrow f(b), f(a) \rightarrow f(b)\}$. Moreover, lambda-calculus with both β-and η-reduction is a weakly orthogonal rewriting system.

Orthogonality is defined in terms of the left-hand sides of the rewrite rules only, whereas weak orthogonality concerns the right-hand sides of rewrite rules. This explains why extensions from the orthogonal to the weakly orthogonal case may cause considerable complications, also for first-order rewriting.

Simultaneous Reduction. The method using simultaneous reduction can be extended to the weakly orthogonal case as follows. The idea is to proceed by induction on a measure that roughly speaking counts the overlap between two co-initial steps $s \twoheadrightarrow t$ and $s \twoheadrightarrow t'$. In case there is no overlap, we proceed as in the orthogonal case. In case there is overlap, we take minimal overlapping redexes u and u' such that u is contracted in $s \twoheadrightarrow t$ and u' is contracted in $s \twoheadrightarrow t'$. Since all trivial pairs are trivial, we have that contracting u yields the same term as contracting u', say s'. The proof proceeds now by showing that $s' \twoheadrightarrow t$ and $s' \twoheadrightarrow t'$, and that the measure of this divergence is smaller than that of the original one. If follows then by induction that $\twoheadrightarrow$ satisfies the diamond property. The proof is illustrated by the following diagram.

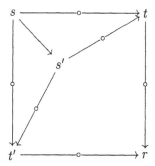

Finite Developments. The proof using developments is adapted to the case of weakly orthogonal systems as follows [25]. First, complete developments are defined for sets of redexes that are pairwise non-overlapping. It turns out that this is sufficient to ensure that complete developments can be defined properly; a global condition on the whole set of redexes under consideration is not necessary. The aim is to show that whenever $s \to t$, by contracting a redex u, and $s \twoheadrightarrow t'$, by contracting a set of redexes V we have that there exists a term r such that $t \twoheadrightarrow r$ and $t' \twoheadrightarrow r$:

There are two possibilities. Either the redex u doesn't overlap with any redex in the set V. In that case both t and t' rewrite by a complete development to the term obtained by performing a complete development of $\{u\} \cup V$, starting in s. In the other case, there is a redex v in the set V such that u and v are overlapping, and hence give rise to a critical pair. Since all critical pairs are trivial, we have that s also rewrites to t by contracting the redex v instead of u. Now we have that t rewrites to t' by performing a complete development of all residuals of V in t.

3.3 Critical Pairs

In the previous subsection we have seen that for left-linear systems, confluence holds if all critical pairs are trivial. Further, we have seen examples showing that a term rewriting system that is not left-linear need not be confluent, even in the absence of critical pairs. A natural next step is to investigate to what extent the condition on critical pairs can be relaxed while maintaining confluence, or a weaker version of it.

Huet [11] (see also [17]) shows that a first-order term rewriting system is locally confluent (that is, $t' \twoheadrightarrow r \twoheadleftarrow t$ whenever $t' \leftarrow s \rightarrow t$) if all its critical pairs are confluent. A critical pair (s, s') is said to be confluent if s and s' have a common reduct. Using Newman's Lemma, confluence of a terminating first-order term rewriting system follows from confluence of its critical pairs.

The idea of the proof as given in [11] is globally as follows. Consider two co-initial rewrite steps $s \rightarrow t$, obtained by contraction of a redex u, and $s \rightarrow t'$, obtained by contraction of a redex u'. If u and u' do not overlap, an analysis of their relative positions yields that a common reduct of t and t' can be found. Now suppose that u and u' are overlapping redexes and let u' be above u. The key auxiliary result states that in that case there must be a critical pair (r, r') such that $t = C[r\sigma]$ and $t' = C[r'\sigma]$. Since critical pairs are confluent, r and r' have a common reduct. As a consequence, t and t' have a common reduct.

The result of [11] is generalised to higher-order term rewriting systems by Nipkow [22, 18]. The proof proceeds basically as in the first-order case, but it is technically significantly more difficult to show the key auxiliary result.

Theorem 3. *A higher-order rewriting system is locally confluent if all its critical pairs are confluent.*

Consider for example the rewrite rules for surjective pairing: $\{\pi_1(\pi(z, z')) \rightarrow z, \pi_2(\pi(z, z')) \rightarrow z', \pi(\pi_1(z), \pi_2(z)) \rightarrow z\}$. There are two critical pairs, namely $(\pi(z, \pi_2(\pi(z, z'))), \pi(z, z'))$ and $(\pi(\pi_1(\pi(z, z')), z'), \pi(z, z'))$. Both are confluent. As a consequence, the rewriting system is locally confluent. Note that also local confluence of lambda-calculus with β-reduction extended with the rules for surjective pairing is obtained as an application of the theorem above.

3.4 Development Closed

For left-linear rewriting systems that are not terminating, confluence of the critical pairs does not guarantee confluence of the rewriting system. This is for instance illustrated by the rewriting system $\{a \rightarrow b, a \rightarrow c, b \rightarrow a, b \rightarrow d\}$.

Huet [11] formulates a criterion on critical pairs that is stronger than the one of being only confluent, and shows that it is a sufficient condition for confluence of left-linear first-order term rewriting systems. This criterion is as follows: if (s, t) is a critical pair, then we have $s \Vdash t$. A critical pair satisfying this criterion is said to be *parallel closed*, and a rewriting system is said to be parallel closed if all its critical pairs are so. This result yields for instance confluence of the term rewriting system $\{f(g(x), b) \to f(g(x), b'), g(a') \to g(a), a \to a', b \to b'\}$, because for the critical pair $(f(g(a), b), f(g(a'), b'))$ we have $f(g(a), b) \mathrel{-\!\!\!\mid\!\!\!\mid\!\!\!\to} f(g(a'), b')$.

Van Oostrom extends in [26] the result that parallel closed rewriting systems are confluent to the higher-order case. The skeleton of the proof is the same as that for first-order rewriting, but the difference is that now another invariant is needed: instead of parallel reduction $(\mathrel{-\!\!\!\mid\!\!\!\mid\!\!\!\to})$ the relation of complete developments $(\mathrel{-\!\!\!\circ\!\!\to})$ is used. Correspondingly, the requirement that all critical pairs are parallel closed can be relaxed: instead, for every critical pair (s, t) it is required that $s \mathrel{-\!\!\!\circ\!\!\to} t$, that is, there is a complete development from s to t. A critical pair satisfying this criterion is said to be *development closed*, and a rewriting system is said to be development closed if all its critical pairs are so. The result of [26] states that a left-linear higher-order rewriting system is confluent if all its critical pairs are development closed; this is an extension of the result that parallel closed rewriting systems are confluent also if only first-order rewriting is considered.

The skeleton of the proof is as follows. We look for a rewriting relation $\Rightarrow$ such that the diamond property of $\Rightarrow$ implies confluence of $\to$. For proving the diamond property of $\Rightarrow$, suppose that $s \Rightarrow t$ and $s \Rightarrow t'$. We proceed by induction on some measure that roughly speaking counts the overlap between those two steps. So first we need to prove the diamond property of $\Rightarrow$ in the case that the measure is zero, that is, there is no overlap between the steps $s \Rightarrow t$ and $s \Rightarrow t'$. In case the measure is greater than zero, we single out a critical divergence between the step $s \to t_0$, obtained by contracting some redex u, and the step $s \to t_0'$, obtained by contracting some redex u'. Suppose that the condition imposed on critical pairs yields that $t_0' \rightsquigarrow t_0$. Now the following is shown:

- We have $t_0 \Rightarrow t$ and $t_0' \Rightarrow t'$.
- The steps $t_0' \rightsquigarrow t_0$ and $t_0 \Rightarrow t$ can be joined to form a step $t_0' \Rightarrow t$.
- The new divergence $t' \Leftarrow t_0' \Rightarrow t$ is smaller than the old one $t' \Leftarrow s \Rightarrow t$.

It then follows then $\Rightarrow$ has the diamond property, which yields confluence of $\to$. The proof idea is illustrated in the following diagram.

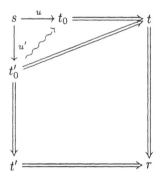

For first-order rewriting [11], parallel reduction $(\mathrel{-\!\!\!\mid\!\!\!\mid\!\!\!\to})$ is taken for $\Rightarrow$. For higher-order rewriting this doesn't work since it doesn't satisfy the diamond property. Instead, the complete development relation $(\mathrel{-\!\!\!\circ\!\!\to})$ is taken for $\Rightarrow$ [26].

Another question is what should be taken for $\rightsquigarrow$. It is needed that $\rightsquigarrow \subseteq \Rightarrow$, and the result is the strongest if $\rightsquigarrow\, =\, \Rightarrow$. The latter is indeed done in both cases.

It is more difficult to show that the $\twoheadrightarrow$-step joining the critical pair and the remainder of the original $\twoheadrightarrow$-step can be combined into a new $\twoheadrightarrow$-step than to show the analogous statement for the parallel rewriting relation $+\!\!\!+\!\!\!\twoheadrightarrow$. Also the argument that the measure decreases is more complex in the higher-order than in the first-order case.

Theorem 4. *A left-linear higher-order rewriting system is confluent if it is development closed.*

As an application of this result, we obtain that the higher-order term rewriting system $\{f(g(x.z(x))) \rightarrow f(z(z(x))), g(x.h'(x)) \rightarrow h(h(x)), h(x) \rightarrow h'(x)\}$ is confluent, because for the critical pair $(f(h(h(x))), f(h'(h'(x))))$ we have that $f(h(h(x))) \twoheadrightarrow f(h'(h'(x)))$.

4 Termination

A rewriting system is *terminating* if it does not admit infinite rewrite sequences; then all rewrite sequences end in a normal form. For first-order rewriting, a lot of techniques and methods to prove termination exist. The theory of termination of higher-order rewriting is so far significantly less well-developed. In this section we discuss two important methods to prove termination that are extended to the higher-order case. The last subsection is concerned with a normalising strategy.

4.1 Termination Models

Termination of a rewriting system can be proved by mapping every term to an element of a non-empty set A that is equipped with a well-founded partial order $>$, in such a way that $[\![s]\!] > [\![t]\!]$ whenever $s \rightarrow t$, where the mapping is denoted by $[\![.]\!]$. For first-order term rewriting, this means that we need to have $[\![C[l\sigma]]\!] > [\![C[r\sigma]]\!]$ for every rewrite rule $l \rightarrow r$, context C, and assignment σ. It is clearly desirable to eliminate some of the quantifications here.

This is done in the method using termination models: here a requirement concerning the rewrite rules guarantees the inequality $[\![C[l\sigma]]\!] > [\![C[r\sigma]]\!]$ to hold for every context C and every assignment σ. A *termination model* for a first-order term rewriting system is an algebra for the signature of the rewriting system in which the terms can be interpreted as usual, with the following properties:

1. The algebra is equipped with a well-founded partial order $>$.
2. The functions $f : A^n \rightarrow A$ of the algebra are required to be *monotonic* in the following sense: $f(\ldots, a, \ldots) > f(\ldots, a', \ldots)$ whenever $a > a'$.
3. For every rewrite rule $l \rightarrow r$ and assignment σ we have $[\![l\sigma]\!] > [\![r\sigma]\!]$.

In fact, instead of the last requirement a condition on the algebraic side is formulated in terms of valuations; that condition implies the last requirement as above. The second requirement yields that $[\![C[s]]\!] > [\![C[t]]\!]$ whenever $[\![s]\!] > [\![t]\!]$.

Huet and Oppen [12] and Zantema [37] prove that a first-order term rewriting system is terminating if and only if it has a termination model. This result provides a complete but not algorithmic method for proving termination.

Van De Pol presents in [28, 29] a generalisation of this result to the higher-order case. This generalisation makes use of ideas that are also present in the work by Gandy [7] and De Vrijer [35, 36] on termination of simply typed λ-calculus. For every type A, the set of *functionals* of type A is defined by induction on the definition of simple types: a base type is interpreted as some fixed non-empty set, and an arrow type $A \to B$ is interpreted as the function space $[\![A]\!] \Rightarrow [\![B]\!]$, where we use the notation $[\![.]\!]$ also for the interpretation of types.

In order to prove termination of simply typed λ-calculus, the λI-terms (λ-terms that cannot erase arguments, for instance $\lambda x.y$ is not a λI-term) are interpreted as certain functionals, called the hereditarily monotonic ones, which come equipped with an ordering. This interpretation cannot be used for proving termination of a higher-order rewriting system for the following reason. First, contraction of a β-redex yields a decrease in the ordering, whereas in higher-order rewriting we work modulo β and hence contraction of a β-redex shouldn't change the interpretation. Second, for proving termination of higher-order rewriting, the ordering should be adapted to make it closed under contexts in $\beta\overline{\eta}$-normal form. These two requirements together yield that the ordering must be reflexive, which makes it useless for proving termination.

The solution presented in [28, 29] is to define two sets of functionals: the set of *weakly monotonic functionals*, and the set of *strict functionals*. Both come equipped with an order. The first is a superset of the set of hereditarily monotonic functionals and can be used for the interpretation of arbitrary λ-terms, the second is a subset of the set of hereditarily monotonic functionals and is used for the interpretation of the function symbols and variables of a higher-order rewriting system. Now all these ingredients are present in the definition of a *termination model* for a higher-order rewriting system, which is given as follows:

1. Every base type is interpreted as a well-founded partial order with some additional property that guarantees the existence of weakly monotonic and strict functionals for all types.
2. Function symbols of the higher-order rewriting system are interpreted as strict functionals.
3. For every rewrite rule $l \to r$, we have $[\![l]\!] > [\![r]\!]$ in the weakly monotonic ordering.

The two key results for the termination method are now as follows. First, whenever $[\![s]\!] > [\![t]\!]$ in the weakly monotonic ordering, we have in the strict ordering that $[\![C[s]]\!] > [\![C[t]]\!]$ with C a context in $\beta\overline{\eta}$-normal form, provided that function symbols and variables are interpreted strictly. Second, it is shown that $[\![s]\!] = [\![t]\!]$ if s and t are β-equal. This is used to prove the following [28, 29].

Theorem 5. *A higher-order rewriting system is terminating if it has a termination model.*

The reverse of the statement does not hold; a simplification of a counterexample in [14] is given in [29]. The restriction that left-hand sides of the rewrite rules should be patterns is not necessary for the proof of the theorem. Further, a corollary is that the combination of a terminating first-order term rewriting system and simply typed λ-calculus with β-reduction is terminating.

To illustrate the use of termination models, we consider as an example (taken from [29]) the higher-order rewriting system $\{\text{and}(z, \text{forall}(x.z'(x))) \rightarrow \text{forall}(x.\text{and}(z, z'(x)))\}$ with and : form $\rightarrow$ form $\rightarrow$ form and forall : (term $\rightarrow$ form) $\rightarrow$ form. Intuitively, term represents the set of terms and form represents the set of formulas. Both are interpreted as the set of natural numbers. Further,

- and is interpreted as $\lambda m, n. \in \mathbb{N}. (2 \cdot m + 2 \cdot n)$,
- forall is interpreted as $\lambda f \in \mathbb{N} \Rightarrow \mathbb{N}. (f(0) + 1)$.

Now we have:

$$
\begin{aligned}
[\![\text{and}(z, \text{forall}(x.z'(x)))]\!] &= 2 \cdot [\![z]\!] + 2 \cdot ([\![z']\!](0) + 1) \\
&> 2 \cdot [\![z]\!] + 2 \cdot [\![z']\!](0) + 1 \\
&= [\![\text{forall}(x.\text{and}(z, z'(x)))]\!].
\end{aligned}
$$

4.2 Recursive Path Ordering

An important method to prove termination of a first-order term rewriting system is the one using the recursive path ordering [6]. This ordering is roughly speaking defined by extending a well-founded ordering on function symbols in a recursive way. The two key results concerning the recursive path ordering are that it is a well-founded ordering on terms, and that $l > r$ implies $C[l\sigma] > C[r\sigma]$ for every context C and every assignment σ. This yields the following method to prove termination of a first-order term rewriting system: define a well-founded ordering on the function symbols and show that in its extension to the recursive path ordering we have $l > r$ for every rewrite rule.

Jouannaud and Rubio define in a recent paper [13] a generalisation of the recursive path ordering to higher-order terms, and show that it provides a method to prove termination of higher-order term rewriting systems. The remarkable feature of their proof is that Kruskal's Tree Theorem is not used to show well-foundedness of the recursive path ordering.

To start with, we show that termination of the first-order term rewriting system $\{f(z, g(z')) \rightarrow g(f(z, z'))\}$ can be proved using the recursive path ordering. Taking $f > g$ as ordering on the function symbols yields that $f(z, g(z')) > g(f(z, z'))$ in the recursive path ordering: first, one is allowed to make copies of $f(z, g(z'))$ under a smaller function symbol (g), yielding $g(f(z, g(z')))$, and second, it is possible to get rid of the innermost symbol g by selecting only its argument z', yielding $f(z, g(z')) > g(f(z, z'))$.

Next, we reconsider the higher-order term rewriting system of the previous subsection $\{\text{and}(z, \text{forall}(x.z'(x))) \rightarrow \text{forall}(x.\text{and}(z, z'(x)))\}$. The structure of this rewrite rule is similar to the one of the first-order example above. Taking as ordering on the function symbols and $>$ forall yields that $\text{and}(z, \text{forall}(x. z'(x))) > \text{forall}(x. \text{and}(z, z'(x)))$ in the recursive path ordering defined in [13].

4.3 Outermost-Fair Rewriting

It may occur that a term can be rewritten to normal form but is also the starting point of an infinite rewrite sequence: consider for example the term $f(a, b)$ in the rewriting system $\{f(a, x) \to c, b \to b\}$. In such a situation it is important to know how to rewrite the term such that eventually a normal form is obtained. In the case of $f(a, b)$, at some point the redex $f(a, b)$ should be contracted in order to reach the normal form c. A *strategy* can be seen as selecting one or more redex occurrences in a term that is not in normal form. A strategy is said to be *normalising* if repeatedly contracting redex occurrences selected by the strategy yields a normal form whenever the initial term has one.

O'Donnell shows in [24] that the *parallel-outermost* strategy, which contracts all redexes that are outermost in a term in one step, is normalising for first-order term rewriting systems that are left-linear, and where trivial critical pairs are only allowed in a certain restricted form. A stronger result obtained in [24] is that *outermost-fair* rewriting is normalising. A rewrite sequence is said to be outermost-fair if every outermost redex occurrence is eliminated eventually. For example, in the rewriting system $\{a \to a, b \to c, f(c, x) \to f(b, x)\}$, the rewrite sequence $f(b, a) \to f(c, a) \to f(b, a) \to \dots$ is outermost-fair, but the rewrite sequence $f(b, a) \to f(b, a) \to f(b, a) \to \dots$ is not. A parallel-outermost rewrite sequence is outermost-fair, since in every step all outermost redexes are eliminated. Hence normalisation of the parallel-outermost strategy is a direct consequence of the normalisation of outermost-fair rewriting.

The idea of the proof that outermost-fair rewriting is normalising as given in [24] is roughly as follows. Consider a term s that has a normal form s'. The intention is to show that all outermost-fair rewrite sequences starting in s end in s'. We fix a rewrite sequence $f : s \twoheadrightarrow s'$ and proceed by induction on its length. If f consists of zero steps, there is nothing to prove. If f consists of one or more rewrite steps, then it is of the form $s \to t \twoheadrightarrow s'$. Let d be an outermost-fair rewrite sequence starting in s. Now the rewrite sequence d is projected over the rewrite step $s \to t$, as in the following picture, where $s_0 = s$ and $t_0 = t$:

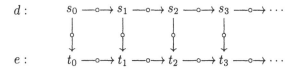

Then the following is shown:

1. If d is outermost-fair then e is outermost-fair.
2. If d is outermost-fair and e ends in a normal form r, then d also ends in r.

It now follows by the induction hypothesis that the outermost-fair rewrite sequence d ends in the normal form s' of s.

We don't consider the proof in more detail here, but just discuss the restriction of left-linearity and the one imposed on critical pairs.

An important observation is that outermost-fair rewriting is not normalising if a redex that is not outermost can create an outermost redex. This can happen if the rewriting system is not left-linear. Consider for instance $\{a \rightarrow b, f(x,x) \rightarrow b, g(x) \rightarrow g(x)\}$. The term $f(g(a), g(b))$ has a normal form, namely b, but it is not reached by the outermost-fair rewrite sequence $f(g(a), g(b)) \rightarrow f(g(a), g(b)) \rightarrow f(g(a), g(b)) \rightarrow \ldots$ in which alternatingly the redexes $g(a)$ and $g(b)$ are contracted. In a rewrite sequence from $f(g(a), g(b))$ to normal form, as for instance $f(g(a), g(b)) \rightarrow f(g(b), g(b)) \rightarrow b$, the contraction of the redex a, which is not outermost, creates the outermost redex $f(g(b), g(b))$.

Further, outermost-fair rewriting is not normalising if redex patterns interfere in an arbitrary way. The term a in the rewriting system $\{a \rightarrow a, a \rightarrow b\}$ has a normal form which is not reached by the outermost-fair rewrite sequence $a \rightarrow a \rightarrow a \rightarrow \ldots$. This interference is ruled out by forbidding the presence of critical pairs. However, it doesn't show that it is necessary to restrict attention to systems without critical pairs. Actually, the result in [24] is obtained for rewriting systems where a special kind of critical pairs is allowed, namely the ones that are trivial and where the overlap is at the top. Rewriting systems that are left-linear and have only this kind of trivial pair, caused by overlap at the top, are said to be *almost orthogonal*. Almost orthogonal systems form a superclass of the orthogonal systems, and a subclass of the weakly orthogonal systems. The rewriting system $\{f(a) \rightarrow f(a), f(x) \rightarrow f(a)\}$ is almost orthogonal but not orthogonal. Note that its trivial critical pair is caused by an overlap at the top. The critical pair in $\{a \rightarrow a, f(a) \rightarrow f(a)\}$ is trivial but the overlap is not at the top; this system is weakly orthogonal but not almost orthogonal.

The above discussion shows that it is necessary to restrict attention to left-linear systems (otherwise outermost-fair rewriting need not be normalising). It is however not yet clear whether the restriction to critical pairs that are trivial and have overlap at the top is necessary. We come back to this point below.

The question arises whether the result of [24] can be extended to the higher-order case. It turns out that this is indeed the case, provided an additional restriction on the rewriting systems is imposed. The point is that in higher-order rewriting, there is, due to the presence of bound variables, yet another way in which a redex that is not outermost can create an outermost redex. As a consequence, outermost-fair rewriting need not be normalising if this additional restriction is not imposed. The following example, due to Van Oostrom, shows the essence of the problem: $\{f(x.z) \rightarrow a, g(z) \rightarrow g(z), h(z) \rightarrow a\}$. The term $f(x.g(h(x)))$ has a normal form: $f(x.g(h(x))) \rightarrow f(x.g(a)) \rightarrow a$. However, this normal form is not reached by the outermost-fair rewrite sequence $f(x.g(h(x))) \rightarrow f(x.g(h(x))) \rightarrow f(x.g(h(x))) \rightarrow \ldots$. Note that contraction of the innermost redex $h(x)$ creates an outermost redex, namely $f(x.g(a))$, by erasing the bound variable; it is only possible to apply the first rewrite rule to a term of the form $f(x.t)$ if t doesn't contain free occurrences of x. This way of creating redexes is ruled out by disallowing the non-occur check as present in the example. Formally, this is done by requiring the rewrite rules to be *fully extended*, a notion which is defined as follows [9].

Definition 4.

1. *Let $z.l$ be a rule-pattern. An occurrence of $z \in z$ in l is* fully extended *if it is of the form $z(x_1, \ldots, x_n)$ with $x_1, \ldots, x_n$ all the variables bound above it.*
2. *A rewrite rule $z.l \to z.r$ is* fully extended *if every occurrence of every $z_i \in z$ in l is fully extended.*
3. *A higher-order rewriting system is* fully extended *if all its rewrite rules are.*

The rewrite rule $z.f(x.z(x)) \to r$ is fully extended. The rewrite rules $z.f(x.z) \to r$ and $\mathsf{abs}(x.\mathsf{app}(z, x)) \to z$ are not fully extended. Recall that $z.f(x.z(x, y))$ and $z.f(x.z(x, x))$ are not even rule-patterns.

Now we come back to the restriction imposed on critical pairs. Recently, it is shown in [27] that outermost-fair rewriting is normalising for *weakly* orthogonal rewriting systems, where arbitrary trivial pairs are allowed. It seems that the restriction to trivial critical pairs cannot be relaxed much more. Consider for instance the rewriting system $\{a \to b, g(x) \to g(x), f(g(b)) \to b\}$. It is parallel closed since we have $f(g(b)) \to b$. The term $f(g(a))$ has a normal form, namely b, but it is not reached by the outermost-fair rewrite sequence $f(g(a)) \to f(g(a)) \to f(g(a)) \to \ldots$. The proof presented in [27] is abstract in nature and applies also to the case of higher-order rewriting. It makes use of ideas that are also present in [34, 20, 8]. For proofs according to the sketch given above, see [24, 4, 32].

Theorem 6. *Outermost-fair rewriting is normalising for higher-order rewriting systems that are weakly orthogonal and fully extended.*

References

1. P. Aczel. A general Church-Rosser theorem. University of Manchester, July 1978.
2. A. Asperti and C. Laneve. Interaction systems I: The theory of optimal reductions. *Mathematical Structures in Computer Science*, 4:457–504, 1994.
3. H.P. Barendregt. *The Lambda Calculus, its Syntax and Semantics*, volume 103 of *Studies in Logic and the Foundations of Mathematics*. North-Holland Publishing Company, revised edition, 1984.
4. J.A. Bergstra and J.W. Klop. Conditional rewrite rules: Confluence and termination. *Journal of Computer and System Sciences*, 32:323–362, 1986.
5. Alonzo Church and J.B. Rosser. Some properties of conversion. *Transactions of the Americal Mathematical Society*, 39:472–482, 1936.
6. N. Dershowitz. Orderings for term rewriting systems. *Theoretical Computer Science*, 17(3):279–301, 1982.
7. R.O. Gandy. Proofs of strong normalization. In J.P. Seldin and J.R. Hindley, editors, *To H.B. Curry: Essays on Combinatory Logic, Lambda Calculus and Formalism*, pages 457–477. Academic Press, 1980.
8. J. Glauert and Z. Khasidashvili. Relative normalization in deterministic residual structures. In *Proceedings of the 19th International Colloquium on Trees in Algebras and Programming (CAAP '96)*, volume 1059 of *Lecture Notes in Computer Science*, pages 180–195, April 1996.

9. M. Hanus and C. Prehofer. Higher-order narrowing with definitional trees. In H. Ganzinger, editor, *Proceedings of the 7th International Conference on Rewriting Techniques and Applications (RTA '96)*, volume 1103 of *Lecture Notes in Computer Science*, pages 138–152, New Brunswick, USA, 1996.

10. R. Hindley. Reductions of residuals are finite. *Transactions of the Americal Mathematical Society*, 240:345–361, June 1978.

11. G. Huet. Confluent reductions: Abstract properties and applications to term rewriting systems. *Journal of the Association for Computing Machinery*, 27(4):797–821, October 1980.

12. G. Huet and D.C. Oppen. Equations and rewrite rules, a survey. In R.V. Book, editor, *Formal Language Theory, Perspectives and Open Problems*, pages 349–405. Academic Press, 1980.

13. J.-P. Jouannaud and A. Rubio. The higher-order recursive path ordering. In *Proceedings of the 14th annual IEEE Symposium on Logic in Computer Science (LICS '99)*, Trento, Italy, 1999. To appear.

14. S. Kahrs. Termination proofs in an abstract setting. To appear in *Mathematical Structures in Computer Science*.

15. Z.O. Khasidashvili. Expression Reduction Systems. In *Proceedings of I. Vekua Institute of Applied Mathematics*, volume 36, pages 200–220, Tblisi, 1990.

16. J.W. Klop. *Combinatory Reduction Systems*, volume 127 of *Mathematical Centre Tracts*. CWI, Amsterdam, 1980. PhD Thesis.

17. D.E. Knuth and P.B. Bendix. Simple word problems in universal algebras. In J. Leech, editor, *Computational Problems in Abstract Algebra*, pages 263–297. Pergamon Press, 1970.

18. R. Mayr and T. Nipkow. Higher-order rewrite systems and their confluence. *Theoretical Computer Science*, 192:3–29, 1998.

19. P.-A. Melliès. *Description Abstraite des Systèmes de Réécriture*. PhD thesis, Université de Paris VII, Paris, France, 1996.

20. A. Middeldorp. Call by need computations to root-stable form. In *Proceedings of the 24th Symposium on Principles of Programming Languages (POPL '97)*, pages 94–105, Paris, France, January 1997. ACM Press.

21. D. Miller. A logic programming language with lambda-abstraction, function variables, and simple unification. *Journal of Logic and Computation*, 1(4):497–536, 1991.

22. T. Nipkow. Higher-order critical pairs. In *Proceedings of the 6th annual IEEE Symposium on Logic in Computer Science (LICS '91)*, pages 342–349, Amsterdam, The Netherlands, July 1991.

23. T. Nipkow and C. Prehofer. Higher-order rewriting and equational reasoning. In W. Bibel and P. Schmitt, editors, *Automated Deduction — A Basis for Applications. Volume I: Foundations*, volume 8 of *Applied Logic Series*, pages 399–430. Kluwer, 1998.

24. M.J. O'Donnell. *Computing in Systems Described by Equations*, volume 58 of *Lecture Notes in Computer Science*. Springer Verlag, 1977.

25. V. van Oostrom. *Confluence for Abstract and Higher-Order Rewriting*. PhD thesis, Vrije Universiteit, Amsterdam, The Netherlands, March 1994.

26. V. van Oostrom. Developing developments. *Theoretical Computer Science*, 175(1):159–181, 1997.

27. V. van Oostrom. Normalisation in weakly orthogonal rewriting. In *Proceedings of the 10th International Conference on Rewriting Techniques and Applications (RTA '99)*, 1999.

28. J.C. van de Pol. Termination proofs for higher-order rewrite systems. In J. Heering, K. Meinke, B. Möller, and T. Nipkow, editors, *Proceedings of the First International Workshop on Higher-Order Algebra, Logic and Term Rewriting (HOA '93)*, volume 816 of *Lecture Notes in Computer Science*, pages 305–325, Amsterdam, The Netherlands, 1994.

29. J.C. van de Pol. *Termination of Higher-order Rewrite Systems*. PhD thesis, Utrecht University, Utrecht, The Netherlands, December 1996.

30. C. Prehofer. *Solving Higher-Order Equations: From Logic to Programming*. PhD thesis, Technische Universität München, 1995.

31. Z. Qian. Unification of higher-order patterns in linear time and space. *Journal of Logic and Computation*, 6:315–341, 1996.

32. F. van Raamsdonk. *Confluence and Normalisation for Higher-Order Rewriting*. PhD thesis, Vrije Universiteit, Amsterdam, The Netherlands, May 1996.

33. B.K. Rosen. Tree-manipulating systems and Church-Rosser theorems. *Journal of the Association for Computing Machinery*, 20(1):160–187, January 1973.

34. R.C. Sekar and I.V. Ramakrishnan. Programming in equational logic: Beyond strong sequentiality. In *Proceedings of the 7th International Symposium on Logic in Computer Science (LICS '90)*, pages 230–241, 1990.

35. R. de Vrijer. Exactly estimating functionals and strong normalization. *Proceedings of the Koninklijke Nederlandse Akademie van Wetenschappen*, 90(4), December 1987.

36. R. de Vrijer. *Surjective Pairing and Strong Normalization: Two Themes in Lambda Calculus*. PhD thesis, Universiteit van Amsterdam, January 1987.

37. H. Zantema. Termination of term rewriting: Interpretation and type elimination. *Journal of Symbolic Computation*, 17:23–50, 1994.

The Maude System*

M. Clavel[1], F. Durán[2], S. Eker[2], P. Lincoln[2], N. Martí-Oliet[3], J. Meseguer[2],
and J. F. Quesada[4]

[1] Department of Philosophy, University of Navarre, Spain
[2] SRI International, Menlo Park, CA 94025, USA
[3] Facultad de Ciencias Matemáticas, Universidad Complutense, Madrid, Spain
[4] CICA (Centro de Informática Científica de Andalucía), Seville, Spain

1 Introduction

Maude is a high-performance language and system supporting both equational
and rewriting logic computation for a wide range of applications, including development of theorem proving tools, language prototyping, executable specification
and analysis of concurrent and distributed systems, and logical framework applications in which other logics are represented, translated, and executed.

Maude's *functional modules* are theories in *membership equational logic* [8, 1],
a Horn logic whose atomic sentences are either equalities $t = t'$ or *membership
assertions* of the form $t : s$, stating that a term t has a certain sort s. Such a logic
extends OBJ3's [4] order-sorted equational logic and supports sorts, subsorts,
subsort polymorphic overloading of operators, and definition of partial functions
with equationally defined domains. Maude's functional modules are assumed
to be Church-Rosser; they are executed by the Maude engine according to the
rewriting techniques and operational semantics developed in [1].

Membership equational logic is a sublogic of *rewriting logic* [6]. A rewrite theory is a pair (T, R) with T a membership equational theory, and R a collection
of labeled and possibly conditional *rewrite rules* involving terms in the signature
of T. Maude's *system modules* are rewrite theories in exactly this sense. The
rewrite rules $r : t \longrightarrow t'$ in R are *not* equations. Computationally, they are interpreted as local *transition rules* in a possibly concurrent system. Logically, they
are interpreted as *inference rules* in a logical system. This makes rewriting logic
both a general *semantic framework* to specify concurrent systems and languages
[7], and a general *logical framework* to represent and execute different logics [5].

Rewriting in (T, R) happens *modulo* the equational axioms in T. Maude supports rewriting modulo different combinations of associativity, commutativity,
identity, and idempotency axioms. The rules in R need not be Church-Rosser
and need not be terminating. Many different rewriting paths are then possible;
therefore, the choice of appropriate *strategies* is crucial for executing rewrite theories. In Maude, such strategies are not an extra-logical part of the language.

* Supported by DARPA through Rome Laboratories Contract F30602-97-C-0312, by
DARPA and NASA through Contract NAS2-98073, by Office of Naval Research Contract N00014-96-C-0114, and by National Science Foundation Grant CCR-9633363.

They are instead *internal strategies* defined by rewrite theories at the metalevel. This is because rewriting logic is *reflective* [2] in the precise sense of having a *universal theory* U that can represent any finitely presented rewrite theory T (including U itself) and any terms t, t' in T as terms $\overline{T}$ and $\bar{t}, \overline{t'}$ in U, so that we have the following equivalence

$$T \vdash t \longrightarrow t' \quad \Leftrightarrow \quad U \vdash \langle \overline{T}, \bar{t} \rangle \longrightarrow \langle \overline{T}, \overline{t'} \rangle.$$

Since U is representable in itself, we can then achieve a "reflective tower" with an arbitrary number of levels of reflection. Maude efficiently supports this reflective tower through its META-LEVEL module, which makes possible not only the declarative definition and execution of user-definable rewriting strategies, but also many other applications, including an extensible *module algebra* of parameterized module operations that is defined and executed within the logic.

This extensibility by reflection is exploited in Maude's design and implementation. Core Maude (Section 2) supports module hierarchies consisting of (unparameterized) functional and system modules and provides the META-LEVEL module. Full Maude (Section 3) is an extension of Core Maude written in Core Maude itself that supports a module algebra of parameterized modules, views, and module expressions in the OBJ style [4] as well as *object-oriented modules* with convenient syntax for object-oriented applications. The paper ends with a summary of different applications (Section 4). The Maude 1.0 system and its documentation have been available for distribution (free of charge) since January 1999 through the Maude web page http://maude.csl.sri.com.

2 Core Maude

The Maude system is built around the Core Maude interpreter, which accepts module hierarchies of (unparameterized) functional and system modules with user-definable mixfix syntax. It is implemented in C++ and consists of two parts: the rewrite engine, and the mixfix frontend.

The rewrite engine is highly modular and does not contain any Maude-specific code. Two key components are the "core" module and the "interface" module. The core module contains classes for objects which are not specific to an equational theory, such as equations, rules, sorts, and connected sort components. The "interface" module contains abstract base classes for objects that may have a different representation in different equational theories, such as symbols, term nodes, dag nodes, and matching automata. New equational theories can be "plugged in" by deriving from the classes in the "interface" module. To date, all combinations of associativity, commutativity, left and right identity and idempotence have been implemented apart from those that contain both associativity and idempotence. New built in symbols with special rewriting (equation or rule) semantics may be easily added.

The engine is designed to provide the look and feel of an interpreter with hooks for source level tracing/debugging and user interrupt handling. These goals prevent a number of optimizations that one would normally implement in a compiler, such as transforming the user's term rewriting system, or keeping pending evaluations on a stack and only building reduced terms. The actual implementation is a semi-compiler where the term rewriting system is compiled to a system of tables and automata, which is then interpreted. Typical performance with the current version is 800K-840K free-theory rewrites per second and 27K-111K associative-commutative (AC) rewrites per second on standard hardware (300MHz Pentium II). The figure for AC rewriting is highly dependent on the complexity of the AC patterns (AC matching is NP-complete) and the size of the AC subjects. The above results were obtained using fairly simple linear and non-linear patterns and large (hundreds of nested AC operators) subjects.

The mixfix frontend consists of a bison/flex-based parser for Maude's surface syntax, a grammar generator (which generates the context free grammar for the user-definable mixfix syntax in a module together with some built-in extensions), a context free parser generator, a mixfix pretty printer, a fully reentrant debugger, the built-in functions for quoted identifiers and the META-LEVEL module, together with a considerable amount of "glue" code holding everything together. Many of the C++ classes are derived from those in the rewrite engine. The Maude parser generator (MSCP) is implemented using SCP as the formal kernel [9]. The techniques used include β-extended GFGs (that is, CFGs extended with "bubbles" (strings of tokens) and precedence/gather patterns). MSCP provides efficient treatment of syntactic reflection, and a basis for flexible syntax definition.

In Maude, key functionality of the universal theory U has been efficiently implemented in the functional module META-LEVEL. In META-LEVEL Maude terms are reified as elements of a data type Term, and Maude modules are reified as terms in a data type Module. The processes of reducing a term to normal form in a functional module and of rewriting a term in a system module using Maude's default interpreter are respectively reified by functions meta-reduce and meta-rewrite. Similarly, the process of applying a rule of a system module to a subject term is reified by a function meta-apply. Furthermore, parsing and pretty printing of a term in a signature, as well as key sort operations are also reified by corresponding metalevel functions.

3 Full Maude

Using reflection Core Maude can be extended to a much richer language with an extensible *module algebra* of module operations that can make Maude modules highly reusable. The basic idea is that the META-LEVEL module can be extended with new data types—extending the Module sort of flat modules to structured and parameterized modules—and with new functions corresponding to new module operations—such as instantiation of parameterized modules by views, flattening of module hierarchies into single modules, desugaring of object-oriented modules into system modules, and so on. All such new types and operations can be defined in Core Maude. Using the meta-parsing and meta-pretty printing functions in META-LEVEL and a simple LOOP-MODE module providing input/output we have developed in Core Maude a user interface for Full Maude. At present, Full Maude supports all of Core Maude plus *object-oriented* modules, *parameterized* modules, *theories* with loose semantics to state formal requirements in parameters, *views* to bind parameter theories to their instances, and *module expressions* instantiating and composing parameterized modules.

4 Applications

Maude is an attractive formal meta-tool for building many advanced applications and formal tools. The largest application so far is Full Maude (7,000 lines of Maude code). Other substantial applications include: an inductive theorem prover; a Church-Rosser checker (both part of a formal environment for Maude and for the CafeOBJ language [3]); an HOL to Nuprl translator; and a translator from J. Millen's CAPSL specification language to the CIL intermediate language. In addition, several language interpreters and strategy languages, several object-oriented specifications—including cryptographic protocols and network applications—and a variety of executable translations mapping logics, architectural description languages and models of computation into the rewriting logic reflective framework have been developed by different authors.

References

1. A. Bouhoula, J.-P. Jouannaud, and J. Meseguer. Specification and proof in membership equational logic. To appear in *Theoretical Computer Science*.
2. M. Clavel. Reflection in general logics and in rewriting logic, with applications to the Maude language. Ph.D. Thesis, University of Navarre, 1998.
3. M. Clavel, F. Durán, S. Eker, and J. Meseguer. Building equational logic tools by reflection in rewriting logic. In *Proc. of the CafeOBJ Symposium '98, Numazu, Japan*. CafeOBJ Project, April 1998.
4. J. Goguen, T. Winkler, J. Meseguer, K. Futatsugi, and J.-P. Jouannaud. Introducing OBJ. Technical Report SRI-CSL-92-03, SRI International, Computer Science Laboratory, 1992. To appear in J.A. Goguen and G.R. Malcolm, editors, *Applications of Algebraic Specification Using OBJ*, Academic Press, 1998.
5. N. Martí-Oliet and J. Meseguer. Rewriting logic as a logical and semantic framework. In J. Meseguer, editor, *Proc. First Intl. Workshop on Rewriting Logic and its Applications*, volume 4 of *Electronic Notes in Theoretical Computer Science*. Elsevier, 1996. http://www1.elsevier.nl/mcs/tcs/pc/volume4.htm.
6. J. Meseguer. Conditional rewriting logic as a unified model of concurrency. *Theoretical Computer Science*, 96(1):73–155, 1992.
7. J. Meseguer. Rewriting logic as a semantic framework for concurrency: a progress report. In *Proc. CONCUR'96, Pisa, August 1996*, pages 331–372. Springer LNCS 1119, 1996.
8. J. Meseguer. Membership algebra as a semantic framework for equational specification. In F. Parisi-Presicce, ed., Proc. WADT'97, 18–61, Springer LNCS 1376, 1998.
9. J. Quesada. *The SCP parsing algorithm based on syntactic constraint propagation*. Ph.D. thesis, University of Seville, 1997.

$\mathcal{TOY}$: A Multiparadigm Declarative System [*]

F.J. López Fraguas and J. Sánchez Hernández

Dep. Sistemas Informáticos y Programación, Univ. Complutense de Madrid
Fac. Matemáticas, Av. Complutense s/n, 28040 Madrid, Spain
email: {fraguas,jaime}@sip.ucm.es   Phone: +34 91 3944429

Abstract. $\mathcal{TOY}$ is the concrete implementation of CRWL, a wide theoretical framework for declarative programming whose basis is a constructor based rewriting logic with lazy non-deterministic functions as the core notion. Other aspects of CRWL supported by $\mathcal{TOY}$ are: polymorphic types; HO features; equality and disequality constraints over terms and linear constraints over real numbers; goal solving by needed narrowing combined with constraint solving. The implementation is based on a compilation of $\mathcal{TOY}$ programs into Prolog.

1   Introduction

$\mathcal{TOY}$ is a system for multiparadigm declarative programming which encompasses (and extends in some interesting ways) functional programming (FP), logic programming (LP) and constraint programming. The system has been made publicly available (at http://mozart.sip.ucm.es/toy), and we have developed for it a variety of interesting programs and programming methodologies [AHL+96,CL98b,CL98a].

As with many other narrowing-based FP+LP proposals (see [Han94] for a survey), the starting point of $\mathcal{TOY}$ can be described as follows: rewrite systems can be seen as functional programs, and rewriting performs evaluation. The use of narrowing instead of rewriting results in goal solving capabilities, turning the rewrite system into a functional logic program. When seen as rewrite systems, $\mathcal{TOY}$ programs have the following characteristics:

1. They follow a *constructor discipline*. Rules use linear constructor-made patterns in left-hand sides.

2. They can be *non-terminating* and *non-confluent*. Therefore $\mathcal{TOY}$ programs serve to compute non-deterministic lazy functions, which turn out to be a useful tool for programming. As a particular source of non-confluence, $\mathcal{TOY}$ allows extra variables to appear in right-hand sides of rules.

3. They use *constrained conditional rules* for defining functions. $\mathcal{TOY}$ can manage *equality* and *disequality constraints* over constructor terms, and *linear constraints* over real numbers.

4. They may be *higher order*. $\mathcal{TOY}$'s approach to HO features is based on an *intensional* view of functions: functions, when partially applied, behave as

---

[*] The authors have been partially supported by the Spanish CICYT ·(project TIC 98-0445-C03-02 'TREND') and the ESPRIT Working Group 22457 (CCL-II).

data constructors, and can be used to form *HO patterns* which, in particular, can appear in left-hand sides of rules and also in answers.

5. They are *polymorphically typed*. Types are inferred (optionally declared) according to Hindley-Milner system.

6. As goal solving mechanism, $\mathcal{TOY}$ uses a suitable combination of lazy narrowing (with a sophisticated strategy, called *demand driven* [LLR93] or *needed narrowing* [AEH94]) and constraint solving. Alternative solutions for a given goal are obtained by backtracking.

2 $\mathcal{TOY}$ in the FLP context

$\mathcal{TOY}$ implements a wide theoretical framework (called CRWL) for declarative programming. The first order untyped core of the framework can be found in [GHL+98]. HO features were included in [GHR97]. Polymorphic types (and also algebraic types, not yet implemented in $\mathcal{TOY}$) are addressed in [AR97a,AR97b], and further extended with constraints in [ALR98a,ALR98b]. The concrete lazy narrowing strategy adopted by $\mathcal{TOY}$ is studied in [LLR93,AEH94], and formally justified within the CRWL-framework in [LS98].

$\mathcal{TOY}$ is an evolution of BABEL [MR92], a functional logic language with a much more restrictive class of programs (confluent, no constraints, more limited HO features). Among other proposals for FLP with publicly available implementations, the most related one is *Curry* [Han98], a still in progress initiative for developing a 'standard' FLP language. $\mathcal{TOY}$ and *Curry* share many characteristics, but there are still remarkable differences:

* *Curry* 's operational model, which combines lazy narrowing and residuation, was proposed and formally justified [Han97] for certain kinds of unconditional confluent programs. For practical reasons, *Curry* now accepts conditional rules and non-deterministic functions, but they lie outside its formal foundations.

* *Curry* contemplates constraints as a special kind of expressions, but in fact only equality constraints are considered. Constraints are also outside the formal foundations of *Curry*.

* *Curry* does not contemplate HO patterns nor HO logical computations.

3 Programming in $\mathcal{TOY}$

The table below contains a small program showing some of $\mathcal{TOY}$'s abilities, together with some goals and the corresponding answers. $\mathcal{TOY}$ borrows from Haskell most of its syntax, with the exception that variables begin with upper-case letters while constructor and function symbols use lower-case.

The goal *(a)* in the table sorts a list, in a pure functional computation. Goal *(b)* sorts a list with a variable as one of its elements; the three given answers include linear constraints over such variable. The answer for goal *(c)* involves a syntactic disequality. In goal *(d)* F is a higher order logic variable; the obtained values for this variable are higher order patterns (`permut`, `sort`,...).

```
% Non-deterministic choice of one of two values
infixr 40 //              X // Y = X                    X // Y = Y

% Non-deterministic insertion of an element into a list
insert X [] = [X]         insert X [Y|Ys] = [X,Y|Ys] // [Y|insert X Ys]

% Non-deterministic generation of permutations of a list
permut [] = []            permut [X|Xs] = insert X (permut Xs)

% Tests if a list of numbers is sorted
sorted [] = true          sorted [X] = true
sorted [X,Y|Ys] = sorted [Y|Ys] <== X <= Y

% Lazy 'generate-and-test' permutation sort. 'check' calls 'sorted' which demands
% its argument, which is lazily non-deterministically generated by 'permut'.
% As soon as the test fails, 'permut' stops the generation and tries another
% alternative for the last choice it made
sort Xs = check (permut Xs)        check Xs = Xs <== sorted Xs == true
```

Goal	Answers
(a) sort [4,2,5,1,3] == L	L == [1,2,3,4,5]; no more solutions
(b) sort [3,2,X] == L	L == [X,2,3] {X=<2}; L == [2,X,3] {X>=2,X=<3};
	L == [2,3,X] {X>=3}; no more solutions
(c) sort [3,2,1] /= L	L /= [1,2,3] ; no more solutions
(d) F [2,1,3] == [1,2,3]	F == permut; F == sort; ...

4 Implementation of $\mathcal{TOY}$

At its outer level, $\mathcal{TOY}$ behaves as a small command interpreter to be executed within a Sicstus Prolog session. There are commands for compiling, running and editing programs, solving goals, showing type information, ...

The core of $\mathcal{TOY}$'s implementation is a process of *compilation to Prolog*: any $\mathcal{TOY}$ program is translated into a set of clauses that reproduce, when executed with Prolog, the expected behaviour of the source program under $\mathcal{TOY}$'s own operational model. Common to all $\mathcal{TOY}$ programs are the clauses for *constraint solving*. For *equality and disequality constraints* (over constructor terms), reduction of arguments is interleaved with occur-check and check for constructor clashes. Disequalities with the form $X/= t$, where t is a constructor term, are in solved form and kept in a *store*, which must be 'awoken' if in a later step X becomes bound. For solving a *linear constraint* $e \diamond e'$ (with $\diamond \in \{<, >, =<, >=, ==, /=\}$), e and e' are reduced to normal forms t and t' and then the Sicstus linear constraint solver is invoked to solve $t \diamond t'$. When a computation is finished, all the stored constraints are conveniently projected over the set of relevant variables for taking part in the answer.

Some other clauses are heavily dependent on the source program. The most important ones are those which control the computation of *head normal forms* for function calls. They must use the rules of the source $\mathcal{TOY}$ program so that $\mathcal{TOY}$'s demand driven strategy is reflected. For this purpose, we built the *def-*

initional tree (see [LLR93]) of each function, from which the Prolog code is extracted.

Although its performance is not very impressive, $\mathcal{TOY}$ easily supports the development of medium size programs. The largest $\mathcal{TOY}$ program of which we are aware (a partial evaluator for functional logic programs) contains about 200 function definitions and 1500 lines of $\mathcal{TOY}$ code.

Acknowledgements:

We thank all members of the Declarative Programming Group at the UCM for their help when developing $\mathcal{TOY}$. Very special thanks merits Rafa Caballero.

References

[AEH94]   S. Antoy, R. Echahed, M. Hanus. A needed narrowing strategy. *Proc. POPL'94*, 268–279, 1994.

[AHL⁺96]   P. Arenas, T. Hortalá, F.J. López, E. Ullán. Real constraints within a functional logic language. *Proc. APPIA-GULP-PRODE'96*, 451–462, 1996.

[ALR98a]   P. Arenas, F.J. López, M. Rodríguez. Embedding multiset constraints into a lazy functional logic language. *Proc. PLILP'98*, Springer LNCS 1490, 429–444, 1998.

[ALR98b]   P. Arenas, F.J. López, M. Rodríguez. Functional plus logic programming with built-in and symbolic constraints, 1998. *Submitted*.

[AR97a]   P. Arenas, M. Rodríguez. A semantic framework for functional logic programming with algebraic polymorphic types. *Proc. TAPSOFT'97*, Springer LNCS 1214, 453–464, 1997.

[AR97b]   P. Arenas, M. Rodríguez. A lazy narrowing calculus for functional logic programming with algebraic polymorphic types. *Proc. ILPS'97*, MIT Press, 53–69, 1997.

[CL98a]   R. Caballero, F.J. López. A functional logic alternative to monads. *Proc. of Workshop on Component-Based Software Development in Computer Logic*, 87–100, 1998.

[CL98b]   R. Caballero, F.J. López. Parsing with non-deterministic functions. *Proc. APPIA-GULP-PRODE'98*, 1–16, 1998.

[GHL⁺98]   J.C. González, T. Hortala, F.J. López, M. Rodríguez. An approach to declarative programming based on a rewriting logic. *To appear in JLP*, 1998. Earlier version in *ESOP'96*.

[GHR97]   J.C. González, M.T. Hortalá, M. Rodríguez. A higher order rewriting logic for functional logic programming. *ICLP'97*, MIT Press, 153–167, 1997.

[Han94]   M. Hanus. The integration of functions into logic programming, *JLP*, 19 & 20, 583–628, 1994.

[Han97]   M. Hanus. A unified computation model for functional and logic programming, *Proc. POPL'97*, 80-93, 1997.

[Han98]   M. Hanus (ed.). Curry: An integrated functional logic language. Available at http://www-i2.informatik.rwth-aachen.de/~hanus/curry, 1998.

[LS98]   F.J. López, J. Sánchez. An efficient narrowing strategy by means of disequality constraints. Tech. Rep. 98/84, Dep. SIP, UCM Madrid, 1998.

[LLR93]   R. Loogen, F.J. López, M. Rodríguez. A demand driven computation strategy for lazy narrowing. *Proc. PLILP'3*, Springer LNCS 714, 184–200, 1993.

[MR92]   J.J. Moreno, M. Rodríguez. Logic programming with functions and predicates: The language BABEL. *JLP*, 12, 191–223, 1992.

UniMoK: A System for Combining Equational Unification Algorithms*

Stephan Kepser[1] and Jörn Richts[2]

[1] SfS, University of Tübingen
Wilhelmstr. 133, 72074 Tübingen Germany
kepser@sfs.nphil.uni-tuebingen.de
[2] Theoretische Informatik, RWTH Aachen
Ahornstr. 55, 52074 Aachen, Germany
richts@informatik.rwth-aachen.de

1 Combining Unification Algorithms

Equational unification algorithms can be used in resolution based theorem provers [9] and rewriting engines [6] to improve their handling of equality. Originally, the requirements of these theorem provers and rewrite engines were such that the unification algorithms had to compute complete sets of unifiers. But with the advent of constraint based approaches to theorem proving [4] and rewriting [8] the interest in unification algorithm that worked merely as decision procedures grew because minimal complete sets of unifiers can be very large – e.g., doubly exponential in the number of variables of the problem in the case of the theory AC – and are hence costly to compute.

Because actual unification problems usually contain function symbols from several different signatures, the following *combination problem* is an important task in unification theory: Given unification algorithms for equational theories $E_1, E_2, \ldots, E_n$ over pairwise disjoint signatures, provide a general method that gives a unification algorithm for the union $E_1 \cup E_2 \cup \ldots \cup E_n$ of these theories. Solutions for this problem were provided by Schmidt-Schauß [10] and Boudet [3] for the combination of algorithms calculating complete sets of unifiers and by Baader and Schulz [1] for combining decision procedures. The combination algorithm presented in [1] is mostly of theoretical interest, it contains many non-deterministic decisions, thus the search space that this algorithm spans is so huge, that it is unusable for practical implementations. Therefore the authors developed optimisation methods [7] for the combination algorithm by Baader and Schulz to gain an implementation that can be used in practise. This implementation is UniMoK.

UniMoK stands for UNIfication MOdule for KEIM. It contains algorithms for unification in certain equational theories and it provides several combination methods for them. All combination algorithms in UniMoK are extensions and optimisations of the combination method by Baader and Schulz [1].

* This work was supported by a DFG grant (SPP "Deduktion") and by the Esprit working group 22457 – CCL II of the EU.

2 Basic Algorithms

The first aim of UNIMOK is to provide an implementation of the combination method by Baader and Schulz and suitable component algorithms for some theories. The combination method of Baader and Schulz requires component algorithms that can solve so-called E_i-unification problems with linear constant restrictions (LCR) for the component theories E_i to be combined. An E_i-unification problem with LCR consists of a set of equations Γ and a linear order $<$ on the variables and constants of Γ where the terms in Γ are built from variables, free constants, and the signature of E_i. A substitution σ is a unifier of $(\Gamma, <)$ if $\sigma(s) =_{E_i} \sigma(t)$ for all equations $s \doteq t \in \Gamma$ and if $\sigma(x)$ does not contain any constant a with $x < a$ for all variables $x \in \Gamma$.

UNIMOK offers algorithms for solving unification problems with LCR for the following equational theories:

- the free theory (syntactic unification),
- the theory A of an associative function symbol
  (with a depth bound, not the Makanin decision procedure),
- the theory AC of an associative and commutative function symbol,
- the theory ACI of an associative, commutative and idempotent function symbol,
- the theory BR of Boolean rings.

For most of these theories, the algorithm could be easily obtained by extending an existing algorithm for unification with constants. For the theory of Boolean rings, a method for constant elimination (see [10]) is used.

All these algorithms were implemented as decision procedures and as algorithms computing complete sets of unifiers. The implementation of the combination algorithm can cope with both kinds of algorithms, i.e., it can work as a decision procedure itself or compute complete sets of unifiers.

The combination algorithm of Baader and Schulz can also be used to combine constraint solvers for so-called quasi-free structures [2]. Unification algorithms are a special case of such constraint solvers. In order to use this property, component algorithms for rational trees and for feature structures were implemented.

3 Optimised Algorithms

The naive implementation of the combination method of Baader and Schulz is mostly for experimental purposes. Due to its large search space of non-deterministic choices this method is not useful for most practical problems. Therefore the authors developed an optimisation technique for this combination method called the deductive method [7]. The implementation of this deductive method is the central and most interesting part of UNIMOK. In short, many decisions in the combination algorithms need not be non-deterministically guessed but can be deduced on the base of one of the component theories involved, the unification problem given, and other decisions already made. Hence the deductive combination algorithm consults the component theories, if they can deduce that certain decisions have to be made deterministically in order for their subproblems to remain solvable. If a component returns such a decision, this decision might enable other components to deduce further decisions. If this process comes to an end before all decisions have been made, the combination algorithm has to make a non-deterministic choice. With this choice it can consult the components again.

The method obviously demands special deductive component algorithms that can deduce such decisions. An equational theory for which only a unification algorithm for problems with LCR exists, can still be used in this combination method but it does not contribute to the deductive process. UNIMOK provides component algorithms computing decisions for the free theory and the theories A, AC, and ACI, for rational trees and feature structures. It also contains general algorithms for collapse-free and regular theories.

Due to its modular, object-oriented approach, UNIMOK is simple to extend. To add a new equational theory E, one has to provide a method for deciding E-unification problems with LCR. In order to contribute to the optimisation, there should also be a method for participating in the deductive process. This is a method that takes a unification problem and a partially specified linear constant restriction as input and returns more information on the decisions to be made.

4 Implementation and Experimental Results

UNIMOK is implemented in Common Lisp on base of the theorem prover development tool box KEIM [5]. KEIM is an open, modular, object-oriented system geared towards ease of use and extensibility, rapid prototyping and universality. It is not designed towards run-time efficiency. To use UNIMOK, an installation of KEIM is required. UNIMOK, thereby, becomes a part of KEIM, and theorem provers developed in KEIM can use UNIMOK for equational unification. In opposite to KEIM, a major design goal behind UNIMOK is the development of efficient code. Basically, all that is needed from KEIM is the module for first order terms which could possibly be replaced with sustainable effort by something more efficient, if needed.

Experimental results show that it is crucial for the deductive combination method in which order the remaining non-deterministic decisions are selected. In [7] the authors presented the so-called iterative strategy, which chooses all non-deterministic decisions for one component first before proceeding to choices for the next component. Run time tests in [7] showed that this strategy is superior for the example problems used there.

However, new run time tests show that this is not true in general. The following table contains sets of example problems solved with the deductive combination method. The problems are randomly generated on the base of a signature containing several A, AC, ACI and free function symbols. Each set contains 200 problems; roughly half of the problems in each set are unifiable. All run times are in seconds. The 'bktrk'-column gives the number of backtracking steps.

	Ded+Iter		Ded			Ded+Iter		Ded	
	time	bktrk	time	bktrk		time	bktrk	time	bktrk
1	816	1953	81	152	11	20	7	21	10
2	232	780	>1h		12	21	8	21	8
3	330	800	1158	1982	13	32	50	31	30
4	58	250	42	110	14	21	47	26	22
5	1362	3971	141	401	15	154	394	3931	12335
6	>1h		103	295	16	26	50	30	31
7	676	2217	189	689	17	319	1116	83	147
8	19	1	19	1	18	1250	2627	44	107
9	67	33	75	33	19	178	462	58	169
10	16	1	15	1	20	99	414	43	159

For some sets, e.g., sets 1, 5, and 6, the iterative strategy ('Ded+Iter') is much worse than the default strategy ('Ded') where all decisions of a certain kind (i.e., identifications; see [7]) are made first; for other sets (2, 3, 15) the iterative strategy is still superior.

Further experiments are needed to develop a strategy combining the strengths of both strategies. Making the component theories choose the next non-deterministic decision might be another option to enhance the selection strategy.

UNIMOK is available at http://www-lti.informatik.rwth-aachen.de/Forschung/unimok.html.

References

1. Franz Baader and Klaus U. Schulz. Unification in the union of disjoint equational theories: Combining decision procedures. *JSC*, 21:211–243, 1996.
2. Franz Baader and Klaus U. Schulz. Combination of constraint solvers for free and quasi-free structures. *Theoretical Computer Science*, 192:107–161, 1998.
3. Alexandre Boudet. Combining unification algorithms. *JSC*, 16(6):597–626, 1993.
4. Hans-Jürgen Bürckert. A resolution principle for clauses with constraints. In Mark E. Stickel, editor, *CADE-10*, pp. 178–192, LNAI 449, Springer, 1990.
5. Xiaorong Huang, Manfred Kerber, Michael Kohlhase, Erica Melis, Dan Nesmith, Jörn Richts, and Jörg H. Siekmann. Keim: A toolkit for automated deduction. In Alan Bundy, editor, *CADE-12*, pp. 807–810, LNAI 814, Springer, 1994.
6. Jean-Pierre Jouannaud and Hélène Kirchner. Completion of a set of rules modulo a set of equations. *SIAM Journal on Computing*, 15:1155–1195, 1986.
7. Stephan Kepser and Jörn Richts. Optimisation techniques for combining constraint solvers. In Maarten de Rijke and Dov M. Gabbay, editors, *Frontiers of Combining Systems, FroCoS'98*. Kluwer Academic Publishers, 1998.
8. Claude Kirchner and Hélène Kirchner. Constrained equational reasoning. In Gaston H. Gonnet, editor, *Proceedings of SIGSAM 1989 International Symposium on Symbolic and Algebraic Computation: ISSAC'89*, pages 382–389. ACM Press, 1989.
9. G.D. Plotkin. Building-in equational theories. *Machine Intelligence*, 7:73–90, 1972.
10. Manfred Schmidt-Schauß. Unification in a combination of arbitrary disjoint equational theories. *Journal of Symbolic Computation*, 8(1,2):51–99, 1989.

$\vec{LR}^2$: A Laboratory for Rapid Term Graph Rewriting[1]

Rakesh Verma and Shalitha Senanayake

Computer Science Department, University of Houston, Houston, TX 77204

1 Introduction

Fast rewriting is needed for equational programming, rewrite based formal verification methods, and symbolic computing systems such as Maple/Mathematica. In any implementation of rewriting techniques efficiency is a critical issue [2]. At the University of Houston we have been developing and evaluating the LR^2 system for fast rewriting for the last several years. There are two motivations for LR^2: we plan to use it for formal verification approaches that include rewriting, and as a testbed for innovating rewriting algorithms that are fast and efficient in practice.

LR^2 consists of a term graph interpreter TGR, and a term graph rewriter that stores the history of its reductions, called Smaran,[2] based on the congruence closure approach [1, 4]. The input to LR^2 is a program representing a convergent rewrite system (a different version allows orthogonal systems) and an input term. Similar to algebraic specification languages like OBJ, ASF+SDF and ELAN a program is composed from modules. Each module defines its own signature and rewriting rules. A module can import other modules. Terms in LR^2 are written in prefix form. The language of LR^2 contains built-in datatypes, viz., integers, floating-point numbers, booleans, characters, sets, multisets, and strings with associated operations. The set datatype supports the usual operations. The string datatype supports membership and indexing operations.

LR^2 also includes a variant detector that can determine if a new term is an alphabetic variant of an existing term, which is currently usable with the history option. If so, the appropriate variant of the result computed for the existing term is used for further rewriting instead of starting from scratch. LR^2 also allows a compact form for storing lists of arithmetic progressions as these occur frequently. Instead of storing the entire list, LR^2 stores the initial value, the final value and the difference. LR^2 provides a set of commands so that it can be called by other systems using UNIX message passing.

LR^2 provides a variety of options to control the amount of history that is stored, if the user chooses the history option. The default using Smaran is to save the results of each rewrite step in a compact data structure. The language of LR^2 allows annotating specific functions with the keyword "memo". This allows to save all the results of rewriting terms that have the specified function symbol at the root. The Delete (history) option in LR^2 allows to delete the entire history of rewrites performed so far except for the given term and its latest reduct after every i^{th} rewrite step, where i can be specified by the user. Another option is to delete the entire history except the given term and the latest reduct as soon as the free user memory drops to a user-specified percentage of total available memory. Options can be combined in any way to suit the application. However, the delete option overrides the other options.

[1] Partially supported by NSF grant CCR-9732186
[2] A preliminary version of Smaran with fewer options was demonstrated at RTA 93.

Operators can be declared AC in LR^2. However, currently only left-linear rules with AC operators can be handled; the matching algorithm for nonlinear rules with AC operators is in the testing phase. In the next section we discuss selected optimizations, and in Section 3 we present some performance results and comparisons.

2 Select Optimizations

In the design and implementation of LR^2 we have attempted to achieve efficiency through both high-level optimizations such as the normalization strategy reported earlier in [3] and also low-level optimizations such as encoding all strings by integers to replace string operations with less expensive integer operators [3]. For lack of space, we mention here only a few of the key optimizations in LR^2.

For the Smaran subsystem, we experimented with alternative representations and chose the most efficient one. We have replaced the earlier use of several large arrays, by a more dynamic data structure that uses a single array, with interconnected lists for each class, and balanced search trees. A signature is stored only in one location and all other data structures address this location. We changed the components of a signature from class numbers to actual pointers to other signatures. A union operation is a constant-time operation implemented by setting a pointer from the root of one class to the root of the other. Of course, after several union operations, the chains of pointers that must be followed can become quite long. Hence we employ the standard technique of path compression on these chains, when obtaining the class number of a signature.

Bottom-up algorithms. The second optimization that is crucial to the efficiency of built-in datatypes is the use of bottom-up algorithms for computations involving the built-in datatypes. The result of this change can be seen for instance in the Fibonacci program, which requires relatively more arithmetic computations than reductions. Here the speed-up over the previous version ranges from 5 to 10 depending on the input term.

Matching and Construction of Signatures. The third major optimization is the use of discrimination trees for matching in LR^2, with a variable list for keeping track of the substitutions for variables to handle consistency checks. The discrimination tree representation is novel in that we do a breadth-first scan of the set of rules and each level of the tree is linked to eliminate expensive recursive calls. Since all strings are encoded as integers in LR^2, the integer encodings for the variables, which are saved in the record for the variables, are used for indexing into the variable list. The use of discrimination trees gives substantial reduction in time for large sets and/or terms requiring over 50000 reductions. For example, for the input term $sieve(from(2, 2000), 500)$, which constructs a list of 2000 numbers starting from 2 and then extracts up to 500 primes from it, the reduction time is almost 25% when discrimination trees are used. We have also adopted a dual representation of rules in LR^2 in which the left-hand sides are stored top-down and right-hand sides are stored bottom-up. With this representation construction of the right-hand side instance for TGR or its signature for Smaran on a successful match is accomplished more efficiently in a bottom-up manner without any expensive recursions. Finally, we have implemented incremental matching algorithms in LR^2 for enhanced efficiency in theorem proving applications. In such applications, new rules may be created or existing rules deleted, hence we have designed the discrimination tree

Program	TGR (s)	Classes	Smaran (s)	Reductions	Computations
$Fib(800.0)$	–	1604	1.99	1602	3198
$huff(700)$	18.38	167803	22.74	327585	163675
$qsort(500)$	16.01	254004	33.77	628253	250501
$primes(800)$	21.46	326836	39.72	651314	656529

Table 1: Experimental Results

structure so that it can efficiently support deletion and insertion operations.

Dependency lists. The fourth optimization that is crucial to the Smaran subsystem in LR^2 is the handling of dependency lists. To update signatures when a class is unioned with another class, we associated a dependency list with each class in Smaran. All signatures to be updated when this class changes are put on its dependency list. The previous version of Smaran handled dependency lists in an eager manner.

In the latest version we have modified the structure and implemented a lazier algorithm. One optimization is that we identify classes on which potentially large number of signatures can depend, e.g., the classes containing the boolean constants $true$ and $false$, and do not create dependency lists for these classes and ensure that these classes are not merged into any other class. More importantly, we take into account the sizes of the dependency lists to decide which class to union into the other. This turns out to be more efficient than weighted union with the sizes of the classes. We also keep track of a signature's location in the dependency list to avoid subsequent searches.

Memory Allocation and Hashing. LR^2 recycles space using free lists for various data structures and it has its own memory allocation and deallocation routines. Finally, for the hashing of strings and signatures we use improved hashing algorithms that are efficiently computable. Practical studies have shown that we get better distribution of values with fewer collisions with the new algorithms. These involve the use of randomization.

3 Results and Comparisons

Because of space limitations, we present only a few experimental results to illustrate the level of efficiency achieved by LR^2. LR^2 is implemented in C and runs on Sun/DEC workstations. Table 1 summarizes results for four benchmarks: Fib - naive fibonacci, $huff$ - to compute huffman codes for a file containing n distinct characters, $qsort$ - quicksort program for a list of n numbers, and $primes$ - naive sieve for first n prime numbers. Input for primes is $sieve(from(2, 6000), 800)$. Construction of the list $from(2, 6000)$ takes 1.42 seconds for Smaran and 0.42 seconds for TGR and is included in the results. Results for quicksort are for the worst case. Note that history is not useful at all for $huff$, $primes$, and $qsort$. To illustrate the options of LR^2, we ran it with the history and delete options but all functions except a dummy were specified "no memo" for $sieve(from(2, 20000), 2500)$ achieving: reductions 5259141, computations 5276867, number of classes 272643, in 219.64 seconds.

Comparison with other systems. We understand the difficulties of comparing software

Example	Reveal	Reveal+Smaran	Reveal+TGR
exgroup	0.7	0.5	0.3
exgroupl	0.5	0.4	0.4

Table 2: Experimental Results (in seconds) of Comparison with Reveal

systems using experimental results, which can be sensitive to the choice of benchmarks, architectures, etc. However, to illustrate the level of efficiency achieved by LR^2 we compared it with other interpreters. In particular, we compared LR^2 with the ELAN interpreter on several benchmarks including those listed in Table 1 and found that LR^2 with history option is between 15 to 80 times faster for linear list reversal with list sizes ranging from 1000 to 4000, for $primes$ LR^2 (with history) is about 15 times faster for first 500 primes, and similar results were obtained for other programs.[3] We also compared LR^2 with the Reveal (version 1.0) prover written in C. Results for some completion examples distributed with Reveal are summarized in Table 2. Since Reveal does not have built-in arithmetic we tried only a linear list reversal program and found that Reveal took total time of 22.91 seconds to reverse a list of 150 elements versus LR^2's (with history) total timings of 1.2 seconds.

4 Discussion and Future Work

In this paper we have presented LR^2 along with some useful extensions and optimizations. We have developed some ideas on how to statically analyze rewrite systems and determine those functions for which history is likely to be useless and those for which history could be useful [5]. We plan to incorporate these in future versions of LR^2 along with lazier handling of dependency lists, more sophisticated reduction strategy [5], and more flexibility to the user in choosing strategies.

Acknowledgements. We thank K.B. Ramesh, S. Kolli for initial work on Smaran.

References

[1] P. Chew. An improved algorithm for computing with equations. In *Proc. IEEE Symp. on Foundations of Computer Science*, volume 21, pages 108–117, 1980.

[2] M. Hermann, C. Kirchner, and H. Kirchner. Implementations of term rewriting systems. *Computer Journal*, 34(1):20–33, 1991.

[3] Rakesh M. Verma. Smaran: A congruence closure based system for equational computations. In C. Kirchner, ed., *Proc. Rewriting Techniques & Applications*, 1993.

[4] Rakesh M. Verma. A theory of using history for equational systems with applications. *Journal of the ACM*, 42(5):984–1020, 1995. Also in IEEE FOCS proc. 1991.

[5] Rakesh M. Verma. Static analysis for high-performance rewriting. Technical report, University of Houston, 1997.

---

[3] We are aware that the focus of the ELAN project is more on compilation

Decidability for Left-Linear Growing Term Rewriting Systems

Takashi Nagaya and Yoshihito Toyama

School of Information Science, JAIST
Tatsunoguchi, Ishikawa 923-1292, Japan
email : {nagaya, toyama}@jaist.ac.jp

Abstract. A term rewriting system is called growing if each variable occurring both the left-hand side and the right-hand side of a rewrite rule occurs at depth zero or one in the left-hand side. Jacquemard showed that the reachability and the sequentiality of linear (i.e., left-right-linear) growing term rewriting systems are decidable. In this paper we show that Jacquemard's result can be extended to left-linear growing rewriting systems that may have *right-non-linear* rewrite rules. This implies that the reachability and the joinability of some class of right-linear term rewriting systems are decidable, which improves the results for right-ground term rewriting systems by Oyamaguchi. Our result extends the class of left-linear term rewriting systems having a decidable call-by-need normalizing strategy. Moreover, we prove that the termination property is decidable for almost orthogonal growing term rewriting systems.

1   Introduction

The original idea of growing term rewriting systems was introduced by Jacquemard [14] for giving a better sufficient condition for sequential rewriting systems. A term rewriting system is called growing if each variable occurring both the left-hand side and the right-hand side of a rewrite rule occurs at depth zero or one in the left-hand side. Jacquemard [14] proved the preservation of recognizability by linear growing term rewriting systems By using this result, he showed that the reachability and the sequentiality of linear (i.e., left-right-linear) growing term rewriting systems are decidable. Similar decidable properties have been shown in [18, 4, 10, 15].

In this paper we extend Jacquemard's result to left-linear growing term rewriting systems that may have *right-non-linear* rewrite rules. The key idea in our proof is to construct deterministic tree automata instead of non-deterministic ones in [14]. The deterministic behavior of tree automata allows us to remove the right-linear restriction from growing term rewriting systems. This implies that the reachability and the joinability of a term rewriting system $\mathcal{R}$ are decidable if the inverse system $\mathcal{R}^{-1}$ is left-linear growing. This result extends the result by Oyamaguchi [17] that the reachability and the joinability of right-ground term rewriting systems are decidable.

Our result gives a better approximation of term rewriting systems, which extends the class of orthogonal term rewriting systems having a decidable call-by-need strategy [2, 14, 7]. Moreover, we prove that termination for almost orthogonal growing term rewriting systems is decidable. Our proof uses Gramlich's theorem [11] that a weakly innermost normalizing TRS $\mathcal{R}$ is terminating if every critical pair of $\mathcal{R}$ is a trivial overlay. Thus the decidability of termination is proven by showing that the set of all ground terms having normal forms by innermost reduction is recognized by a tree automaton for left-linear growing term rewriting systems.

This paper is organized as follows. Section 2 gives the definitions of term rewriting systems and tree automata. In Section 3, we show the recognizability concerning left-linear growing term rewriting systems. Using this result, Section 4 shows that the reachability and the joinability of right-linear term rewriting systems are decidable if the inverse of them are growing. In Section 5, we extend the class of orthogonal term rewriting systems having a decidable call-by-need strategy. Section 6 proves that termination for almost orthogonal growing term rewriting systems is decidable.

2 Preliminaries

2.1 Term Rewriting Systems

We mainly follow the notation of [1, 6, 16]. Let $\mathcal{F}$ be a finite set of *function symbols* denoted by $f, g, h, \ldots$, and let $\mathcal{V}$ be a countably infinite set of *variables* denoted by $x, y, z, \ldots$, where $\mathcal{F} \cap \mathcal{V} = \phi$. The set of all *terms* built from $\mathcal{F}$ and $\mathcal{V}$ is denoted by $\mathcal{T}(\mathcal{F}, \mathcal{V})$. The set of variables occurring in a term t is denoted by $\mathcal{V}(t)$. Terms not containing variables are called *ground* terms. The set of all ground terms built from $\mathcal{F}$ is denoted by $\mathcal{T}(\mathcal{F})$. A term t is *linear* if every variable in t occurs only once in t. Identity of terms is denoted by $\equiv$.

If p is a position in t then $t|_p$ denotes the *subterm* of t at p. A subterm s of t is *proper* if $s \not\equiv t$. We write $s \subset t$ to indicate that s is a proper subterm of t. $t[s]_p$ denotes the term obtain from t by replacing the subterm $t|_p$ with s. If t has an occurrence of some variable x then we write $x \in t$.

A *substitution* σ is a mapping from $\mathcal{V}$ into $\mathcal{T}(\mathcal{F}, \mathcal{V})$. Substitutions are extended into homomorphisms from $\mathcal{T}(\mathcal{F}, \mathcal{V})$ into $\mathcal{T}(\mathcal{F}, \mathcal{V})$. We write $t\sigma$ instead of $\sigma(t)$. A term s is an *instance* of a term t if there exists a substitution σ such that $s \equiv t\sigma$.

A *term rewriting system* (TRS) $\mathcal{R}$ is a finite set of rewrite rules. A *rewrite rule* is a pair $\langle l, r \rangle$ of terms. (We do not assume that $l \notin \mathcal{V}$ and any variable in r also occurs in l.) We write $l \to r$ for $\langle l, r \rangle$. An instance of the left-hand side of a rewrite rule is a *redex*. The rewrite rules of a TRS $\mathcal{R}$ define a reduction relation $\to_{\mathcal{R}}$ on $\mathcal{T}(\mathcal{F}, \mathcal{V})$ as follow: $t \to_{\mathcal{R}} s$ iff there exist a rewrite rule $l \to r \in \mathcal{R}$, a position p in t and a substitution σ such that $t|_p \equiv l\sigma$ and $s \equiv t[r\sigma]_p$. The transitive-reflexive closure of $\to_{\mathcal{R}}$ is denoted by $\xrightarrow{*}_{\mathcal{R}}$. The inverse relation of $\xrightarrow{*}_{\mathcal{R}}$ is denoted by $\xleftarrow{*}_{\mathcal{R}}$. A *normal form* is a term without redexes. We say that t *has a normal form* if $t \xrightarrow{*}_{\mathcal{R}} s$ for some normal form s. A TRS $\mathcal{R}$ is *terminating* (*strongly normalizing*) if there exists no infinite reduction sequence $t_0 \to_{\mathcal{R}} t_1 \to_{\mathcal{R}} t_2 \to_{\mathcal{R}} \cdots$. A TRS $\mathcal{R}$ is *weakly normalizing* if every term has a normal form.

A rewrite rule $l \to r$ is *ground* (*linear*) if l and r are ground (linear). A rewrite rule $l \to r$ is *left-linear* (*right-linear*) if l (r) is linear. A TRS $\mathcal{R}$ is *ground* (*linear, left-linear, right-linear*) if every rewrite rule in $\mathcal{R}$ is ground (linear, left-linear, right-linear).

Let $l \rightarrow r$ and $l' \rightarrow r'$ be two rules of $\mathcal{R}$. We assume that they are renamed to have no common variables. Suppose that p is a position of l such that $l|_p \notin \mathcal{V}$ and l' are unifiable with a most general unifier σ. Then the pair $\langle l[r']_p\sigma, r'\sigma\rangle$ is called a *critical pair* of $\mathcal{R}$. If $l \rightarrow r$ and $l' \rightarrow r'$ are the same rule, then we do not consider the case $p = \varepsilon$. A critical pair $\langle l[r']_p\sigma, r'\sigma\rangle$ with $p = \varepsilon$ is an *overlay*. A critical pair $\langle t, s\rangle$ is *trivial* if $t \equiv s$. An *orthogonal* TRS is a left-liner TRS without critical pairs. A left-linear TRS is *almost orthogonal* if all its critical pairs are trivial overlays.

2.2 Ω-terms

Let $\mathcal{R}$ be a TRS. We add a new constant Ω to $\mathcal{F}$. The set $\mathcal{T}(\mathcal{F} \cup \{\Omega\}, \mathcal{V})$ is abbreviated to $\mathcal{T}_\Omega$. Elements of $\mathcal{T}_\Omega$ are called Ω-*terms*. We say that an Ω-term t is a *normal form* if t does not contain neither redexes nor Ω's. The set of all normal forms is denoted by $\mathrm{NF}_\mathcal{R}$. t_Ω denotes the Ω-term obtained from t by replacing all variables in t with Ω. The prefix ordering $\leq$ on $\mathcal{T}_\Omega$ is defined as follows: (i) $\Omega \leq t$ for all $t \in \mathcal{T}_\Omega$, (ii) $f(s_1, \ldots, s_n) \leq f(t_1, \ldots, t_n)$ if $s_i \leq t_i$ for any $1 \leq i \leq n$, (iii) $x \leq x$ for all $x \in \mathcal{V}$. Two Ω-terms t and s are compatible, written $t \uparrow s$, if there exists an Ω-term r such that $t \leq r$ and $s \leq r$. In this case the least upper bound of t and s is denoted by $t \sqcup s$.

2.3   Tree Automata

A *tree automaton* is a tuple $\mathcal{A} = (\mathcal{F}, Q, Q^f, \Delta)$ where $\mathcal{F}$ is a finite set of function symbols, Q is a finite set of *states*, $Q^f \subseteq Q$ is a set of *final* states and Δ is a set of ground rewrite rules of the form $f(q_1, \ldots, q_n) \rightarrow q$ or $q \rightarrow q'$ where $f \in \mathcal{F}$, $q_1, \ldots, q_n, q, q' \in Q$. The latter rules are called ϵ-*rules*. We use $\rightarrow_\mathcal{A}$ for the reduction relation $\rightarrow_\Delta$ on $\mathcal{T}(\mathcal{F} \cup Q)$. A term $t \in \mathcal{T}(\mathcal{F})$ is *accepted* by $\mathcal{A}$ if $t \xrightarrow{*}_\mathcal{A} q$ for some $q \in Q_f$. The tree language $L(\mathcal{A})$ recognized by $\mathcal{A}$ is the set of all terms accepted by $\mathcal{A}$. A set L is *recognizable* if there exists a tree automaton $\mathcal{A}$ such that $L = L(\mathcal{A})$. A tree automaton $\mathcal{A}$ is *deterministic* if there are neither ϵ-rules nor different rules with the same left-hand side. A tree automaton $\mathcal{A}$ is *complete* if there is at least one rule $f(q_1, \ldots, q_n) \rightarrow q$ in Δ for all $f \in \mathcal{F}$ and $q_1, \ldots, q_n \in Q$. The following properties of tree automata are well-known [3, 8].

Lemma 1. The class of recognizable tree languages is closed under union, intersection and complementation. $\qquad\square$

Lemma 2. The emptiness problem for tree automata is decidable. $\qquad\square$

Note. In this paper, we regard pairs of terms as rewrite rules without restrictions. Hence the left-hand side of a rewrite rule may be a variable and the right-hand side of a rewrite rule can have a variable not occurring in the left-hand side. This is convenient for approximations of TRSs. Moreover, we consider rewriting on ground terms only. Replacing every variable in terms with a fresh constant, rewriting on non-ground terms can be simulated by that on ground terms. Thus this restriction entails no loss of generality and would simplify matters.

3 Left-Linear Growing TRSs

The definition of growing was given by Jacquemard in [14]. He showed that if $\mathcal{R}$ is a linear growing TRS then the set $\{\, t \in \mathcal{T}(\mathcal{F}) \mid \exists s \in L \;\; t \xrightarrow{*}_{\mathcal{R}} s \,\}$ is recognizable for every recognizable tree language L. In this section we improve this result by replacing *linear growing* with *left-linear growing*.

In the following definition, unlike Jacquemard, we do not assume the linearity for growing TRSs.

Definition 3. A rewrite rule $l \to r$ is *growing* if all variables in $\mathcal{V}(l) \cap \mathcal{V}(r)$ occur at depth 0 or 1 in l. A TRS $\mathcal{R}$ is *growing* if every rewrite rule in $\mathcal{R}$ is growing.

Example 1. Let

$$\mathcal{R} = \begin{cases} f(f(x,y), z) \to f(z, g(z)) \\ g(x) \to f(g(y), z). \end{cases}$$

Then $\mathcal{R}$ is growing. But the following $\mathcal{R}'$ is not growing.

$$\mathcal{R}' = \begin{cases} f(f(x,y), z) \to f(x, g(z)) \\ g(x) \to f(g(y), z). \end{cases}$$

Let R be a binary relation on a set A and let $B \subseteq A$. Then we define $R(B)$ as $\{\, y \in A \mid \exists x \in B \;\; (x, y) \in R \,\}$. Now, we are ready to prove our main result that if $\mathcal{R}$ is a left-linear growing TRS then the set $(\xleftarrow{*}_{\mathcal{R}})(L) = \{\, t \in \mathcal{T}(\mathcal{F}) \mid \exists s \in L \;\; t \xrightarrow{*}_{\mathcal{R}} s \,\}$ is recognizable for every recognizable tree language L.

Let $\mathcal{R}$ be a left-linear growing TRS and let L be a tree language recognized by $\mathcal{A}_L = (\mathcal{F}, Q_L, Q_L^f, \Delta_L)$. We now construct a tree automaton recognizing $(\xleftarrow{*}_{\mathcal{R}})(L)$ from $\mathcal{R}$ and $\mathcal{A}_L$. Let $\mathcal{L} = \{\, l \in \mathcal{T}(\mathcal{F}, \mathcal{V}) \mid l \notin \mathcal{V}, f(\ldots, l, \ldots) \to r \in \mathcal{R} \,\}$. Then every term in $\mathcal{L}$ is linear because of the left-linearity of $\mathcal{R}$. Since the set of all ground instances of a linear term is recognizable [3, 8], we have an automaton $\mathcal{A}_l = (\mathcal{F}, Q_l, Q_l^f, \Delta_l)$ with $L(\mathcal{A}_l) = \{\, l\sigma \mid \sigma : \mathcal{V} \to \mathcal{T}(\mathcal{F}) \,\}$ for each $l \in \mathcal{L}$. Without loss of generality, we assume $Q_a \cap Q_b = \phi$ for any $a, b \in \{L\} \cup \mathcal{L}$ with $a \neq b$. The tree automaton $\mathcal{A}_\cup = (\mathcal{F}, Q_\cup, Q_\cup^f, \Delta_\cup)$ is defined by $Q_\cup = \bigcup_{l \in \mathcal{L}} Q_l \cup Q_L$, $Q_\cup^f = Q_L^f$ and $\Delta_\cup = \bigcup_{l \in \mathcal{L}} \Delta_l \cup \Delta_L$.

Starting from $\mathcal{A}_0 = \mathcal{A}_\cup$, Jacquemard's method in [14] constructs *non-deterministic* tree automata $\mathcal{A}_0, \mathcal{A}_1, \mathcal{A}_2, \ldots$, which can define a *non-deterministic* tree automaton $\mathcal{A}_k$ as $\lim \mathcal{A}_i$ since the number of states is bounded. Then the obtained $\mathcal{A}_k$ accepts $(\xleftarrow{*}_{\mathcal{R}})(L)$ [14]. However this method does not work for *left-liner* growing TRSs. Since the right-hand sides of rewrite rules of left-linear growing TRSs may have multiple occurrences of variables, a subterm in a redex can be duplicated through rewriting. However the *non-deterministic* tree automaton $\mathcal{A}_k$ does not guarantee to reduce the same duplicated subterm to the same state. Thus it cannot trace rewriting by non-right-linear rewrite rules.

The above observation naturally leads us to deterministic tree automata construction for tracing the behavior of left-linear growing TRSs. A naive construction method is to transform an induced non-deterministic automaton into the deterministic automaton at each step in Jacquemard' one [14]. However, this method can not guarantee $\lim \mathcal{A}_i$ because the transformation explodes the number of states, in fact it requires exponentially many states. To prevent this state explosion we carefully construct deterministic tree automata $\mathcal{A}_0, \mathcal{A}_1, \mathcal{A}_2, \ldots \mathcal{A}_k$ as follows, using a fixed set of states.

Let $\mathcal{A}_0 = (\mathcal{F}, Q, Q^f, \Delta_0)$ where $Q = 2^{Q_\cup}$, $Q^f = \{ A \in Q \mid A \cap Q_\cup^f \neq \phi \}$ and Δ_0 contains the following rules:

$$f(A_1, \ldots, A_n) \to A$$
$$\text{if } A = \{ q \in Q_\cup \mid \exists q_1 \in A_1, \ldots, \exists q_n \in A_n \ f(q_1, \ldots, q_n) \xrightarrow{*}_{\mathcal{A}_\cup} q \}.$$

For $0 \leq i \leq k$, $\mathcal{A}_{i+1} = (\mathcal{F}, Q, Q^f, \Delta_{i+1})$ $(\mathcal{A}_k = (\mathcal{F}, Q, Q^f, \Delta_k))$ is obtained from $\mathcal{A}_i$ as follows:

- If there exist $f(A_1, \ldots, A_n) \to A \in \Delta_i$, $l \to r \in \mathcal{R}$ and $A' \in Q$ satisfying the following **Condition 1** or **2**:
  - **Condition 1:**
    1. $l \equiv f(l_1, \ldots, l_n)$,
    2. for each $1 \leq j \leq n$, $l_j \notin \mathcal{V}$ implies $A \cap Q_{l_j}^f \neq \phi$,
    3. there exists $\theta : \mathcal{V} \to Q$ such that
       - (a) $r\theta \xrightarrow{*}_{\mathcal{A}_i} A'$,
       - (b) for each $x \in r$, if $x \equiv l_j$ for some j then $x\theta = A_j$,
         $$\text{otherwise } t \xrightarrow{*}_{\mathcal{A}_i} x\theta \text{ for some } t \in \mathcal{T}(\mathcal{F}),$$
    4. $A \subset A \cup A'$ (i.e., $A \cup A'$ properly includes A),
  - **Condition 2:**
    1'. $l \in \mathcal{V}$,
    2'. there exists $\theta : \mathcal{V} \to Q$ such that
       - (a') $r\theta \xrightarrow{*}_{\mathcal{A}_i} A'$,
       - (b') for each $x \in r$, if $x \equiv l$ then $x\theta = A$,
         $$\text{otherwise } t \xrightarrow{*}_{\mathcal{A}_i} x\theta \text{ for some } t \in \mathcal{T}(\mathcal{F}),$$
    3'. $A \subset A \cup A'$ (i.e., $A \cup A'$ properly includes A),
  then $\Delta_{i+1} = (\Delta_i \backslash \{f(A_1, \ldots, A_n) \to A\}) \cup \{f(A_1, \ldots, A_n) \to A \cup A'\}$.
- Otherwise, $\Delta_k = \Delta_i$.

From 4 of **Condition 1** and 3' of **Condition 2**, it is clear that the process of construction terminates.

Example 2. Let
$$\mathcal{R} = \begin{cases} f(x) \to g(x, x) \\ a \to b. \end{cases}$$

Let $L = \{g(a, b)\}$ and $\mathcal{A}_L = (\mathcal{F}, Q_L, Q_L^f, \Delta_L)$ where $Q_L = \{q_a, q_b, q_f\}$, $Q_L^f = \{q_f\}$ and $\Delta_L = \{a \to q_a, b \to q_b, f(q_a, q_b) \to q_f\}$. Then $f(\{q_a, q_b\}) \to \phi \in \Delta_0$, $f(x) \to g(x, x) \in \mathcal{R}$ and $\{q_f\} \in Q$ satisfy **Condition 1** because we have $g(\{q_a, q_b\}, \{q_a, q_b\}) \to_{\mathcal{A}_0} \{q_f\}$. Thus we can first replace the right-hand side of $f(\{q_a, q_b\}) \to \phi$ with $\{q_f\}$. Next the right-hand side of $a \to \{q_a\} \in \Delta_1$ can be replaced with $\{q_a, q_b\}$. Consequently, we obtain $\Delta_k = \Delta_2$. The term $f(a)$ in $(\xrightarrow{*}_{\mathcal{R}})(L)$ is accepted by $\mathcal{A}_k$ because $f(a) \to_{\mathcal{A}_k} f(\{q_a, q_b\}) \to_{\mathcal{A}_k} \{q_f\} \in Q^f$. Note that $f(a)$ is not accepted by the automaton generated by the method in [14].

In the following we prove that $L(A_k) = (\xleftarrow{*}_{\mathcal{R}})(L)$. We write $t \xrightarrow{*}_{\mathcal{R}} \cdot \xrightarrow{*}_{\mathcal{A}} q$ if $t \xrightarrow{*}_{\mathcal{R}} s \xrightarrow{*}_{\mathcal{A}} q$ for some $s \in T(\mathcal{F})$.

Lemma 4. Let $t \in T(\mathcal{F}, V)$, $\theta : V \to Q$ and $\sigma : V \to T(\mathcal{F})$ such that $x\sigma \xrightarrow{*}_{\mathcal{R}} \cdot \xrightarrow{*}_{\mathcal{A}_\cup} q'$ for any $x \in t$ and $q' \in x\theta$. For each $0 \le i \le k$, if $t\theta \xrightarrow{*}_{\mathcal{A}_i} A \in Q$ then $t\sigma \xrightarrow{*}_{\mathcal{R}} \cdot \xrightarrow{*}_{\mathcal{A}_\cup} q$ for any $q \in A$.

Note. In the above claim the condition "$x\sigma \xrightarrow{*}_{\mathcal{R}} \cdot \xrightarrow{*}_{\mathcal{A}_\cup} q'$ for any $x \in t$ and $q' \in x\theta$" cannot be replaced with a simpler form "$x\sigma \xrightarrow{*}_{\mathcal{R}} \cdot \xrightarrow{*}_{\mathcal{A}_0} x\theta$ for any $x \in t$", because the first condition means $\forall x \in t, \forall q' \in x\theta, \exists s \in T(\mathcal{F})[x\sigma \xrightarrow{*}_{\mathcal{R}} s \xrightarrow{*}_{\mathcal{A}_\cup} q']$ but the second one means $\forall x \in t, \exists s \in T(\mathcal{F}), \forall q' \in x\theta[x\sigma \xrightarrow{*}_{\mathcal{R}} s \xrightarrow{*}_{\mathcal{A}_\cup} q']$, which is different from the first one.

Proof. We prove the lemma by induction on i.
   Base step. We use induction on the structure of t. The case $t \equiv x$ is trivial. Let $t \equiv f(t_1, \ldots, t_n)$. Assume $t\theta \equiv f(t_1, \ldots, t_n)\theta \xrightarrow{*}_{\mathcal{A}_0} f(A_1, \ldots, A_n) \to_{\mathcal{A}_0} A$. Let $q \in A$. Then by the definition of Δ_0 there exist $q_1 \in A_1, \ldots, q_n \in A_n$ such that $f(q_1, \ldots, q_n) \xrightarrow{*}_{\mathcal{A}_\cup} q$. By induction hypothesis, for each $1 \le j \le n$ there exists s_j such that $t_j\sigma \xrightarrow{*}_{\mathcal{R}} s_j \xrightarrow{*}_{\mathcal{A}_\cup} q_j$. Thus we have $t\sigma \equiv f(t_1\sigma, \ldots, t_n\sigma) \xrightarrow{*}_{\mathcal{R}} f(s_1, \ldots, s_n) \xrightarrow{*}_{\mathcal{A}_\cup} f(q_1, \ldots, q_n) \xrightarrow{*}_{\mathcal{A}_\cup} q$.
   Induction step. Let $f(A_1, \ldots, A_n) \to A' \in \Delta_i \backslash \Delta_{i-1}$. We use induction on the number m of reduction steps using this rule in the reduction $t\theta \xrightarrow{*}_{\mathcal{A}_i} A$. If $m = 0$ then $t\theta \xrightarrow{*}_{\mathcal{A}_{i-1}} A$. Thus it follows from induction hypothesis on i that $t\sigma \xrightarrow{*}_{\mathcal{R}} \cdot \xrightarrow{*}_{\mathcal{A}_\cup} q$ for any $q \in A$. Let $m > 0$. Suppose

$$t\theta \equiv t\theta[f(t_1, \ldots, t_n)\theta]_p \xrightarrow{*}_{\mathcal{A}_{i-1}} t\theta[f(A_1, \ldots, A_n)]_p \to_{\mathcal{A}_i} t\theta[A']_p \xrightarrow{*}_{\mathcal{A}_i} A.$$

Let $\tilde{t} \equiv t[z]_p$ where $z \notin t$. We define $\tilde\theta : V \to Q$ and $\tilde\sigma : V \to T(\mathcal{F})$ as follows: if $x \equiv z$ then $x\tilde\theta = A'$ and $x\tilde\sigma \equiv f(t_1, \ldots, t_n)\sigma$, otherwise $x\tilde\theta \equiv x\theta$ and $x\tilde\sigma \equiv x\sigma$. Clearly $\tilde{t}\tilde\theta \equiv t\theta[A']_p$ and $\tilde{t}\tilde\sigma \equiv t\sigma$. We will show the following claim:

$$x\tilde\sigma \xrightarrow{*}_{\mathcal{R}} \cdot \xrightarrow{*}_{\mathcal{A}_\cup} q \text{ for any } x \in \tilde{t} \text{ and } q \in x\tilde\theta.$$

Then by applying induction hypothesis on m to $\tilde{t}\tilde\theta \equiv t\theta[A']_p \xrightarrow{*}_{\mathcal{A}_i} A$, we can obtain $\tilde{t}\tilde\sigma \equiv t\sigma \xrightarrow{*}_{\mathcal{R}} \cdot \xrightarrow{*}_{\mathcal{A}_\cup} q$ for any $q \in A$. Thus the lemma holds.
   Proof of the claim. Let $x \in \tilde{t}$. If $x \not\equiv z$ then it follows from the assumption of the lemma that $x\tilde\sigma \xrightarrow{*}_{\mathcal{R}} \cdot \xrightarrow{*}_{\mathcal{A}_\cup} q$ for any $q \in x\tilde\theta$. We consider the case $x \equiv z$. Assume that $f(A_1, \ldots, A_n) \to A'_1 \in \Delta_{i-1}$, $l \to r \in \mathcal{R}$ and $A'_2 \in Q$ satisfy **Condition 1** or **2** and $A' = A'_1 \cup A'_2$. Since $f(t_1, \ldots, t_n)\theta \xrightarrow{*}_{\mathcal{A}_{i-1}} f(A_1, \ldots, A_n) \to_{\mathcal{A}_{i-1}} A'_1$, it follows from induction hypothesis on i that

$$f(t_1, \ldots, t_n)\sigma \xrightarrow{*}_{\mathcal{R}} \cdot \xrightarrow{*}_{\mathcal{A}_\cup} q \text{ for any } q \in A'_1. \tag{1}$$

We distinguish two cases.
   Case 1. **Condition 1** is satisfied. Let $l \equiv f(l_1, \ldots, l_n)$. By applying induction hypothesis on i to $t_j\theta \xrightarrow{*}_{\mathcal{A}_{i-1}} A_j$ for $1 \le j \le n$, we obtain $t_j\sigma \xrightarrow{*}_{\mathcal{R}} \cdot \xrightarrow{*}_{\mathcal{A}_\cup} q$ for

any $q \in A_j$. For each $1 \leq j \leq n$, let s_j be a term such that if $l_j \in \mathcal{V}$ then $s_j \equiv t_j\sigma$, otherwise $t_j\sigma \xrightarrow{*}_{\mathcal{R}} s_j \xrightarrow{*}_{\mathcal{A}_\cup} q \in Q^f_{l_j}$. From the disjointness of the sets of states, $s_j \xrightarrow{*}_{\mathcal{A}_\cup} q \in Q^f_{l_j}$ implies $s_j \xrightarrow{*}_{\mathcal{A}_{l_j}} q \in Q^f_{l_j}$. Hence $f(s_1, \ldots, s_n)$ is an instance of l by the linearity of l. Let $\theta' : \mathcal{V} \to Q$ be a substitution defined by 3 of **Condition 1**. Let $\sigma' : \mathcal{V} \to \mathcal{T}(\mathcal{F})$ be a substitution such that for any $y \in r$ if $y \equiv l_j$ for some j then $y\sigma' \equiv s_j$, otherwise $y\sigma' \xrightarrow{*}_{\mathcal{A}_{i-1}} y\theta'$. Then from the growingness of $\mathcal{R}$ we have the reduction $f(s_1, \ldots, s_n) \to_{\mathcal{R}} r\sigma'$. Furthermore, we can see $y\sigma' \xrightarrow{*}_{\mathcal{A}_{i-1}} y\theta'$ for any $y \in r$. Therefore, by induction hypothesis on i, $y\sigma' \xrightarrow{*}_{\mathcal{R}} \cdot \xrightarrow{*}_{\mathcal{A}_\cup} q$ for any $y \in r$ and $q \in y\theta'$. Applying induction hypothesis on i to $r\theta' \xrightarrow{*}_{\mathcal{A}_{i-1}} A'_2$, it is obtained that $r\sigma' \xrightarrow{*}_{\mathcal{R}} \cdot \xrightarrow{*}_{\mathcal{A}_\cup} q$ for any $q \in A'_2$. Thus, since $f(t_1, \ldots, t_n)\sigma \xrightarrow{*}_{\mathcal{R}} r\sigma'$, we have

$$f(t_1, \ldots, t_n)\sigma \xrightarrow{*}_{\mathcal{R}} \cdot \xrightarrow{*}_{\mathcal{A}_\cup} q \quad \text{for any } q \in A'_2. \tag{2}$$

Because $z\tilde{\theta} = A' = A'_1 \cup A'_2$ and $z\tilde{\sigma} \equiv f(t_1, \ldots, t_n)\sigma$, it follows from (1) and (2) that $z\tilde{\sigma} \xrightarrow{*}_{\mathcal{R}} \cdot \xrightarrow{*}_{\mathcal{A}_\cup} q$ for any $q \in z\tilde{\theta}$. Therefore the claim holds.

Case 2. **Condition 2** is satisfied. Let $\theta' : \mathcal{V} \to Q$ be a substitution defined by 2' of **Condition 2**. Let $\sigma' : \mathcal{V} \to \mathcal{T}(\mathcal{F})$ be a substitution such that for any $y \in r$ if $y \equiv l$ then $y\sigma' \equiv f(t_1, \ldots, t_n)\sigma$, otherwise $y\sigma' \xrightarrow{*}_{\mathcal{A}_{i-1}} y\theta'$. Using (1) and induction hypothesis on i, we obtain $y\sigma' \xrightarrow{*}_{\mathcal{R}} \cdot \xrightarrow{*}_{\mathcal{A}_\cup} q$ for any $y \in r$ and $q \in y\theta'$. Applying induction hypothesis on i to $r\theta' \xrightarrow{*}_{\mathcal{A}_{i-1}} A'_2$, it is obtained that $r\sigma' \xrightarrow{*}_{\mathcal{R}} \cdot \xrightarrow{*}_{\mathcal{A}_\cup} q$ for any $q \in A'_2$. Since $f(t_1, \ldots, t_n)\sigma \to_{\mathcal{R}} r\sigma'$,

$$f(t_1, \ldots, t_n)\sigma \xrightarrow{*}_{\mathcal{R}} \cdot \xrightarrow{*}_{\mathcal{A}_\cup} q \quad \text{for any } q \in A'_2. \tag{3}$$

Therefore, it follows from (1) and (3) that $z\tilde{\sigma} \xrightarrow{*}_{\mathcal{R}} \cdot \xrightarrow{*}_{\mathcal{A}_\cup} q$ for any $q \in z\tilde{\theta}$. Hence the claim holds. $\square$

Lemma 5. $L(\mathcal{A}_k) \subseteq (\xleftarrow{*}_{\mathcal{R}})(L)$.

Proof. Let $t \in L(\mathcal{A}_k)$ i.e., $t \xrightarrow{*}_{\mathcal{A}_k} A$ for some $A \in Q^f$. By the definition of Q^f, A has a final state q of $\mathcal{A}_L$. From Lemma 4, there exists $s \in \mathcal{T}(\mathcal{F})$ such that $t \xrightarrow{*}_{\mathcal{R}} s \xrightarrow{*}_{\mathcal{A}_\cup} q$. By the disjointness of the sets of states, we have $s \xrightarrow{*}_{\mathcal{A}_L} q \in Q^f_L$. Thus $t \in (\xleftarrow{*}_{\mathcal{R}})(L)$. $\square$

Lemma 6. Let $t \in \mathcal{T}(\mathcal{F}, \mathcal{V})$. Let $\theta, \theta' : \mathcal{V} \to Q$ with $x\theta \subseteq x\theta'$ for any $x \in t$. If $t\theta \xrightarrow{*}_{\mathcal{A}_i} A \in Q$ then $t\theta' \xrightarrow{*}_{\mathcal{A}_k} A'$ for some $A' \in Q$ with $A \subseteq A'$.

Proof. We prove the lemma by induction on i.

Base step. We use the induction on the structure of t. The case $t \equiv x$ is trivial. Let $t \equiv f(t_1, \ldots, t_n)$. Then we assume $f(t_1, \ldots, t_n)\theta \xrightarrow{*}_{\mathcal{A}_0} f(A_1, \ldots, A_n) \to_{\mathcal{A}_0} A \in Q$. By induction hypothesis, for each $1 \leq j \leq n$ there exists $A'_j \in Q$ such that $t_j\theta' \xrightarrow{*}_{\mathcal{A}_k} A'_j$ and $A_j \subseteq A'_j$. By the definition of $\mathcal{A}_0$, Δ_0 has a rule $f(A'_1, \ldots, A'_n) \to A'$ with $A \subseteq A'$. Then by the construction of $\mathcal{A}_k$, Δ_k has a

rule $f(A'_1, \ldots, A'_n) \to A''$ with $A' \subseteq A''$. Thus we obtain $f(t_1, \ldots, t_n)\theta' \xrightarrow{*}_{\mathcal{A}_k}$
$f(A'_1, \ldots, A'_n) \to_{\mathcal{A}_k} A''$ and $A \subseteq A''$.

Induction step. We use the induction on the structure of t. The case $t \equiv x$
is trivial. Let $t \equiv f(t_1, \ldots, t_n)$. Assume $f(t_1, \ldots, t_n)\theta \xrightarrow{*}_{\mathcal{A}_i} f(A_1, \ldots, A_n) \to_{\mathcal{A}_i}$
$A \in Q$. By induction hypothesis on the structure of t, for each $1 \leq j \leq n$ there
exists $A'_j \in Q$ such that $t_j\theta' \xrightarrow{*}_{\mathcal{A}_k} A'_j$ and $A_j \subseteq A'_j$. Since $\mathcal{A}_k$ is deterministic and
complete, there exists exactly one $A' \in Q$ such that $f(A'_1, \ldots, A'_n) \to A' \in \Delta_k$.
We will show $A \subseteq A'$. If $f(A_1, \ldots, A_n) \to A \in \Delta_{i-1}$ then from induction hypothesis on i it follows that $A \subseteq A'$. Otherwise, we assume that $f(A_1, \ldots, A_n) \to$
$B_1 \in \Delta_{i-1}$, $l \to r \in \mathcal{R}$ and $B_2 \in Q$ satisfy **Condition 1** or **2** and $A = B_1 \cup B_2$.
From induction hypothesis on i, we get $B_1 \subseteq A'$. We distinguish two cases.

Case 1. **Condition 1** is satisfied. Let $l \equiv f(l_1, \ldots, l_n)$ and let $\theta_1 : \mathcal{V} \to Q$
be a substitution defined by 3 of **Condition 1**. Then let θ_2 be a substitution
from $\mathcal{V}$ to Q such that for every $x \in r$ if $x \equiv l_j$ then $x\theta_2 = A'_j$, otherwise
$t \xrightarrow{*}_{\mathcal{A}_k} x\theta_2$ for some $t \in \mathcal{T}(\mathcal{F})$ with $t \xrightarrow{*}_{\mathcal{A}_{i-1}} x\theta_1$. Using induction hypothesis on
i, we can show that $x\theta_1 \subseteq x\theta_2$ for every $x \in r$. Applying induction hypothesis
on i to $r\theta_1 \xrightarrow{*}_{\mathcal{A}_{i-1}} B_2$, we obtain $r\theta_2 \xrightarrow{*}_{\mathcal{A}_k} B'_2$ for some $B'_2 \in Q$ with $B_2 \subseteq B'_2$.
Therefore $f(A'_1, \ldots, A'_n) \to A' \in \Delta_k$, $l \to r \in \mathcal{R}$ and $B'_2 \in Q$ satisfy $1, 2$
and 3 of **Condition 1**. By the construction of $\mathcal{A}_k$, they must not satisfy 4 of
Condition 1. Thus we have $A' = A' \cup B'_2$. Hence $A = B_1 \cup B_2 \subseteq A' \cup B'_2 = A'$.

Case 2. **Condition 2** is satisfied. Let $\theta_1 : \mathcal{V} \to Q$ be a substitution defined
by $2'$ of **Condition 2**. Then let $\theta_2 : \mathcal{V} \to Q$ be a substitution such that for every
$x \in r$ if $x \equiv l$ then $x\theta_2 = A'$, otherwise $t \xrightarrow{*}_{\mathcal{A}_k} x\theta_2$ for some $t \in \mathcal{T}(\mathcal{F})$ with
$t \xrightarrow{*}_{\mathcal{A}_{i-1}} x\theta_1$. Using induction hypothesis on i, we can show that $y\theta_1 \subseteq y\theta_2$ for
every $y \in r$. Applying induction hypothesis on i to $r\theta_1 \xrightarrow{*}_{\mathcal{A}_{i-1}} B_2$, we obtain
$r\theta_2 \xrightarrow{*}_{\mathcal{A}_k} B'_2$ for some $B'_2 \in Q$ with $B_2 \subseteq B'_2$. Thus $f(A'_1, \ldots, A'_n) \to A' \in \Delta_k$,
$l \to r \in \mathcal{R}$ and $B'_2 \in Q$ satisfy $1'$ and $2'$ of **Condition 2**. By the construction
of $\mathcal{A}_k$ they must not satisfy $3'$ of **Condition 2**, i.e., $A' = A' \cup B'_2$. Hence $A =$
$B_1 \cup B_2 \subseteq A' \cup B'_2 = A'$. □

Lemma 7. Let $t \in \mathcal{T}(\mathcal{F})$ and $t \xrightarrow{*}_{\mathcal{A}_k} A \in Q$. If $t \xrightarrow{*}_{\mathcal{A}_\cup} q \in Q_\cup$ then $q \in A$.

Proof. Since $\mathcal{A}_0$ is complete, there exists $A' \in Q$ such that $t \xrightarrow{*}_{\mathcal{A}_0} A'$. By induction of the structure of t, we can show that $A' = \{ q \in Q_\cup \mid t \xrightarrow{*}_{\mathcal{A}_\cup} q \}$. Thus,
if $t \xrightarrow{*}_{\mathcal{A}_\cup} q \in Q_\cup$ then $q \in A'$. Because $\mathcal{A}_k$ is deterministic, we get $A' \subseteq A$ by
Lemma 6. Hence $q \in A$. □

Lemma 8. $L(\mathcal{A}_k) \supseteq (\xleftarrow{*}_{\mathcal{R}})(L)$.

Proof. Assume that $t \xrightarrow{*}_{\mathcal{R}} s$ for some $s \in L$. We show that $t \in L(\mathcal{A}_k)$ by
induction on the length m of this reduction. If $m = 0$ then $t \in L$. Thus $t \xrightarrow{*}_{\mathcal{A}_\cup} q$
for some $q \in Q^f_L$. Since $\mathcal{A}_k$ is complete, there exists $A \in Q$ such that $t \xrightarrow{*}_{\mathcal{A}_k} A$.
According to Lemma 7, $q \in A$ and therefore $A \in Q^f$. Hence $t \in L(\mathcal{A}_k)$. Let
$m > 0$. Then we assume that

$$t \equiv t[l\sigma]_p \to_{\mathcal{R}} t[r\sigma]_p \xrightarrow{*}_{\mathcal{R}} s \in L$$

with $l \to r \in \mathcal{R}$. By induction hypothesis, $t[r\sigma]_p$ is accepted by $\mathcal{A}_k$. Since $\mathcal{A}_k$ is deterministic, there exists $\theta : \mathcal{V} \to Q$ such that

$$t[r\sigma]_p \xrightarrow{*}_{\mathcal{A}_k} t[r\theta]_p \xrightarrow{*}_{\mathcal{A}_k} t[A]_p \xrightarrow{*}_{\mathcal{A}_k} B \in Q^f$$

where $A \in Q$. By completeness of $\mathcal{A}_k$, we assume that

$$t \equiv t[f(t_1, \ldots, t_n)]_p \xrightarrow{*}_{\mathcal{A}_k} t[f(A_1, \ldots, A_n)]_p \to_{\mathcal{A}_k} t[A']_p \xrightarrow{*}_{\mathcal{A}_k} B' \in Q$$

where $f(A_1, \ldots, A_n) \to A' \in \Delta_k$ and $n \geq 0$. We consider the following two cases.

Case 1. $l \equiv f(l_1, \ldots, l_n)$. If $l_j \notin \mathcal{V}$ then t_j is accepted by $\mathcal{A}_{l_j}$ and thus A_j has $q \in Q^f_{l_j}$ by Lemma 7. Because $\mathcal{A}_k$ is deterministic, for any $x \in r$, $x \equiv l_j$ implies $x\theta \equiv A_j$. Therefore $f(A_1, \ldots, A_n) \to A' \in \Delta_k$, $l \to r \in \mathcal{R}$ and $A \in Q$ fulfill 1, 2 and 3 of **Condition 1**. By the construction of $\mathcal{A}_k$, they must not satisfy 4 of **Condition 1**. Thus $A \subseteq A'$. Since Lemma 6 yields $B \subseteq B'$, we obtain $B' \in Q^f$. Therefore $t \in L(\mathcal{A}_k)$.

Case 2. $l \equiv x$ for some $x \in \mathcal{V}$. Because $\mathcal{A}_k$ is deterministic, if $x \in r$ then $x\theta \equiv A'$. Therefore $f(A_1, \ldots, A_n) \to A' \in \Delta_k$, $l \to r \in \mathcal{R}$ and $A \in Q$ fulfill $1'$ and $2'$ of **Condition 2**. By the construction of $\mathcal{A}_k$, they must not satisfy $3'$ of **Condition 2** and thus $A \subseteq A'$. According to Lemma 6, $B \subseteq B'$ and therefore $B' \in Q^f$. Hence $t \in L(\mathcal{A}_k)$. $\qquad \square$

Thus we obtain the following theorem.

Theorem 9. Let $\mathcal{R}$ be a left-linear growing TRS and let L be a recognizable tree language. Then the set $(\xleftarrow{*}_{\mathcal{R}})(L)$ is recognized by a tree automaton. $\qquad \square$

If $\mathcal{R}$ is left-linear TRS then $\mathrm{NF}_{\mathcal{R}}$ is a recognizable set. Thus, from Theorem 9 the set $(\xleftarrow{*}_{\mathcal{R}})(\mathrm{NF}_{\mathcal{R}})$ is recognizable for a left-linear growing $\mathcal{R}$. This means that the weakly normalizing property of left-linear growing TRSs is decidable.

4 Reachability and Joinability

The reachability problem for $\mathcal{R}$ is the problem of deciding whether $t \xrightarrow{*}_{\mathcal{R}} s$ given two terms t and s. It is well-known that this problem is undecidable for general TRSs. Oyamaguchi [17] has shown that this problem is decidable for right-ground TRSs. Decidability for linear growing TRSs was shown by Jacquemard [14]. Since a singleton set of a term is recognizable, we can extend these results by using Theorem 9.

Theorem 10. The reachability problem for left-linear growing TRSs is decidable. $\qquad \square$

For a TRS $\mathcal{R}$, we define $\mathcal{R}^{-1}$ by $\mathcal{R}^{-1} = \{ r \to l \mid l \to r \in \mathcal{R} \}$. Clearly $t \xrightarrow{*}_{\mathcal{R}} s$ iff $s \xrightarrow{*}_{\mathcal{R}^{-1}} t$. By Theorem 10, we obtain the following theorem.

Theorem 11. Let $\mathcal{R}$ be a TRS such that $\mathcal{R}^{-1}$ is left-linear and growing. The reachability problem for $\mathcal{R}$ is decidable. $\qquad \square$

If $\mathcal{R}$ is right-ground TRS then $\mathcal{R}^{-1}$ is left-linear and growing. Thus, the above theorem is a generalization of Oyamaguchi's result.

The joinability problem for $\mathcal{R}$ is the problem of deciding given finite number of terms $t_1, \ldots, t_n$, whether there exists a term s such that $t_i \xrightarrow{*} s$ for any $1 \leq i \leq n$. Oyamaguchi [17] has shown that this problem is decidable for right-ground TRSs. This result is extended as follows.

Theorem 12. Let $\mathcal{R}$ be a TRS such that $\mathcal{R}^{-1}$ is left-linear and growing. The joinability problem for $\mathcal{R}$ is decidable.

Proof. Let $t_1, \ldots, t_n$ be terms. Then $t_1, \ldots, t_n$ are joinable iff

$$(\xrightarrow{*}_{\mathcal{R}})(\{t_1\}) \cap \cdots \cap (\xrightarrow{*}_{\mathcal{R}})(\{t_n\}) \neq \phi.$$

By Theorem 9, $(\xrightarrow{*}_{\mathcal{R}})(\{t_i\}) = (\xleftarrow{*}_{\mathcal{R}^{-1}})(\{t_i\})$ is recognizable for any $1 \leq i \leq n$. Thus from Lemmas 1 and 2 the theorem follows. $\qquad\square$

5 Decidable Approximations

The recognizability result of Section 3 gives us better decidable approximations of TRSs than Jacquemard's ones in [14].

A TRS $\mathcal{R}'$ is an *approximation* of a TRS $\mathcal{R}$ if $\xrightarrow{*}_{\mathcal{R}} \subseteq \xrightarrow{*}_{\mathcal{R}'}$. An *approximation mapping* τ is a mapping from TRSs to TRSs such that $\tau(\mathcal{R})$ is an approximation of $\mathcal{R}$ for every TRS $\mathcal{R}$.

Definition 13. Let $\mathcal{R} = \{\, l_i \rightarrow r_i \mid 1 \leq i \leq n \,\}$ be a TRS. A *left-linear growing approximation* of $\mathcal{R}$ is a left-linear growing TRS $\{\, l_i' \rightarrow r_i \mid 1 \leq i \leq n \,\}$ where for any $1 \leq i \leq n$, l_i' is obtained from l_i by replacing variables which do not match both conditions of left-linearity and growingness with fresh variables. An approximation mapping τ is *left-linear growing* if $\tau(\mathcal{R})$ is a left-linear growing approximation of $\mathcal{R}$ for every TRS $\mathcal{R}$.

If $\mathcal{R}$ is a left-linear growing TRS then the left-linear growing approximation of $\mathcal{R}$ is $\mathcal{R}$ itself. If τ is a left-linear growing mapping then $\mathrm{NF}_{\mathcal{R}} = \mathrm{NF}_{\tau(\mathcal{R})}$ for every left-linear TRS $\mathcal{R}$.

Example 3. Let

$$\mathcal{R} = \begin{cases} f(g(x), y) \rightarrow f(x, f(y, x)) \\ g(x) \rightarrow f(x, x). \end{cases}$$

Then a left-linear growing approximation of $\mathcal{R}$ is

$$\mathcal{R}' = \begin{cases} f(g(z), y) \rightarrow f(x, f(y, x)) \\ g(x) \rightarrow f(x, x). \end{cases}$$

Jacquemard [14] introduced *linear* (i.e., *left-right-linear*) *growing approximation mappings* which return a *linear growing approximation* of $\mathcal{R}$ for every TRS $\mathcal{R}$. Here a linear growing approximation of $\mathcal{R}$ is a *linear* growing TRS obtained from $\mathcal{R}$ by replacing variables of rewrite rules. For example, the TRS $\{\, f(g(z), y) \to f(x, f(y, w)),\ g(x) \to f(x, y) \,\}$ is a linear growing approximation of $\mathcal{R}$ of Example 3. Our approximations are better than Jacquemard's ones because they are not assumed to be right-linear. See [9] for another approximations technique based on tree automata induced by TRSs.

Durand and Middeldorp [7] studied approximations of TRSs to call-by-need computations. They presented the framework for decidable call-by-need computations without notions of index and sequentiality. The following notions originate from [7]. Let τ be an approximation mapping. The redex at a position p in $t \in \mathcal{T}(\mathcal{F})$ is τ-*needed* if there exists no $s \in \mathrm{NF}_{\mathcal{R}}$ such that $t[\Omega]_p \xrightarrow{*}_{\tau(\mathcal{R})} s$. Note that a normal form s does not contain Ω's. It is well-known that every τ-needed redex is *needed* defined by Huet and Lévy [13] if $\mathcal{R}$ is an orthogonal TRS. The class $\mathcal{C}_\tau$ of TRSs is defined as follows: $\mathcal{R} \in \mathcal{C}_\tau$ iff every term not in normal form has a τ-needed redex.

Theorem 14. Let $\mathcal{R}$ be a left-linear TRS. Let τ be a left-linear growing approximation mapping.

(1) It is decidable whether a redex in a term is τ-needed.
(2) It is decidable whether $\mathcal{R} \in \mathcal{C}_\tau$.

Proof. The set $\mathrm{NF}_{\mathcal{R}}$ is recognizable since $\mathcal{R}$ is left-linear. Using Theorem 9, we obtain the recognizability of the set $\{\, t \in \mathcal{T}(\mathcal{F} \cup \{\Omega\}) \mid \exists s \in \mathrm{NF}_{\mathcal{R}}\ t \xrightarrow{*}_{\tau(\mathcal{R})} s \,\}$. Therefore, the theorem follows from Theorems 15 and 29 in [7]. $\qquad\square$

Let $\mathcal{R}$ be an orthogonal TRS satisfying the variable restriction that $l \notin \mathcal{V}$ and $\mathcal{V}(r) \subseteq \mathcal{V}(l)$ for every $l \to r \in \mathcal{R}$. If $\tau(\mathcal{R}) = \mathcal{R}$ then τ-neededness coincides with neededness [7]. It was also shown by Huet and Lévy [13] that every term not in normal from has a needed redex. Thus we have the following corollary.

Corollary 15. Let $\mathcal{R}$ be an orthogonal growing TRS satisfying the variable restriction. Then the neededness is decidable and we have $\mathcal{R} \in \mathcal{C}_\tau$ for every left-linear growing approximation mapping τ. $\qquad\square$

The following theorem shows that left-linear growing approximations extend the class of orthogonal TRSs having a decidable call-by-need normalizing strategy.

Theorem 16. Let τ be a left-liner growing approximation mapping and let τ' be a linear growing approximation mapping. Then $\mathcal{C}_{\tau'} \subset \mathcal{C}_\tau$ even if these classes are restricted to orthogonal TRSs.

Proof. For every TRS $\mathcal{R}$, τ'-neededness implies τ-neededness because we have $\xrightarrow{*}_{\tau(\mathcal{R})} \subseteq \xrightarrow{*}_{\tau'(\mathcal{R})}$. Thus $\mathcal{C}_{\tau'} \subseteq \mathcal{C}_\tau$. Let $\mathcal{R} = \{g(x) \to f(x, x, x)\} \cup \mathcal{R}'$ where $\mathcal{R}' = \{f(a, b, x) \to a, f(b, x, a) \to a, f(x, a, b) \to b\}$. From Corollary 15 we have $\mathcal{R} \in \mathcal{C}_\tau$. We will show that $\mathcal{R} \notin \mathcal{C}_{\tau'}$. If $\tau'(\mathcal{R}) = \{g(x) \to f(y, z, x)\} \cup \mathcal{R}'$ then

$g(b) \xrightarrow{*}_{\tau'(\mathcal{R})} a$ and $g(b) \xrightarrow{*}_{\tau'(\mathcal{R})} b$. Therefore, the term $f(g(b), g(b), g(b))$ does not have τ'-needed redexes. Similarly, we can show that $f(g(a), g(a), g(a))$ does not have τ'-needed redexes for other linear growing approximations of $\mathcal{R}$. Hence $\mathcal{R} \notin \mathcal{C}_{\tau'}$. $\qquad \square$

6 Termination of Almost Orthogonal Growing TRSs

Termination is decidable for ground TRSs [12], right-ground TRSs [5] and right-linear monadic TRSs [19]. In this section, we show that termination of almost orthogonal growing TRSs is decidable. If a TRS $\mathcal{R}$ contains a rewrite rule which does not satisfy the variable restriction then $\mathcal{R}$ is not terminating. Thus we may assume that $\mathcal{R}$ satisfies the variable restriction. We first explain the theorem of Gramlich [11], which is used in our proof.

A reduction $t \rightarrow_{\mathcal{R}} s$ by applying a rule at position p is *innermost* if every proper subterm of $t|_p$ is normal form. The innermost reduction is denoted by $\rightarrow_{\mathcal{I}}$. We say that a term t is *weakly innermost normalizing* if $t \xrightarrow{*}_{\mathcal{I}} s$ for some normal form s. A TRS $\mathcal{R}$ is *weakly innermost normalizing* if every term t is weakly innermost normalizing.

Theorem 17 [11]. Let $\mathcal{R}$ be a TRS such that every critical pair of $\mathcal{R}$ is a trivial overlay.

(a) $\mathcal{R}$ is terminating iff $\mathcal{R}$ is weakly innermost normalizing.
(b) For any term t, t is terminating iff t is weakly innermost normalizing. $\quad \square$

According to Theorem 17, if we can prove the decidability of weakly innermost normalizing then termination is decidable. We show that the set of all ground terms being weakly innermost normalizing is recognizable. From here on we assume that $\mathcal{R}$ is a left-linear growing TRS.

We must construct a tree automaton which recognizes the set of all ground terms being weakly innermost normalizing. We start with the deterministic and complete tree automaton $\mathcal{A}_{NF}$ by Comon [2] which accepts ground normal forms. The set $\mathcal{S}_{\mathcal{R}}$ is defined as follows: $\mathcal{S}_{\mathcal{R}} = \{ t \in \mathcal{T}_{\Omega} \mid t \sqsubset l_{\Omega}, l \rightarrow r \in \mathcal{R} \}$. $\mathcal{S}_{\mathcal{R}}^*$ is the smallest set such that $\mathcal{S}_{\mathcal{R}} \subseteq \mathcal{S}_{\mathcal{R}}^*$ and if $t, s \in \mathcal{S}_{\mathcal{R}}^*$ and $t \uparrow s$ then $t \sqcup s \in \mathcal{S}_{\mathcal{R}}^*$. $\mathcal{A}_{NF} = (\mathcal{F}, Q_{NF}, Q_{NF}^f, \Delta_{NF})$ is defined by $Q_{NF} = \{ q_t \mid t \in \mathcal{S}_{\mathcal{R}}^*$ and t does not contain redexes $\} \cup \{q_{\Omega}, q_{\text{red}}\}$, $Q_{NF}^f = Q_{NF} \backslash \{q_{\text{red}}\}$ and Δ_{NF} consists of the following rules:

- $f(q_{t_1}, \ldots, q_{t_n}) \rightarrow q_t$
  if $f(t_1, \ldots, t_n)$ is not a redex and
  t is maximal Ω-term w.r.t. $\leq$ such that $t \leq f(t_1, \ldots, t_n)$ and $q_t \in Q_{NF}^f$,
- $f(q_{t_1}, \ldots, q_{t_n}) \rightarrow q_{\text{red}}$ if $f(t_1, \ldots, t_n)$ is a redex,
- $f(q_1, \ldots, q_n) \rightarrow q_{\text{red}}$ if $q_{\text{red}} \in \{q_1, \ldots, q_n\}$.

The following lemma shows that $\mathcal{A}_{NF}$ recognizes the set of ground normal forms.

Lemma 18 [2]. Let $t \in \mathcal{T}(\mathcal{F})$.

(i) $\mathcal{A}_{NF}$ is deterministic and complete.

(ii) If $t \xrightarrow{*}_{\mathcal{A}_{NF}} q_s \in Q_{NF}^f$ then t is a normal form, $s \leq t$ and $u \leq s$ for any $q_u \in Q_{NF}^f$ with $u \leq t$.

(iii) If $t \xrightarrow{*}_{\mathcal{A}_{NF}} q_{\text{red}}$ then t is not a normal form. $\qquad\qquad\square$

We inductively construct tree automata $\mathcal{A}_0, \mathcal{A}_1, \ldots, \mathcal{A}_k$ as follows. Let $\mathcal{A}_0 = (\mathcal{F}, Q, Q^f, \Delta_0) = (\mathcal{F}, Q_{NF}, Q_{NF}^f, \Delta_{NF}) = \mathcal{A}_{NF}$. $\mathcal{A}_{i+1} = (\mathcal{F}, Q, Q^f, \Delta_{i+1})$ (or $\mathcal{A}_k = (\mathcal{F}, Q, Q^f, \Delta_k)$) is obtained from $\mathcal{A}_i$ as follows:

- If there exist $q_{t_1} \in Q^f, \ldots, q_{t_n} \in Q^f$, $f(l_1, \ldots, l_n) \to r \in \mathcal{R}$ and $q \in Q$ such that

  (1) $f(l_1, \ldots, l_n)_\Omega \leq f(t_1, \ldots, t_n)$

  (2) there exists $\theta : \mathcal{V} \to Q$ such that $r\theta \xrightarrow{*}_{\mathcal{A}_i} q$ and $x \equiv l_j$ implies $x\theta = q_{t_j}$ for every $x \in r$ and $1 \leq j \leq n$,

  (3) $f(q_{t_1}, \ldots, q_{t_n}) \to q \notin \Delta_i$,

  then $\Delta_{i+1} = \Delta_i \cup \{f(q_{t_1}, \ldots, q_{t_n}) \to q\}$.
- Otherwise, $\Delta_k = \Delta_i$.

Since the set of states is fixed, the number of new rules is bounded. Thus, the process of construction terminates. Note that $\mathcal{A}_1, \ldots, \mathcal{A}_k$ are non-deterministic. In the following we prove that

$$L(\mathcal{A}_k) = \{\, t \in T(\mathcal{F}) \mid t \text{ is weakly innermost normalizing}\, \}.$$

Lemma 19. Let $t \in T(\mathcal{F})$. For any $0 \leq i \leq k$, if $t \xrightarrow{*}_{\mathcal{A}_i} q \in Q$ then $t \xrightarrow{*}_\mathcal{I} s \xrightarrow{*}_{\mathcal{A}_{NF}} q$ for some $s \in T(\mathcal{F})$.

Proof. We prove the lemma by induction on i. *Base step.* Trivial. *Induction step.* Assume that $q_{s_1} \in Q^f, \ldots, q_{s_n} \in Q^f$, $f(l_1, \ldots, l_n) \to r \in \mathcal{R}$ and $q_1 \in Q$ satisfy the conditions of construction and Δ_i is obtained by adding the rule $f(q_{s_1}, \ldots, q_{s_n}) \to q_1$ to Δ_{i-1}. We use induction on the number m of applications of the rule $f(q_{s_1}, \ldots, q_{s_n}) \to q_1$ in the reduction $t \xrightarrow{*}_{\mathcal{A}_i} q$. If $m = 0$ then $t \xrightarrow{*}_{\mathcal{A}_{i-1}} q$. Thus it follows from induction hypothesis on i that $t \xrightarrow{*}_\mathcal{I} s \xrightarrow{*}_{\mathcal{A}_{NF}} q$ for some $s \in T(\mathcal{F})$. Let $m > 0$. Suppose that

$$t \equiv t[f(t_1, \ldots, t_n)]_p \xrightarrow{*}_{\mathcal{A}_{i-1}} t[f(q_{s_1}, \ldots, q_{s_n})]_p \to_{\mathcal{A}_i} t[q_1]_p \xrightarrow{*}_{\mathcal{A}_i} q.$$

For every $1 \leq j \leq n$, we obtain $u_j \in T(\mathcal{F})$ such that $t_j \xrightarrow{*}_\mathcal{I} u_j \xrightarrow{*}_{\mathcal{A}_{NF}} q_{s_j}$ by applying induction hypothesis on i to $t_j \xrightarrow{*}_{\mathcal{A}_{i-1}} q_{s_j}$. According to Lemma 18 (ii), $f(s_1, \ldots, s_n) \leq f(u_1, \ldots, u_n)$ and $u_1, \ldots, u_n$ are normal forms. Because we have $f(l_1, \cdots, l_n)_\Omega \leq f(s_1, \ldots, s_n)$ by the condition (1), we obtain the following reduction sequence:

$$f(t_1, \ldots, t_n) \xrightarrow{*}_\mathcal{I} f(u_1, \ldots, u_n) \equiv f(l_1, \ldots, l_n)\sigma \to_\mathcal{I} r\sigma.$$

Let θ be a substitution which is satisfied in the condition (2) of construction. Then from the growingness of $\mathcal{R}$ we have $r\sigma \xrightarrow{*}_{\mathcal{A}_{NF}} r\theta$ and hence $r\sigma \xrightarrow{*}_{\mathcal{A}_{i-1}} q_1$.

Applying induction hypothesis on m to $t[r\sigma]_p \xrightarrow{*}_{\mathcal{A}_{i-1}} t[q_1]_p \xrightarrow{*}_{\mathcal{A}_i} q$, we obtain $s \in T(\mathcal{F})$ such that $t[r\sigma]_p \xrightarrow{*}_{\mathcal{I}} s \xrightarrow{*}_{\mathcal{A}_{NF}} q$. Thus we have $t \xrightarrow{*}_{\mathcal{I}} s \xrightarrow{*}_{\mathcal{A}_{NF}} q$ since $t \rightarrow_{\mathcal{I}} t[r\sigma]_p$. □

Lemma 20. $L(\mathcal{A}_k) \subseteq \{ t \in T(\mathcal{F}) \mid t \text{ is weakly innermost normalizing } \}$.

Proof. From Lemmas 18 and 19. □

Lemma 21. Let $t \in T(\mathcal{F})$ be a normal form. Then there exists exactly one q in Q such that $t \xrightarrow{*}_{\mathcal{A}_k} q$. Furthermore, q is the state q_s in Q^f such that $s \leq t$ and $u \leq s$ for any $q_u \in Q^f$ with $u \leq t$.

Proof. By Lemma 19, $t \xrightarrow{*}_{\mathcal{A}_k} q$ iff $t \xrightarrow{*}_{\mathcal{A}_{NF}} q$. Thus, from Lemma 18 the claim follows. □

Lemma 22. $L(\mathcal{A}_k) \supseteq \{ t \in T(\mathcal{F}) \mid t \text{ is weakly innermost normalizing } \}$.

Proof. Assume that $t \xrightarrow{*}_{\mathcal{I}} s$ for some normal form s. We show that $t \in L(\mathcal{A}_k)$ by induction on the length m of this reduction. Let $m = 0$. Then t is a normal form and hence $t \in L(\mathcal{A}_{NF}) \subseteq L(\mathcal{A}_k)$. Let $m > 0$. We assume that

$$t \equiv t[f(l_1, \ldots, l_n)\sigma]_p \rightarrow_{\mathcal{I}} t[r\sigma]_p \xrightarrow{*}_{\mathcal{I}} s$$

with $f(l_1, \ldots, l_n) \rightarrow r \in \mathcal{R}$. By induction hypothesis, $t[r\sigma]_p$ is accepted by $\mathcal{A}_k$, i.e., $t[r\sigma]_p \xrightarrow{*}_{\mathcal{A}_k} q$ for some $q \in Q^f$. Because $x\sigma$ is a normal form for every $x \in r$, Lemma 21 yields $\theta : \mathcal{V} \rightarrow Q$ such that

$$t[r\sigma]_p \xrightarrow{*}_{\mathcal{A}_k} t[r\theta]_p \xrightarrow{*}_{\mathcal{A}_k} t[q_1]_p \xrightarrow{*}_{\mathcal{A}_k} q$$

where $q_1 \in Q$. For any $1 \leq j \leq n$, by Lemma 21 we have exactly one $q_{s_j} \in Q$ with $l_j\sigma \xrightarrow{*}_{\mathcal{A}_k} q_{s_j}$ because $l_j\sigma$ is a normal form. Note that if $l_j \equiv x$ and $x \in r$ then $x\theta = q_{s_j}$. For any $1 \leq j \leq n$, $q_{l_{j\Omega}} \in Q^f$ since $l_{j\Omega} \in \mathcal{S}_{\mathcal{R}}^*$ and $l_{j\Omega}$ does not contain redexes. According to Lemma 21 $f(l_1, \ldots, l_n)_\Omega \leq f(s_1, \ldots, s_n)$. Therefore $q_{s_1} \in Q^f, \ldots, q_{s_n} \in Q^f$, $f(l_1, \ldots, l_n) \rightarrow r \in \mathcal{R}$ and $q_1 \in Q$ satisfy the conditions (1) and (2) of construction. By the construction of $\mathcal{A}_k$, Δ_k has the rule $f(q_{s_1}, \ldots, q_{s_n}) \rightarrow q_1$. Thus, since

$$t \equiv t[f(l_1, \ldots, l_n)\sigma]_p \xrightarrow{*}_{\mathcal{A}_k} t[f(q_{s_1}, \ldots, q_{s_n})]_p \rightarrow_{\mathcal{A}_k} t[q_1]_p \xrightarrow{*}_{\mathcal{A}_k} q \in Q^f,$$

t is accepted by $\mathcal{A}_k$. □

Thus we obtain the following result.

Lemma 23. Let $\mathcal{R}$ be a left-linear growing TRS. The set of ground terms being weakly innermost normalizing is recognized by a tree automaton. □

Theorem 24. Termination is decidable for almost orthogonal growing TRSs.

Proof. Let $\mathcal{R}$ be an almost orthogonal growing TRS. According to Lemma 17, $\mathcal{R}$ is strongly normalizing iff every ground term is weakly innermost normalizing. From Lemmas 1, 2 and 23, it is decidable whether every ground term is weakly innermost normalizing. □

Acknowledgments

The authors would like to thank the anonymous referees for their comments and suggestions. This work is partially supported by Grants from Ministry of Education, Science and Culture of Japan, 10139214 and 10680346.

References

1. F. Baader and T. Nipkow, Term Rewriting and All That, Cambridge University Press, 1998.
2. H. Comon, Sequentiality, Second Order Monadic Logic and Tree Automata, *Proceedings of the 10th IEEE Symposium on Logic in Computer Science, San Diego*, pp.508-517, 1995.
3. H. Comon, M. Dauchet, R. Gilleron, D. Lugiez, S. Tison, and M. Tommasi, Tree Automata Techniques and Applications, *Preliminary Version*.
4. J.-L. Coquidé, M. Dauchet, R. Gilleron, and S. Vágvölgyi, Bottom-Up Tree Pushdown Automata: Classification and Connection with Rewrite Systems, *Theoretical Computer Science* **127**, pp.69-98, 1994.
5. N. Dershowitz, Termination of Linear Rewriting Systems, *LNCS* **115**, pp.448-458, 1981.
6. N. Dershowitz and J.-P. Jouannaud, Rewrite Systems, in *Handbook of Theoretical Computer Science*, Vol.B, ed. J. van Leeuwen, North-Holland, pp.243-320, 1990.
7. I. Durand and A. Middeldorp, Decidable Call by Need Computations in Term Rewriting (Extended Abstract), *LNAI* **1249**, pp.4-18,1997.
8. F. Gécseg and M. Steinby, Tree Automata, Akadémiai Kiadó, Budapest, 1984.
9. T. Genet, Decidable Approximations of Sets of Descendants and Sets of Normal Forms, *LNCS* **1379**, pp.151-165, 1998.
10. R. Gilleron and S. Tison, Regular Tree Languages and Rewrite Systems, *Fundamenta Informaticae* **24**, pp.157-176, 1995.
11. B. Gramlich, Abstract Relations between Restricted Termination and Confluence Properties of Rewrite Systems, *Fundamenta Informaticae* **24**, pp.3-23, 1995.
12. G. Huet and D. Lankford, On the Uniform Halting Problem for Term Rewriting Systems, *INRIA Technical Report* **283**, 1978.
13. G. Huet and J.-J. Lévy, Computations in Orthogonal Rewriting Systems, I and II, in *Computational Logic, Essays in Honor of Alan Robinson*, eds. J.-L. Lassez and G. Plotkin, MIT Press, pp.396-443, 1991.
14. F. Jacquemard, Decidable Approximations of Term Rewriting Systems, *LNCS* **1103**, pp.362-376, 1996.
15. K. Kitaoka, T. Takai, Y. Kaji, T. Tanaka, H. Seki, Finite Overlapping Term Rewriting Systems Effectively Preserve Recognizablility, *IEICE Technical Report* **COMP98**-45, Vol.98, No.380, pp.57-64, 1998 (*in Japanese*).
16. J.W. Klop, Term Rewriting Systems, in *Handbook of Logic in Computer Science*, Vol.2, eds. S. Abramasky, D. Gabbay, and T. Maibaum, pp.1-116, Oxford University Press, 1992.
17. M. Oyamaguchi, The Reachability and Joinability Problems for Right-Ground Term-Rewriting Systems, *Journal of Information Processing* **13**, pp.347-354, 1990.
18. K. Salomaa, Deterministic Tree Pushdown Automata and Monadic Tree Rewriting Systems, *J. Comput. and Syst. Sci.* **37**, pp.367-394, 1988.
19. K. Salomaa, Decidability of Confluence and Termination of Monadic Term Rewriting Systems, *LNCS* **488**, pp.275-286, 1991.

Transforming Context-Sensitive Rewrite Systems

Jürgen Giesl[1] and Aart Middeldorp[2]

[1] Dept. of Computer Science, Darmstadt University of Technology
Alexanderstr. 10, 64283 Darmstadt, Germany
`giesl@informatik.tu-darmstadt.de`

[2] Institute of Information Sciences and Electronics
University of Tsukuba, Tsukuba 305-8573, Japan
`ami@is.tsukuba.ac.jp`

Abstract. We present two new transformation techniques for proving termination of context-sensitive rewriting. Our first method is simple, sound, and more powerful than previously suggested transformations. However, it is not complete, i.e., there are terminating context-sensitive rewrite systems that are transformed into non-terminating term rewrite systems. The second method that we present in this paper is both sound and complete. This latter result can be interpreted as stating that from a termination perspective there is no reason to study context-sensitive rewriting.

1 Introduction

In the presence of infinite reductions in term rewriting, the search for normal forms is usually guided by adopting a suitable reduction strategy. Consider the following rewrite rules which form a part of a term rewrite system that implements the Sieve of Eratosthenes for generating the infinite list of all prime numbers (we did not include the rules defining divides):

$$
\begin{array}{ll}
\mathsf{primes} & \to \mathsf{sieve}(\mathsf{from}(\mathsf{s}(\mathsf{s}(0)))) \quad \mathsf{head}(x : y) \to x \\
\mathsf{from}(x) & \to x : \mathsf{from}(\mathsf{s}(x)) \quad \mathsf{tail}(x : y) \ \to y \\
\mathsf{if}(\mathsf{true}, x, y) & \to x \quad \mathsf{sieve}(x : y) \to x : \mathsf{filter}(x, \mathsf{sieve}(y)) \\
\mathsf{if}(\mathsf{false}, x, y) & \to y \\
\mathsf{filter}(\mathsf{s}(\mathsf{s}(x)), y : z) & \to \mathsf{if}(\mathsf{divides}(\mathsf{s}(\mathsf{s}(x)), y), \mathsf{filter}(\mathsf{s}(\mathsf{s}(x)), z), y : \mathsf{filter}(\mathsf{s}(\mathsf{s}(x)), z))
\end{array}
$$

A term like $\mathsf{head}(\mathsf{tail}(\mathsf{tail}(\mathsf{primes}))))$ admits a finite reduction to the normal form $\mathsf{s}^5(0)$ (the third prime number) as well as infinite reductions. The infinite reductions can for instance be avoided by always contracting the leftmost-outermost redex. Context-sensitive rewriting (Lucas [10, 11]) provides an alternative way of solving the non-termination problem. Rather than specifying which redexes may be contracted, in context-sensitive rewriting for every function symbol one indicates which arguments may not be evaluated and a contraction of a redex is allowed only if it is does not take place in a forbidden argument of a function symbol above it. For instance, by forbidding all contractions in the argument t of a term of the form $s : t$, infinite reductions are no longer possible while normal forms can still be computed. This example illustrates that this restricted form of rewriting has strong connections with lazy evaluation strategies used in functional programming languages, because it allows us to deal with non-terminating programs and infinite data structures, cf. [11].

In this paper we are concerned with the problem of showing termination of context-sensitive rewriting. More precisely, we consider transformations from context-sensitive rewrite systems to ordinary term rewrite systems that are *sound* with respect to termination: termination of the transformed term rewrite system implies termination of the original context-sensitive rewrite system. The advantage of such an approach is that all techniques for proving termination of term rewriting (e.g., [3, 6, 8, 14]) can be used to infer termination of context-sensitive rewriting. Two such transformations are reported in the literature, by Lucas [10] and by Zantema [17]. We add two more. Our first transformation is simple, its soundness is easily established, and it improves upon the transformations of [10, 17]. To be precise, we prove that the class of terminating context-sensitive rewrite systems for which our transformation succeeds is larger than that of Lucas' transformation and we claim that the same holds for Zantema's transformation. None of these three transformations succeeds in transforming every terminating context-sensitive rewrite system into a terminating term rewrite system. In other words, they all lack *completeness*. We analyze the failure of completeness for our first transformation, resulting in a second transformation with is both sound and complete. Hence it appears that from a termination point of view there is no reason to study context-sensitive rewriting further. We come back to this issue in the final part of the paper.

The remainder of the paper is organized as follows. In the next section we recall the definition of context-sensitive rewriting as well as the previous transformations of Lucas and Zantema. In Section 3 we present our first transformation and prove that it is sound. Despite being incomplete, we argue that it can handle more systems than the transformations of Lucas and Zantema. In Section 4 we refine our first transformation into a sound and complete one. The bulk of this section is devoted to the completeness proof. We make some concluding remarks in Section 5.

2 Preliminaries and Related Work

Familiarity with the basics of term rewriting ([4, 7, 9]) is assumed. Let $\mathcal{F}$ be a signature. A function $\mu \colon \mathcal{F} \to \mathcal{P}(\mathbb{N})$ is called a *replacement map* if $1 \leqslant i \leqslant \text{arity}(f)$ for all $f \in \mathcal{F}$ and $i \in \mu(f)$. A *context-sensitive rewrite system* (CSRS for short) is a term rewrite system (TRS) $\mathcal{R}$ over a signature $\mathcal{F}$ that is equipped with a replacement map μ. We always assume that $\mathcal{F}$ contains a constant. The context-sensitive rewrite relation $\to_{\mathcal{R},\mu}$ is defined as the restriction of the usual rewrite relation $\to_{\mathcal{R}}$ to contractions of redexes at *active* positions. A position π in a term t is (μ-)active if $\pi = \varepsilon$ (the root position), or $t = f(t_1, \ldots, t_n)$,

$\pi = i \cdot \pi'$, $i \in \mu(f)$, and π' is active in t_i. So $s \rightarrow_{\mathcal{R},\mu} t$ if and only if there exist a rewrite rule $l \rightarrow r$ in $\mathcal{R}$, a substitution σ, and an active position π in s such that $s|_{\pi} = l\sigma$ and $t = s[r\sigma]_{\pi}$.

Consider the TRS of the introduction. By taking $\mu(:) = \mu(\mathsf{if}) = \mu(\mathsf{sieve}) = \mu(\mathsf{from}) = \mu(\mathsf{s}) = \mu(\mathsf{head}) = \mu(\mathsf{tail}) = \{1\}$, and $\mu(\mathsf{filter}) = \mu(\mathsf{divides}) = \{1, 2\}$ we obtain a terminating CSRS. The term $0 : \mathsf{from}(\mathsf{s}(0))$, which has an infinite reduction in the TRS, is a normal form of the CSRS because the reduction step to $0 : (\mathsf{s}(0) : \mathsf{from}(\mathsf{s}(\mathsf{s}(0))))$ is no longer possible as the contracted redex occurs at a forbidden position $(2 \notin \mu(:))$.

Context-sensitive rewriting subsumes ordinary rewriting (by taking $\mu(f) = \{1, ..., n\}$ for every n-ary function symbol f). The interesting case is when $\mathcal{R}$ admits infinite reductions and μ is defined in such a way that $\rightarrow_{\mathcal{R},\mu}$ is terminating but still capable of computing $(\mathcal{R}\text{-})$normal forms. For the latter aspect we refer to Lucas [11]; in this paper we are only concerned with termination of context-sensitive rewriting.

Lucas [10] presented a simple transformation from CSRSs to TRSs which is sound with respect to termination. Let $(\mathcal{R}, \mu)$ be a CSRS over a signature $\mathcal{F}$. The idea of the transformation is to replace every function symbol $f \in \mathcal{F}$ by a new function symbol f_{μ} where all arguments except the active ones are removed. Thus, the arity of f_{μ} is $|\mu(f)|$. The transformed system $\mathcal{R}^{\mathsf{L}}_{\mu}$ results from $\mathcal{R}$ by normalising all terms in its rewrite rules using the (terminating and confluent) TRS consisting of all rules

$$f(x_1, \ldots, x_n) \rightarrow f_{\mu}(x_{i_1}, \ldots, x_{i_k})$$

such that $\mu(f) = \{i_1, \ldots, i_k\}$ with $i_1 < \cdots < i_k$. For instance, if $\mathcal{R}$ is the TRS of the introduction and μ is defined as above, then $\mathcal{R}^{\mathsf{L}}_{\mu}$ consists of the following rewrite rules:

$$
\begin{array}{lll}
\mathsf{primes}_{\mu} & \rightarrow \mathsf{sieve}_{\mu}(\mathsf{from}_{\mu}(\mathsf{s}_{\mu}(\mathsf{s}_{\mu}(0_{\mu})))) & \mathsf{head}_{\mu}(:_{\mu}(x)) \rightarrow x \\
\mathsf{from}_{\mu}(x) & \rightarrow :_{\mu}(x) & \mathsf{tail}_{\mu}(:_{\mu}(x)) \rightarrow y \\
\mathsf{sieve}_{\mu}(:_{\mu}(x)) & \rightarrow :_{\mu}(x) & \\
\mathsf{filter}_{\mu}(\mathsf{s}_{\mu}(\mathsf{s}_{\mu}(x)), :_{\mu}(y)) & \rightarrow \mathsf{if}_{\mu}(\mathsf{divides}_{\mu}(\mathsf{s}_{\mu}(\mathsf{s}_{\mu}(x)), y)) & \\
\mathsf{if}_{\mu}(\mathsf{true}_{\mu}) & \rightarrow x & \\
\mathsf{if}_{\mu}(\mathsf{false}_{\mu}) & \rightarrow y & \\
\end{array}
$$

Note that $\mathcal{R}^{\mathsf{L}}_{\mu}$ is not terminating due to the extra variables in the right-hand sides of the rules for tail_{μ} and if_{μ}.

Zantema [17] presented a more complicated transformation in which subterms at forbidden positions are marked rather than discarded. The transformed system $\mathcal{R}^{\mathsf{Z}}_{\mu}$ consists of two parts. The first part results from a translation of the rewrite rules of $\mathcal{R}$, as follows. Every function symbol f occurring in a left or right-hand side is replaced by $\underline{f}$ (a fresh function symbol of the same arity as f) if it occurs in a forbidden argument of the function symbol directly above it. These new function symbols are used to block further reductions at this position.

In addition, if a variable x occurs in a forbidden position in the left-hand side l of a rewrite rule $l \rightarrow r$ then all occurrences of x in r are replaced by $\mathsf{a}(x)$.

Here a is a new unary function symbol which is used to activate blocked function symbols again. The second part of $\mathcal{R}_\mu^Z$ consists of rewrite rules that are needed for blocking and unblocking function symbols:

$$f(x_1, \ldots, x_n) \to \underline{f}(x_1, \ldots, x_n)$$
$$\mathsf{a}(\underline{f}(x_1, \ldots, x_n)) \to f(x_1, \ldots, x_n)$$

for every n-ary f for which $\underline{f}$ appears in the first part of $\mathcal{R}_\mu^Z$, together with the rule $\mathsf{a}(x) \to x$. The example CSRS $(\mathcal{R}, \mu)$ is transformed into

$$
\begin{array}{ll}
\text{primes} & \to \text{sieve}(\text{from}(\text{s}(\text{s}(0)))) \\
\text{from}(x) & \to x : \underline{\text{from}}(\text{s}(x)) \\
\text{sieve}(x : y) & \to x : \underline{\text{filter}}(x, \text{sieve}(\mathsf{a}(y))) \\
\text{filter}(\text{s}(\text{s}(x)), y : z) & \to \text{if}(\text{divides}(\text{s}(\text{s}(x)), y), \underline{\text{filter}}(\text{s}(\text{s}(x)), \mathsf{a}(z)), \\
& \qquad\qquad\qquad\qquad y : \underline{\text{filter}}(\text{s}(\text{s}(x)), \mathsf{a}(z)))
\end{array}
$$

$$
\begin{array}{llll}
\text{if}(\text{true}, x, y) & \to \mathsf{a}(x) & & \\
\text{if}(\text{false}, x, y) & \to \mathsf{a}(y) & \text{head}(x : y) & \to x \\
\text{from}(x) & \to \underline{\text{from}}(x) & \text{tail}(x : y) & \to \mathsf{a}(y) \\
\text{sieve}(x) & \to \underline{\text{sieve}}(x) & \mathsf{a}(\underline{\text{from}}(x)) & \to \text{from}(x) \\
\text{filter}(x, y) & \to \underline{\text{filter}}(x, y) & \mathsf{a}(\underline{\text{sieve}}(x)) & \to \text{sieve}(x) \\
x : y & \to x \underline{:} y & \mathsf{a}(\underline{\text{filter}}(x, y)) & \to \text{filter}(x, y) \\
\mathsf{a}(x) & \to x & \mathsf{a}(x \underline{:} y) & \to x : y
\end{array}
$$

This transformation is sound but not complete as we have the infinite reduction

$$\text{sieve}(\mathsf{a}(\underline{\text{from}}(0))) \to_{\mathcal{R}_\mu^Z}^+ 0 : \underline{\text{filter}}(0, \text{sieve}(\mathsf{a}(\underline{\text{from}}(\text{s}(0)))))$$
$$\to_{\mathcal{R}_\mu^Z}^+ 0 : \underline{\text{filter}}(0, \text{s}(0) : \underline{\text{filter}}(\text{s}(0), \text{sieve}(\mathsf{a}(\underline{\text{from}}(\text{s}(\text{s}(0)))))))$$
$$\to_{\mathcal{R}_\mu^Z}^+ \cdots$$

in the TRS $\mathcal{R}_\mu^Z$.

Zantema's method appears to be more powerful than Lucas' transformation but actually the two methods are incomparable (cf. the TRS consisting of the single rule $\mathsf{c} \to \mathsf{f}(\mathsf{g}(\mathsf{c}))$ with $\mu(\mathsf{f}) = \varnothing$ and $\mu(\mathsf{g}) = \{1\}$).

3 A Sound Transformation

In this section we present our first transformation from CSRSs to TRSs. The advantage of this transformation is that it is very easy and more powerful than the transformations of Lucas and Zantema defined in the preceding section. In the transformation we will extend the original signature $\mathcal{F}$ of the TRS by two additional unary function symbols active and mark.

Essentially, the idea for the transformation is to mark the active positions in a term on the object level, because those positions are the only ones where context-sensitive rewriting may take place. For this purpose we use the new function symbol active. Thus, instead of a rule $l \to r$ the transformed TRS should contain a rule whose left-hand side is active(l). Moreover, after rewriting an instance of l to the corresponding instance of r, we have to mark the new active positions in the resulting term. For that purpose we use the function mark. So we replace every rule $l \to r$ by active(l) $\to$ mark(r). To mark all active positions in a term, the rules for mark must have the form

$$\mathsf{mark}(f(x_1, \ldots, x_n)) \to \mathsf{active}(f([x_1], \ldots, [x_n]))$$

where the form of the argument $[x_i]$ depends on whether i is an active argument of f: If $i \in \mu(f)$ then x_i must also be marked active and thus $[x_i] = \mathsf{mark}(x_i)$, otherwise the ith argument of f is not active and we define $[x_i] = x_i$. Finally, we also need a rule to deactivate terms. For example, consider the TRS consisting of the following rewrite rules:

$$\begin{aligned}
\mathsf{a} &\to \mathsf{f}(\mathsf{b}) \\
\mathsf{f}(\mathsf{b}) &\to \mathsf{a} \\
\mathsf{b} &\to \mathsf{c}
\end{aligned}$$

No matter how the replacement map μ is defined, the resulting CSRS is not terminating. Suppose $\mu(f) = \{1\}$. In the transformed system we would have the rules

$$\begin{aligned}
\mathsf{active(a)} &\to \mathsf{mark(f(b))} & \mathsf{mark(a)} &\to \mathsf{active(a)} \\
\mathsf{active(f(b))} &\to \mathsf{mark(a)} & \mathsf{mark(b)} &\to \mathsf{active(b)} \\
\mathsf{active(b)} &\to \mathsf{mark(c)} & \mathsf{mark(c)} &\to \mathsf{active(c)} \\
& & \mathsf{mark(f(x))} &\to \mathsf{active(f(mark(x)))}
\end{aligned}$$

This TRS is terminating because active(a) can be reduced to active(f(active(b))), but if we cannot deactivate the subterm active(b) then the second rule is not applicable. Thus, we have to add the rule active(x) $\to x$. To summarize, we obtain the following transformation.

Definition 1. *Let $(\mathcal{R}, \mu)$ be a CSRS over a signature $\mathcal{F}$. The TRS $\mathcal{R}^1_\mu$ over the signature $\mathcal{F} \cup \{\mathsf{active}, \mathsf{mark}\}$ consists of the following rewrite rules:*

$$\begin{aligned}
\mathsf{active}(l) &\to \mathsf{mark}(r) & &\text{for all } l \to r \in \mathcal{R} \\
\mathsf{mark}(f(x_1, \ldots, x_n)) &\to \mathsf{active}(f([x_1]_f, \ldots, [x_n]_f)) & &\text{for all } f \in \mathcal{F} \\
\mathsf{active}(x) &\to x
\end{aligned}$$

Here $[x_i]_f = \mathsf{mark}(x_i)$ if $i \in \mu(f)$ and $[x_i]_f = x_i$ otherwise. The subset of $\mathcal{R}^1_\mu$ consisting of all rules of the form

$$\mathsf{mark}(f(x_1, \ldots, x_n)) \to \mathsf{active}(f([x_1]_f, \ldots, [x_n]_f))$$

will be denoted by $\mathcal{M}$.

Soundness of our transformation is an easy consequence of the following lemma which shows how context-sensitive reduction steps are simulated in the transformed system.

Lemma 1. *Let $(\mathcal{R}, \mu)$ be a CSRS over a signature $\mathcal{F}$ and let $s, t \in \mathcal{T}(\mathcal{F})$. If $s \to_{\mathcal{R}, \mu} t$ then $\mathsf{mark}(s){\downarrow}_{\mathcal{M}} \to^{+}_{\mathcal{R}^1_\mu} \mathsf{mark}(t){\downarrow}_{\mathcal{M}}$.*

Proof. First note that $\mathcal{M}$ is confluent and terminating, so $u{\downarrow}_{\mathcal{M}}$ exists for every term u. There exist a rewrite rule $l \to r \in \mathcal{R}$, a substitution σ, and an active position π in s such that $s|_\pi = l\sigma$ and $t = s[r\sigma]_\pi$. We prove the lemma by induction on π. If $\pi = \varepsilon$ then $s = l\sigma$ and $t = r\sigma$. An easy induction on the structure of s reveals that $\mathsf{mark}(s){\downarrow}_{\mathcal{M}} \to^{*}_{\mathcal{R}^1_\mu} \mathsf{active}(s)$ (one just has to eliminate all inner occurrences of active in $\mathsf{mark}(s){\downarrow}_{\mathcal{M}}$). Since $\mathsf{active}(s) \to \mathsf{mark}(t)$ is an instance of a rule in $\mathcal{R}^1_\mu$ we obtain

$$\mathsf{mark}(s){\downarrow}_{\mathcal{M}} \to^{*}_{\mathcal{R}^1_\mu} \mathsf{active}(s) \to_{\mathcal{R}^1_\mu} \mathsf{mark}(t) \to^{+}_{\mathcal{R}^1_\mu} \mathsf{mark}(t){\downarrow}_{\mathcal{M}}.$$

If $\pi = i \cdot \pi'$ then we have $s = f(s_1, \ldots, s_i, \ldots, s_n)$ and $t = f(s_1, \ldots, t_i, \ldots, s_n)$ with $s_i \to_{\mathcal{R}, \mu} t_i$. Note that $i \in \mu(f)$ due to the definition of context-sensitive rewriting. For $1 \leqslant j \leqslant n$ define $s'_j = \mathsf{mark}(s_j){\downarrow}_{\mathcal{M}}$ if $j \in \mu(f)$ and $s'_j = s_j$ if $j \notin \mu(f)$. The induction hypothesis yields $s'_i \to^{+}_{\mathcal{R}^1_\mu} \mathsf{mark}(t_i){\downarrow}_{\mathcal{M}}$. Since

$$\mathsf{mark}(s){\downarrow}_{\mathcal{M}} = \mathsf{active}(f(s'_1, \ldots, s'_i, \ldots, s'_n))$$

and

$$\mathsf{mark}(t){\downarrow}_{\mathcal{M}} = \mathsf{active}(f(s'_1, \ldots, \mathsf{mark}(t_i){\downarrow}_{\mathcal{M}}, \ldots, s'_n)),$$

the result follows. □

Theorem 1. *Let $(\mathcal{R}, \mu)$ be a CSRS over a signature $\mathcal{F}$. If $\mathcal{R}^1_\mu$ is terminating then $(\mathcal{R}, \mu)$ is terminating.*

Proof. If $(\mathcal{R}, \mu)$ is not terminating then there exists an infinite reduction of ground terms. Any such sequence is transformed by the previous lemma into an infinite reduction in $\mathcal{R}^1_\mu$. □

The converse of the above theorem does not hold, i.e., the transformation is incomplete.

Example 1. As an example of a terminating CSRS that is transformed into a non-terminating TRS by our transformation, consider the following variant $\mathcal{R}$ of a well-known example from Toyama [15]:

$$f(b, c, x) \to f(x, x, x) \qquad d \to b \qquad d \to c$$

If we define $\mu(f) = \{3\}$ then the resulting CSRS is terminating because the usual cyclic reduction of $f(b, c, d)$ to $f(d, d, d)$ and further to $f(b, c, d)$ cannot be done any more, as one would have to reduce the first and second argument of f. However, the transformed TRS $\mathcal{R}^1_\mu$

$$
\begin{aligned}
\mathsf{active}(f(b, c, x)) &\to \mathsf{mark}(f(x, x, x)) & \mathsf{mark}(f(x, y, z)) &\to \mathsf{active}(f(x, y, \mathsf{mark}(z))) \\
\mathsf{active}(d) &\to \mathsf{mark}(b) & \mathsf{mark}(b) &\to \mathsf{active}(b) \\
\mathsf{active}(d) &\to \mathsf{mark}(c) & \mathsf{mark}(c) &\to \mathsf{active}(c) \\
\mathsf{active}(x) &\to x & \mathsf{mark}(d) &\to \mathsf{active}(d)
\end{aligned}
$$

is not terminating:

$$\begin{aligned}
\text{mark}(f(b, c, d)) &\rightarrow \text{active}(f(b, c, \text{mark}(d))) \\
&\rightarrow \text{active}(f(b, c, \text{active}(d))) \\
&\rightarrow \text{mark}(f(\text{active}(d), \text{active}(d), \text{active}(d))) \\
&\rightarrow^+ \text{mark}(f(\text{mark}(b), \text{mark}(c), d)) \\
&\rightarrow^+ \text{mark}(f(\text{active}(b), \text{active}(c), d)) \\
&\rightarrow^+ \text{mark}(f(b, c, d))
\end{aligned}$$

Note that $\mathcal{R}^L_\mu$:

$$f_\mu(x) \rightarrow f_\mu(x) \qquad d_\mu \rightarrow b_\mu \qquad d_\mu \rightarrow c_\mu$$

and $\mathcal{R}^Z_\mu$:

$$\begin{aligned}
f(\underline{b}, \underline{c}, x) &\rightarrow f(x, x, x) & a(\underline{b}) &\rightarrow b \\
d &\rightarrow b & a(\underline{c}) &\rightarrow c \\
d &\rightarrow c & b &\rightarrow \underline{b} \\
a(x) &\rightarrow x & c &\rightarrow \underline{c}
\end{aligned}$$

also fail to be terminating ($\mathcal{R}^Z_\mu$ admits the cycle $f(\underline{b}, \underline{c}, d) \rightarrow f(d, d, d) \rightarrow^+$ $f(b, c, d) \rightarrow^+ f(\underline{b}, \underline{c}, d)$).

Nevertheless, compared to the transformations of Lucas and Zantema, our easy transformation appears to be very powerful. There are numerous CSRSs where our transformation succeeds and which cannot be handled by the other two transformations.

Example 2. As a simple example, consider the terminating CSRS $\mathcal{R}$

$$\begin{aligned}
g(x) &\rightarrow h(x) \\
c &\rightarrow d \\
h(d) &\rightarrow g(c)
\end{aligned}$$

with $\mu(g) = \mu(h) = \varnothing$ from [17]. The TRSs $\mathcal{R}^L_\mu$:

$$g_\mu \rightarrow h_\mu \qquad c_\mu \rightarrow d_\mu \qquad h_\mu \rightarrow g_\mu$$

and $\mathcal{R}^Z_\mu$:

$$\begin{aligned}
g(x) &\rightarrow h(a(x)) & a(\underline{c}) &\rightarrow c \\
c &\rightarrow d & a(\underline{d}) &\rightarrow d \\
h(\underline{d}) &\rightarrow g(\underline{c}) & c &\rightarrow \underline{c} \\
a(x) &\rightarrow x & d &\rightarrow \underline{d}
\end{aligned}$$

are non-terminating ($\mathcal{R}^Z_\mu$ admits the cycle $g(\underline{c}) \rightarrow h(a(\underline{c})) \rightarrow h(c) \rightarrow h(d) \rightarrow$ $h(\underline{d}) \rightarrow g(\underline{c})$). In contrast, our simple transformation generates the TRS

$$\begin{aligned}
\text{active}(g(x)) &\rightarrow \text{mark}(h(x)) & \text{mark}(g(x)) &\rightarrow \text{active}(g(x)) \\
\text{active}(c) &\rightarrow \text{mark}(d) & \text{mark}(h(x)) &\rightarrow \text{active}(h(x)) \\
\text{active}(h(d)) &\rightarrow \text{mark}(g(c)) & \text{mark}(c) &\rightarrow \text{active}(c) \\
\text{active}(x) &\rightarrow x & \text{mark}(d) &\rightarrow \text{active}(d)
\end{aligned}$$

which is terminating.[1]

Moreover, while the techniques of Lucas and Zantema fail for the Sieve of Eratosthenes example from the introduction, our transformation generates a terminating TRS. In fact, we do not know of any example where the method of Lucas or Zantema works but our method fails. (In particular, our transformation succeeds for all terminating CSRSs presented in [17].) This strongly suggests that our proposal is more powerful than the previous two approaches. For the transformation of Lucas this can indeed be proved.

Theorem 2. *Let $(\mathcal{R}, \mu)$ be a CSRS over a signature $\mathcal{F}$. If $\mathcal{R}_\mu^L$ is terminating then $\mathcal{R}_\mu^1$ is terminating.*

Proof. We prove termination of $\mathcal{R}_\mu^1$ using the *dependency pair* approach of Arts and Giesl [1–3]. The dependency pairs of $\mathcal{R}_\mu^1$ are

$$\langle \text{ACTIVE}(l), \text{MARK}(r)\rangle \qquad \text{for all } l \to r \text{ in } \mathcal{R} \qquad \text{(i)}$$
$$\langle \text{MARK}(f(x_1,\ldots,x_n)), \text{ACTIVE}(f([x_1]_f,\ldots,[x_n]_f))\rangle \text{ for all } f \in \mathcal{F} \qquad \text{(ii)}$$
$$\langle \text{MARK}(f(x_1,\ldots,x_n)), \text{MARK}(x_i)\rangle \qquad \text{for } f \in \mathcal{F}, i \in \mu(f) \qquad \text{(iii)}$$

To prove termination of $\mathcal{R}_\mu^1$ we have to find a weakly monotonic quasi-order $\succsim$ and a well-founded order $\succ$ which is compatible with $\succsim$ (i.e., $\succ \circ \succsim \subseteq \succ$) such that both $\succ$ and $\succsim$ are closed under substitution. Then it is sufficient if the following constraints are satisfied. Dependency pairs of kind (i) and (iii) should be strictly decreasing and for dependency pairs of kind (ii) it is enough if they are weakly decreasing. Moreover, all rules of $\mathcal{R}_\mu^1$ should be weakly decreasing. Thus, we only have to demand

$$\text{ACTIVE}(l) \succ \text{MARK}(r) \qquad \text{for all } l \to r \text{ in } \mathcal{R}$$
$$\text{MARK}(f(x_1,\ldots,x_n)) \succsim \text{ACTIVE}(f([x_1]_f,\ldots,[x_n]_f)) \text{ for all } f \in \mathcal{F}$$
$$\text{MARK}(f(x_1,\ldots,x_n)) \succ \text{MARK}(x_i) \qquad \text{for all } f \in \mathcal{F}, i \in \mu(f)$$
$$\text{active}(l) \succsim \text{mark}(r) \qquad \text{for all } l \to r \text{ in } \mathcal{R}$$
$$\text{mark}(f(x_1,\ldots,x_n)) \succsim \text{active}(f([x_1]_f,\ldots,[x_n]_f)) \qquad \text{for all } f \in \mathcal{F}$$
$$\text{active}(x) \succsim x$$

Let $\mathcal{A}$ be the (confluent and terminating) TRS consisting of the rewrite rules

$$\text{ACTIVE}(x) \to x$$
$$\text{MARK}(x) \to x$$
$$\text{active}(x) \to x$$
$$\text{mark}(x) \to x$$
$$f(x_1,\ldots,x_n) \to f_\mu(x_{i_1},\ldots,x_{i_k})$$

for all $f \in \mathcal{F}$ where $\mu(f) = \{i_1,\ldots,i_k\}$ with $i_1 < \cdots < i_k$. Define $s \succ t$ if and only if $s\!\downarrow_{\mathcal{A}} (\to_{\mathcal{R}_\mu^L} \cup \rhd)^+ t\!\downarrow_{\mathcal{A}}$. Here $\rhd$ denotes the proper subterm relation.

---

[1] This can be proved using the *dependency pair* approach ([3]): Since the pair $\langle \text{ACTIVE}(h(d)), \text{MARK}(g(c))\rangle$ can occur at most once in any *chain* of dependency pairs, it follows that there are no infinite chains and hence the TRS is terminating.

Moreover, let $s \succsim t$ hold if and only if $s\downarrow_{\mathcal{A}} \to_{\mathcal{R}_\mu^L}^* t\downarrow_{\mathcal{A}}$. One easily verifies that $\succ$ and $\succsim$ satisfy the above demands ($\succ$ is well founded by the termination of $\mathcal{R}_\mu^L$). Hence, due to the soundness of the dependency pair approach, the termination of $\mathcal{R}_\mu^1$ is established. □

This theorem can also be proved using the *self-labelling* technique of [12].

4   A Sound and Complete Transformation

In this section we present a transformation of context-sensitive rewrite systems which is not only sound but also complete with respect to termination. To appreciate the non-triviality of this result, the reader may want to try to construct a sound and complete transformation (together with a proof of completeness) before reading any further.

Let us first investigate why the transformation of Sect. 3 lacks completeness. Consider again the CSRS $(\mathcal{R}, \mu)$ of Example 1. The reason for the non-termination of $\mathcal{R}_\mu^1$ is that terms may have occurrences of active at forbidden positions, even if we start with a "proper" term (like mark(f(b, c, d))). The "forbidden" occurrences of active in the first two arguments of f (in the term mark(f(active(d), active(d), active(d)))) lead to contractions which are impossible in the underlying CSRS. Thus, the key to achieving a complete transformation is to control the number of occurrences of active. We do this in a rather drastic manner: we will work with a single occurrence of active. Of course, we cannot forbid the existence of terms with multiple occurrences of active but we can make sure that no new active symbols are introduced during the contraction of an active redex.

Working with a single active occurrence entails that we have to shift it in a non-deterministic fashion downwards to any active position. This is achieved by the rules

$$\text{active}(f(x_1, \ldots, x_i, \ldots, x_n)) \to f'(x_1, \ldots, \text{active}(x_i), \ldots, x_n)$$

for every $i \in \mu(f)$. When shifting the active symbol to an argument of f, the original function symbol f is replaced by a new function symbol f'. This is to ensure that no reductions can take place above the current position of active. By this shifting of the symbol active, our TRS implements an algorithm to search for redexes subject to the constraints of the replacement map μ. Once we have shifted active to the position of the desired redex, we can apply one of the rules

$$\text{active}(l) \to \text{mark}(r)$$

as in the previous transformation. The function symbol mark is used to mark the contractum of the selected redex. In order to continue the reduction it has to be replaced by active again. Since the next reduction step may of course take place at a position above the previously contracted redex, we first have to shift mark upwards through the term, i.e., we use rules of the form

$$f'(x_1, \ldots, \text{mark}(x_i), \ldots, x_n) \to \text{mark}(f(x_1, \ldots, x_i, \ldots, x_n))$$

for every $i \in \mu(f)$. We want to replace mark by active if there are no f' symbols left above it. Since the absence of f' symbols cannot be determined, we introduce a new unary function symbol top to mark the position below which reductions may take place. Thus, the reduction of a term s with respect to a CSRS is modelled by the reduction of the term top(active(s)) in the transformed TRS. If top(active(s)) is reduced to a term top(mark(t)), we are ready to replace mark by active. This suggests adding the rule

$$\mathsf{top}(\mathsf{mark}(x)) \to \mathsf{top}(\mathsf{active}(x)).$$

However, as illustrated with the counterexample in Sect. 3, we have to avoid making infinite reductions with terms which contain inner occurrences of new symbols like active and mark. For that reason we want to make sure that this rule is only applicable to terms that do not contain any other occurrences of the new function symbols. Thus, before reducing top(mark(t)) to top(active(t)) we check whether the term t is *proper*, i.e., whether it contains only function symbols from the original signature $\mathcal{F}$. This is easily achieved by new unary function symbols proper and ok. For any ground term $t \in \mathcal{T}(\mathcal{F})$, proper($t$) reduces to ok($t$), but if t contains one of the newly introduced function symbols then the reduction of proper(t) is blocked. This is done by the rules

$$\mathsf{proper}(c) \to \mathsf{ok}(c)$$

for every constant $c \in \mathcal{F}$ and

$$\mathsf{proper}(f(x_1, \ldots, x_n)) \to f(\mathsf{proper}(x_1), \ldots, \mathsf{proper}(x_n))$$
$$f(\mathsf{ok}(x_1), \ldots, \mathsf{ok}(x_n)) \to \mathsf{ok}(f(x_1, \ldots, x_n))$$

for every function symbol $f \in \mathcal{F}$ of arity $n > 0$. Now, instead of the rule top(mark(x)) $\to$ top(active(x)) we adopt the rules

$$\mathsf{top}(\mathsf{mark}(x)) \to \mathsf{top}(\mathsf{proper}(x))$$
$$\mathsf{top}(\mathsf{ok}(x)) \to \mathsf{top}(\mathsf{active}(x)).$$

This concludes our informal explanation of the new transformation, whose formal definition is summarized below.

Definition 2. *Let* $(\mathcal{R}, \mu)$ *be a CSRS over a signature* $\mathcal{F}$. *The TRS* $\mathcal{R}_\mu^2$ *over the signature* $\mathcal{F}' = \mathcal{F} \cup \{\mathsf{active}, \mathsf{mark}, \mathsf{top}, \mathsf{proper}, \mathsf{ok}\} \cup \{f' \mid f \in \mathcal{F} \text{ is not a constant}\}$ *consists of the following rewrite rules (for all* $l \to r \in \mathcal{R}$, $f \in \mathcal{F}$ *of arity* $n > 0$, $i \in \mu(f)$, *and constants* $c \in \mathcal{F}$):

$$\mathsf{active}(l) \to \mathsf{mark}(r)$$
$$\mathsf{active}(f(x_1, \ldots, x_i, \ldots, x_n)) \to f'(x_1, \ldots, \mathsf{active}(x_i), \ldots, x_n)$$
$$f'(x_1, \ldots, \mathsf{mark}(x_i), \ldots, x_n) \to \mathsf{mark}(f(x_1, \ldots, x_i, \ldots, x_n))$$
$$\mathsf{proper}(c) \to \mathsf{ok}(c)$$
$$\mathsf{proper}(f(x_1, \ldots, x_n)) \to f(\mathsf{proper}(x_1), \ldots, \mathsf{proper}(x_n))$$
$$f(\mathsf{ok}(x_1), \ldots, \mathsf{ok}(x_n)) \to \mathsf{ok}(f(x_1, \ldots, x_n))$$
$$\mathsf{top}(\mathsf{mark}(x)) \to \mathsf{top}(\mathsf{proper}(x))$$
$$\mathsf{top}(\mathsf{ok}(x)) \to \mathsf{top}(\mathsf{active}(x))$$

In the remainder of this section we show that our second transformation is both sound and complete. We start with a preliminary lemma, which states that proper has indeed the desired effect.

Lemma 2. *Let $(\mathcal{R}, \mu)$ be a CSRS over a signature $\mathcal{F}$. Let $s, t \in \mathcal{T}(\mathcal{F}')$. We have* $\mathsf{proper}(s) \to_{\mathcal{R}_\mu^2}^+ \mathsf{ok}(t)$ *if and only if $s = t$ and $s \in \mathcal{T}(\mathcal{F})$.*

Proof. The "if" direction is an easy induction proof on the structure of s. The "only if" direction can be proved by induction on the number of symbols in s.

If the root of s is a function symbol $g \in \mathcal{F}' \setminus (\mathcal{F} \cup \{\mathsf{proper}\})$ then $\mathsf{proper}(s)$ cannot be rewritten at the root. Thus, any one-step reduction of $\mathsf{proper}(s)$ would yield a term of the form $\mathsf{proper}(s')$ where $s \to_{\mathcal{R}_\mu^2} s'$. If $g \in \{\mathsf{active}, \mathsf{mark}\} \cup \{f' \mid f \in \mathcal{F}$ is not a constant$\}$ then the root symbol of s' must also be from that set. Similarly, if g is ok or top, then the root symbol of s' is g as well. This implies that no reduct of $\mathsf{proper}(s)$ can be reduced at the root position either. Hence $\mathsf{proper}(s) \to_{\mathcal{R}_\mu^2}^+ \mathsf{ok}(t)$ cannot hold and the claim holds vacuously.

In the remaining case the root symbol of s is from $\mathcal{F} \cup \{\mathsf{proper}\}$. Thus, s has the form $\mathsf{proper}^m(u)$ for some $m \geqslant 0$ where the root of u is different from proper. In order to reduce $\mathsf{proper}(s)$ at the root, we first have to reduce $s = \mathsf{proper}^m(u)$ to a term with a root symbol from $\mathcal{F}$. Similar to the observations above, the root symbol of u cannot be from $\mathcal{F}' \setminus \mathcal{F}$. If u is a constant from $\mathcal{F}$ then the only applicable rule is $\mathsf{proper}(u) \to \mathsf{ok}(u)$. Thus, $\mathsf{proper}(s) = \mathsf{proper}^{m+1}(u)$ is reduced to the normal form $\mathsf{proper}^m(\mathsf{ok}(u))$. So in this case $\mathsf{proper}(s)$ can only rewrite to a term of the form $\mathsf{ok}(t)$ if $m = 0$ and thus the claim of the lemma holds trivially.

Otherwise, $u = f(u_1, \ldots, u_n)$ with $f \in \mathcal{F}$ of arity $n > 0$. The reduction from $\mathsf{proper}(s)$ to $\mathsf{ok}(t)$ must start as follows:

$$
\begin{aligned}
\mathsf{proper}(s) \;&=\; \mathsf{proper}(\mathsf{proper}^m(f(u_1, \ldots, u_n))) \\
&\to_{\mathcal{R}_\mu^2}^* \mathsf{proper}(\mathsf{proper}^m(f(u_1', \ldots, u_n'))) \\
&\to_{\mathcal{R}_\mu^2} \mathsf{proper}(\mathsf{proper}^{m-1}(f(\mathsf{proper}(u_1'), \ldots, \mathsf{proper}(u_n')))) \\
&\to_{\mathcal{R}_\mu^2}^* \cdots \\
&\to_{\mathcal{R}_\mu^2} \mathsf{proper}(f(u_1'', \ldots, u_n'')) \\
&\to_{\mathcal{R}_\mu^2}^* f(\mathsf{proper}(u_1''), \ldots, \mathsf{proper}(u_n''))
\end{aligned}
$$

where $\mathsf{proper}^m(u_i) \to_{\mathcal{R}_\mu^2}^* u_i''$ for all $1 \leqslant i \leqslant n$. (Note that the root symbol f of u must not be rewritten to ok, for otherwise no reduction step at the root can take place.) To reduce $f(\mathsf{proper}(u_1''), \ldots, \mathsf{proper}(u_n''))$ to a term of the form $\mathsf{ok}(t)$, every argument $\mathsf{proper}(u_i'')$ must be reduced to a term of the form $\mathsf{ok}(t_i)$ and then $f(\mathsf{ok}(t_1), \ldots, \mathsf{ok}(t_n))$ can be reduced to $\mathsf{ok}(f(t_1, \ldots, t_n))$. But if $\mathsf{proper}(u_i'') \to_{\mathcal{R}_\mu^2}^* \mathsf{ok}(t_i)$ then we also have $\mathsf{proper}(\mathsf{proper}^m(u_i)) \to_{\mathcal{R}_\mu^2}^* \mathsf{ok}(t_i)$. The induction hypothesis yields $\mathsf{proper}^m(u_i) = t_i$ and $\mathsf{proper}^m(u_i) \in \mathcal{T}(\mathcal{F})$ for all $1 \leqslant i \leqslant n$. So in this case we have $m = 0$ as well, i.e., s cannot contain any occurrence of proper. Consequently, $\mathsf{ok}(f(t_1, \ldots, t_n))$ is in normal form and hence $s = u = f(u_1, \ldots, u_n) = f(t_1, \ldots, t_n) = t \in \mathcal{T}(\mathcal{F})$. $\square$

The next lemma shows how context-sensitive reduction steps are simulated by the second transformation. The "if" part is used in the completeness proof.

Lemma 3. *Let $(\mathcal{R}, \mu)$ be a CSRS over a signature $\mathcal{F}$ and let $s \in \mathcal{T}(\mathcal{F})$. We have $s \to_{\mathcal{R}, \mu} t$ if and only if $\mathsf{active}(s) \to^+_{\mathcal{R}^2_\mu} \mathsf{mark}(t)$.*

Proof. The "only if" direction is easily proved by induction on the depth of the position of the redex contracted in $s \to_{\mathcal{R}, \mu} t$. We prove here the "if" direction by induction on s. There are two possibilities for the rewrite rule of $\mathcal{R}^2_\mu$ that is applied in the first step of the reduction from $\mathsf{active}(s)$ to $\mathsf{mark}(t)$. If a rule of the form $\mathsf{active}(l) \to \mathsf{mark}(r)$ is used, then $s = l\sigma$ for some substitution σ. Since $r\sigma$ contains only symbols from $\mathcal{F}$, $\mathsf{mark}(r\sigma)$ is in normal form and thus $t = r\sigma$. Clearly $s \to_{\mathcal{R}, \mu} t$.

Otherwise, s must have the form $f(s_1, \ldots, s_i, \ldots, s_n)$ and in the first reduction step $\mathsf{active}(s)$ is reduced to $f'(s_1, \ldots, \mathsf{active}(s_i), \ldots, s_n)$ for some $i \in \mu(f)$. Note that all reductions of the latter term to a term of the form $\mathsf{mark}(t)$ have the form

$$f'(s_1, \ldots, \mathsf{active}(s_i), \ldots, s_n) \to^+_{\mathcal{R}^2_\mu} f'(s_1, \ldots, \mathsf{mark}(t_i), \ldots, s_n)$$
$$\to_{\mathcal{R}^2_\mu} \mathsf{mark}(f(s_1, \ldots, t_i, \ldots, s_n)).$$

Hence $t = f(s_1, \ldots, t_i, \ldots, s_n)$. The induction hypothesis yields $s_i \to_{\mathcal{R}, \mu} t_i$ and as $i \in \mu(f)$ we also have $s \to_{\mathcal{R}, \mu} t$. $\qquad\square$

Soundness of our second transformation is now easily shown.

Theorem 3. *Let $(\mathcal{R}, \mu)$ be a CSRS over a signature $\mathcal{F}$. If $\mathcal{R}^2_\mu$ is terminating then $(\mathcal{R}, \mu)$ is terminating.*

Proof. If $(\mathcal{R}, \mu)$ is not terminating then there exists an infinite reduction of ground terms in $\mathcal{T}(\mathcal{F})$. Note that $s \to_{\mathcal{R}, \mu} t$ implies $\mathsf{active}(s) \to^+_{\mathcal{R}^2_\mu} \mathsf{mark}(t)$ by Lemma 3. Hence it also implies

$$\mathsf{top}(\mathsf{active}(s)) \to^+_{\mathcal{R}^2_\mu} \mathsf{top}(\mathsf{mark}(t)) \to_{\mathcal{R}^2_\mu} \mathsf{top}(\mathsf{proper}(t)).$$

Moreover, by Lemma 2 we have $\mathsf{proper}(t) \to^+_{\mathcal{R}^2_\mu} \mathsf{ok}(t)$ and thus

$$\mathsf{top}(\mathsf{proper}(t)) \to^+_{\mathcal{R}^2_\mu} \mathsf{top}(\mathsf{ok}(t)) \to_{\mathcal{R}^2_\mu} \mathsf{top}(\mathsf{active}(t)).$$

Concatenating these two reductions shows that $\mathsf{top}(\mathsf{active}(s)) \to^+_{\mathcal{R}^2_\mu} \mathsf{top}(\mathsf{active}(t))$ whenever $s \to_{\mathcal{R}, \mu} t$. Hence any infinite reduction of ground terms in $(\mathcal{R}, \mu)$ is transformed into an infinite reduction in $\mathcal{R}^2_\mu$. $\qquad\square$

To prove that the converse of Theorem 3 holds as well, we define $\mathcal{S}^2_\mu$ as the TRS $\mathcal{R}^2_\mu$ without the two rewrite rules for top. The following lemma states that we do not have to worry about $\mathcal{S}^2_\mu$.

Lemma 4. *The TRS S_μ^2 is terminating for any CSRS $(\mathcal{R}, \mu)$.*

Proof. Let $\mathcal{F}$ be the signature of $(\mathcal{R}, \mu)$. The rewrite rules of S_μ^2 are oriented from left to right by $\succ_{\text{rpo}}$, the recursive path order [5] induced by the following precedence $\succ$ on $\mathcal{F}'$:

$$\text{active} \succ f' \succ \text{mark} \succ \text{proper} \succ f \succ c \succ \text{ok}$$

for every non-constant $f \in \mathcal{F}$ and every constant $c \in \mathcal{F}$. Since $\succ$ is well-founded, it follows that S_μ^2 is terminating. □

Now we are ready to present the main theorem of the paper.

Theorem 4. *Let $(\mathcal{R}, \mu)$ be a CSRS over a signature $\mathcal{F}$. If $(\mathcal{R}, \mu)$ is terminating then $\mathcal{R}_\mu^2$ is terminating.*

Proof. First note that the precedence used in the proof of Lemma 4 cannot be extended to deal with the whole of $\mathcal{R}_\mu^2$ as the second rewrite rule for top requires ok $\succ$ active. Since $\mathcal{R}_\mu^2$ lacks collapsing rules, it is sufficient to prove termination of any *typed* version of $\mathcal{R}_\mu^2$, cf. [16, 13]. Thus we may assume that the function symbols of $\mathcal{R}_\mu^2$ come from a many-sorted signature, where the only restriction is that the left and right-hand side of any rewrite rule are well-typed and of the same type. We use two sorts α and β, with top of type $\alpha \to \beta$ and all other symbols of type $\alpha \times \ldots \times \alpha \to \alpha$. So if $\mathcal{R}_\mu^2$ allows an infinite reduction then there exists an infinite reduction of well-typed terms. Since both types contain a ground term, we may assume for a proof by contradiction that there exists an infinite reduction starting from a well-typed *ground* term t. Terms of type α are terminating by Lemma 4 since they cannot contain the symbol top and thus the only applicable rules stem from S_μ^2. So t is a ground term of type β, which implies that $t = \text{top}(t')$ with t' of type α. Since t' is terminating, the infinite reduction starting from t must contain a root reduction step. So t' reduces to mark(t_1) or ok(t_1) for some term t_1 (of type α). We consider the former possibility, the latter possibility is treated in a very similar way. The infinite reduction starts with

$$t \to_{\mathcal{R}_\mu^2}^* \text{top}(\text{mark}(t_1)) \to_{\mathcal{R}_\mu^2} \text{top}(\text{proper}(t_1)).$$

Since proper(t_1) is of type α and thus terminating, after some further reduction steps another step takes place at the root. This is only possible if proper(t_1) reduces to ok(t_2) for some term t_2. According to Lemma 2 we must have $t_1 = t_2 \in \mathcal{T}(\mathcal{F})$. Hence the presupposed infinite reduction continues as follows:

$$\text{top}(\text{proper}(t_1)) \to_{\mathcal{R}_\mu^2}^+ \text{top}(\text{ok}(t_1)) \to_{\mathcal{R}_\mu^2} \text{top}(\text{active}(t_1)).$$

Repeating this kind of reasoning reveals that the infinite reduction must be of the following form, where all root reduction steps between top(proper(t_1)) and top(mark(t_3)) are made explicit:

$$t \to_{\mathcal{R}_\mu^2}^* \text{top}(\text{proper}(t_1)) \to_{\mathcal{R}_\mu^2}^+ \text{top}(\text{ok}(t_1)) \to_{\mathcal{R}_\mu^2} \text{top}(\text{active}(t_1)) \to_{\mathcal{R}_\mu^2}^+ \text{top}(\text{mark}(t_2))$$
$$\to_{\mathcal{R}_\mu^2} \text{top}(\text{proper}(t_2)) \to_{\mathcal{R}_\mu^2}^+ \text{top}(\text{ok}(t_2)) \to_{\mathcal{R}_\mu^2} \text{top}(\text{active}(t_2)) \to_{\mathcal{R}_\mu^2}^+ \text{top}(\text{mark}(t_3))$$
$$\to_{\mathcal{R}_\mu^2} \cdots$$

Hence $\mathsf{active}(t_i) \to_{\mathcal{R}_\mu^2}^+ \mathsf{mark}(t_{i+1})$ and $t_i \in \mathcal{T}(\mathcal{F})$ for all $i \geqslant 1$. We obtain

$$t_1 \to_{\mathcal{R},\mu} t_2 \to_{\mathcal{R},\mu} t_3 \to_{\mathcal{R},\mu} \cdots$$

from Lemma 3, contradicting the termination of $(\mathcal{R}, \mu)$. □

5 Conclusion and Further Work

In this paper we presented two new transformations from CSRSs to TRSs whose purpose is to reduce the problem of proving termination of CSRSs to the problem of proving termination of TRSs. The advantage of such an approach is that all termination techniques for ordinary term rewriting (including future developments) become available for context-sensitive rewriting as well. So in particular, these techniques can now also be used to analyze the termination behaviour of lazy functional programs which may be modelled by CSRSs. Our first transformation is simple, sound, and appears to be more powerful than previously suggested transformations. Our second transformation is not only sound but also complete, so it transforms every terminating CSRS into a terminating TRS.

Our transformations also form a basis for *automated* termination proofs of CSRSs. Of course, a direct termination proof of $\mathcal{R}_\mu^2$ cannot be obtained by a path order amenable to automation and even a powerful method like the dependency pair approach often will not succeed in finding a fully automated termination proof. To a lesser extent this is already true for our first transformation. However, our transformations are suitable for changes in their presentation which do not result in any significant change in their behaviour, but which ease the termination proofs of the resulting TRSs considerably.

For instance, for the first transformation an obvious idea is to normalize the right-hand sides of the $\mathsf{active}(l) \to \mathsf{mark}(r)$ rules with respect to the subsystem $\mathcal{M}$. Another natural idea is to replace the single symbol active by fresh symbols f_{active} for every $f \in \mathcal{F}$. This amounts to replacing every occurrence of the pattern $\mathsf{active}(f(\cdots))$ in the rewrite rules by $f_{\mathsf{active}}(\cdots)$ as well as expanding the rule $\mathsf{active}(x) \to x$ into all rules of the form $f_{\mathsf{active}}(x_1, \ldots, x_n) \to f(x_1, \ldots, x_n)$. If we apply both ideas to the TRS $\mathcal{R}_\mu^1$ of Example 2 we obtain the TRS

$\mathsf{g}_{\mathsf{active}}(x) \to \mathsf{h}_{\mathsf{active}}(x)$	$\mathsf{mark}(\mathsf{g}(x)) \to \mathsf{g}_{\mathsf{active}}(x)$	$\mathsf{g}_{\mathsf{active}}(x) \to \mathsf{g}(x)$
$\mathsf{c}_{\mathsf{active}} \quad\to \mathsf{d}_{\mathsf{active}}$	$\mathsf{mark}(\mathsf{h}(x)) \to \mathsf{h}_{\mathsf{active}}(x)$	$\mathsf{h}_{\mathsf{active}}(x) \to \mathsf{h}(x)$
$\mathsf{h}_{\mathsf{active}}(\mathsf{d}) \to \mathsf{g}_{\mathsf{active}}(\mathsf{c})$	$\mathsf{mark}(\mathsf{c}) \quad\to \mathsf{c}_{\mathsf{active}}$	$\mathsf{c}_{\mathsf{active}} \quad\to \mathsf{c}$
	$\mathsf{mark}(\mathsf{d}) \quad\to \mathsf{d}_{\mathsf{active}}$	$\mathsf{d}_{\mathsf{active}} \quad\to \mathsf{d}$

which is compatible with $\succ_{\mathrm{rpo}}$ for the precedence $\mathsf{mark} \succ \mathsf{c}_{\mathsf{active}} \succ \mathsf{d}_{\mathsf{active}} \succ \mathsf{d} \succ \mathsf{c} \succ \mathsf{g}_{\mathsf{active}} \succ \mathsf{g} \succ \mathsf{h}_{\mathsf{active}} \succ \mathsf{h}$.

Refinements like those mentioned above should be studied further. Termination of the TRS resulting from our first (incomplete) transformation is sometimes easier to prove than termination of the TRS resulting from our second (complete) one. Thus, we conclude by stating that while our second transformation is superior to all previous incomplete ones, at present our incomplete transformation of Sect. 3 as well as the ones of Lucas [10] and Zantema [17] may still be useful for the purpose of automation. In addition, the latter paper contains a complete semantic characterization of context-sensitive rewriting which can be used in a direct termination proof attempt.

References

1. T. Arts and J. Giesl. Automatically proving termination where simplification orderings fail. In *Proceedings of the 7th International Joint Conference on the Theory and Practice of Software Development*, volume 1214 of *LNCS*, pages 261–273, 1997.
2. T. Arts and J. Giesl. Proving innermost normalisation automatically. In *Proceedings of the 8th International Conference on Rewriting Techniques and Applications*, volume 1232 of *LNCS*, pages 157–172, 1997.
3. T. Arts and J. Giesl. Termination of term rewriting using dependency pairs. *Theoretical Computer Science*, 1999. To appear.
4. F. Baader and T. Nipkow. *Term Rewriting and All That*. Cambridge University Press, 1998.
5. N. Dershowitz. Orderings for term-rewriting systems. *Theoretical Computer Science*, 17:279–301, 1982.
6. N. Dershowitz. Termination of rewriting. *Journal of Symbolic Computation*, 3:69–116, 1987.
7. N. Dershowitz and J.-P. Jouannaud. Rewrite systems. In *Handbook of Theoretical Computer Science*, volume B, pages 243–320. Elsevier, 1990.
8. B. Gramlich. Abstract relations between restricted termination and confluence properties of rewrite systems. *Fundamenta Informaticae*, 24:3–23, 1995.
9. J. W. Klop. Term rewriting systems. In *Handbook of Logic in Computer Science, Vol. 2*, pages 1–116. Oxford University Press, 1992.
10. S. Lucas. Termination of context-sensitive rewriting by rewriting. In *Proceedings of the 23rd International Colloquium on Automata, Languages and Programming*, volume 1099 of *LNCS*, pages 122–133, 1996.
11. S. Lucas. Context-sensitive computations in functional and functional logic programs. *Journal of Functional and Logic Programming*, 1:1–61, 1998.
12. A. Middeldorp, H. Ohsaki, and H. Zantema. Transforming termination by self-labelling. In *Proceedings of the 13th International Conference on Automated Deduction*, volume 1104 of *LNAI*, pages 373–387, 1996.
13. H. Ohsaki and A. Middeldorp. Type introduction for equational rewriting. In *Proceedings of the 4th Symposium on Logical Foundations of Computer Science*, volume 1234 of *LNCS*, pages 283–293, 1997.
14. J. Steinbach. Simplification orderings: History of results. *Fundamenta Informaticae*, 24:47–87, 1995.
15. Y. Toyama. Counterexamples to the termination for the direct sum of term rewriting systems. *Information Processing Letters*, 25:141–143, 1987.
16. H. Zantema. Termination of term rewriting: Interpretation and type elimination. *Journal of Symbolic Computation*, 17:23–50, 1994.
17. H. Zantema. Termination of context-sensitive rewriting. In *Proceedings of the 8th International Conference on Rewriting Techniques and Applications*, volume 1232 of *LNCS*, pages 172–186, 1997.

Context-Sensitive AC-Rewriting

M. C. F. Ferreira and A. L. Ribeiro

Dep. de Informática, Fac. de Ciências e Tecnologia,
Univ. Nova de Lisboa, 2825-114 Caparica, Portugal
{cf,ar}@di.fct.unl.pt

Abstract. *Context-sensitive rewriting* was introduced in [7] and consists of syntactical restrictions imposed on a Term Rewriting System indicating how reductions *can* be performed. So *context-sensitive rewriting* is a restriction of the usual rewrite relation which reduces the reduction space and allows for a finer control of the reductions of a term. In this paper we extend the concept of context-sensitive rewriting to the framework *rewriting modulo an associative-commutative theory* in two ways: by restricting reductions and restricting AC-steps, and we then study this new relation with respect to the property of termination.

1   Introduction

The concept of *context-sensitive rewriting* was introduced by Lucas [7] and consists of syntactical restrictions imposed on a Term Rewriting System (TRS) indicating how reductions *can* be performed; this is achieved by associating to each function symbol a set of positions, namely the positions where reductions can be performed. So *context-sensitive rewriting* is a restriction of the usual rewrite relation which reduces the reduction space and allows for a finer control of the reductions of a term.

Context-sensitive rewriting is also related to lazy evaluation strategies in functional programming: the same mechanism used for indicating where reductions can (and cannot) take place can also be used for indicating which reductions are not needed. As an example consider the definition of the if-then-else [8]:

$$\text{if-then-else}(true, t, e) \rightarrow t \qquad \text{if-then-else}(false, t, e) \rightarrow e$$

where we only want to eventually evaluate the "then"or "else"-branch once we know what the result of evaluating the condition is. This restriction can be achieved in context-sensitive rewriting by stating that we are only allowed to evaluate the first argument of terms headed by the symbol if-then-else. Then when reducing any term of the form if-then-else(c, t, e), unwanted reductions can be avoided.

Of the properties of TRS's, *confluence* and *termination* are two of the most relevant, ensuring that a computation always yields the same result and that that result exists, respectively. Context-sensitive rewriting has been studied with respect to termination in [9, 15] and with respect to confluence in [6, 8]; for a thorough account on this topic see [10]. However, many interesting and useful systems have operators which are *associative* and *commutative* (AC), and if one wants to fully capture the functionality of such operators, rewriting has to be replaced by *AC-rewriting*, i. e., rewriting modulo an associative-commutative theory. For example, if one wants to incorporate the "and" and "or" operators to the system describing the if-then-else, given that these are AC operators, one should now consider AC-rewriting.

In this paper we extend the concept of context-sensitive rewriting to AC-rewriting by restricting both reduction steps and AC-steps and study the property of termination for this new rewrite relation following the approach of [15] for termination of context-sensitive rewriting. As in the case of rewriting, context-sensitive AC-rewriting preserves AC-termination but goes further since there are systems which are context-sensitive AC-terminating but not AC-terminating. This makes context-sensitive (AC) rewriting particularly interesting for dealing with systems with infinite data structures since they are inherently non-terminating.

The rest of the paper is organized as follows. In section 2 we review some concepts relative to AC-rewriting and context-sensitive rewriting. In section 3, we introduce the notion of context-sensitive AC-rewriting, context-sensitive quasi-orders and interpretations and we show how to formulate termination of a context-sensitive AC-relation in terms of these entities. In section 4 we present a transformation which allows to translate context-sensitive AC-termination to AC-termination, and prove the soundness of this (not complete) transformation. In section 5 we present some examples and we conclude in section 6. Due to space restrictions, most proofs are omitted (see [2] for full proofs).

2 Preliminaries

We introduce some notation and give some basic notions over orders, TRS's and AC-rewriting. For more information the reader is referred to [1, 4, 13].

A *poset* $(S, \succ)$ is a set S together with a partial order, $\succ$; we say that $\succ$ is well-founded if there are no infinite sequences of the form $s_0 \succ s_1 \succ \cdots$. *Quasi-orders* over a set S are denoted, in general, by $\succeq$. Any quasi-order defines an equivalence relation, namely $\succeq \cap \preceq$ (denoted by $\sim$), and a partial order, which we consider to be $\succeq \setminus \preceq$, denoted by $\succ$. Conversely, the union of a given partial order $\succ$ and equivalence $\sim$ is a quasi-order only if they satisfy $(\sim \circ \succ \circ \sim) = \succ$, where $\circ$ represents relation composition; in this case we say that $\succ$ and $\sim$ are *compatible*. A quasi-order $\succeq$ is well-founded if $\succ$ is.

$\mathcal{T}(\mathcal{F}, \mathcal{X})$ denotes the set of terms over $\mathcal{F}$, a non-empty (possibly infinite) set of function symbols, and $\mathcal{X}$, a denumerable set of variables such that $\mathcal{F} \cap \mathcal{X} = \emptyset$. For technical and practical reasons we assume that our set of terms is never empty. For $t \in \mathcal{T}(\mathcal{F}, \mathcal{X})$, the set $var(t)$ contains the variables occurring in t. For $f \in \mathcal{F}$, $ar(f)$ denotes the arity of f.

A *TRS* is a tuple $(\mathcal{F}, \mathcal{X}, R)$, with $R \subseteq \mathcal{T}(\mathcal{F}, \mathcal{X}) \times \mathcal{T}(\mathcal{F}, \mathcal{X})$. The elements (l, r) of R are the rules of the TRS and are usually denoted by $l \to r$; we require that they satisfy $l \notin \mathcal{X}$ and $var(r) \subseteq var(l)$. In the following, unless otherwise specified, we identify the TRS with R, being $\mathcal{F}$ the set of function symbols occurring in R. A TRS $(\mathcal{F}, \mathcal{X}, R)$ induces a *reduction relation* on $\mathcal{T}(\mathcal{F}, \mathcal{X})$, denoted by $\rightarrow_R$, as follows: $s \rightarrow_R t$ (s *reduces* or *rewrites* to t) if and only if $s = C[l^\sigma]$ and $t = C[r^\sigma]$, for some linear context C^1, substitution $\sigma : \mathcal{X} \rightarrow \mathcal{T}(\mathcal{F}, \mathcal{X})$ and rule $l \rightarrow r \in R$. The transitive closure of $\rightarrow_R$ is denoted by $\rightarrow_R^+$ and its reflexive-transitive closure by $\rightarrow_R^*$. By $\rightarrow_R^n$, $n \in \mathbb{N}$, we denote the composition of $\rightarrow_R$ with itself n times (if $n = 0$ then $\rightarrow_R^n$ is the identity). A *rewrite sequence* is a sequence of reduction steps $t_0 \rightarrow_R t_1 \rightarrow_R \cdots$.

A binary relation Θ over $\mathcal{T}(\mathcal{F}, \mathcal{X})$ is: *terminating* if there are no infinite sequences $s_0 \Theta s_1 \Theta s_2 \ldots$; *closed under substitutions* if $s\Theta t \Rightarrow s^\sigma \Theta t^\sigma$, for any substitution $\sigma : \mathcal{X} \rightarrow \mathcal{T}(\mathcal{F}, \mathcal{X})$; *closed under contexts* if $s\Theta t \Rightarrow C[s]\Theta C[t]$, for any linear context C; a *congruence* if it is an equivalence relation closed under contexts; *subterm compatible*, or having the *subterm property* if $C[s]\Theta s$, for any non-trivial context C.

A partial order $\succ$ (resp. quasi-order $\succeq$) on $\mathcal{T}(\mathcal{F}, \mathcal{X})$ is a *rewrite order* (resp. *rewrite quasi-order*) if it is closed under contexts and substitutions (resp. both $\succ$ and $\sim$ are closed under contexts and substitutions); a *reduction (quasi-)order* is a well-founded rewrite (quasi-)order.

An *equation* (or *axiom*) over $\mathcal{T}(\mathcal{F}, \mathcal{X})$ is a pair of terms (s, t); an *equational system* over $\mathcal{T}(\mathcal{F}, \mathcal{X})$ is a set of equations. An important class of equational systems are the *permutative* or length-preserving theories [12]. This class comprises the so-called *AC-theories*, where A stands for associativity and C for commutativity. AC-theories contain only *associative* and *commutative* axioms which have the respective forms $f(x, f(y, z)) = f(f(x, y), z)$ and $f(x, y) = f(y, x)$, for binary function symbol f. Given a signature $\mathcal{F}$, we denote by $\mathcal{F}_{AC}$ the subset of $\mathcal{F}$ containing the function symbols which are AC.

Any equational system EQ generates a congruence on the set of terms; we denote by $=_{EQ}$ the least congruence closed under substitutions containing EQ. Without loss of generality, we assume that any equational system is symmetric, i. e., if $(s, t) \in EQ$ then also $(t, s) \in EQ$; however for the sake of simplicity, when expressing EQ extensively we omit symmetric equations. With this assumption, the equational theory generated by a set of equations becomes:

Definition 1. *The equational theory generated by an equational system EQ is denoted by $=_{EQ}$ and is the least congruence on $\mathcal{T}(\mathcal{F}, \mathcal{X})$ containing EQ and closed under substitutions, i. e., $s =_{EQ} t$ iff either $s = t$, or $s = C[e_1^\sigma]$ and $t = C[e_2^\sigma]$, for some equation $(e_1, e_2) \in EQ$, context C and substitution σ, or $s =_{EQ} u$ and $u =_{EQ} t$, for some term u.*

Definition 2. *An equational rewrite system R/EQ consists of a TRS R and an equational system EQ, both defined over the same set of terms. Its associated equational rewrite relation $\rightarrow_{R/EQ}$ is given by: $s \rightarrow_{R/EQ} t$ iff there are terms u, v such that $s =_{EQ} u \rightarrow_R v =_{EQ} t$. We speak of* equational rewriting *or* rewriting modulo a set of equations.

---

[1] A linear context is a context with a single occurrence of the trivial context $\square$.

A TRS is *terminating* if its rewrite relation $\rightarrow_R$ is terminating (i.e., there are no infinite rewrite sequences). If EQ is an equational system and R a TRS, we say that R is E-terminating (or that R/EQ is terminating) if the relation $\rightarrow_{R/EQ}$ is terminating, i. e., if there are no infinite sequences of the form: $s_0 =_{EQ} s_0' \rightarrow_R s_1 =_{EQ} s_1' \rightarrow_R s_2 =_{EQ} s_2' \rightarrow_R s_3 \ldots$

An equational rewrite system R/EQ is compatible with a quasi-order $\succeq = \succ \cup \sim$ (on $\mathcal{T}(\mathcal{F}, \mathcal{X})$) if $=_{EQ} \subseteq \sim$ and $\rightarrow_R \subseteq \succ$. We have that [3, 5, 11]:

Theorem 1. *A rewrite system is terminating if and only if it is compatible with a reduction order. An equational rewrite system is terminating if and only if it is compatible with a reduction quasi-order.*

Definition 3. *[7] Given a signature $\mathcal{F}$, a replacement map for $\mathcal{F}$ is a function $\mu : \mathcal{F} \rightarrow \mathcal{P}(\mathbb{N})$ such that $\mu(f) \subseteq \{1, \ldots, ar(f)\}$, for any $f \in \mathcal{F}$.*

The replacement function indicates which positions of a term may be reduced. For example if $\mu(f) = \{1, 3\}$ then the term $f(t_1, t_2, t_3)$ may be reduced by reducing t_1, t_3[2], or the whole term; reductions in t_2 are not allowed.

The replacement function is the basis for the definition of context-sensitive reduction. We take the definition given in [15], instead of the original one: they are nevertheless equivalent[3].

Definition 4. *Let μ be a replacement map, and let R be a TRS. The context-sensitive rewrite relation $\hookrightarrow_{R,\mu}$ over $\mathcal{T}(\mathcal{F}, \mathcal{X})$ is the least relation satisfying*

- $l^\sigma \hookrightarrow_{R,\mu} r^\sigma$, *for any rule $l \rightarrow r \in R$ and substitution $\sigma : \mathcal{X} \rightarrow \mathcal{T}(\mathcal{F}, \mathcal{X})$*
- $f(t_1, \ldots, t_{i-1}, t, t_{i+1}, \ldots, t_n) \hookrightarrow_{R,\mu} f(t_1, \ldots, t_{i-1}, u, t_{i+1}, \ldots, t_n)$, *if $t \hookrightarrow_{R,\mu} u$, $f \in \mathcal{F}$ and $i \in \mu(f)$, for all $t_1, \ldots, t_{i-1}, t_{i+1}, \ldots, t_n \in \mathcal{T}(\mathcal{F}, \mathcal{X})$*

Analogous to the notion of termination and E-termination, we can define μ-termination: given a TRS R and a replacement map μ, we say that R is μ-terminating if the relation $\hookrightarrow_{R,\mu}$ is terminating.

The following notion restricts the property closedness under contexts.

Definition 5. *Given a binary relation Θ over $\mathcal{T}(\mathcal{F}, \mathcal{X})$ and a replacement map μ, if $s \Theta t$ implies that $f(\ldots, u_{i-1}, s, u_{i+1}, \ldots) \Theta f(\ldots, u_{i-1}, t, u_{i+1}, \ldots)$, whenever $i \in \mu(f)$, we say that Θ is closed under μ-contexts.*

3 Context-Sensitive Reduction Quasi-Orders and Interpretations

We start by extending the definition of context-sensitive rewriting (Def. 4) to the AC-framework. The first thing that needs to be done is to define what the replacement map can do to AC-symbols. Recall that the only requirement

---

[2] Of course reductions in the subterms may themselves be restricted.

[3] In [7, 6] the possibility of defining context-sensitive reduction as in Definition 4 is also pointed out.

we have on a replacement map μ is that $\mu(f) \subseteq \{1,\ldots,ar(f)\}$. If the function symbol happens to be an AC-symbol, then some values of $\mu(f)$ do not make sense (this is essentially due to commutativity, since we can permute the subterms and bring terms in forbidden positions to positions which are not forbidden). Therefore for AC-symbols f either $\mu(f) = \emptyset$ or $\mu(f) = \{1,2\}$.

As mentioned before we will restrict both the rewrite and the AC-steps so we will work with two replacement maps μ_r and μ_{ac} denoting, respectively, the replacement map for rewriting and the replacement map for AC-steps; in the following and unless otherwise specified, we will use a generic μ to denote a pair of replacement maps (μ_r, μ_{ac}).

Definition 6. *The restricted equational theory generated by an equational system EQ and replacement map μ is denoted by $=_{EQ,\mu}$ and is the least relation over $\mathcal{T}(\mathcal{F}, \mathcal{X})$ that satisfies*

- $s =_{EQ,\mu} s$,
- $e_1^\sigma =_{EQ,\mu} e_2^\sigma$, *for any equation* $(e_1, e_2) \in EQ$,
- $f(\ldots, t_{i-1}, u, t_{i+1}, \ldots) =_{EQ,\mu} f(\ldots, t_{i-1}, v, t_{i+1} \ldots)$, *for any terms u, v such that $u =_{EQ,\mu} v$, and any position $i \in \mu(f)$,*
- $s =_{EQ,\mu} t$, *if there exists some u such that $s =_{EQ,\mu} u$, and $u =_{EQ,\mu} t$.*

Obviously for any equational system EQ and replacement map μ we have $=_{EQ,\mu} \subseteq =_{EQ}$, so in particular $=_{AC,\mu_{ac}} \subseteq =_{AC}$.

Definition 7. *Given an AC-rewrite system R/AC and replacement maps $\mu = (\mu_r, \mu_{ac})$, the context-sensitive AC-rewrite relation $\hookrightarrow_{R/AC,\mu}$ over $\mathcal{T}(\mathcal{F}, \mathcal{X})$ is given by: $s \hookrightarrow_{R/AC,\mu} t \Leftrightarrow \exists s', t' : s =_{AC,\mu_{ac}} s' \hookrightarrow_{R,\mu_r} t' =_{AC,\mu_{ac}} t$*

The classical way of proving termination of TRS's is to find a suitable reduction order compatible with the rewrite rules (Theorem 1); equivalently one can find a suitable $\mathcal{F}$-algebra $(\mathcal{A}, \succ)$, such that the operations interpreting the elements of $\mathcal{F}$ are monotone and the order $\succ$ is well-founded. In [15], the notions of reduction order and interpretation were extended for context-sensitive rewriting. Following this approach, we are going to define μ-reduction quasi-orders and interpretations for context-sensitive AC-rewriting; we will then show that μ-AC-termination can be characterized in terms of μ-reduction quasi-orders and interpretations. First we define μ-AC-termination.

Definition 8. *Given an AC-rewrite system R/AC and replacement maps $\mu = (\mu_r, \mu_{ac})$, we say that R is μ-AC-terminating, or R/AC is μ-terminating, if the relation $\hookrightarrow_{R/AC,\mu}$ is terminating, i. e., if there are no infinite context-sensitive AC-rewrite sequences, $s_0 \hookrightarrow_{R/AC,\mu} s_1 \hookrightarrow_{R/AC,\mu} s_2 \hookrightarrow_{R/AC,\mu} \cdots$*

We now introduce the notion of μ-reduction quasi-order which is an extension of the μ-reduction order in [15], but also accounting for the equivalence part.

Definition 9. *Let $\mu = (\mu_r, \mu_{ac})$ be replacement maps. A μ-reduction quasi-order on $\mathcal{T}(\mathcal{F}, \mathcal{X})$ is defined to be a well-founded quasi-order $\succeq$ on $\mathcal{T}(\mathcal{F}, \mathcal{X})$ such that $\succ$ is closed under substitutions and μ_r-contexts, and $\sim$ is closed under substitutions and μ_{ac}-contexts.*

Next we are going to define the concept of compatibility of a μ-reduction quasi-order with an AC-term rewriting system.

Definition 10. *A μ-reduction quasi-order $\succeq$ on $\mathcal{T}(\mathcal{F}, \mathcal{X})$ is called* compatible *with an AC-TRS R/AC iff $l \succ r$ for every rewrite rule $l \to r \in R$, and $e_1 \sim e_2$ for every equation $(e_1, e_2) \in AC$.*

The lemma that follows relates a μ-reduction quasi-order compatible with an AC-TRS with the context-sensitive AC-rewrite relation associated with it.

Lemma 1. *Let $\succeq$ be a μ-reduction quasi-order compatible with R/AC, and let $u, v \in \mathcal{T}(\mathcal{F}, \mathcal{X})$ be arbitrary. Then (1) if $u \hookrightarrow_{R,\mu_r} v$ then $u \succ v$, and (2) if $u =_{AC,\mu_{ac}} v$ then $u \sim v$.*

We can now state a result analogous to Theorem 1 but for context-sensitive AC-rewriting, i. e., we show that μ-termination of an AC-TRS can be shown by finding a compatible μ-reduction quasi-order.

Theorem 2. *An AC-TRS R/AC is μ-terminating iff it admits a compatible μ-reduction quasi-order $\succeq$ on $\mathcal{T}(\mathcal{F}, \mathcal{X})$.*

As was mentioned before, another classical way of proving termination of rewrite relations is to define a monotone $\mathcal{F}$-algebra together with a well-founded order such that reductions are translated via the interpretation to decreasing steps in the order. This notion of interpretation was adapted for context-sensitive rewrite relations in [15]; we adapt this notion for context-sensitive AC-rewriting.

Given a signature $\mathcal{F}$, an $\mathcal{F}$-algebra is defined as usual, i. e., a structure consisting of a non-empty set A and an interpretation function $f_A : A^n \to A$, $(n \geq 0)$ for each function symbol $f \in \mathcal{F}$. The definition of term evaluation is also standard. Given an assignment $\alpha : \mathcal{X} \to A$, we define the term evaluation $[\alpha] : \mathcal{T}(\mathcal{F}, \mathcal{X}) \to A$, as: (1) $[\alpha]x = \alpha(x)$, for any $x \in \mathcal{X}$, and (2) $[\alpha]f(t_1, \ldots, t_n) = f_A([\alpha]t_1, \ldots, [\alpha]t_n)$, for any $f \in \mathcal{F}$, and terms $t_1, \ldots, t_n \in \mathcal{T}(\mathcal{F}, \mathcal{X})$.

In order to use $\mathcal{F}$-algebras for proving termination we still need to require some more properties. In general, one requires that the underlying set of the algebra is well-ordered by some well-founded partial order and that the interpretation functions f_A are monotone in all arguments. For context-sensitive relations this last requirement can be relaxed since we only need to require monotonicity for arguments that may be reduced (μ_r-monotonicity; this is the approach taken in [15]), and for context-sensitive AC-relations one needs to impose extra requirements, namely that the interpretation and the order are somehow compatible with the restricted AC-theory (this actually means that a partial order is not suitable for our purposes and we need to work with *quasi-orders*).

Definition 11. *Let $\mu = (\mu_r, \mu_{ac})$. A μ-monotone $\mathcal{F}$-algebra $(A, \mathcal{F}, \succeq)$ is an $\mathcal{F}$-algebra $(A, \mathcal{F})$ provided with a quasi-order $\succeq$ in A, such that*

- *every interpretation function f_A is μ_r-monotone, i. e., if $f \in \mathcal{F}$, $i \in \mu_r(f)$, $a, b \in A$ and $a \succ b$ then $f_A(\ldots, a_{i-1}, a, a_{i+1}, \ldots) \succ f_A(\ldots, a_{i-1}, b, a_{i+1}, \ldots)$, for all $a_j \in A, 1 \leq j \leq n, j \neq i$.*

    − *every interpretation function f_A is stable on μ_{ac}-arguments with respect to the equivalence $\sim$, i.e., if $f \in \mathcal{F}$, $i \in \mu_{ac}(f)$, $a, b \in A$ and $a \sim b$ then $f_A(\ldots, a_{i-1}, a, a_{i+1}, \ldots) \sim f_A(\ldots, a_{i-1}, b, a_{i+1}, \ldots)$, for all $a_j \in A$, $1 \leq j \leq n$, $j \neq i$.*

If additionally $\succeq$ is well-founded, we say that the algebra is well-founded.

As usual a monotone algebra induces an order relation on the set of terms.

Definition 12. *Let $\mu = (\mu_r, \mu_{ac})$. Let $\mathcal{A} = (A, \mathcal{F}, \succeq)$ be a μ-monotone $\mathcal{F}$-algebra. The quasi-order $\succeq_A$ induced by $\mathcal{A}$ on $\mathcal{T}(\mathcal{F}, \mathcal{X})$ is defined by: $t \succ_A t'$ if and only if $[\alpha](t) \succ [\alpha](t')$, for all $\alpha : \mathcal{X} \rightarrow A$; and $t \sim_A t'$ if and only if $[\alpha](t) \sim [\alpha](t')$, for all $\alpha : \mathcal{X} \rightarrow A$.*

It is not difficult to see that $\succeq_A = \succ_A \cup \sim_A$ is indeed well-defined and a quasi-order.

    The following lemma establishes that a (μ_r, μ_{ac})-monotone $\mathcal{F}$-algebra induces a (μ_r, μ_{ac})-reduction quasi-order on terms.

Lemma 2. *Let $\mu = (\mu_r, \mu_{ac})$. Let $\mathcal{A} = (A, \mathcal{F}, \succeq)$ be a non-empty well founded μ-monotone $\mathcal{F}$-algebra. Then the relation $\succeq_A$ on $\mathcal{T}(\mathcal{F}, \mathcal{X})$ is a μ-reduction quasi-order.*

As a consequence of the last statement and Theorem 2, we are now able to say that finding a non-empty well-founded μ-monotone $\mathcal{F}$-algebra, which induces a quasi-order on terms, is another way of proving μ-termination of an AC-TRS.

Theorem 3. *Let μ_r and μ_{ac} be replacement maps for $\mathcal{F}$, and let $\mu = (\mu_r, \mu_{ac})$. An AC-TRS R/AC is μ-terminating if and only if a non-empty well-founded μ-monotone $\mathcal{F}$-algebra $\mathcal{A} = (A, \mathcal{F}, \succeq)$ exists for which $\succeq_A$ is compatible with R/AC.*

Proof. If there exists a non-empty well-founded μ-monotone $\mathcal{F}$-algebra compatible with R/AC, the conclusion that R/AC is μ-terminating comes from applying Lemma 2 and then Theorem 2. Conversely, if R/AC is μ-terminating then the algebra obtained by taking $A = \mathcal{T}(\mathcal{F}, \mathcal{X})$, interpreting each function symbol by itself, and by taking $\succeq = (\hookrightarrow^+_{R/AC, \mu} \cup =_{AC, \mu_{ac}})$ is a non-empty[4] well-founded, μ-monotone and compatible with R/AC.

4   Relating Context-Sensitive AC-Rewriting and AC-Rewriting

In [15] a transformation is given that allows to study context-sensitive termination in terms of termination of a non-context-sensitive relation. In this section we modify and extend the method for context-sensitive AC-rewriting: we present a new transformation which encompasses the one given in [15] and adapt it to

---

[4] Recall that $\mathcal{T}(\mathcal{F}, \mathcal{X})$ is non-empty.

AC-rewriting. Before doing so, we give an intuitive idea of the transformation. The goal is to introduce the effects of the replacement map in the TRS directly so that one obtains a different TRS such that its termination implies μ-termination of the original. The simplest way of doing so is to replace each forbidden position in a term by a new constant, and apply this transformation to the left and right-hand sides of every rule. We show this transformation with an example.

Example 1. [8] Let R be given by the rules

$$sel(0, h : t) \rightarrow h \quad sel(s(n), h : t) \rightarrow sel(n, t) \quad from(n) \rightarrow n : from(s(n))$$

These rules describe the selection of an element of order n from an arbitrarily long list. The list constructor is represented by the infix function symbol ":". Clearly the system is not terminating, however using a lazy reduction strategy we would only evaluate $from(s(n))$ whenever that evaluation would be necessary for obtaining the result. This can be simulated with context-sensitive rewriting by stating that $\mu_r(:) = \{1\}$ ($\mu_r(f) = \{1, \ldots, ar(f)\}$ for all other function symbols), disallowing reductions in the second argument of ":". If we would apply our naïve transformation to the system, we would get (where $\diamond$ is a fresh constant):

$$sel(0, h : \diamond) \rightarrow h \quad sel(s(n), h : \diamond) \rightarrow sel(n, t) \quad from(n) \rightarrow n : \diamond$$

and this is not even a TRS.

We can already see what is wrong with this transformation: we may have variables which occur in forbidden positions in the left-hand side but in non-forbidden positions in the right-hand side and this transformation then loses those variables resulting in a system which is not even a TRS.

In order to avoid the problem illustrated in the example, we have to keep track of the variables which go from forbidden to visible positions when we go from l to r in a rule $l \rightarrow r$. The method given in [15] to achieve this consists in marking function symbols and variables occurring in forbidden positions. The marking has to be such that when a variable goes from forbidden to allowed positions we may be able to undo the marking. We will use the following marking: we will underline function symbols, and a variable x occurring in a forbidden position in the left-hand side of a rule will be replaced by a term $a(x)$ in the right-hand side of the rule. In this notation we depart slightly from the notation of [15] for marking terms, there function symbols are overlined instead of underlined. However in order to avoid confusion with the operation of flattening, commonly used in AC-systems and usually denoted by overlining, we replace overline by underline. We illustrate this in the previous example.

Example 2. The system corresponding to the marking of R is the following:

$$sel(0, h : t) \rightarrow h \quad sel(s(n), h : t) \rightarrow sel(n, a(t)) \quad from(n) \rightarrow n : \underline{from}(\underline{s}(n))$$

The previous example also illustrates the difference between our transformation and the one presented in [15]. Using the method from [15] the third rule would be $from(n) \rightarrow n : \underline{from}(s(n))$. With this method only head function symbols are marked, and not all function symbols.

We now give the definition of marking (underlining).

Definition 13. *For every term in $\mathcal{T}(\mathcal{F}, \mathcal{X})$, underlining is defined by*

$$\underline{x} = x \qquad\qquad for\ x \in \mathcal{X}$$
$$\underline{f(t_1,\ldots,t_n)} = \underline{f}(\underline{t_1},\ldots,\underline{t_n})\ for\ f \in \mathcal{F}$$

We formalize now the concept of variable appearing in a forbidden position; forbidden positions are always relative to a replacement map.

Definition 14. *[15] Given a term $t \in \mathcal{T}(\mathcal{F}, \mathcal{X})$ and a replacement map μ, the set $Forb(t)$ of forbidden variables of t is defined inductively as:*

- $Forb(x) = \emptyset$, *for any $x \in \mathcal{X}$,*
- $Forb(f(t_1,\ldots,t_n)) = \left(\bigcup_{i \notin \mu(f)} var(t_i)\right) \cup \left(\bigcup_{i \in \mu(f)} Forb(t_i)\right)$, *for $f \in \mathcal{F}$,*
  $t_1,\ldots,t_n \in \mathcal{T}(\mathcal{F}, \mathcal{X})$.

We say that a variable x is forbidden *or* appears in a forbidden position *wrt a term t if $x \in Forb(t)$.*

As we said before, given a rewrite rule $l \rightarrow r$, a variable x appearing in a forbidden position in l has its non-forbidden occurrences in r replaced by a term $a(x)$. This can be achieved by defining a special substitution.

Definition 15. *[15] Let μ be a replacement map. For any term t, we define the substitution $\tau(t)$ by $x^{\tau(t)} = a(x)$, if $x \in Forb(t)$ and $x^{\tau(t)} = x$, otherwise.*

For example in the rule $f(x,y,y) \rightarrow g(x,x,y)$, with $\mu_r(f) = \{2\}, \mu_r(g) = \{2\}$, we have $Forb(l) = \{x,y\}$ and applying the substitution $\tau(l)$ to r we get the term $g(a(x), a(x), a(y))$.

Notation 1. *In the following, let $\underline{\mathcal{F}} = \{\underline{f} \mid f \in \mathcal{F}\}$, $\mathcal{F}' = \mathcal{F} \cup \underline{\mathcal{F}} \cup \{a\}$, where a is a fresh unary function symbol, and $\bar{\mathcal{F}}_{AC'} = \mathcal{F}_{AC} \cup \{\underline{f} : f \in \mathcal{F}_{AC}\}$. We will denote by AC' the associative-commutative equations for the symbols in $\mathcal{F}_{AC'}$.*

The actual marking of the rewrite rules is done with the function Φ given next; this function differs from the one given in [15] in the way forbidden subterms are handled.

Definition 16. *The function $\Phi : \mathcal{T}(\mathcal{F}, \mathcal{X}) \rightarrow \mathcal{T}(\mathcal{F}', \mathcal{X})$ is defined as follows:*

- $\Phi(x) = x$, *for any $x \in \mathcal{X}$,*
- $\Phi(f(t_1,\ldots,t_n)) = \underline{f}(u_1,\ldots,u_n)$, *for any $f \in \mathcal{F}$, where $u_i = \Phi(t_i)$ if $i \in \mu(f)$, and $u_i = \underline{t_i}$, otherwise.*

Given a TRS R, the TRS $\Phi(R)$ is defined by: $\Phi(R) = \{\Phi(l) \rightarrow (\Phi(r))^{\tau(l)} \mid l \rightarrow r \in R\}$.

In order to relate μ-(AC-)termination with (AC-)termination, the function Φ is not enough. Note that this function transforms the original TRS in another one, but introduces new terms in the right-hand side of rules, corresponding to the marking of variables. To be able to simulate the μ-reductions of the original context-sensitive system, we must have the possibility of eliminating the marking of variables. This is done with another TRS that we define next.

Definition 17. *The TRS $Bar(\mathcal{F})$ over $\mathcal{F}'$ is defined by*

$$Bar(\mathcal{F}) = \begin{cases} a(\underline{f}(x_1,\ldots,x_n)) \to f(y_1,\ldots,y_n) \text{ for all } f \in \mathcal{F}, \text{ where} \\ \qquad\qquad\qquad\qquad y_i = a(x_i), \text{ if } i \in \mu(f), \text{ and} \\ \qquad\qquad\qquad\qquad y_i = x_i, \text{ if } i \notin \mu(f) \\ f(x_1,\ldots,x_n) \quad \to \underline{f}(x_1,\ldots,x_n) \text{ for all } f \in \mathcal{F} \\ a(x) \qquad\qquad\quad \to x \end{cases}$$

The difference between the system $Bar(\mathcal{F})$ given in Definition 17 and the system given in [15] is in the first rule. As we mark all function symbols appearing in forbidden positions, we must also be able to unmark them recursively when they change to non-forbidden positions.

Recall that we are working with two replacement maps μ_r and μ_{ac}; in the following all functions/definitions which depend on a replacement map (eg. $Bar(\mathcal{F})$, $\Phi(\ )$) will be taken *with respect to* the replacement map restricting the rewrite steps μ_r.

We can now state the main result of this section (like Proposition 4 in [15]).

Theorem 4. *Let R/AC be an AC-TRS over $\mathcal{F}$. Let μ_r, μ_{ac} be replacement maps for $\mathcal{F}$ such that $\mu_r(f) \neq \emptyset$, for all AC-symbols f. Let $AC'_{\mu_{ac}}$ be the AC'-theory restricted by the replacement map μ_{ac}. If $(\Phi(R) \cup Bar(\mathcal{F}))/AC'_{\mu_{ac}}$ terminates then R/AC μ-terminates.*

The converse of this result does not hold; the example given in [15] for the analogous result, with an AC-symbol added can be used to show that the converse of Theorem 4 does not hold either. As for the condition imposed on the replacement map μ_r with respect to AC-symbols, this condition is necessary in order to ensure that the function Φ is compatible with the AC-theory, as we discuss later.

Before proving Theorem 4, we present a small example showing that our transformation subsumes the one from [15].

Example 3. [15] The system $f(x) \to g(h(f(x)))$ (without AC-symbols) is μ_r-terminating for a replacement map $\mu_r(f) = \mu_r(h) = \{1\}$ and $\mu_r(g) = \emptyset$; according to the transformation from [15], the system $\Phi(R) \cup Bar(\mathcal{F})$ would contain the rule $f(x) \to g(\underline{h}(f(x)))$ which is not terminating, so μ_r-termination of the original system could not be inferred in this case; our system $\Phi(R) \cup Bar(\mathcal{F})$ would contain instead the rule $f(x) \to g(\underline{h}(\underline{f}(x)))$ and will be terminating. In general better results will be obtained by marking the subterms recursively instead of marking only the head function symbol of a subterm occurring in a position not indexed by the replacement map.

To prove Theorem 4 we show that if $s =_{AC,\mu_{ac}} t$ then $\Phi(s) =_{AC',\mu_{ac}} \Phi(t)$ and if $s \hookrightarrow_{R,\mu_r} t$ then $\Phi(s) \to^+_{\Phi(R)\cup Bar(\mathcal{F})} \Phi(t)$.

Lemma 3. *Let μ_r, μ_{ac} be replacement maps for $\mathcal{F}$ such that $\mu_r(f) \neq \emptyset$, for all AC-symbols f, let $s, t \in \mathcal{T}(\mathcal{F}, \mathcal{X})$. If $s =_{AC,\mu_{ac}} t$ then $\Phi(s) =_{AC',\mu_{ac}} \Phi(t)$ and $\underline{s} =_{AC',\mu_{ac}} \underline{t}$.*

The previous result, is not valid if we consider AC-symbols with forbidden positions. Consider the following terms $s = f(f(x,y),z))$ and $t = f(x, f(y,z))$. Then obviously $s =_{AC} t$ and $\Phi(s) = f(\underline{f}(x,y),z)$, $\Phi(t) = f(x,\underline{f}(y,z))$, and using only the AC-equations for $f, \underline{f}$, there is no way of relating $\Phi(s)$ and $\Phi(t)$. This is not surprising and can be solved at least in two different ways. One can incorporate equations of the form $f(\underline{f}(x,y),z) = f(x,\underline{f}(y,z))$ to the equational theory; this has the unpleasant consequence that the equational system no longer is an AC-system and therefore cannot be dealt with methods typically used for AC-systems. One can also work with the flattened form of terms. In that case we never have two consecutive occurrences (one rooted just below the other) of the same AC-symbol in the same term, so this problem does not arise. But one has then to develop all the theory using flattened terms and flattened rewriting, a more difficult notion to work with (see [2]).

The following result is adapted from [15].

Lemma 4. *Let $t, u \in \mathcal{T}(\mathcal{F}, \mathcal{X})$. If $t \hookrightarrow_{R,\mu_r} u$ then $\Phi(t) \to^+_{\Phi(R)\cup Bar(\mathcal{F})} \Phi(u)$.*

We can now prove the main result, Theorem 4.

Proof. **(Theorem 4)** Assume that R/AC is not μ-terminating. Then there is an infinite reduction $s_0 =_{AC,\mu_{ac}} s_1 \hookrightarrow_{R,\mu_r} s_1' =_{AC,\mu_{ac}} s_2 \hookrightarrow_{R,\mu_r} s_2' \dots$. According to Lemmas 3 and 4, respectively, if $s_i =_{AC,\mu_{ac}} s_{i+1}$ then $\Phi(s_i) =_{AC',\mu_{ac}} \Phi(s_{i+1})$, and if $s_i \hookrightarrow_{R,\mu_r} s_i'$ then $\Phi(s_i) \to^+_{\Phi(R)\cup Bar(\mathcal{F})} \Phi(s_i')$. Then $\Phi(s_0) =_{AC',\mu_{ac}} \Phi(s_1) \to^+_{\Phi(R)\cup Bar(\mathcal{F})} \Phi(s_1') =_{AC',\mu_{ac}} \Phi(s_2) \to^+_{\Phi(R)\cup Bar(\mathcal{F})} \Phi(s_2') \dots$, which contradicts $\Phi(R) \cup Bar(\mathcal{F})/AC'_{\mu_{ac}}$ being terminating.

If one wants to use automatic methods to prove termination of $(\Phi(R) \cup Bar(\mathcal{F}))/AC'_{\mu_{ac}}$ then it is better to use the full AC'-theory instead of its restricted version $AC'_{\mu_{ac}}$. The following corollary can then be of practical use.

Corollary 1. *Let R/AC be an AC-TRS over $\mathcal{F}$. Let μ_r, μ_{ac} be replacement maps for $\mathcal{F}$ such that $\mu_r(f) \neq \emptyset$, for all AC-symbols f. If $(\Phi(R)\cup Bar(\mathcal{F}))/AC'$ terminates then R/AC μ-terminates.*

4.1 Optimizing the Method

In [15], an optimization of the transformation method was given. This optimization was based on the observation that it was unnecessary to mark function symbols which did not appear in the rewrite rules of $\Phi(R)$ and therefore the set of marked symbols can be reduced (as well as the set of rules for handling the marking). We will also apply that optimization in our setting (this optimization will also affect AC-symbols). First we introduce the following notation:

Notation 2. *In the following, let*

$$\widehat{\mathcal{F}} = \{f \mid f \in \mathcal{F} \text{ and } \underline{f} \text{ occurs in } \Phi(l) \text{ or } \Phi(r), \text{ for some } l \to r \in R \}$$

and let $\widehat{\underline{\mathcal{F}}} = \{\underline{f} \mid f \in \widehat{\mathcal{F}}\}^5$. *$\widehat{AC}$ represents the AC-theory generated by $\mathcal{F}_{AC}$ plus the AC-equations for the symbols $\underline{g} \in \widehat{\underline{\mathcal{F}}}$, such that g is an AC-symbol, and $\widehat{AC}_{\mu_{ac}}$ represents the μ_{ac} restricted $\widehat{AC}$ theory.*

We define the TRS Λ which will be used for removing redundant marks.

Definition 18. *The TRS Λ over $\mathcal{T}(\mathcal{F}', \mathcal{X})$ contains the rewrite rules:*

$$\underline{f}(x_1, \ldots, x_n) \to f(x_1, \ldots, x_n), \text{ for all } f \in \mathcal{F} \setminus \widehat{\mathcal{F}}$$

Theorem 5. *The TRS Λ is complete, i. e., confluent and terminating.*

The previous theorem tell us that Λ actually defines a function, namely the function that given a term associates to it its (unique and always existing) normal form wrt the TRS Λ. It is not difficult to see that this function coincides with the function λ from [15].

In [15], the rewrite system $\Psi(R, \mu)$ was defined as $\Phi(R) \cup Bar(\widehat{\mathcal{F}})$ and shown to have the same termination properties as $\Phi(R) \cup Bar(\mathcal{F})$; we will proceed similarly. We define the TRS $\Psi(R, \mu_r)$ which is obtained by *filtering* the TRS $\Phi(R) \cup Bar(\mathcal{F})$ through the function Λ defined above. This filtering is achieved by taking the rewrite rules $\Lambda(l) \to \Lambda(r)$, where $l \to r$ is a rule in $\Phi(R) \cup Bar(\mathcal{F})$ and $\Lambda(l) \neq \Lambda(r)$. Indeed this filtering induces a partition of $\Phi(R) \cup Bar(\mathcal{F}) = S \cup T$, such that T contains all rules $l \to r$ for which $\Lambda(l) = \Lambda(r)$, and S all the other rules. It is not difficult to see that the TRS T contains the rules: $f(x_1, \ldots, x_n) \to \underline{f}(x_1, \ldots, x_n)$ for all $f \in \mathcal{F} \setminus \widehat{\mathcal{F}}$ and the normalized S, which coincides with $\Psi(\overline{R}, \mu_r)$, contains rules:

$$\Psi(R, \mu_r) = \begin{cases} \Phi(l) & \to (\Phi(r))^{\tau(l)} & \text{for all } l \to r \in R \\ a(\underline{f}(x_1, \ldots, x_n)) \to f(u_1, \ldots, u_n) & \text{for all } f \in \widehat{\mathcal{F}}, \text{ where} \\ & \quad u_i = a(x_i), \text{ if } i \in \mu_r(f) \text{ and} \\ & \quad u_i = x_i, \text{ if } i \notin \mu_r(f) \\ a(f(x_1, \ldots, x_n)) \to f(u_1, \ldots, u_n) & \text{for all } f \in \mathcal{F} \setminus \widehat{\mathcal{F}}, \text{ where} \\ & \quad u_i = a(x_i), \text{ if } i \in \mu_r(f) \text{ and} \\ & \quad u_i = x_i, \text{ if } i \notin \mu_r(f) \\ f(x_1, \ldots, x_n) & \to \underline{f}(x_1, \ldots, x_n) \text{ where } f \in \widehat{\mathcal{F}} \\ a(x) & \to x \end{cases}$$

Now we state that the TRS thus obtained has the same termination properties as $\Phi(R) \cup Bar(\mathcal{F})$.

Theorem 6. *$\Psi(R, \mu_r)/\widehat{AC}_{\mu_{ac}}$ is terminating iff $\Phi(R) \cup Bar(\mathcal{F})/AC'_{\mu_{ac}}$ is terminating.*

---

[5] The set $\widehat{\mathcal{F}}$ corresponds to $\mathcal{F}_0$ in [15].

5 Examples

Example 4. The following system defines the operational semantics of a process language which has sequential (;) and parallel (||) composition and a choice operator (+) to account for non-determinism of computations. The language has a non-terminating iterator constructor (It) and instructions for terminating execution (*abort*) or do-nothing (*skip*).

In order to avoid infinite reductions of a term containing the iterator, we only allow for reductions on the first argument of ";", that is $\mu_r(;) = \{1\}$ (for all other function symbols f, $\mu_r(f)$ contains all positions). This strategy also corresponds to the operational semantics of the sequential operator, which is to execute the second instruction only after the preceding one is complete. The AC-operators are || and + and we do not restrict AC-steps.

$$
\begin{array}{ll}
It(p) \quad\quad\ \ \rightarrow p \ ; \ It(p) & p + q \quad\quad \rightarrow p \\
(p \ + \ q) \ ; \ r \rightarrow (p \ ; \ r) \ + \ (q \ ; \ r) & p + q \quad\quad \rightarrow q \\
(p \ ; \ q) \ ; \ r \ \ \rightarrow p \ ; \ (q \ ; \ r) & p \ || \ abort \ \rightarrow abort \\
p \ ; \ skip \quad\ \rightarrow p & p \ || \ skip \ \ \rightarrow p \\
abort \ ; \ p \quad \rightarrow abort & p \ || \ (q \ + \ r) \rightarrow (p \ || \ q) \ + \ (p \ || \ r)
\end{array}
$$

In order to see that R/AC is μ-terminating we define a μ-interpretation as follows. We take the set $\mathbb{N}_2$ of all naturals greater than one, ordered by the usual order on $\mathbb{N}$ and define the interpretation functions: $It_{\mathbb{N}}(x) = 2x + 1, x \ ;_{\mathbb{N}} \ y = 2x, x \ +_{\mathbb{N}} \ y = x + y + 1, x \ ||_{\mathbb{N}} \ y = xy, abort_{\mathbb{N}} = 2, skip_{\mathbb{N}} = 2$.

It is not difficult to see that with these operations $\mathbb{N}_2$ constitutes a (non-empty) μ-monotone well-founded $\mathcal{F}$-algebra compatible with the AC-theory for symbols ||, + and compatible with the rewrite rules, so according to Theorem 3, R/AC is μ-terminating; note that R/AC is not terminating due to the first rewrite rule.

Example 5. Let + be AC, and suppose we have the system:

$$
\begin{array}{ll}
f(x) \quad\ \rightarrow x \ : \ f(d(x)) & 0 + y \quad\ \rightarrow y \\
d(0) \quad\ \rightarrow 0 & s(x) \ + \ y \rightarrow s(x \ + \ y) \\
d(s(x)) \rightarrow d(x) \ + \ s(s(0))
\end{array}
$$

where ":" stands for the list constructor and the function f computes the list with elements $x, 2x, 4x, \ldots$. Suppose that $\mu_r(:) = \{1\}$ and $\mu_r(g) = \{1, \ldots, ar(g)\}$, for any $g \neq :$. We do not restrict AC-steps. The system $\Psi(R, \mu_r)$ consists of the rules above but with the first rule replaced by $f(x) \rightarrow x : \underline{f}(\underline{d}(x))$; furthermore it also contains the rules $a(x \ + \ y) \rightarrow a(x) \ + \ a(y), a(\underline{f}(x)\overline{)} \rightarrow f(a(x)), f(x) \rightarrow \underline{f}(x)$, $a(\underline{d}(x)) \rightarrow d(a(x)), d(x) \rightarrow \underline{d}(x), a(s(x)) \rightarrow s(a\overline{(}x)), a(x : y) \rightarrow a(x) : y$, and $a(x) \rightarrow x$. + is the only AC-operator.

μ-termination of R/AC is proven if we prove AC-termination of $\Psi(R, \mu_r)$. This can be done using an AC-compatible order like the one in [14], with the precedence satisfying $a > f > \underline{f}, :, \underline{d}, a > d > \underline{d}, 0, +, + > s$.

Example 6. The following example represents a function that checks whether two trees representing arithmetic expressions are the mirror image of each other and in this case computes the value of the expression (for the sake of simplicity we only consider the operator +). AC-operators are + and *and*.

$$
\begin{aligned}
mirror(x, y) \quad &\rightarrow if(m(x, y), x, false) & 0 + y \quad &\rightarrow y \\
m(0, 0) \quad &\rightarrow true & s(x) + y \quad &\rightarrow s(x + y) \\
m(0, s(x)) \quad &\rightarrow false & true \text{ and } x \quad &\rightarrow x \\
m(s(x), s(y)) \quad &\rightarrow m(x, y) & false \text{ and } x \quad &\rightarrow false \\
m(x + y, z + w) \quad &\rightarrow m(x, w) \text{ and } m(y, z) \\
if(true, x, y) \quad &\rightarrow x & if(false, x, y) \quad &\rightarrow y
\end{aligned}
$$

This TRS is actually AC-terminating as can be seen using the order from [14], for example. But one can avoid unwanted reductions and AC-steps by stating that $\mu_r(if) = \{1\}$ and $\mu_{ac}(mirror) = \mu_{ac}(m) = \emptyset$.

6 Conclusions

The concept of context-sensitive rewriting was introduced by Lucas [7]. It is a restriction of the rewrite relation which allows for a finer control of reductions and is related to lazy evaluation strategies in functional languages. In this paper we extended the concept of context-sensitive rewriting to rewriting modulo an AC-theory by restricting both rewrite steps and AC-steps, and we studied the termination properties of context-sensitive AC-rewriting following the approach of Zantema [15] for context-sensitive rewriting. Context-sensitive (AC) termination, or μ-termination is a more general property than (AC) termination since this last property implies the former but not vice-versa, thus context-sensitive relations can also be used to study termination properties of reduction strategies for systems that are not terminating. We presented two techniques for proving μ-AC-termination: the first method provides a complete technique and consists in the definition of a suitable interpretation algebra for the (restricted) AC-rewrite system and the context-sensitive AC-rewrite relation, then μ-AC-termination of a system R/AC can be concluded from the existence of such a suitable interpretation. The second, not complete, method consists in transforming the (restricted) AC-rewrite system and the context-sensitive relation in an (restricted) AC-rewrite system where the reduction relation considered is either the usual AC-rewrite relation or a restricted version thereof (the restriction is on the application of AC-steps); μ-AC-termination of the original system can be inferred from (restricted) AC-termination of the transformed system. The advantage of this method over the previous one is that usual well-known techniques for proving AC-termination can be used to infer μ-AC-termination. It is also possible to develop a transformation for relating restricted and unrestricted AC-theories and lifting the restriction $\mu_r(f) \neq \emptyset$, for $f \in AC$; for that the use of flattened μ-rewriting is needed (see [2] for details).

The extension of context-sensitive rewriting presented restricts the applications of reductions steps and the application of AC-steps; it is not difficult to find examples where restriction of application of AC-steps leads to termination while unrestricted application does not (consider for example the rule $f(a + b) \rightarrow f(b + a)$ with $\mu_{ac}(f) = \emptyset$), however we do not have yet interesting applications for which it is useful to also restrict AC-steps.

In the future we intend to focus in these kind of applications and also study the confluence properties of the relation presented.

Acknowledgements The authors would like to thank Salvador Lucas for his helpful comments.

References

[1] N. Dershowitz and J.-P. Jouannaud. Rewrite systems. In *Handbook of Theoretical Computer Science*, volume B, chapter 6, pages 243–320. Elsevier, 1990.

[2] M. C. F. Ferreira and A. L. Ribeiro. Context-sensitive AC-rewriting (extended version). Draft, 1999.

[3] J.-P. Jouannaud and M. Muñoz. Termination of a set of rules modulo a set of equations. In *Proc. of the 7th Int. Conf. on Automated Deduction*, volume 170 of *LNCS*, pages 175 – 193. Springer, 1984.

[4] J. W. Klop. Term rewriting systems. In *Handbook of Logic in Computer Science*, volume II, pages 1–116. Oxford University Press, 1992.

[5] D. S. Lankford. Canonical inference. Technical Report Memo ATP-36, Automatic Theorem Proving Project, University of Texas, Austin, 1975.

[6] S. Lucas. Context-sensitive computations in functional and functional logic programs. *The Journal of Functional and Logic Programming*, 1998(1):1–61.

[7] S. Lucas. Fundamentals of context-sensitive rewriting. In *Proc. of the XXII Seminar on Current Trends in Theory and Practice of Informatics, SOFSEM'95*, volume 1012 of *LNCS*, pages 405–412. Springer, 1995.

[8] S. Lucas. Context-sensitive computations in confluent programs. In *Proc. of the 8th Int. Symposium PLILP'96 - Programming Languages: Implementations, Logics, and Programs*, volume 1140 of *LNCS*, pages 408–422. Springer, 1996.

[9] S. Lucas. Termination of context-sensitive rewriting by rewriting. In *Proc. of the 23rd Int. Colloquium on Automata, Languages and Programming, ICALP96*, volume 1099 of *LNCS*, pages 122–133. Springer, 1996.

[10] S. Lucas. *Reescritura con Restricciones de Reemplazamiento*. PhD thesis, Universidad Politecnica de Valencia, October 1998. In Spanish.

[11] Z. Manna and S. Ness. On the termination of Markov algorithms. In *Proc. of the 3rd Hawaii Int. Conf. on System Science*, pages 789–792, Honolulu, 1970.

[12] G. Peterson and M. Stickel. Complete sets of reductions for some equational theories. *Journal of the ACM*, 28:233–264, 1981.

[13] D. A. Plaisted. Equational reasoning and term rewriting systems. In *Handbook of Logic in Artificial Intelligence and Logic Programming*, volume 1, pages 273–364. Oxford Science Publications, Clarendon Press - Oxford, 1993.

[14] A. Rubio. A fully syntactic AC-RPO. In this volume.

[15] H. Zantema. Termination of context-sensitive rewriting. In *Proc. of the 8th Int. Conf. on Rewriting Techniques and Applications, RTA'97*, Lecture Notes in Computer Science. Springer, 1997.

The Calculus of Algebraic Constructions[*]

Frédéric Blanqui[†], Jean-Pierre Jouannaud[†] and Mitsuhiro Okada[‡]

[†] LRI, CNRS and Université de Paris-Sud

Bât. 405, 91405 Orsay Cedex, France

[‡] Department of Philosophy, Keio University,

108 Minatoku, Tokyo, Japan

Tel: +33-1-69156905   FAX: +33-1-69156586   Tel-FAX:+33-1-43212975

Abstract : This paper is concerned with the foundations of the Calculus of Algebraic Constructions (CAC), an extension of the Calculus of Constructions by inductive data types. CAC generalizes inductive types equipped with higher-order primitive recursion, by providing definitions of functions by pattern-matching which capture recursor definitions for arbitrary non-dependent and non-polymorphic inductive types satisfying a strictly positivity condition. CAC also generalizes the first-order framework of abstract data types by providing dependent types and higher-order rewrite rules. Full proofs are available at http://www.lri.fr/~blanqui/publis/rta99full.ps.gz.

1   Introduction

Proof assistants allow one to build complex proofs by using macros, called tactics, which generate proof terms representing the sequence of deduction rules used in the proof. These proof terms are then "type-checked" in order to ensure the correct use of each deduction step. As a consequence, the correctness of the proof assistant, hence of the verification itself, relies solely on the correctness of the type-checker, but not on the tactics themselves. This approach has a major problem: proof objects may become very large. For example, proving that $0+100$ equals its normal form 100 in some encoding of Peano arithmetic will generate a proof of a hundred steps, assuming + is defined by induction on its second argument. Such proofs occur in terms, as well as in subterms of a dependent type. Our long term goal is to cure this situation by restoring the balance between computations and deductions, as argued in [14]. The work presented in this paper intends to be a first important step towards this goal. To this end, we will avoid encodings by incorporating to the Calculus of Constructions (CC) [9] user-defined function symbols defined by sets of first and higher-order rewrite

---

[*]This work was partly supported by the Centre National de la Recherche Scientifique of France, Grants-in-aid for Scientific Research of Ministry of Education, Science and Culture of Japan, and the Oogata-kenkyuu-jyosei grant of Keio University.

rules. These rules will be used in conjunction with the usual proof reduction rule that reduces subterms in dependent types:

$$\frac{\Gamma \vdash M : T \quad T \longleftrightarrow^*_{R \cup \beta} T'}{\Gamma \vdash M : T'}$$

Since the pioneer work by Breazu-Tannen in 1988 [5] on the confluence of the combination of the simply-typed λ-calculus with first-order algebraic rewriting, soon followed, as for the strong normalization, by Breazu-Tannen and Gallier [6] and, independently, by Okada [21], this question has been very active. We started our program at the beginning of the decade, by developing the notion of abstract data type system [18], in which the user defined computations could be described by using rewrite rules belonging to the so-called *General Schema*, a generalization of higher-order primitive recursion. This work was done in the context of a bounded polymorphic type discipline, and was later extended to CC [1].

In [4], we introduced, in the context of the simply-typed λ-calculus, a new and more flexible definition of the General Schema to capture the rewrite rules defining recursors for strictly positive inductive types [10], problem left open in [18]. In this paper, we similarly equip CC with non-dependent and non-polymorphic inductive types, and first and higher-order rewriting. Our main result is that this extension is compatible with CC.

In [10], strictly positive inductive types can be dependent and polymorphic. Hence, further work will be needed to reach the expressive power of the Calculus of Inductive Constructions [22], implemented in the Coq proof assistant [3], all the more so since it handles strong elimination, that is the possibility to define types by induction. But our new General Schema seems powerful and flexible enough to be further extended to such a calculus, hence resulting in to a simpler strong normalization proof.

As a consequence of our result, it will become possible to develop a new version of the Coq proof assistant, in which the user may define functions by pattern-matching and then develop libraries of decision procedures using this kind of functional style. Ensuring the consistency of the underlying proof theory requires a proof that the user-defined rules obey the General Schema, a task that can be easily automated. Note also that, since most of the time, when one develops proofs, the efficiency of rewriting does not really matter, the type-checker of the proof development system can be kept small and not too difficult to certify, hence conforming to the idea of relying on a small easy-to-check kernel.

2 Definition of the calculus

2.1 Syntax

Definition 1 (Algebraic types) *Given a set S of sorts, we define the sets $\mathcal{T}_S$ of algebraic types:*

$$s := \mathbf{s} \mid (s \rightarrow s)$$

where s ranges over S and →associates to the right such that $s_1 \rightarrow (s_2 \rightarrow s_3)$ can be written $s_1 \rightarrow s_2 \rightarrow s_3$. An algebraic type $s_1 \rightarrow \ldots \rightarrow s_n$ is first-order *if each s_i is a sort, otherwise it is* higher-order.

Definition 2 (Constructors) *We assume that each sort s has an associated set $\mathcal{C}(s)$ of* constructors. *Each constructor C is equipped with an algebraic type $\tau(C)$ of the form $s_1 \rightarrow \ldots \rightarrow s_n \rightarrow s$; n is called the* arity *of C, and s its* output *type. We denote by C^n the set of constructors of arity n.*

A constructor C is first-order *if its type is first-order, otherwise it is* higher-order. *Constructor declarations define a quasi-ordering on sorts: $s \geq_S t$ if and only if t occurs in the type of a constructor belonging to $\mathcal{C}(s)$. In the following, we will assume that $>_S$ is well-founded, ruling out mutually inductive sorts.*

Definition 3 (Algebraic signature) *Given a non empty sequence $s_1, \ldots, s_n, s$ of algebraic types, we denote by $\mathcal{F}_{s_1, \ldots, s_n, s}$ the set of* function symbols *of arity n, of type $\tau(f) = s_1 \rightarrow \ldots \rightarrow s_n \rightarrow s$ and of* output type *s. We will denote by $\mathcal{F}^n$ the set of function symbols of arity n, and by $\mathcal{F}$ the set of all function symbols. Function symbols with a first-order (resp. higher-order) type are called* first-order *(resp.* higher-order*).*

Here are familiar examples of sorts and functions:

(i) the sort **bool** of booleans whose constructors are **true** : **bool** and **false** : **bool**; $\mathtt{if}_t$ of arity 3 is a defined function of type $\mathtt{bool} \rightarrow t \rightarrow t \rightarrow t$, for any algebraic type t;

(ii) the sort **nat** of natural numbers whose constructors are **0** : **nat** and **s** : **nat** →**nat**; **+** of arity 2 is a defined function of type **nat** →**nat** →**nat**;

(iii) the sort $\mathtt{list}_t$ of lists of elements of an algebraic type t whose constructors are $\mathtt{nil}_t$: $\mathtt{list}_t$ and $\mathtt{cons}_t$: $t \rightarrow \mathtt{list}_t \rightarrow \mathtt{list}_t$; $\mathtt{append}_t$ of arity 2 is a defined function of type $\mathtt{list}_t \rightarrow \mathtt{list}_t \rightarrow \mathtt{list}_t$, while $\mathtt{map}_{t,t'}$ of arity 2 is a defined function of type $(t \rightarrow t') \rightarrow \mathtt{list}_t \rightarrow \mathtt{list}_{t'}$;

(iv) the sort **ord** of ordinals whose constructors are $\mathtt{0_{ord}}$: **ord**, $\mathtt{s_{ord}}$: **ord** → **ord** and $\mathtt{lim_{ord}}$: (**nat** →**ord**) →**ord**.

Definition 4 (Terms) *The set Term of CAC terms is inductively defined as:*

$$a := x \mid s \mid \star \mid \Box \mid \lambda x{:}a.a \mid \Pi x{:}a.a \mid (a\ a) \mid C(a_1, \ldots, a_n) \mid f(a_1, \ldots, a_n)$$

where s ranges over $\mathcal{S}$, C over C^n, f over $\mathcal{F}^n$ and x over Var, a set of variables made of two disjoint infinite sets $Var^\Box$ and $Var^\star$. The application $(a\ b)$ associates to the left such that $(a_1\ a_2)\ a_3$ can be written $a_1\ a_2\ a_3$. The sequence of terms $a_1 \ldots a_n$ is denoted by the vector $\vec{a}$ of length $|\vec{a}| = n$. A term $C(\vec{a})$ (resp. $f(\vec{a})$) is said to be constructor headed *(resp.* function headed*).*

After Dewey, the set $Pos(a)$ of *positions* in a term a is a language over the alphabet $\mathbb{N}^+$ of strictly positive natural numbers. Note that abstraction and product have two arguments, the type and the body. The *subterm* of a term a at position $p \in Pos(a)$ is denoted by $a|_p$ and the term obtained by replacing $a|_p$ by a term b is written $a[b]_p$. We write $a \unrhd b$ if b is a subterm of a.

Figure 1: Typing rules of CAC

(ax) $\qquad \vdash \star : \square$

(sort) $\qquad \vdash \mathbf{s} : \star \qquad\qquad\qquad (\mathbf{s} \in \mathcal{S})$

(var) $\qquad \dfrac{\Gamma \vdash c : p}{\Gamma, x : c \vdash x : c} \qquad\qquad (x \in Var^p \setminus dom(\Gamma),\ p \in \{\star, \square\})$

(weak) $\qquad \dfrac{\Gamma \vdash a : b \quad \Gamma \vdash c : p}{\Gamma, x : c \vdash a : b} \qquad (x \in Var^p \setminus dom(\Gamma),\ p \in \{\star, \square\})$

(cons) $\qquad \dfrac{\Gamma \vdash a_1 : s_1 \quad \ldots \quad \Gamma \vdash a_n : s_n}{\Gamma \vdash C(a_1, \ldots, a_n) : \mathbf{s}} \qquad (C \in \mathcal{C}^n,\ \tau(C) = s_1 \to \ldots \to s_n \to \mathbf{s})$

(fun) $\qquad \dfrac{\Gamma \vdash a_1 : s_1 \quad \ldots \quad \Gamma \vdash a_n : s_n}{\Gamma \vdash f(a_1, \ldots, a_n) : s} \qquad (f \in \mathcal{F}_{s_1, \ldots, s_n, s},\ n \geq 0)$

(abs) $\qquad \dfrac{\Gamma, x : a \vdash b : c \quad \Gamma \vdash (\Pi x{:}a.c) : q}{\Gamma \vdash (\lambda x{:}a.b) : (\Pi x{:}a.c)} \qquad (x \notin dom(\Gamma),\ q \in \{\star, \square\})$

(app) $\qquad \dfrac{\Gamma \vdash a : (\Pi x{:}b.c) \quad \Gamma \vdash d : b}{\Gamma \vdash (a\ d) : c\{x \mapsto d\}}$

(conv) $\qquad \dfrac{\Gamma \vdash a : b \quad \Gamma \vdash b' : p}{\Gamma \vdash a : b'} \qquad \begin{array}{l}(p \in \{\star, \square\},\ b \longrightarrow^*_\beta b' \text{ or } b' \longrightarrow^*_\beta b \\ \qquad \text{or } b \longrightarrow^*_R b' \text{ or } b' \longrightarrow^*_R b)\end{array}$

(prod) $\qquad \dfrac{\Gamma \vdash a : p \quad \Gamma, x : a \vdash b : q}{\Gamma \vdash (\Pi x{:}a.b) : q} \qquad (x \notin dom(\Gamma),\ p, q \in \{\star, \square\})$

We note by $FV(a)$ and $BV(a)$ the sets of respectively free and bound variables occurring in a term a, and by $Var(a)$ their union. By convention, *bound and free variables will always be assumed different*. As in the untyped λ-calculus, terms that only differ from each other in their bound variables will be identified, an operation called *α-conversion*. A *substitution* θ of domain $dom(\theta) = \{\vec{x}\}$ is written $\{\vec{x} \mapsto \vec{b}\}$. Substitutions are written in postfix notation, as in $a\theta$.

Finally, we traditionally consider that $(b\ \vec{a})$, $\lambda \vec{x}{:}\vec{a}.b$ and $\Pi \vec{x}{:}\vec{a}.b$, denote all three the term b if $\vec{a}$ is the empty sequence, and the respective terms $(\ldots((b\ a_1)\ a_2) \ldots a_n)$, $\lambda x_1 : a_1.(\lambda x_2 : a_2.(\ldots(\lambda x_n : a_n.b)\ldots))$ and $\Pi x_1 : a_1.(\Pi x_2 : a_2.(\ldots(\Pi x_n : a_n.b)\ldots))$ otherwise. We also write $a \to b$ for the term $\Pi x{:}a.b$ when $x \notin FV(b)$. This abbreviation allows us to see algebraic types as terms of our calculus.

2.2 Typing rules

Definition 5 (Typing rules) *A* declaration *is a pair $x{:}a$ made of a variable x and a term a. An* environment *Γ is a (possibly empty) ordered sequence of declarations of the form $x_1{:}a_1, \ldots, x_n{:}a_n$, where all x_i are distinct; $dom(\Gamma) = \{x_1, \ldots, x_n\}$ is its* domain, *$FV(\Gamma) = \bigcup_{x:a \in \Gamma} FV(a)$ is its set of free variables,* *and $\Gamma(x_i) = a_i$. A typing judgement is a triple $\Gamma \vdash a : b$ made of an environment Γ and two terms a, b. A term a has type b in an environment Γ if the judgement $\Gamma \vdash a : b$ can be deduced by the rules of Figure 1. An environment is valid if $\star$ can be typed in it. An environment is algebraic if every declaration has the form $x{:}c$, where c is an algebraic type.*

The rules (sort), (cons) and (fun) are added to the rules of CC [9]. The (conv) rule expresses that types depend on reductions via terms. In CC, the relation used in the side condition is the monotonic, symmetric, reflexive, transitive closure of the β-rewrite relation $(\lambda x{:}a.b)\, c \longrightarrow_\beta b\{x \mapsto c\}$.

In our calculus, there are two kinds of computation rules: β- (or proof-) reduction and the user-defined rewrite rules, denoted by $\longrightarrow_R$. This contrasts with the other calculi of constructions, in which the meaning of (conv) is fixed by the designer of the language, while it depends on the user in our system. The unusual form of the side condition of our conversion rule is due to the fact that no proof of subject reduction is known for a conversion rule with the more natural side condition $b \longleftrightarrow^*_{\beta R} b'$. See [1] for details.

The structural properties of CC are also true in CAC. See [1] and [2] for details. We just recall the different term classes that compose the calculus.

Definition 6 *Let Kind be the set $\{K \in Term \mid \exists \Gamma, \Gamma \vdash K : \Box\}$ of kinds, Constr be the set $\{T \in Term \mid \exists \Gamma, \exists K \in Kind, \Gamma \vdash T : K\}$ of type constructors, Type be the set $\{\tau \in Term \mid \exists \Gamma, \Gamma \vdash \tau : \star\}$ of types, Obj be the set $\{u \in Term \mid \exists \Gamma, \exists \tau \in Type, \Gamma \vdash u : \tau\}$ of objects, and Thm be the set $Constr \cup Kind$ of theorems.*

Lemma 7 *Kinds, type constructors and objects can be characterized as follows:*

- $K := \star \mid \Pi x{:}\tau.K \mid \Pi \alpha{:}K.K$
- $T := \mathbf{s} \mid \alpha \mid \Pi x{:}\tau.\tau \mid \Pi \alpha{:}K.\tau \mid \lambda x{:}\tau.T \mid \lambda \alpha{:}K.T \mid (T\, u) \mid (T\, T)$
- $u := x \mid C(u_1, \ldots, u_n) \mid f(u_1, \ldots, u_n) \mid \lambda x{:}\tau.u \mid \lambda \alpha{:}K.u \mid (u\, u) \mid (u\, T)$

where $\alpha \in Var^\Box$ and $x \in Var^\star$.

2.3 Inductive types

Inductive types have been introduced in CC for at least two reasons: firstly, to ease the user's description of his/her specification by avoiding the complicated impredicative encodings which were necessary before; secondly, to transform inductive proofs into inductive procedures via the Curry-Howard isomorphism. The logical consistency of the calculus follows from the existence of a least fixpoint, a property which is ensured syntactically in the Calculus of Inductive Constructions by restricting oneself to strictly positive types [10].

Definition 8 (Positive and negative type positions) *Given an algebraic type s, its sets of positive and negative positions are inductively defined as follows:*

$$Pos^+(s \in \mathcal{S}) = \epsilon \qquad\qquad Pos^-(s \in \mathcal{S}) = \emptyset$$
$$Pos^+(s \to t) = 1{\cdot}Pos^-(s) \,\cup\, 2{\cdot}Pos^+(t) \quad Pos^-(s \to t) = 1{\cdot}Pos^+(s) \,\cup\, 2{\cdot}Pos^-(t)$$

Given an algebraic type t, we say that s does occur positively in t if s occurs in t, and each occurrence of s in t is at a positive position.

Definition 9 (Inductive sorts) *Let* s *be a sort whose constructors are* $C_1, \ldots,$ C_n *and suppose that* C_i *has type* $s_1^i \to \ldots \to s_{n_i}^i \to$ s. *Then we say that:*

(i) s *is a* basic inductive sort *if each* s_j^i *is* s *or a basic inductive sort smaller than* s *in* $<_{\mathcal{S}}$,

(ii) s *is a* strictly positive inductive sort *if each* s_j^i *is either a strictly positive inductive sort smaller than* s *in* $<_{\mathcal{S}}$, *or of the form* $s_1' \to \ldots \to s_p' \to$ s *where each* s_k' *is built from strictly positive inductive sorts smaller than* s *in* $<_{\mathcal{S}}$.

In the following, we will assume that every inductive sort of a user specification is strictly positive.

The sort nat whose constructors are 0 : nat and s : nat $\to$ nat is a basic sort. The sort ord whose constructors are 0_{ord} : ord, s_{ord} : ord $\to$ ord and $\lim_{\mathrm{ord}}$: (nat $\to$ ord) $\to$ ord is a strictly positive sort, since ord $>_{\mathcal{S}}$ nat.

Definition 10 (Strictly positive recursors) *Let* s *be a strictly positive inductive sort generated by the constructors* $C_1, \ldots, C_n$ *of respective types* $s_1^i \to \ldots \to s_{n_i}^i \to$ s. *The associated recursor* rec_t^s *of algebraic output type* t *is a function symbol of arity* $n + 1$, *and type* s $\to t_1 \to \ldots \to t_n \to t$ *where* $t_i = s_1^i \to \ldots \to s_{n_i}^i \to s_1^i \{s \mapsto t\} \to \ldots \to s_{n_i}^i \{s \mapsto t\} \to t$. *It is defined by the rewrite rules:*

$$\mathrm{rec}_t^s(C_i(\vec{a}), \vec{b}) \longrightarrow b_i \, \vec{a} \, \vec{d} \quad \text{where}$$

$d_j = a_j$ *if* s *is not in* s_i^j, *otherwise* $s_j^i = s_1' \to \ldots \to s_p' \to$ s *and* $d_j = \lambda \vec{x} : \vec{s'} \{s \mapsto t\} . \mathrm{rec}_t^s(a_j \, \vec{x}, \vec{b})$.

Via the Curry-Howard isomorphism, a recursor of a sort s corresponds to the structural induction principle associated to the set of elements built from the constructors of s. Strictly positive types are found in many proof assistants based on the Curry-Howard isomorphism, e.g. in Coq [3]. Here are a few recursors:

$$\mathrm{rec}_{\mathrm{bool}}^t(\mathrm{true}, u, v) \longrightarrow u \qquad \mathrm{rec}_{\mathrm{nat}}^t(0, u, v) \longrightarrow u$$
$$\mathrm{rec}_{\mathrm{bool}}^t(\mathrm{false}, u, v) \longrightarrow v \qquad \mathrm{rec}_{\mathrm{nat}}^t(s(n), u, v) \longrightarrow v \, n \, \mathrm{rec}_{\mathrm{nat}}^t(n, u, v)$$

$$\mathrm{rec}_{\mathrm{ord}}^t(0_{\mathrm{ord}}, u, v, w) \longrightarrow u$$
$$\mathrm{rec}_{\mathrm{ord}}^t(s_{\mathrm{ord}}(n), u, v, w) \longrightarrow v \, n \, \mathrm{rec}_{\mathrm{ord}}^t(n, u, v, w)$$
$$\mathrm{rec}_{\mathrm{ord}}^t(\lim_{\mathrm{ord}}(f), u, v, w) \longrightarrow w \, f \, \lambda n{:}\,\mathrm{nat}.\mathrm{rec}_{\mathrm{ord}}^t(f \, n, u, v, w)$$

$\mathrm{rec}_{\mathrm{bool}}^t$ is if_t, and $\mathrm{rec}_{\mathrm{nat}}^t$ is Gödel's higher-order primitive recursion operator.

2.4   User-defined rules

First, we define the syntax of terms that may be used for rewrite rules:

Definition 11 (Rule terms) *Terms built up solely from constructors, function symbols and variables of* $Var^\star$, *are called* algebraic. *Their set is defined by the following grammar:*

$$a := x^\star \mid C(a_1, \ldots, a_n) \mid f(a_1, \ldots, a_n)$$

where $x^\star$ *ranges over* $Var^\star$, C *over* $\mathcal{C}^n$ *and* f *over* $\mathcal{F}^n$. *An algebraic term is* first-order *if its function symbols and constructors are first-order, and* higher-order *otherwise. The set of* rule terms *is defined by the following grammar:*

$$a := x^\star \mid \lambda x^\star{:}s.a \mid (a\ a) \mid C(a_1, \ldots, a_n) \mid f(a_1, \ldots, a_n)$$

where $x^\star$ ranges over $Var^\star$, s over $\mathcal{T}_S$, C over $\mathcal{C}^n$ and f over $\mathcal{F}^n$. A rule term is first-order *if it is a* first-order algebraic *term, otherwise it is* higher-order.

Definition 12 (Rewrite rules) *A* rewrite rule *is a pair* $l \longrightarrow r$ *of rule terms such that* l *is headed by a function symbol* f *which is said to be* defined, *and* $FV(r) \subseteq FV(l)$. *Given a set* R *of rewrite rules, a term* a *rewrites to a term* b *at position* $m \in Pos(a)$ *with the rule* $l \longrightarrow r \in R$, *written* $a \longrightarrow^m_R b$ *if* $a|_m = l\theta$ *and* $b = a[r\theta]_m$ *for some substitution* θ.

A rewrite rule is first-order *if* l *and* r *are both first-order, otherwise it is* higher-order. *A first-order rewrite rule* $l \longrightarrow r$ *is* conservative *if no (free) variable has more occurrences in* r *than in* l. *The rules induce the following quasi-ordering on function symbols:* $f \geq_{\mathcal{F}} g$ *iff* g *occurs in a defining rule of* f.

We assume that first-order function symbols are defined only by first-order rewrite rules. Of course, it is always possible to treat a first-order function symbol as an higher-order one. Here are examples of rules:

$$\mathrm{if}_t(\mathbf{true}, u, v) \longrightarrow u \qquad\qquad \mathrm{map}_{t,t'}(f, \mathbf{nil}_t) \longrightarrow \mathbf{nil}_{t'}$$
$$\mathrm{if}_t(\mathbf{false}, u, v) \longrightarrow v \qquad \mathrm{map}_{t,t'}(f, \mathbf{cons}_t(x, l)) \longrightarrow \mathbf{cons}_{t'}(f\ x, \mathrm{map}_{t,t'}(f, l))$$

$$+(x, 0) \longrightarrow x \qquad\qquad\qquad \mathrm{ack}(0, y) \longrightarrow \mathbf{s}(y)$$
$$+(x, \mathbf{s}(y)) \longrightarrow \mathbf{s}(+(x, y)) \qquad\qquad \mathrm{ack}(\mathbf{s}(x), 0) \longrightarrow \mathrm{ack}(x, \mathbf{s}(0))$$
$$+(+(x, y), z) \longrightarrow +(x, +(y, z)) \quad \mathrm{ack}(\mathbf{s}(x), \mathbf{s}(y)) \longrightarrow \mathrm{ack}(x, \mathrm{ack}(\mathbf{s}(x), y))$$

Having rewrite rules in our calculus brings many benefits, in addition to obtaining proofs in which computational steps are transparent. In particular, it enhances the declarativeness of the language, as examplified by the Ackermann's function, for which the definition in Coq [3] must use two mutually recursive functions. For subject reduction, the following properties will be needed:

Definition 13 (Admissible rewrite rules) *A* rewrite rule $l \longrightarrow r$, *where* l *is headed by a function symbol whose output type is* s, *is* admissible *if and only if it satisfies the following conditions:*

- *there exists an algebraic environment* Γ_l *in which* l *is well-typed,*
- *for any environment* Γ, $\Gamma \vdash l{:}s \Rightarrow \Gamma \vdash r{:}s$.

We assume that rules use distinct variables and note by Γ_R *the union of the* Γ_l's.

2.5 Definition of the General Schema

Let us consider the example of a strictly positive recursor rule, for the sort $\mathbf{ord}$:

$$\mathrm{rec}^t_{\mathbf{ord}}(\mathbf{lim}_{\mathbf{ord}}(f), u, v, w) \longrightarrow w\ f\ \lambda n{:}\mathbf{nat}.\mathrm{rec}^t_{\mathbf{ord}}(f\ n, u, v, w)$$

To prove the decreasingness of the recursive call arguments, one would like to compare $\mathbf{lim}_{\mathbf{ord}}(f)$ with f, and not $\mathbf{lim}_{\mathbf{ord}}(f)$ with $(f\ n)$. To this end, we introduce the notion of the *critical subterm* of an application, and then interpret a function call by the critical subterms of its arguments. Here, f will be the critical subterm of $(f\ n)$, hence resulting in the desired comparison.

Definition 14 (Γ,s-critical subterm) *Given an algebraic type s and an environment Γ, a term a is a Γ,s-term if it is typable in Γ by an algebraic type in which s occurs positively. A term b is a Γ,s-subterm of a term a, a $\trianglerighteq_{\Gamma,s} b$, if b is a subterm of a, of which each superterm is a Γ,s-term. Writing a Γ,s-term a in its application form $a_1 \ldots a_n$, where a_1 is not an application, its Γ,s-critical subterm $\chi_\Gamma^s(a)$ is the smallest Γ,s-subterm $a_1 \ldots a_k$ (see Figure 2).*

For a higher-order function symbol, the arguments that have to be compared via their critical subterm, are said to be at *inductive positions*. They correspond to the arguments on which the function is inductively defined. Next, we define a notion of status that allows users to precise how to compare the arguments of recursive calls. Roughly speaking, it is a simple combination of multiset and lexicographic comparisons.

Definition 15 (Status orderings) *A status of arity n is a term of the form $lex(t_1, \ldots, t_p)$ where t_i is either x_j for some $j \in [1..n]$, or a term of the form $mul(x_{k_1}, \ldots, x_{k_q})$ such that each variable x_i, $1 \le i \le n$, occurs at most once. A position i is lexicographic if there exists j such that $t_j = x_i$. A status term is a status whose variables are substituted by arbitrary terms of CAC.*

Let stat be a status of arity n, I be a subset of the lexicographic positions of stat, called inductive positions, $S = \{>^i\}_{i \in I}$ a set of orders on terms indexed by I, and $>$ an order on terms. We define the corresponding status ordering, $>_{stat}^S$ on sequences of terms as follows:

- $(a_1, \ldots, a_n) >_{stat}^S (b_1, \ldots, b_n)$ iff $stat\{\vec{x} \mapsto \vec{a}\} >_{stat}^S stat\{\vec{x} \mapsto \vec{b}\}$,
- $lex(c_1, \ldots, c_p) >_{lex(t_1, \ldots, t_p)}^S lex(d_1, \ldots, d_p)$ iff $(c_1, \ldots, c_p) (>_{t_1}^S, \ldots, >_{t_p}^S)_{lex} (d_1, \ldots, d_p)$,
- $>_{x_i}^S$ is $>^i$ if $i \in I$, otherwise it is $>$,
- $mul(c_1, \ldots, c_q) >_{mul(x_{k_1}, \ldots, x_{k_q})} mul(d_1, \ldots, d_q)$ iff $\{c_1, \ldots, c_q\} >_{mul} \{d_1, \ldots, d_q\}$.

Note that it boils down to the usual lexicographic ordering if $stat = lex(x_1, \ldots, x_n)$ or to the multiset ordering if $stat = lex(mul(x_1, \ldots, x_n))$. $>_{stat}^S$ is well-founded if so is $>$ and each $>^i$.

For example, let $>$ and $\succ$ be some orders, $stat = lex(x_2, mul(x_1, x_3))$, $I = \{1\}$, and $S = (\succ)$. Then, $(a_1, a_2, a_3) >_{stat}^S (b_1, b_2, b_3)$ iff $a_2 \succ b_2$, or else $a_2 = b_2$ and $\{a_1, a_3\} >_{mul} \{b_1, b_3\}$.

Definition 16 (Critical interpretation) *Given an environment Γ, the critical interpretation function $\phi_{f,\Gamma}$ of a function symbol $f \in \mathcal{F}_{s_1, \ldots, s_n, s}$ is:*

- $\phi_{f,\Gamma}(a_1, \ldots, a_n) = (\phi_{f,\Gamma}^1(a_1), \ldots, \phi_{f,\Gamma}^n(a_n))$,
- $\phi_{f,\Gamma}^i(a_i) = a_i$ if $i \notin Ind(f)$,
- $\phi_{f,\Gamma}^i(a_i) = \chi_\Gamma^{s_i}(a_i)$ if $i \in Ind(f)$.

The critical ordering associated to f is $>_{f,\Gamma} = \triangleright_{stat_f}^S$, where $S = (\triangleright_{\Gamma,s_i}^S)_{i \in Ind(f)}$.

According to Definition 15, the critical ordering is nothing but the usual subterm ordering at non-inductive positions, and the critical subterm ordering of Definition 14 at inductive positions.

Figure 2: Critical subterm

We are now ready for describing the schema for higher-order rewrite rules. Given some lefthand side rule, we define a set of acceptable righthand sides, called computable closure. In the next section, we prove that it preserves strong normalization.

Definition 17 (Accessible subterms) *A term b is said to be* accessible *in a well-typed term c if it is a subterm of c which is typable by a basic inductive sort, or if there exists $p \in \mathcal{P}os(c)$ such that $c|_p = b$, and $\forall q < p$, $c|_q$ is headed by a constructor. b is said to be* accessible *in $\vec{c}$ if it is so in some $c \in \vec{c}$.*

Definition 18 (Computable closure) *Given an algebraic environment Γ containing Γ_R and a term $f(\vec{c})$ typable in Γ, the* computable closure $\mathcal{CC}_{f,\Gamma}(\vec{c})$ *of $f(\vec{c})$ in Γ is defined as the least set of Γ-terms containing all terms accessible in $\vec{c}$, all variables in $dom(\Gamma) \setminus FV(\vec{c})$, and closed under the following operations:*
(i) constructor application: let C be a constructor of type $s_1 \to \ldots \to s_n \to s$; then $C(\vec{u}) \in \mathcal{CC}_{f,\Gamma}(\vec{c})$ iff $u_i : s_i \in \mathcal{CC}_{f,\Gamma}(\vec{c})$ for all $i \in [1..n]$,
(ii) defined application: let $g \in \mathcal{F}_{s_1,\ldots,s_n,t}$ such that $g <_{\mathcal{F}} f$; then $g(\vec{u}) \in \mathcal{CC}_{f,\Gamma}(\vec{c})$ iff $u_i : s_i \in \mathcal{CC}_{f,\Gamma}(\vec{c})$ for all $i \in [1..n]$,
(iii) application: let $u : s \to t \in \mathcal{CC}_{f,\Gamma}(\vec{c})$ and $v : s \in \mathcal{CC}_{f,\Gamma}(\vec{c})$; then $(uv) \in \mathcal{CC}_{f,\Gamma}(\vec{c})$,
(iv) abstraction: let $u \in \mathcal{CC}_{f,\Gamma}(\vec{c})$ and $x : s \in \Gamma$; then $\lambda x : s.u \in \mathcal{CC}_{f,\Gamma}(\vec{c})$,
(v) reduction: let $u \in \mathcal{CC}_{f,\Gamma}(\vec{c})$, and v be a reduct of u using a β-rewrite step or a higher-order rewrite rule for a function symbol $g <_{\mathcal{F}} f$; then $v \in \mathcal{CC}_{f,\Gamma}(\vec{c})$,
(vi) recursive call: let $\vec{c'}$ be a vector of n terms in $\mathcal{CC}_{f,\Gamma}(\vec{c})$ of respective types $s_1, \ldots, s_n$, such that $\phi_{f,\Gamma}(\vec{c}) = \vec{c} >_{f,\Gamma} \phi_{f,\Gamma}(\vec{c'})$; then $f(\vec{c'}) \in \mathcal{CC}_{f,\Gamma}(\vec{c})$.

A useful finite approximation of this infinite set is defined by the Coquand's notion of *structurally smaller* [7], where only cases (i), (iii), (v) (one β-step only) and (vi) are used, with a multiset status which forbids the use of nested recursions. Our definition is therefore richer for two independent reasons. Note further that Coquand restricts himself to the cases for which his ordering is well-founded, a property that we think related to the positivity condition.

This can also be compared with the current criterion used in Coq for accepting function definitions by fixpoint and constructor matching [11]. Functions are defined by induction on one argument at a time, this argument must be constructor headed, and recursive calls can be made only with its immediate subterms. We are now ready for defining the schema:

Definition 19 (General Schema) *A set R of rewrite rules satisfies the General Schema if*

(i) its first-order part is conservative and strongly normalizing,

(ii) each higher-order function $f \in \mathcal{F}_{s_1,\ldots,s_n,t}$ is defined by a set of admissible rewrite rules of the form $f(\bar{c}) \longrightarrow e$ such that $e \in CC_{f,\Gamma}(\bar{c})$ for some algebraic Γ containing Γ_R (the environment in which the rules of R are defined).

All pattern-matching definitions given so far satisfy the General Schema, including the first-order ones. We could have imposed that the first-order rules also satisfy the General Schema: this would have simplified our definition, but at the price of restricting the expressivity for the first-order rules. In our formulation, the strong normalization property of the first-order rules has to be proved beforehand. Tools exist that do the job automatically for many practical examples. Note that recursor rules of any strictly positive inductive type satisfy the General Schema:

Lemma 20 *The recursor rules for strictly positive inductive sorts satisfy the General Schema.*

2.6 CAC computations

Definition 21 (Reduction relation) *Given a set R of rewrite rules satisfying the General Schema, including the set Rec of recursor rules of a given user specification, the CAC rewrite relation is $\longrightarrow\ =\ \longrightarrow_\beta \cup \longrightarrow_R$. The CAC reduction relation is its reflexive and transitive closure denoted by $\longrightarrow^*$. Its transitive closure is denoted by $\longrightarrow^+$. Its reflexive, symmetric and transitive closure is denoted by $\longleftrightarrow^*$. A term is in normal form if it cannot be β-reduced, Rec-reduced or R-reduced. An expansion is the inverse of a reduction: a expanses to b if b reduces to a.*

Our calculus enjoys the subject reduction property, that is, preservation of types under reductions. The proof uses a weak version of confluence, see [1].

Full confluence is proved after strong-normalization, by using Newman's Lemma, and by assuming there are no critical pairs between any two higher-order rules, and between the higher-order rules, the first-order rules and the β-reduction rule (by considering that the abstraction is an unary function symbol, and the application a binary one).

3 Strong-normalization

A term is *strongly normalizable* if any reduction issuing from it terminates. Strong-normalization and confluence together imply the logical soundness of the system as well as the decidability of type-checking. In this section, we investigate only the former. Let SN be the set of strongly normalizable terms.

To prove the strong normalization property for well-typed terms, we use the well known proof technique of Girard dubbed "reducibility candidates" [17], further extended by Coquand and Gallier to the Calculus of Constructions [8]. Note that these proofs use well-typed candidates, that is, sets of well-typed terms.

There exists proofs with lighter notations based on untyped candidates [16], but which do not allow one to reason about the type of the elements of a reducibility candidate, as it will be necessary to do with our extension of the General Schema. For a comprehensive survey of the method, see [15].

The strong normalization proof of Coquand and Gallier can easily be tailored to our need. It suffices to define an adequate interpretation for the inductive types, and to prove that, if the arguments of a function call belong to the interpretation of their type, then the function call itself belongs to the interpretation of its output type. We recall the definitions that are necessary for the understanding of our extension, and refer the reader to [8] for a complete exposition.

3.1 Interpretation of theorems

Definition 22 (Reducibility candidates) *We define the set Neutr of neutral terms as being the set of terms that are not an abstraction or constructor headed. Let $\mathcal{T}_{\Delta,A} = \{\Delta' \vdash a \mid \Delta' \vdash a : A, \ \Delta' \supseteq \Delta\}$, $SN_{\Delta,A} = \{\Delta' \vdash a \in \mathcal{T}_{\Delta,A} \mid a \in SN\}$.*

Given a valid environment Δ, the family $\mathcal{C}$ of saturated sets $\mathcal{C}_{\Delta,A}$ where A is a Δ-theorem, is defined by the properties listed below.

1. *If $A = \square$, then $\mathcal{C}_{\Delta,A}$ is the set $\{SN_{\Delta,\square}\}$.*
2. *If A is a Δ-type or a Δ-kind, then $\mathcal{C}_{\Delta,A}$ is the set of non empty sets $S \subseteq SN_{\Delta,A}$ such that the following properties hold:*

    (S1) $S \supseteq \{\Delta' \vdash x\vec{a} \in \mathcal{T}_{\Delta,A} \mid x \in Var \text{ and } \vec{a} \in SN\}$.

    (S2) For every neutral term t such that $\Delta' \vdash t \in \mathcal{T}_{\Delta,A}$, if, for every immediate reduct t' of t, $\Delta' \vdash t' \in S$, then $\Delta' \vdash t \in S$.

    (S3) Whenever $\Delta' \vdash t \in S$ and $\Delta' \subseteq \Delta''$, then $\Delta'' \vdash t \in S$.

    (S4) Whenever $\Delta' \vdash t \in S$ and t' is a reduct of t, then $\Delta' \vdash t' \in S$.

3. *If A is a type constructor of type $\Pi x : B.C$ in Δ, then $\mathcal{C}_{\Delta,A}$ is the set of functions with the following properties:*

    (a) If B is a kind, then
    - *$f \in \mathcal{C}_{\Delta,A}$ is a function with domain $\{(\Delta' \vdash T, S) \mid \Delta' \vdash T \in \mathcal{T}_{\Delta,B} \text{ and } S \in \mathcal{C}_{\Delta',T}\}$ such that $f(\Delta' \vdash T, S) \in \mathcal{C}_{\Delta',AT}$,*
    - *$f(\Delta' \vdash T_1, S_1) = f(\Delta' \vdash T_2, S_2)$ whenever $T_1 \longleftrightarrow^* T_2$.*

    (b) If B is a type, then
    - *$f \in \mathcal{C}_{\Delta,A}$ is a function with domain $\mathcal{T}_{\Delta,B}$ such that $f(\Delta' \vdash t) \in \mathcal{C}_{\Delta',At}$,*
    - *$f(\Delta' \vdash t_1) = f(\Delta' \vdash t_2)$ whenever $t_1 \longleftrightarrow^* t_2$.*

Compared to [8], we extended (S2) to neutral terms to take care of functions, and added (S4) to insure that reducibility candidates are stable by reduction.

Definition 23 (Interpretation of algebraic types) *Given a valid environment Δ, we define the interpretation of algebraic types as follows:*

- *$can_{\Delta,s} = \{\Delta' \vdash a \in SN_{\Delta,s} \mid \text{ if } a \longrightarrow^* C(\vec{b}) \text{ and } \tau(C) = s_1 \rightarrow \ldots \rightarrow s_n \rightarrow s,$ then $\Delta' \vdash b_i \in can_{\Delta,s_i}$ for every $i \in [1..n]\}$,*
- *$can_{\Delta,s \rightarrow t} = \{\Delta' \vdash a \in \mathcal{T}_{\Delta,s \rightarrow t} \mid \forall \Delta'' \subseteq \Delta', \forall \Delta'' \vdash b \in can_{\Delta,s}, \Delta'' \vdash ab \in can_{\Delta,t}\}$.*

Let us justify the definition. Since $>_S$ is assumed to be well-founded, our hypothesis is that the definition makes sense for every algebraic type built from sorts strictly smaller than a given sort s. Let P be the set of subsets of $SN_{\Delta,s}$ that contains all strongly normalizable terms that do not reduce to a term headed by a constructor of s. P is a complete lattice for set inclusion. Given an element $X \in P$, we define the following function on algebraic types built from sorts smaller than s: $R_X(s) = X$, $R_X(t) = can_{\Delta,t}$ and $R_X(s \to t) = can_{\Delta,s \to t}$. Now, let $F : P \to P$, $X \mapsto X \cup Y$ where $Y = \{a \in SN_{\Delta,s} \mid$ if $a \twoheadrightarrow C(\vec{b})$ and $\tau(C) = s_1 \to \ldots \to s_n \to s$ then $b_i \in R_X(s_i)$ for every $i \in [1..n]\}$. Since inductive sorts are assumed to be positive, one can show that F is monotone. Hence, from Tarski's Theorem, it has a least fixed point $can_{\Delta,s} \in C_{\Delta,s}$.

Definition 24 (Well-typed substitutions) *Given two valid environments Δ and Γ, a substitution θ is a* well-typed substitution *from Γ to Δ if $dom(\theta) \subseteq dom(\Gamma)$ and, for every variable $x \in dom(\Gamma)$, $\Delta \vdash x\theta : \Gamma(x)\theta$.*

Definition 25 (Candidate assignments) *Given two valid environments Δ and Γ, and a well-typed substitution θ from Γ to Δ, a* candidate assignment *compatible with θ is a function ξ from $Var^\square$ to the set of saturated sets such that, for every variable $\alpha \in dom(\Gamma) \cap Var^\square$, $\xi(\alpha) \in C_{\Delta,\alpha\theta}$.*

Compared to [8] where well-typed substitutions and candidate assignments are packaged together, we prefer to separate them since the former is introduced to deal with abstractions, while the latter is introduced to deal with polymorphism. We are now ready to give the definition of the interpretation of theorems.

Definition 26 (Interpretation of theorems) *Given two valid environments Δ and Γ, a well-typed substitution θ from Γ to Δ, and a candidate assignment ξ compatible with θ, we define the interpretation of Γ-theorems as follows:*

- $[\![\Gamma \vdash \square]\!]_{\Delta,\theta,\xi} = SN_{\Delta,\square}$,
- $[\![\Gamma \vdash \star]\!]_{\Delta,\theta,\xi} = SN_{\Delta,\star}$,
- $[\![\Gamma \vdash s]\!]_{\Delta,\theta,\xi} = can_{\Delta,s}$,
- $[\![\Gamma \vdash \alpha]\!]_{\Delta,\theta,\xi} = \xi(\alpha)$,
- $[\![\Gamma \vdash \lambda x{:}\tau.T]\!]_{\Delta,\theta,\xi} =$ *the function which associates* $[\![\Gamma, x{:}\tau \vdash T]\!]_{\Delta',\theta\{x \mapsto t\},\xi}$ *to every* $\Delta' \vdash t \in \mathcal{T}_{\Delta,\tau\theta}$,
- $[\![\Gamma \vdash \lambda \alpha{:}K'.T]\!]_{\Delta,\theta,\xi} =$ *the function which associates* $[\![\Gamma, \alpha{:}K' \vdash T]\!]_{\Delta',\theta\{\alpha \mapsto T'\},\xi\{\alpha \mapsto S\}}$ *to every* $(\Delta' \vdash T', S) \in \{(\Delta' \vdash T', S) \mid \Delta' \vdash T' : K'\theta, \Delta' \supseteq \Delta, S \in C_{\Delta',T'}\}$,
- $[\![\Gamma \vdash T\ t]\!]_{\Delta,\theta,\xi} = [\![\Gamma \vdash T]\!]_{\Delta,\theta,\xi}(\Delta \vdash t\theta)$
- $[\![\Gamma \vdash T\ T']\!]_{\Delta,\theta,\xi} = [\![\Gamma \vdash T]\!]_{\Delta,\theta,\xi}(\Delta \vdash T'\theta, [\![\Gamma \vdash T']\!]_{\Delta,\theta,\xi})$
- $[\![\Gamma \vdash \Pi x{:}\tau.A]\!]_{\Delta,\theta,\xi} = \{\Delta' \vdash a \in \mathcal{T}_{\Delta,\Pi x{:}\tau\theta.A\theta} \mid \forall \Delta'' \supseteq \Delta', \forall \Delta'' \vdash t \in [\![\Gamma \vdash \tau]\!]_{\Delta'',\theta,\xi}, \Delta'' \vdash at \in [\![\Gamma, x{:}\tau \vdash A]\!]_{\Delta'',\theta\{x \mapsto t\},\xi}\}$,
- $[\![\Gamma \vdash \Pi \alpha{:}K.A]\!]_{\Delta,\theta,\xi} = \{\Delta' \vdash a \in \mathcal{T}_{\Delta,\Pi \alpha{:}K\theta.A\theta} \mid \forall \Delta'' \supseteq \Delta', \forall \Delta'' \vdash T \in [\![\Gamma \vdash K]\!]_{\Delta'',\theta,\xi}, \forall S \in C_{\Delta'',T}, \Delta'' \vdash aT \in [\![\Gamma, \alpha{:}K \vdash A]\!]_{\Delta'',\theta\{\alpha \mapsto T\},\xi\{\alpha \mapsto S\}}\}$.

The last two cases correspond to the "stability by application". The well-definedness of this definition is insured by the following lemma.

Lemma 27 (Interpretation correctness) *Assume that Δ and Γ are two valid environments, θ is a well-typed substitution from Γ to Δ, and ξ is a candidate assignment compatible with θ. Then, for every Γ-theorem A, $[\![\Gamma \vdash A]\!]_{\Delta,\theta,\xi} \in \mathcal{C}_{\Delta,A\theta}$.*

We are now able to state the main lemma for the strong normalization theorem.

Definition 28 (Reducible substitutions) *Given two valid environments Δ and Γ, a well-typed substitution θ from Γ to Δ, and a candidate assignment ξ compatible with θ, θ is said to be valid with respect to ξ if, for every variable $x \in dom(\Gamma)$, $\Delta \vdash x\theta \in [\![\Gamma \vdash \Gamma(x)]\!]_{\Delta,\theta,\xi}$.*

Lemma 29 (Main lemma) *Assume that $\Gamma \vdash a : b$, Δ is a valid environment, θ is a well-typed substitution from Γ to Δ, and ξ is a candidate assignment compatible with θ. If θ is valid with respect to ξ, then $\Delta \vdash a\theta \in [\![\Gamma \vdash b]\!]_{\Delta,\theta,\xi}$.*

Proof: As in [8], by induction on the structure of the derivation. We give only the additional cases. The case (cons) is straightforward. The case (fun) is proved by Theorem 33 to come for the case of higher-order function symbols, and by [18] for the case of first-order function symbols. □

Theorem 30 (Strong normalization) *Assume that the higher-order rules satisfy the General Schema. Then, any well-typed term is strongly normalizable.*

Proof: Application of the Main Lemma, see [8] for details.

3.2 Reducibility of higher-order function symbols

One can see that the critical interpretation is not compatible with the reduction relation, and not stable by substitution either. We solve this problem by using yet another interpretation function for terms enjoying both properties and relating to the previous one as follows:

Definition 31 (Admissible recursive call interpretation) *A recursive call interpretation for a function symbol f is given by:*
(i) a function $\Phi_{f,\Gamma}$ operating on arguments of f, for each environment Γ,
(ii) a status ordering $\geq^{S}_{stat_f}$ where S is a set of orders indexed by $Ind(f)$.
A recursive call interpretation is admissible *if it satisfies the following properties:*
(Stability) *Assume that $f(\vec{c'}) \in \mathcal{CC}_{f,\Gamma}(\vec{c})$, hence $\phi_{f,\Gamma}(\vec{c}) = \vec{c} >_{f,\Gamma} \phi_{f,\Gamma}(\vec{c'})$, Δ is a valid environment, and θ is a well-typed substitution from Γ to Δ such that $\vec{c}\theta$ are strongly normalizable terms. Then, $\Phi_{f,\Delta}(\vec{c}\theta) >^{S}_{stat_f} \Phi_{f,\Delta}(\vec{c'}\theta)$.*
(Compatibility) *Assume that s is the output type of f, $\vec{a}$ and $\vec{a'}$ are two sequences of strongly normalizable terms such that $\Delta \vdash f(\vec{a}) : s$ and $\vec{a} \longrightarrow^{*} \vec{a'}$. Then, $\Phi_{f,\Delta}(\vec{a}) \geq^{S}_{stat_f} \Phi_{f,\Delta}(\vec{a'})$.*

The definition of the actual interpretation function, which is intricate, can be found in the full version of the paper. Before to prove the reducibility of higher-order function symbols, we need the following result.

Lemma 32 (Compatibility of accessibility with reducibility)
If $\Delta \vdash a \in can_{\Delta,A}$ and $b \in \mathcal{T}_{\Delta,B}$ is accessible in a, then $\Delta \vdash b \in can_{\Delta,B}$.

Theorem 33 (Reducibility of higher-order function symbols)
Assume that the higher-order rules satisfy the General Schema. Then, for every higher-order function symbol $f \in \mathcal{F}_{s_1,\ldots,s_n,s}$, $\Delta \vdash f(\vec{a}) \in can_{\Delta,s}$ provided that $\Delta \vdash f(\vec{a}) : s$ and $\Delta \vdash a_i \in can_{\Delta,s_i}$ for every $i \in [1..n]$.

Proof: The proof uses three levels of induction: on the function symbols ordered by $>_\mathcal{F}$, on the sequence of terms to which f is applied, and on the righthand side structure of the rules defining f. By induction hypothesis (1), any g occurring in the rules defining f satisfies the lemma.

We proceed to prove that $\Delta \vdash f(\vec{a}) \in can_{\Delta,s}$ by induction (2) on $(\Phi_{f,\Delta}(\vec{a}), \vec{a})$ with $(\geq^S_{stat_f}, (\longrightarrow^*)_{lex})_{lex}$ as well-founded order. Since $b = f(\vec{a})$ is a neutral term, by definition of reducibility candidates, it suffices to prove that every reduct b' of b belongs to $can_{\Delta,s}$.

If b is not reduced at its root then one a_i is reduced. Thus, $b' = f(\vec{a'})$ such that $\vec{a} \longrightarrow \vec{a'}$. As reducibility candidates are stable by reduction, $\Delta \vdash a'_i \in can_{\Delta,s_i}$, hence the induction hypothesis (2) applies since the interpretation is compatible with reductions.

If b is reduced at its root then $\vec{a} = \vec{c}\theta$ and $b' = e\theta$ for some terms $\vec{c}, e$ and substitution θ such that $f(\vec{c}) \longrightarrow e$ is the applied rule. θ is a well-typed substitution from Γ_R to Δ, and ξ is compatible with θ since $dom(\Gamma_R) \cap Var^\square = \emptyset$. We now show that θ is compatible with ξ. Let x be a free variable of e of type t. By definition of the General Schema, x is an accessible subterm of $\vec{c}$. Hence, by Lemma 32, $\Delta \vdash x\theta \in can_{\Delta,t}$ since, for every $i \in [1..n]$, $\Delta \vdash c_i\theta \in can_{\Delta,s_i}$.

Given an algebraic environment Γ containing Γ_R, let us show by induction (3) on the structure of $e \in \mathcal{CC}_{f,\Gamma}(\vec{c})$ that, for any well-typed substitution θ from Γ to Δ compatible with ξ, $e\theta \in can_{\Delta,t}$, provided that $c_i\theta \in can_{\Delta,s_i}$ for every $i \in [1..n]$.

Base case: either e is accessible in c_i, or e is a variable of $dom(\Gamma) \setminus FV(\vec{c})$. In the first case, this results from Lemma 32, and in the second case, this results from the fact that θ is compatible with ξ. Now, let us go through the different closure operations of the definition of $\mathcal{CC}_{f,\Gamma}()$.

(i) construction: $e = C(e_1, \ldots, e_p)$ and $\tau(C) = t_1 \to \ldots \to t_p \to \mathbf{t}$. $e\theta \in can_{\Delta,t}$ since, by induction hypothesis (3), $e_i\theta \in can_{\Delta,t_i}$.

(ii) defined application: $e = g(e_1, \ldots, e_p)$ with $\tau(g) = t_1 \to \ldots \to t_p \to t$ and $g <_\mathcal{F} f$. By induction hypothesis (3), $e_i\theta \in can_{\Delta,t_i}$. Hence, $e\theta \in can_{\Delta,t}$, by [18] for first-order function symbols, or by induction hypothesis (1) for higher-order ones, since $g <_\mathcal{F} f$.

(iii) application: $e = u\ v$. $e\theta \in can_{\Delta,t}$ since, by induction hypothesis (3), $u\theta \in can_{\Delta,t' \to t}$ and $v\theta \in can_{\Delta,t'}$.

(iv) abstraction: $e = \lambda x \!:\! t_1.u$ and $t = t_1 \to t_2$ such that $\Gamma, x \!:\! t_1 \vdash u \!:\! t_2$. Let $v \in can_{\Delta, t_1}$. By induction hypothesis (3), $u\theta\{x \mapsto v\} \in can_{\Delta, x:t_1, t_2}$. Hence, $(\lambda x \!:\! t_1.u\theta)v \in can_{\Delta, x:t_1, t_2}$ and $e\theta \in can_{\Delta, t}$.

(v) reduction: e is a reduct of a term $u \in \mathcal{CC}_{f, \Gamma}(\vec{c})$. Since $\Gamma \vdash u \!:\! t$, by induction hypothesis (3), $u\theta \in can_{\Delta, t}$. Since reducibility candidates are stable by reduction, $e\theta \in can_{\Delta, t}$.

(vi) admissible recursive call: $e = f(\vec{c'})$ and $\phi_{f, \Gamma}(\vec{c}) = \vec{c} >_{f, \Gamma} \phi_{f, \Gamma}(\vec{c'})$. The induction hypothesis (1) applies since the interpretation is stable. $\square$

This achieves the proof of the strong normalization property.

4 Conclusion and future work

We have defined an extension of the Calculus of Constructions by higher-order rewrite rules defining uncurried function symbols via the so called *General Schema* [4], which will allow a smooth integration in proof assistants like Coq, of function definitions by pattern-matching on the one hand, and decision procedures on the other hand. This result extends previous work by Barbanera et al. [1], by allowing for non-dependent and non-polymorphic inductive types. In our strong normalization proof based on Girard's reducibility candidates, we have indeed used a powerful generalization of the General Schema, of which the recursors for strictly positive inductive types are an instance, which is an important step of its own.

Several problems need to be solved to achieve our program, that is to extend the Coq proof assistant [3] with rewriting facilities. Firstly, to generalize our results to arbitrary positive inductive types, for which the type being defined may occur at any positive position of the argument types of its constructors. Secondly, to extend the results to dependent and polymorphic inductive types as defined by Coquand and Paulin in [10]. This is indeed the same problem, of defining and proving a generalization of the schema. Thirdly, to allow rewriting at the type level, enabling one to define types by induction. The corresponding recursor rules are called strong elimination [22]. We have already preliminary results in the latter two directions. Lastly, to accommodate the η-rule. By following [12], we plan to try the use of the η-rule as an expansion, instead of as a reduction. In this context, it would also be interesting to see to which extent the works by Nipkow [20] and Klop [19] on higher-order rewriting systems could be integrated in our framework. Fourthly, following [13], we also want to introduce modules in our calculus to be able to develop libraries of reusable parameterized proofs.

Acknowledgements: We want to thank Maribel Fernández for her careful reading, and the useful remarks by the anonymous referees.

References

[1] F. Barbanera, M. Fernández, and H. Geuvers. Modularity of strong normalization in the algebraic-λ-cube. *Journal of Functional Programming*, 7(6), 1997.

[2] H. Barendregt. Introduction to generalized type systems. *Journal of Functional Programming*, 1992.

[3] *The Coq Proof Assistant Reference Manual Version 6.2*. INRIA-Rocquencourt-CNRS-Université Paris Sud-ENS Lyon, 1998.

[4] F. Blanqui, J.-P. Jouannaud, and M. Okada. Inductive Data Type Systems, 1998.

[5] V. Breazu-Tannen. Combining algebra and higher-order types. In *Third IEEE Annual Symposium on Logic in Computer Science*, pages 82–90. 1988.

[6] V. Breazu-Tannen and J. Gallier. Polymorphic rewriting conserves algebraic strong normalization. *Theoretical Computer Science*, 83(1):3–28, June 1991.

[7] T. Coquand. Pattern matching with dependent types. In B. Nordström, K. Pettersson, G. Plotkin, editors, *Workshop on Types for Proofs and Programs*, 1992.

[8] T. Coquand and J. Gallier. A proof of strong normalization for the Theory of Constructions using a Kripke-like interpretation. *1st Intl. Workshop on Logical Frameworks*. 1990.

[9] T. Coquand and G. Huet. The Calculus of Constructions. *Information and Computation*, 76:96–120, 1988.

[10] T. Coquand and C. Paulin-Mohring. Inductively defined types. In P. Martin-Löf and G. Mints, editors, *Proceedings of Colog'88*, LNCS 417. Springer-Verlag, 1990.

[11] C. Cornes. *Conception d'un langage de haut niveau de representation de preuves: Récurrence par filtrage de motifs; Unification en présence de types inductifs primitifs; Synthèse de lemmes d'inversion*. PhD thesis, Université de Paris 7, 1997.

[12] R. Di Cosmo and D. Kesner. Combining algebraic rewriting, extensional lambda calculi, and fixpoints. *Theoretical Computer Science*, 169(2):201–220, 1996.

[13] J. Courant. A module calculus for Pure Type Systems. *TLCA'97*.

[14] G. Dowek, T. Hardin, and C. Kirchner. Theorem proving modulo. Technical Report 3400, INRIA, 1998.

[15] J. Gallier. On Girard's "Candidats de Réductibilité". In P.-G. Odifreddi, editor, *Logic and Computer Science*. North Holland, 1990.

[16] H. Geuvers. A short and flexible proof of strong normalization for the Calculus of Constructions. In P. Dybjer, B. Nordström, and J. Smith, editors, *Selected Papers 2nd Intl. Workshop on Types for Proofs and Programs, TYPES'94, Båstad, Sweden, 6–10 June 1994*, volume 996 of *LNCS*, pages 14–38. 1995.

[17] J.-Y. Girard, Y. Lafont, and P. Taylor. *Proofs and Types*. Cambridge Tracts in Theoretical Computer Science. Cambridge University Press, 1988.

[18] J.-P. Jouannaud and M. Okada. Abstract Data Type Systems. *Theoretical Computer Science*, 173(2):349–391, February 1997.

[19] J. W. Klop, V. van Oostrom, and F. van Raamsdonk. Combinatory reduction systems: introduction and survey. *Theoretical Computer Science*, 121(1-2):279–308, December 1993.

[20] T. Nipkow. Higher-order critical pairs. In *Proc. 6th IEEE Symp. Logic in Computer Science, Amsterdam*, pages 342–349, 1991.

[21] M. Okada. Strong normalizability for the combined system of the typed lambda calculus and an arbitrary convergent term rewrite system. In G. H. Gonnet, editor, *Proceedings of the ACM-SIGSAM 1989 International Symposium on Symbolic and Algebraic Computation*, pages 357–363. ACM Press, July 1989.

[22] B. Werner. *Une Théorie des Constructions Inductives*. Thèse, Université Paris 7, 1994.

HOL-$\lambda\sigma$: An Intentional First-Order Expression of Higher-Order Logic

Gilles Dowek[1], Thérèse Hardin[2], and Claude Kirchner[3]

[1] INRIA-Rocquencourt, B.P. 105, 78153 Le Chesnay Cedex, France,
Gilles.Dowek@inria.fr, http://coq.inria.fr/~dowek
[2] LIP6, UPMC, 4 place Jussieu, 75252 Paris Cedex 05, France,
Therese.Hardin@lip6.fr, http://www.lip6.fr/~hardin
[3] LORIA & INRIA, 615, rue du Jardin Botanique, 54600 Villers-lès-Nancy, France
Claude.Kirchner@loria.fr, http://www.loria.fr/~ckirchne

Abstract. We propose a first-order presentation of higher-order logic based on explicit substitutions. It is intentionally equivalent to the usual presentation of higher-order logic based on λ-calculus, i.e. a proposition can be proved without the extensionality axioms in one theory if and only if it can in the other. The *Extended Narrowing and Resolution* first-order proof-search method can be applied to this theory. This allows to simulate higher-order resolution step by step and furthermore leaves room for further optimizations and extensions.

Introduction

Higher-order logic is a formalism that allows a natural expression of program specifications and of mathematics. It is used in many theorem provers — either automatic or tactic driven — such as HOL, Isabelle, PVS, λ-Prolog, TPS, etc. Higher-order logic can be expressed in many different ways using combinators, λ-calculus, etc. Some of these formulations present higher-order logic as a first-order theory, some other do not. Expressing higher-order logic as a first-order theory permits to use standard first-order proof-search methods. Extensions, for example integrating algebraic axioms, are easier to study and handle in this simple framework.

There are several ways to encode higher-order logic as a first-order theory and several proof search methods for each encoding, which are more or less efficient with respect to the standard higher-order resolution. For instance the encoding of higher-order logic using combinators is not intentionally equivalent to the standard presentation using λ-calculus, because some proofs require the extensionality axioms in this presentation, but not in the standard one. This leads to inefficiencies.

In this paper we give a first-order presentation of higher-order logic called HOL-$\lambda\sigma$ using the so called, *calculus of explicit substitutions* [ACCL91]. We show that this presentation is intentionally equivalent to the usual presentation of higher-order logic based on λ-calculus, i.e. the theories are still equivalent when we drop the extensionality axioms in both cases. A rather surprising side effect of this presentation of higher-order logic is that it provides a clarification of the intricate skolemization rule of higher-order logic [Mil83,Mil87].

On the proof search side, we show that this theory can be mechanized with the *Extended Narrowing and Resolution* (ENAR) method introduced in [DHK98b]. We retrieve higher-order resolution as a particular case as it can be simulated step by step by the ENAR method applied to HOL-$\lambda\sigma$. But ENAR permits more optimizations as it permits for instance to delay application of substitutions. It keeps also the simplicity of first-order frameworks and can easily be extended, for instance with equational axioms. A first step in this direction can be found in [KR97].

The ENAR proof search method relies upon a presentation of first-order logic called *deduction modulo* that allows to build-in a congruence identifying terms and also propositions. This leads to shorter and more direct proofs by making congruent propositions equivalent instead of requiring explicit proof arguments. Hence, we shall express HOL-$\lambda\sigma$ in deduction modulo.

In order to remain self contained, we recall the principal ideas of *deduction modulo* in section 1. Then, we recall in section 2 the usual presentation of higher-order logic based on λ-calculus (HOL-λ) and in section 3 its first-order presentation based on Curry combinators. Section 4 introduces HOL-$\lambda\sigma$, establishes its main properties (termination, confluence, consistency and cut elimination) and presents the equivalence theorem between HOL-λ and HOL-$\lambda\sigma$ which rests upon cut elimination. In section 5 we show that the rather intricate Skolem theorem for higher-order logic can be deduced from the first-order one. At last section 6 presents briefly the ENAR proof search method (whose completeness rests upon cut elimination) and its application to HOL-$\lambda\sigma$.

1 Deduction modulo

In this paper we shall use a presentation of first-order logic, called *deduction modulo* [DHK98b], that permits to identify propositions modulo a congruence.

In deduction modulo, the notions of language, term and proposition are that of (many sorted) first-order logic. We consider theories to be formed with a set of axioms Γ *and a congruence*, denoted $\equiv$, defined on propositions. As a consequence, the deduction rules must take into account this equivalence. For instance, the *modus ponens* cannot be stated as usual

$$\frac{A \Rightarrow B \quad A}{B}$$

but, as the two occurrences of A need not be identical, but need only to be congruent, it is stated as

$$\frac{A' \Rightarrow B \quad A}{B} \text{ if } A \equiv A'$$

In fact, as the congruence may identify implications with other propositions, a slightly more general formulation is needed

$$\frac{C \quad A}{B} \text{ if } C \equiv A \Rightarrow B$$

All the rules of natural deduction or sequent calculus may be stated in a similar way. Figure 1 gives a formulation of *sequent calculus modulo*.

$$\frac{}{P \vdash Q}\text{axiom} \text{ if } P \equiv Q \qquad\qquad \frac{\Gamma, P \vdash \Delta \quad \Gamma \vdash Q, \Delta}{\Gamma \vdash \Delta}\text{cut} \text{ if } P \equiv Q$$

$$\frac{\Gamma, Q_1, Q_2 \vdash \Delta}{\Gamma, P \vdash \Delta}\text{contr-l} \text{ if } P \equiv Q_1 \equiv Q_2 \qquad \frac{\Gamma \vdash Q_1, Q_2, \Delta}{\Gamma \vdash P, \Delta}\text{contr-r} \text{ if } P \equiv Q_1 \equiv Q_2$$

$$\frac{\Gamma \vdash \Delta}{\Gamma, P \vdash \Delta}\text{weak-l} \qquad\qquad \frac{\Gamma \vdash \Delta}{\Gamma \vdash P, \Delta}\text{weak-r}$$

$$\frac{\Gamma \vdash P, \Delta \quad \Gamma, Q \vdash \Delta}{\Gamma, R \vdash \Delta}\Rightarrow\text{-l} \text{ if } R \equiv (P \Rightarrow Q) \qquad \frac{P, \Gamma \vdash Q, \Delta}{\Gamma \vdash R, \Delta}\Rightarrow\text{-r} \text{ if } R \equiv (P \Rightarrow Q)$$

$$\frac{\Gamma, P, Q \vdash \Delta}{\Gamma, R \vdash \Delta}\wedge\text{-l} \text{ if } R \equiv (P \wedge Q) \qquad \frac{\Gamma \vdash P, \Delta \quad \Gamma \vdash Q, \Delta}{\Gamma \vdash R, \Delta}\wedge\text{-r} \text{ if } R \equiv (P \wedge Q)$$

$$\frac{\Gamma, P \vdash \Delta \quad \Gamma, Q \vdash \Delta}{\Gamma, R \vdash \Delta}\vee\text{-l} \text{ if } R \equiv (P \vee Q) \qquad \frac{\Gamma \vdash P, Q, \Delta}{\Gamma \vdash R, \Delta}\vee\text{-r} \text{ if } R \equiv (P \vee Q)$$

$$\frac{\Gamma \vdash P, \Delta}{\Gamma, R \vdash \Delta}\neg\text{-l} \text{ if } R \equiv \neg P \qquad\qquad \frac{\Gamma, P \vdash \Delta}{\Gamma \vdash R, \Delta}\neg\text{-r} \text{ if } R \equiv \neg P$$

$$\frac{}{\Gamma, P \vdash \Delta}\bot\text{-l} \text{ if } P \equiv \bot$$

$$\frac{\Gamma, P\{x \leftarrow t\} \vdash \Delta}{\Gamma, Q \vdash \Delta}(x, P, t) \,\forall\text{-l} \text{ if } Q \equiv \forall x\, P \qquad \frac{\Gamma \vdash P\{x \leftarrow y\}, \Delta}{\Gamma \vdash Q, \Delta}(x, P, y) \,\forall\text{-r} \text{ if } Q \equiv \forall x\, P$$

$$\frac{\Gamma, P\{x \leftarrow y\} \vdash \Delta}{\Gamma, Q \vdash \Delta}(x, P, y) \,\exists\text{-l} \text{ if } Q \equiv \exists x\, P \qquad \frac{\Gamma \vdash P\{x \leftarrow t\}, \Delta}{\Gamma \vdash Q, \Delta}(x, P, t) \,\exists\text{-r} \text{ if } Q \equiv \exists x\, P$$

where the rules ∀-r and ∃-l assume that $y \notin FV(\Gamma \Delta)$

Fig. 1. The sequent calculus modulo

As an example, the proposition $\exists x\; 2 \times x = 4$ is rather cumbersome to prove in sequent calculus with the axioms of arithmetic. Indeed to prove the proposition $2 \times 2 = 4$ we have to say that $2 \times 2 = 1 \times 2 + 2$, $1 \times 2 + 2 = 0 \times 2 + 2 + 2$, ... and thus to use the axioms of arithmetic and equality many and many times.

In contrast, in sequent calculus modulo, we have the following proof

$$\dfrac{\dfrac{\dfrac{\overline{4 = 4 \vdash 2 \times 2 = 4}\ \text{axiom}}{\forall x\ x = x \vdash 2 \times 2 = 4}\ (x, x = x, 4)\ \forall\text{-l}}{\forall x\ x = x \vdash \exists x\ 2 \times x = 4}\ (x, 2 \times x = 4, 2)\ \exists\text{-r}}{}$$

Substituting the variable x by the term 2 in the proposition $2 \times x = 4$ yields the proposition $2 \times 2 = 4$, that is congruent to $4 = 4$. The transformation of one proposition into the other, that requires several proof steps in sequent calculus, is dropped from the proof in deduction modulo. It is a mere computation that need not be written, because everybody can re-do it by him/herself.

In this case, the congruence can be defined by a rewrite system defined on terms:

$$0 + y \longrightarrow y, \quad S(x) + y \longrightarrow S(x + y), \quad 0 \times y \longrightarrow 0, \quad S(x) \times y \longrightarrow x \times y + y$$

Notice that, in the proof above, we do not need the axioms of addition and multiplication. Indeed, these axioms are now redundant: since the terms $0 + y$ and y are congruent, the axiom $\forall y\ 0 + y = y$ is congruent to the equality axiom $\forall y\ y = y$. Hence, it can be dropped. In other words, these equivalences on terms have been built-in [Plo72,And71,PS81,Sti85,JK86,Mar94,Vir95,Vir98].

But in many situations, it is also natural to define a congruence at the proposition level. For instance, we may add to the previous system the rule of integral rings

$$x \times y = 0 \longrightarrow x = 0 \vee y = 0$$

that rewrites an atomic proposition to a disjunction. The main originality of deduction modulo is that it allows to define such a congruence directly on propositions with rules rewriting atomic propositions to arbitrary ones.

In this paper, all congruences will be defined by confluent rewrite systems. As these rewrite systems are defined on propositions and propositions contain binders, these rewrite systems are in fact *Combinatory Reduction Systems* [KvOvR93].

Notice that deduction modulo is not a proper extension of first-order logic. It is proved in [DHK98b] that for every congruence $\equiv$, we can find a theory $\mathcal{T}$ such that $\Gamma \vdash P$ is provable modulo $\equiv$ if and only if $\mathcal{T}\Gamma \vdash P$ is provable in ordinary first-order logic. Of course, the provable propositions are the same, but the proofs are very different, indeed much shorter in deduction modulo.

2 HOL-λ

We recall very quickly the usual presentation of higher-order logic [Chu40,And86]. Terms are those of a simply typed λ-calculus with two base types ι and o and the following constants $\dot{\Rightarrow}$, $\dot{\wedge}$ and $\dot{\vee}$, of type $o \to o \to o$, $\dot{\neg}$ of type $o \to o$, $\dot{\bot}$ of type o, $\dot{\forall}_T$ and $\dot{\exists}_T$ of type $(T \to o) \to o$ (we use a notation with a dot for the constants to distinguish them from the connectors and quantifiers of first-order logic). Propositions are terms of type o. The unique $\beta\eta$-normal form of a term a is written $a\downarrow$. The deduction rules are given in figure 2 where all propositions are supposed to be normal.

An alternative presentation does not normalize the propositions after the quantifier rules but takes β and η as axioms.

This system is well-known to be consistent and to enjoy cut elimination [Gir70,Gir72].

$$\frac{}{P \vdash P}\text{axiom} \qquad \frac{\Gamma, P \vdash \Delta \quad \Gamma \vdash P, \Delta}{\Gamma \vdash \Delta}\text{cut}$$

$$\frac{\Gamma, P, P \vdash \Delta}{\Gamma, P \vdash \Delta}\text{contr-l} \qquad \frac{\Gamma \vdash P, P, \Delta}{\Gamma \vdash P, \Delta}\text{contr-r}$$

$$\frac{\Gamma \vdash \Delta}{\Gamma, P \vdash \Delta}\text{weak-l} \qquad \frac{\Gamma \vdash \Delta}{\Gamma \vdash P, \Delta}\text{weak-r}$$

$$\frac{\Gamma \vdash P, \Delta \quad \Gamma, Q \vdash \Delta}{\Gamma, (\dot{\Rightarrow} P Q) \vdash \Delta}\dot{\Rightarrow}\text{-l} \qquad \frac{P, \Gamma \vdash Q, \Delta}{\Gamma \vdash (\dot{\Rightarrow} P Q), \Delta}\dot{\Rightarrow}\text{-r}$$

$$\frac{\Gamma, P, Q \vdash \Delta}{\Gamma, (\dot{\wedge} P Q) \vdash \Delta}\dot{\wedge}\text{-l} \qquad \frac{\Gamma \vdash P, \Delta \quad \Gamma \vdash Q, \Delta}{\Gamma \vdash (\dot{\wedge} P Q), \Delta}\dot{\wedge}\text{-r}$$

$$\frac{\Gamma, P \vdash \Delta \quad \Gamma, Q \vdash \Delta}{\Gamma, (\dot{\vee} P Q) \vdash \Delta}\dot{\vee}\text{-l} \qquad \frac{\Gamma \vdash P, Q, \Delta}{\Gamma \vdash (\dot{\vee} P Q), \Delta}\dot{\vee}\text{-r}$$

$$\frac{\Gamma \vdash P, \Delta}{\Gamma, (\dot{\neg} P) \vdash \Delta}\dot{\neg}\text{-l} \qquad \frac{\Gamma, P \vdash \Delta}{\Gamma \vdash (\dot{\neg} P), \Delta}\dot{\neg}\text{-r}$$

$$\frac{}{\Gamma, \dot{\perp} \vdash \Delta}\dot{\perp}\text{-l}$$

$$\frac{\Gamma, (P\, t)\downarrow \vdash \Delta}{\Gamma, (\dot{\forall} P) \vdash \Delta}t\, \dot{\forall}\text{-l} \qquad \frac{\Gamma \vdash (P\, y)\downarrow, \Delta}{\Gamma \vdash (\dot{\forall} P), \Delta}\dot{\forall}\text{-r}$$

$$\frac{\Gamma, (P\, y)\downarrow \vdash \Delta}{\Gamma, (\dot{\exists} P) \vdash \Delta}\dot{\exists}\text{-l} \qquad \frac{\Gamma \vdash (P\, t)\downarrow, \Delta}{\Gamma \vdash (\dot{\exists} P), \Delta}t\, \dot{\exists}\text{-r}$$

where the rules $\dot{\forall}$-r and $\dot{\exists}$-l assume that $y \notin FV(\Gamma\Delta)$

Fig. 2. HOL-λ: The deduction rules of HOL-λ

3 HOL-C

Higher-order logic can be expressed as a many-sorted first-order theory whose sorts are all simple types. In such a presentation, when t is a term of type $T \to U$

and u a term of type T we cannot write the application of the term t to the term u as $(t\ u)$, but we need to introduce a function symbol $\alpha_{T,U}$ and write this term $\alpha_{T,U}(t,u)$. The rank of the function symbol $\alpha_{T,U}$ is $(T \to U, T)U$. Of course we shall continue to write $(t\ u)$ for the term $\alpha_{T,U}(t,u)$.

To express function terms and predicate terms, instead of using λ-calculus, we introduce for each applicative term t whose variables are among $x_1, \ldots, x_n$ a constant symbol written $x_1, \ldots, x_n \longmapsto t$ and an axiom

$$((x_1, \ldots, x_n \longmapsto t)\ x_1\ \ldots\ x_n) = t$$

Such constant symbols are called *combinators*.

At last, we introduce a predicate symbol ε of rank (o) that transforms a term t of type o into the proposition $\varepsilon(t)$. We add axioms that relate the connectors and quantifiers (e.g. $\wedge$) and their replication as constant symbols (e.g. $\dot\wedge$), for instance:

$$\varepsilon(\dot\wedge\ x\ y) \Leftrightarrow (\varepsilon(x) \wedge \varepsilon(y))$$

Thus, the language contains:

- for each applicative term t of type U whose variables are among $x_1, \ldots, x_n$ of type $T_1, \ldots, T_n$, a constant symbol $x_1, \ldots, x_n \longmapsto t$ of type $T_1 \to \ldots \to T_n \to U$,
- constant symbols $\dot\Rightarrow$, $\dot\wedge$ and $\dot\vee$ of sort $o \to o \to o$, $\dot\neg$ of type $o \to o$, $\dot\perp$ of type o, for each type T constant symbols $\dot\forall_T$ and $\dot\exists_T$ of type $(T \to o) \to o$,
- for each pair of types (T, U), a binary function symbol $\alpha_{T,U}$ of rank $(T \to U, T)\ U$,
- a unary predicate symbol ε of rank (o).

The axioms are:

$$\varepsilon(((x_1, \ldots, x_n \longmapsto t)\ x_1\ \ldots\ x_n) = t)$$

$$\varepsilon(\dot\Rightarrow\ x\ y) \Leftrightarrow (\varepsilon(x) \Rightarrow \varepsilon(y))$$

$$\varepsilon(\dot\wedge\ x\ y) \Leftrightarrow (\varepsilon(x) \wedge \varepsilon(y))$$

$$\varepsilon(\dot\vee\ x\ y) \Leftrightarrow (\varepsilon(x) \vee \varepsilon(y))$$

$$\varepsilon(\dot\neg\ x) \Leftrightarrow \neg\varepsilon(x)$$

$$\varepsilon(\dot\perp) \Leftrightarrow \perp$$

$$\varepsilon(\dot\forall\ x) \Leftrightarrow \forall y\ \varepsilon(x\ y)$$

$$\varepsilon(\dot\exists\ x) \Leftrightarrow \exists y\ \varepsilon(x\ y)$$

These axioms can be dropped if we work modulo the congruence defined by the rewrite system formed with the rule

$$((x_1, \ldots, x_n \longmapsto t)\ x_1\ \ldots\ x_n) \longrightarrow t$$

and those of figure 4.

Translation from λ-terms to combinators is usually called λ-*lifting*. This translation can be modified in order to use only the combinators $S = x, y, z \longmapsto ((x\ z)\ (y\ z))$ and $K = x, y \longmapsto x$. As well known, this translation does not preserve the term structure.

This presentation of higher-order logic can be shown to be equivalent to the presentation with λ-calculus if we take the extensionality axioms

$$\forall f\ \forall g\ ((\forall x\ \varepsilon((f\ x) = (g\ x))) \Rightarrow \varepsilon(f = g))$$

$$\forall x\ \forall y\ (\varepsilon(x \Leftrightarrow y) \Rightarrow \varepsilon(x = y))^1$$

in both cases, i.e. a proposition P is provable in the presentation of higher-order logic with λ-calculus if and only if the proposition $\varepsilon(P')$ is provable in the first-order theory above. But, if we drop the extensionality axioms, then the two presentations are not equivalent anymore. For instance, the proposition

$$((\lambda x\ \lambda y\ x)\ (u\ v)) = \lambda y\ (u\ v)$$

is provable in presentation with λ-calculus while its translation

$$((f, x \longmapsto (f\ x))\ (x, y \longmapsto x)\ (u\ v)) = ((u, v, y \longmapsto (u\ v))\ u\ v)$$

requires extensionality. Even when the extensionality axioms are taken, the formulations with λ-calculus and combinators are only weakly equivalent: provable propositions are the same, but the proofs are very different, some proofs requiring only $\beta\eta$-conversion when expressed with λ-calculus and the use of extensionality when expressed with combinators. This leads to inefficiencies when searching for proofs.

4 HOL-$\lambda\sigma$

We present in this section another first-order formulation of higher-order logic. It is not based on combinators as previously, but on de Bruijn indices and explicit substitutions. It allows to avoid the previously mentioned drawbacks of combinators.

4.1 The theory

In λ-calculus with de Bruijn indices, bound variables are replaced by an index indicating the binding height of this variable, i.e. the number of λ's between this occurrence and its binder. For instance the term $\lambda x\ (x\ (\lambda y\ x))$ is written $\lambda\ (1\ (\lambda\ 2))$. This notation is also a first-order language with a binary function symbol α, a unary function symbol λ and constant symbols $1, 2, 3 \ldots$. Simple sorts are not sufficient anymore with de Bruijn indices. Indeed, we need to give a sort not only to terms like $(\lambda_A\ 1)$ (that gets the sort $A \to A$), but also to terms of the form 1. Thus, as detailed in [DHK95], we have to consider sorts of

β-reduction and η-reduction:

$$(\lambda a)b \longrightarrow a[b.id]$$

$$\lambda(a\ 1) \longrightarrow b \text{ if } a =_\sigma b[\uparrow]$$

σ-reduction:

$$(a\ b)[s] \longrightarrow (a[s]\ b[s])$$

$$1[a.s] \longrightarrow a$$

$$a[id] \longrightarrow a$$

$$(\lambda a)[s] \longrightarrow \lambda(a[1.(s \circ \uparrow)])$$

$$(a[s])[t] \longrightarrow a[s \circ t]$$

$$id \circ s \longrightarrow s$$

$$\uparrow \circ (a.s) \longrightarrow s$$

$$(s_1 \circ s_2) \circ s_3 \longrightarrow s_1 \circ (s_2 \circ s_3)$$

$$(a.s) \circ t \longrightarrow a[t].(s \circ t)$$

$$s \circ id \longrightarrow s$$

$$1.\uparrow \longrightarrow id$$

$$1[s].(\uparrow \circ s) \longrightarrow s$$

Fig. 3. The rewrite rules of $\lambda\sigma$-calculus

the form $\Gamma \vdash T$ where T is a simple type and Γ a context, i.e. a list of simple types.

With de Bruijn indices conversion axioms use an external definition for substitution. Moreover this substitution is not well-defined on open terms of this first-order language. This is solved by considering an extension of this calculus: the *calculus of explicit substitutions* [ACCL91] also called $\lambda\sigma$-calculus. This calculus introduces also sorts of the form $\Gamma \vdash \Delta$ for substitutions that are lists of terms and symbols to build such substitutions id, ., $\uparrow$ and $\circ$. Then a new term constructor is introduced _[_] that permits to apply an explicit substitution to a term. The rewrite rules describing the evaluation of the $\lambda\sigma$-calculus are given in figure 3.

HOL-$\lambda\sigma$ is a many-sorted first-order theory with sorts of the form $\Gamma \vdash T$ and $\Gamma \vdash \Delta$ where Γ and Δ are sequences of simple types and T is a simple type.

Definition 1. (Language) The language contains the following function symbols:

---

[1] Here, equality is Leibniz' equality, i.e. $\lambda x\ \lambda y\ \dot{\forall}\lambda p\ ((p\ x)\dot{\Rightarrow}(p\ y))$

1_A^Γ constant of sort $A.\Gamma \vdash A$

$\alpha_{A\to B,A}^\Gamma$ binary function of rank $(\Gamma \vdash A \to B, \Gamma \vdash A)\Gamma \vdash B$

$\lambda_{A,B}^\Gamma$ unary function of rank $(A.\Gamma \vdash B)\Gamma \vdash A \to B$

$[\ ]_A^{\Gamma,\Gamma'}$ binary function of rank $(\Gamma' \vdash A, \Gamma \vdash \Gamma')\Gamma \vdash A$

id^Γ constant of sort $\Gamma \vdash \Gamma$

$\uparrow_A^\Gamma$ constant of sort $A.\Gamma \vdash \Gamma$

$\cdot_A^{\Gamma,\Gamma'}$ binary function of rank $(\Gamma \vdash A, \Gamma \vdash \Gamma')\Gamma \vdash A.\Gamma'$

$\circ^{\Gamma,\Gamma',\Gamma''}$ binary function of rank $(\Gamma \vdash \Gamma'', \Gamma'' \vdash \Gamma')\Gamma \vdash \Gamma'$

$\dot\Rightarrow$ constant of sort $\vdash o \to o \to o$

$\dot\wedge$ constant of sort $\vdash o \to o \to o$

$\dot\vee$ constant of sort $\vdash o \to o \to o$

$\dot\neg$ constant of sort $\vdash o \to o$

$\dot\bot$ constant of sort $\vdash o$

$\dot\forall_A$ constant of sort $\vdash (A \to o) \to o$

$\dot\exists_A$ constant of sort $\vdash (A \to o) \to o$

and a single unary predicate symbol:

$$\varepsilon \text{ of rank } (\vdash o)$$

We denote $\lambda\sigma\mathcal{L}$ the rewrite rules of $\lambda\sigma$-calculus together with the logical rules $\mathcal{L}$ given in figure 4 and we write $A \equiv B$ when A and B are congruent modulo $\lambda\sigma\mathcal{L}$.

$$\varepsilon(\dot\Rightarrow x\ y) \longrightarrow \varepsilon(x) \Rightarrow \varepsilon(y)$$
$$\varepsilon(\dot\wedge x\ y) \longrightarrow \varepsilon(x) \wedge \varepsilon(y)$$
$$\varepsilon(\dot\vee x\ y) \longrightarrow \varepsilon(x) \vee \varepsilon(y)$$
$$\varepsilon(\dot\neg x) \longrightarrow \neg\varepsilon(x)$$
$$\varepsilon(\dot\bot) \longrightarrow \bot$$
$$\varepsilon(\dot\forall_T x) \longrightarrow \forall y\ \varepsilon(x\ y)$$
$$\varepsilon(\dot\exists_T x) \longrightarrow \exists y\ \varepsilon(x\ y)$$

Fig. 4. The $\mathcal{L}$-rewrite rules

4.2 Properties

We prove in the full version of the paper [DHK98a] that the system $\lambda\sigma\mathcal{L}$ is weakly terminating and confluent on terms containing only term variables. The weak termination property is proved by encoding the system $\lambda\sigma\mathcal{L}$ into the typed

$\lambda\sigma$-calculus. The confluence property is proved by showing that $\lambda\sigma^*$ and $\mathcal{L}$ are confluent and strongly commute. We can then apply Hindley-Rosen lemma.

We also prove in the full version of the paper that the theory HOL-$\lambda\sigma$ is consistent and has the cut elimination property. Consistence can be proved by constructing a model. Following the method introduced in [DW98] we prove cut-elimination by constructing a so called *pre-model* of the theory. The cut elimination property is used in the proof of the embedding theorem 3 and in the completeness proof of the *Extended Narrowing and Resolution* method applied to HOL-$\lambda\sigma$.

4.3 Embedding HOL-λ into HOL-$\lambda\sigma$

We now want to prove that HOL-$\lambda\sigma$ is intentionally equivalent to the usual presentation of higher-order logic HOL-λ.

Following [DHK95], we define a translation from λ-calculus to $\lambda\sigma$-calculus called *pre-cooking*. This translation replaces the bound variables by the appropriate indices and adds an appropriate $[\uparrow^n]$ operator to free variables and constants according to the context in which they occur.

To each variable x of type T, we associate the sort $\vdash T$ in $\lambda\sigma$-calculus.

Definition 2. Let a be a λ-term. The pre-cooking of a is the $\lambda\sigma$-term defined by $a_F = F(a, [\,])$ where $F(a, l)$ is defined using the list of variable l ($[\,]$ being the empty list) by:

- $F((\lambda x.a), l) = \lambda(F(a, x.l))$,
- $F((a\ b), l) = F(a, l)F(b, l)$,
- $F(x, l) = 1[\uparrow^{k-1}]$, if x is the k-th variable of l
- $F(x, l) = x[\uparrow^n]$ where n is the length of l if x is a variable not occurring in l or a constant.

Theorem 3. *If $p_1, \ldots, p_n, q_1, ..., q_m$ are propositions in HOL-λ then the sequent $p_1, \ldots, p_n \vdash q_1, ..., q_m$ is provable in HOL-λ if and only if the sequent $\varepsilon(p_{1F}), \ldots, \varepsilon(p_{nF}) \vdash \varepsilon(q_{1F}), ..., \varepsilon(q_{mF})$ is provable in HOL-$\lambda\sigma$.*

The proof of this result is given in [DHK98a] and rely on the next propositions.

Proposition 4. *– If t has the type T then t_F has the sort $\vdash T$,*
- *$(\{a/x\}b)_F = \{x \mapsto a_F\}b_F$,*
- *$a =_{\beta\eta} b$ in λ-calculus if and only if $a_F =_{\lambda\sigma} b_F$ in $\lambda\sigma$-calculus.*

The purpose of the following definition and propositions is to characterize the image of the pre-cooking mapping.

Definition 5. A *F-term* is a $\lambda\sigma$-term containing only variables which sort has an empty context. A *F-proposition* is a proposition of the form $\varepsilon(P)$ where P is a *F-term*.

Proposition 6. *If t is a $\lambda\sigma\mathcal{L}$-normal F-term well-typed in the empty context then there is a λ-term u such that $t = u_F$.*

Proposition 7. *Let $\Gamma \vdash \Delta$ be a sequent containing only F-propositions. Then, if this sequent has a proof, it also has a proof where all propositions are F-propositions and all the witnesses F-terms.*

5 Skolemization in HOL-$\lambda\sigma$

Skolemization in higher-order logic is known to be more complicated than in first-order logic. Indeed, the naive skolemization in higher-order logic permits to transform some unprovable formulations of the axiom of choice into provable propositions. Thus the naive skolemization rule has to be restricted in such a way that skolemizing a proposition of the form

$$\forall x_1 \ldots \forall x_n \, \exists y \, (P \; x_1 \; \ldots \; x_n \; y)$$

introduces a skolem symbol f^n that can only be used if applied to at least n terms and moreover the variables free in those terms cannot be bound in any context. For instance the term $\lambda y \, (f^1 \; x \; y)$ is correct, while the terms f^1, $(F \; f^1)$ and $\lambda x \, (f^1 \; x \; y)$ are not (Miller's conditions) [Mil83,Mil87].

A further motivation for expressing higher-order logic as a first-order theory is to avoid this cumbersome rule by reusing the usual first-order skolemization rule. We show below that when we apply the first-order skolemization rule to HOL-$\lambda\sigma$ we get Miller's conditions.

5.1 Miller's conditions in HOL-λ

The naive skolemization in higher-order logic, that transforms $\forall x \, \exists y \, (P \; x \; y)$ in $\forall x \, (P \; x \; (f \; x))$ with f constant of type $T \to U$ (where T is the type of x and U that of y) is unsound. Indeed the axiom of choice

$$\forall x \, \exists y \, (P \; x \; y) \Rightarrow \exists g \, \forall x \, (P \; x \; (g \; x))$$

is not provable in type theory [And72]. Thus from the proposition $\forall x \, \exists y \, (P \; x \; y)$ we cannot deduce $\exists g \, \forall x \, (P \; x \; (g \; x))$ while naively skolemizing this proposition yields $\forall x \, (P \; x \; (f \; x))$ from which we can obviously deduce $\exists g \, \forall x \, (P \; x \; (g \; x))$. Miller [Mil83,Mil87] has proposed an alternative skolemization rule that transforms a proposition of the form

$$\forall x_1 \, \forall x_2 \, \ldots \, \forall x_n \, \exists y \, (P \; x_1 \; x_2 \; \ldots \; x_n \; y)$$

into

$$\forall x_1 \, \forall x_2 \, \ldots \, \forall x_n \, (P \; x_1 \; x_2 \; \ldots \; x_n \; (f^n \; x_1 \; x_2 \; \ldots \; x_n)).$$

Two conditions are added:

- the symbol f^n can be used only when applied to at least n arguments (e.g. $(f^1\ x)$ is a term, but f^1 alone is not).
- the variables free in the necessary arguments cannot be bound by a λ higher in the term (e.g. $\lambda x\ (f^1\ y)$ is a term, but $\lambda x\ (f^1\ x)$ is not).

Recall however that, as it is usual in higher-order logic, $\forall x\ P$ is a notation for the term $\dot{\forall}\ (\lambda x\ P)$ where $\dot{\forall}$ is a constant. Then the skolemized proposition $\forall x\ (P\ x\ (f^1\ x))$ itself does not verify the second condition since x is bound by the external quantifier. Hence, we must either introduce quantifiers as new binders or give a more restricted form to Skolem theorem. If we use skolemization to put a proposition to be refuted in clausal form, then the universal quantifier will be suppressed yielding the proposition $(P\ X\ (f^1\ X))$ where X is a free variable. So we can state Skolem theorem as the correctness of this transformation with respect to resolution i.e. the clausal form of a proposition can be refuted by resolution if and only if the proposition itself is provable in sequent calculus. In [Mil83,Mil87] Miller formulates his theorem as the correctness of this transformation with respect to the connection method.

5.2 HOL-$\lambda\sigma$

Skolem theorem applies to HOL-$\lambda\sigma$ as it applies to any first-order theory. A proposition of the form $\forall x\ \exists y\ (P\ x\ y)$ is skolemized as $\forall x\ (P\ x\ f(x))$ where f is a unary function symbol. Hence we get back Miller's first condition. The rank of this symbol is $(\Gamma \vdash T)\Delta \vdash U$, i.e. it maps an argument of sort $\Gamma \vdash T$ into a term of sort $\Delta \vdash U$. The sort of the argument expresses Miller's second condition as it restricts the free variables in this term.

For example the proposition $\forall x\ \exists y\ \varepsilon(P\ x\ y)$ is skolemized as $\forall x\ \varepsilon(P\ x\ f(x))$ where f has rank $(\vdash T) \vdash U$, which requires the argument of f to be well typed in the empty context. For instance the λ-term $\lambda x\ (f^1\ x)$ which violates Miller's second condition, is expressed by the term $\lambda f(1)$ that is not well typed, while the term $\lambda x\ (f^1\ y)$ that verifies Miller's second condition is expressed by the term $\lambda f(y)$ that is well typed. We thus reap the benefit of using the $\lambda\sigma$-calculus where sorts explicit the scope of terms.

6 Automated theorem proving in HOL-$\lambda\sigma$

We are now able to wrap-up together the above ingredients to get a first-order presentation of higher-order resolution. To this end, as with any first-order theory modulo, we can use the ENAR method developed in [DHK98b] to search proofs in HOL-$\lambda\sigma$.

6.1 The ENAR method

The ENAR method applies to congruences described by class rewrite systems, i.e. pairs composed of a rewrite system $\mathcal{R}$ rewriting atomic propositions to propositions and a set of equational axioms $\mathcal{E}$ equating terms with terms and defining a congruence denoted $=_\varepsilon$.

As compared to first-order resolution, the ENAR method first replaces unification by equational unification modulo $\mathcal{E}$. The unification problems are kept as constraints written $t =^?_{\mathcal{E}} u$ and a clause C constrained by a set of equations E is written $C[E]$. Hence, we construct refutations with the **Extended Resolution** rule presented in figure 5. Then, as $\mathcal{R}$ rewrites atomic propositions to non atomic ones, we need another rule that instantiates, rewrites and puts in clausal form the result using the operator cl. This rule is called **Extended Narrowing** by analogy with the narrowing rule of equational unification.

$$\frac{\{A_1,\ldots,A_n,B_1,\ldots,B_m\}[E_1] \quad \{\neg C_1,\ldots,\neg C_p,D_1,\ldots,D_q\}[E_2]}{\{B_1,\ldots,B_m,D_1,\ldots,D_q\}[E_1 \cup E_2 \cup \{A_1 \ldots =^?_{\mathcal{E}} A_n =^?_{\mathcal{E}} C_1 \ldots =^?_{\mathcal{E}} C_p\}]} \quad \textbf{Ext. Res.}$$

$$\frac{C[E]}{cl(C[r]_p)[E \cup \{C_{|p} =^?_{\mathcal{E}} l\}]} \quad \textbf{Ext. Narrowing} \quad \text{if } l \to r \in \mathcal{R} \text{ and } C_{|p} \notin \mathcal{X}$$

Fig. 5. Extended narrowing and resolution (ENAR)

Theorem 8. *[DHK98b] Let $\mathcal{RE}$ be a confluent and weakly terminating class rewrite system such that the cut rule is redundant in sequent calculus modulo $\mathcal{RE}$. Then, the sequent*

$$A_1,\ldots,A_n \vdash B_1,\ldots,B_m$$

is provable in sequent calculus modulo if and only if from the constrained clauses

$$cl(\{\{A_1\},\ldots,\{A_n\},\{\neg B_1\},\ldots,\{\neg B_m\}\})[\emptyset]$$

we can derive the empty clause constrained by a $\mathcal{E}$-unifiable set of equations.

6.2 Applying ENAR to HOL-$\lambda\sigma$

In [DHK98b], we have applied ENAR to a first-order presentation of higher-order logic using combinators and we have shown that the **Extended Narrowing** rule specializes to the **Splitting** rule of higher-order resolution [Hue72,Hue73]. Unfortunately equational unification modulo the conversion axioms of combinators is not higher-order unification.

If we apply this method to HOL-$\lambda\sigma$, we obtain another proof search method for higher-order logic. As shown in the previous sections, HOL-$\lambda\sigma$ fulfills the hypotheses of theorem 8, so this method is complete. The **Extended Narrowing** rule still specializes to the **Splitting** rule of higher-order resolution, but the unification required is the unification modulo the system $\lambda\sigma$ that we have shown to be equivalent to higher-order unification in [DHK95]. Thus, the method obtained this way simulates higher-order resolution step by step.

Conclusion

In this paper we have given a first-order presentation of higher-order logic. This presentation is intentionally equivalent to the presentation of higher-order logic based on λ-calculus. Applying the Extended Narrowing and Resolution method to this theory gives higher-order resolution. Hence we show this way that expressing higher-order logic as a first-order theory and applying a first-order proof search method is at least as efficient as a direct implementation, provided we take the good first-order expression of higher-order logic and the good proof search method.

Expressing higher-order resolution in a first-order framework permits to clarify its features: higher-order unification, the splitting rule and higher-order resolution. Higher-order unification is equational unification in an appropriate theory. The splitting rule is an instance of the extended narrowing rule introduced in [DHK98b], it is needed because the rewrite system of higher-order logic transforms atomic propositions into non atomic ones. The higher-order skolemization rule is an instance of the first-order one. Its scoping particularities are consequences of the sort system of higher-order logic.

As we stay in a first-order setting, we can also reuse optimizations of first-order theorem proving such as redundancy criteria and subsumption. Another consequence is that extending the method to equational higher-order resolution requires only to add more reduction rules to the rewrite system $\lambda \sigma \mathcal{L}$, then narrowing provides an equational higher-order unification algorithm [KR97] and the proof search method is complete provided deduction modulo the extended theory verifies the cut elimination property.

References

[ACCL91] M. Abadi, L. Cardelli, P.-L. Curien, and J.-J. Lévy. Explicit substitutions. *Journal of Functional Programming*, 1(4):375–416, 1991.

[And71] P. Andrews. Resolution in type theory. *Journal of Symbolic Logic*, 36:414–432, 1971.

[And72] P. Andrews. General models, descriptions and choice in type theory. *The Journal of Symbolic Logic*, 37(2):385–394, 1972.

[And86] P. Andrews. *An Introduction to Mathematical Logic and Type Theory: To Truth through Proof.* Academic Press inc., New York, 1986.

[Chu40] A. Church. A formulation of the simple theory of types. *Journal of Symbolic Logic*, 5:56–68, 1940.

[DHK95] G. Dowek, T. Hardin, and C. Kirchner. Higher-order unification via explicit substitutions, extended abstract. In D. Kozen, editor, *Proceedings of LICS'95*, pages 366–374, San Diego, June 1995.

[DHK98a] G. Dowek, T. Hardin, and C. Kirchner. HOL-λσ an intentional first-order expression of higher-order logic. Rapport de Recherche 3556, Institut National de Recherche en Informatique et en Automatique, November 1998. http://coq.inria.fr/~dowek/3556.ps.gz.

[DHK98b] G. Dowek, T. Hardin, and C. Kirchner. Theorem proving modulo. Rapport de Recherche 3400, Institut National de Recherche en Informatique et en Automatique, April 1998. http://coq.inria.fr/~dowek/RR-3400.ps.gz.

[DW98]   G. Dowek and B. Werner. Proof normalization modulo. Rapport de Recherche 3542, Institut National de Recherche en Informatique et en Automatique, November 1998. http://coq.inria.fr/~dowek/RR-3542.ps.gz.

[Gir70]   J.-Y. Girard. Une extension de l'interprétation de Gödel à l'analyse et son application à l'élimination des coupures dans l'analyse et la théorie des types. In *J.E. Fenstad (Ed.), Second Scandinavian Logic Symposium.* North-Holland, 1970.

[Gir72]   J.-Y. Girard. *Interprétation fonctionnelle et élimination des coupures dans l'arithmétique d'ordre supérieur.* PhD thesis, Paris VII, 1972.

[Hue72]   G. Huet. *Constrained Resolution: A Complete Method for Type Theory.* PhD thesis, Case Western Reserve University, 1972.

[Hue73]   G. Huet. The undecidability of unification in third order logic. *Information and Control*, 22:257–267, 1973.

[JK86]   J.-P. Jouannaud and H. Kirchner. Completion of a set of rules modulo a set of equations. *SIAM Journal of Computing*, 15(4):1155–1194, 1986. Preliminary version in Proceedings 11th ACM Symposium on Principles of Programming Languages, Salt Lake City (USA), 1984.

[KR97]   C. Kirchner and C. Ringeissen. Higher-Order Equational Unification via Explicit Substitutions. In *Proceedings 6th International Joint Conference ALP'97-HOA'97, Southampton (UK)*, volume 1298 of *Lecture Notes in Computer Science*, pages 61–75. Springer-Verlag, 1997.

[KvOvR93] J. Klop, V. van Oostrom, and F. van Raamsdonk. Combinatory reduction systems: introduction and survey. *Theoretical Computer Science*, 121:279–308, 1993.

[Mar94]   C. Marché. Normalised rewriting and normalised completion. In S. Abramsky, editor, *Proceedings 9th IEEE Symposium on Logic in Computer Science, Paris (France)*, pages 394–403, 1994.

[Mil83]   D. Miller. *Proofs in higher order logic.* PhD thesis, Carnegie Mellon University, 1983.

[Mil87]   D. Miller. A compact representation of proofs. *Studia Logica*, XLVI(4):347–370, 1987.

[Plo72]   G. Plotkin. Building-in equational theories. *Machine Intelligence*, 7:73–90, 1972.

[PS81]   G. Peterson and M. Stickel. Complete sets of reductions for some equational theories. *Journal of the ACM*, 28:233–264, 1981.

[Sti85]   M. Stickel. Automated deduction by theory resolution. *Journal of Automated Reasoning*, 1(4):285–289, 1985.

[Vir95]   P. Viry. Rewriting modulo a rewrite system. Technical report TR-20/95, Dipartimento di informatica, Università di Pisa, December 1995.

[Vir98]   P. Viry. Adventures in sequent calculus modulo equations. In C. Kirchner and H. Kirchner, editors, *Proceedings of the 2nd International Workshop on Rewriting Logic and its Applications, WRLA'98*, volume 15, Pont-à-Mousson (France), September 1998. Electronic Notes in Theoretical Computer Science.

On the Connections between Rewriting and Formal Language Theory

Friedrich Otto

Fachbereich Mathematik/Informatik, Universität Kassel, D–34109 Kassel, Germany
`otto@theory.informatik.uni-kassel.de`

Abstract. Formal language theory, and in particular the theory of automata, has provided many tools that have been found extremely useful in rewriting theory, since automata can be used for deciding certain properties of rewriting systems as well as for constructing (weakly) confluent rewriting systems. On the other hand, rewriting theory has had some influence on the development of formal language theory, since based on certain rewriting systems some interesting classes of formal languages have been defined. Here a survey on some connections between rewriting and formal language theory is given, starting from the classical string languages and string-rewriting systems and continuing with tree automata and term-rewriting systems.

1   Introduction

String-rewriting systems (or *semi-Thue systems*) are intimately connected with formal language theory, since under the name of *sets of productions* they form an essential part of Chomsky's phrase-structure grammars (see, for example, [17]). In particular, the various classes of the Chomsky hierarchy are defined by placing certain restrictions on the form of the productions (that is, the rewrite rules) that are admitted in a grammar. Hence, it is not surprising that techniques and results that have been developed in formal language theory are often very helpful in investigating certain properties of string-rewriting systems.

In fact, many sets associated with string-rewriting systems are context-free or even regular languages, and in fact corresponding descriptions, for example accepting automata, can often be constructed effectively from the string-rewriting system under consideration. This is the case for the sets of reducible and irreducible strings with respect to a finite (or left-regular) system, and the same is true for certain sets of descendants and unions of congruence classes with respect to some restricted systems. Based on these language-theoretical properties, some decision problems can be solved effectively, in some cases even efficiently. An example is Book's reduction algorithm for deciding the word problem for finite convergent string-rewriting systems that is based on a realization of a left-most reduction through a two-pushdown automaton [1]. Another example is Book's decision algorithm for linear sentences that express properties of Thue congruences generated by finite monadic and confluent string-rewriting systems [2, 3].

The test for confluence of a finite noetherian system reduces to checking emptiness of finitely many intersections of finite sets [5]. However, if we want to verify that a finite noetherian system is confluent on a certain congruence class only, then this test is much more complicated. In fact, this task turns out to be undecidable in general, and even for finite monadic systems it reduces to checking equality for finitely many pairs of one-turn languages [51]. String-rewriting systems will be addressed in Section 2.

Prefix-rewriting systems can be used to describe left-congruences in monoids. Hence, in the case of groups they yield descriptions of subgroups [41]. Under certain restrictions a prefix-rewriting system can be completed using a Knuth-Bendix-style completion procedure [40], thus giving a rewrite-based algorithm for deciding membership in the subgroup considered. However, there is a simpler method for completing prefix-rewriting systems that is based on finite-state acceptors, and that is applicable to certain classes of finite convergent presentations of groups [43]. Prefix-rewriting systems will be discussed in Section 3.

The finite, length-reducing, and confluent string-rewriting systems have been used to define the class CRL of Church-Rosser languages in [45]. From the definition it follows immediately that the membership problem for each Church-Rosser language is decidable in linear time. Hence, CRL is contained in the class CSL of context-sensitive languages, and it is shown in the original paper that CRL contains the class $DCFL$ of deterministic context-free languages. However, only very recently the exact relationship between the class CRL and the class CFL of context-free languages could be settled [7, 8, 48]. The Church-Rosser languages will be the contents of Section 4.

The concepts of formal language and automata theory have been generalized to first-order terms and term languages (see, for example, [22, 23]), while on the other hand term-rewriting systems can be seen as a corresponding generalization of string-rewriting systems. Accordingly, automata-theoretical notions and techniques have been applied successfully to describe certain sets of terms that are associated with term-rewriting systems, and to solve certain decision problems.

Here, however, some technical complications arise that have no counterparts in the string case. A string-rewriting system S on an alphabet $\Sigma = \{a_1, \ldots, a_m\}$ can be interpreted as a term-rewriting system $R_S = \{\ell(x) \to r(x) \mid (\ell \to r) \in S\}$ on the signature $F_\Sigma = \{a_1, \ldots, a_m, \not\in\}$, where each letter a_i is considered as a function symbol of arity one, and $\not\in$ is a constant. Thus, *linear* term-rewriting systems form a generalization of string-rewriting systems in that function symbols of arity larger than one are admitted. However, general term-rewriting systems embody a further generalization in that they may contain *non-linear* terms. Hence, the problem of linearity versus non-linearity plays an important role in the study of term-rewriting systems.

On the other hand, *ground term-rewriting systems* have particularly nice properties due to the serious restriction to the applicability of their rules. These systems will be discussed in Section 5, while Section 6 is devoted to the various generalizations of techniques and results of automata theory to term languages and term-rewriting systems. Here we will in particular address the question of presenting the set of irreducible (ground) terms of a finite term-rewriting system through a finite tree-automaton, and the property of preserving regularity.

Due to space limitations only some fundamental definitions will be given in the paper. For further information regarding the notions introduced and for proofs of the results presented, the interested reader is asked to consult the literature, where [5] serves as our main reference on string-rewriting systems, [18] is our main reference on term-rewriting systems, and [23] is our main reference on tree automata.

Being a contribution to the proceedings of the 10th International Conference on Rewriting Techniques and Applications (RTA'99) this survey paper cannot possibly cover all the various aspects of the many connections between rewriting and formal language theory. Therefore, this article only presents some of the more fundamental connections that I have chosen based on my personal taste and experience. Others may feel that some important connections have been neglected. I apologize to all of them.

2   String-rewriting systems

Let Σ be a finite alphabet. Then Σ^* denotes the set of strings over Σ including the empty string λ. As usual the concatenation of two strings u and v will be denoted as uv, and numerical exponents will be used to abbreviate strings.

A *string-rewriting system* S on Σ is a subset of $\Sigma^* \times \Sigma^*$, the elements of which are called *(rewrite) rules*. By dom(S) we denote the set of all left-hand sides of rules of S, and by range(S) we denote the set of all right-hand sides. The *reduction relation* $\to_S^*$ defined by S is the reflexive and transitive closure of the *single-step reduction relation* $\to_S := \{(u\ell v, urv) \mid u, v \in \Sigma^*, (\ell, r) \in S\}$. A string $w \in \Sigma^*$ is called *reducible* if $w \to_S z$ holds for some string $z \in \Sigma^*$, otherwise w is called *irreducible*. By RED(S) (IRR(S)) we denote the set of all strings that are reducible (irreducible) modulo S. Obviously, RED$(S) = \Sigma^* \cdot \text{dom}(S) \cdot \Sigma^*$ and IRR$(S) = \Sigma^* \smallsetminus \text{RED}(S)$. Thus, if S is a finite system, then RED(S) and IRR(S) are regular languages. Actually, we have the following result.

Proposition 1. [27]
Given a finite string-rewriting system S, deterministic finite-state acceptors for the sets RED(S) and IRR(S) can be constructed in polynomial time.

For $w \in \Sigma^*$, $\Delta_S^*(w) := \{z \in \Sigma^* \mid w \to_S^* z\}$ is the set of *descendants* of w, $\nabla_S^*(w) := \{z \in \Sigma^* \mid z \to_S^* w\}$ is the set of *ancestors* of w, and $[w]_S := \{z \in \Sigma^* \mid w \leftrightarrow_S^* z\}$ is the *congruence class* of w. Here $\leftrightarrow_S^*$ denotes the *Thue congruence* generated by S, which is simply the reflexive, symmetric, and transitive closure of the relation $\to_S$. For a language $L \subseteq \Sigma^*$, the sets $\Delta_S^*(L)$, $\nabla_S^*(L)$, and $[L]_S$ are defined accordingly.

For a finite system S the sets of the form $\Delta_S^*(w)$, $\nabla_S^*(w)$, and $[w]_S$ are clearly recursively enumerable, but in general they are not even recursive. For certain restricted classes of string-rewriting systems however, we obtain much stronger results.

A string-rewriting system S is called

- *length-reducing* if $|\ell| > |r|$ holds for each rule (ℓ, r) of S, where $|w|$ denotes the *length* of the string w,
- *monadic* if it is length-reducing, and range$(S) \subseteq \Sigma \cup \{\lambda\}$,
- *special* if it is length-reducing, and range$(S) = \{\lambda\}$.

Obviously, a length-reducing system is *noetherian*. In fact, a system of this form has a linear upper bound on the length of reduction sequences. Although there are much more general classes of noetherian string-rewriting systems, we will not consider them in this paper.

If S is a length-reducing system, then $\Delta_S^*(w)$ is a finite set for each string w. However, already for finite confluent systems of this form we obtain very general languages once we consider sets of the form $\Delta_S^*(L)$, where $L \subseteq \Sigma^*$ is a regular language.

Proposition 2. [50]
Let $E \subseteq \Sigma^*$ be a recursively enumerable language. Then there exist a finite, length-reducing, and confluent string-rewriting system S on some alphabet Γ properly containing Σ and two regular languages $L_1, L_2 \subseteq \Gamma^*$ such that

$$\pi_\Sigma(\Delta_S^*(L_1) \cap L_2) = E = \pi_\Sigma([L_1]_S \cap L_2),$$

where π_Σ denotes the projection from Γ^* onto Σ^*.

On the other hand we have the following positive result.

Proposition 3. [4, 5, 37, 63]
Let S be a monadic string-rewriting system on Σ, and let $L \subseteq \Sigma^*$ be a regular language. Then the set $\Delta_S^*(L)$ is again a regular language. If S is finite, then a finite-state acceptor for this language can be constructed in polynomial time from a finite-state acceptor for L.

The acceptor for $\Delta_S^*(L)$ is simply obtained from the one for L by adding transitions. Accordingly the polynomial time-bound carries over even to certain classes of infinite monadic systems.

If S is a finite monadic system, then $S^{-1} := \{(r, \ell) \mid (\ell, r) \in S\}$ can be interpreted as the set of productions of a context-free grammar. Hence, it is easily seen that the set $\nabla_S^*(L)$ is a context-free language for each finite monadic string-rewriting system S and each context-free language L. If S is confluent, then $[w]_S = \nabla_S^*(w)$ for each irreducible string w. Hence, we obtain the following result.

Proposition 4. [4]
Let S be a finite string-rewriting system on Σ that is monadic and confluent.

(a) For each context-free set $L \subseteq \mathrm{IRR}(S)$ of irreducible strings, $[L]_S$ is a context-free language.

(b) For each regular set $L \subseteq \Sigma^*$, $[L]_S$ is a deterministic context-free language.

Underlying part (b) of Proposition 4 is the following general result.

Proposition 5. [1, 5]
Let S be a finite and noetherian string-rewriting system on Σ. Then there exists a deterministic automaton with two-pushdown stores that, given a string $w \in \Sigma^*$ as input, computes the irreducible descendant of w modulo S with respect to left-most reductions. If S is monadic, then this computation can be performed by a standard deterministic pushdown automaton.

Based on the positive results for monadic systems above Book has developed a decision procedure for a restricted class of sentences of first-order predicate calculus without equality, where the set of nonlogical symbols consists of a binary predicate symbol $\equiv$, a binary function symbol $\cdot$, a constant symbol a for each letter a from a fixed finite alphabet Σ, and a constant symbol 1.

Let S be a string-rewriting system on Σ. By interpreting the function symbol $\cdot$ as the multiplication in the monoid $M_S := \Sigma^* / \leftrightarrow_S^*$, by interpreting each constant a as the monoid element $[a]_S$ and the constant 1 as the identity $[\lambda]_S$ of the monoid M_S, and by interpreting the predicate symbol $\equiv$ as the congruence $\leftrightarrow_S^*$, we obtain an interpretation for these sentences expressing some properties of M_S.

Let Σ be a finite alphabet, and let V_E and V_U be two disjoint countable sets of symbols such that $(V_E \cup V_U) \cap \Sigma = \emptyset$. The symbols of V_E are *existential variables*, while those of V_U are *universal variables*. A string in $(\Sigma \cup V_U)^*$ is a *universal term*, and a string in $(\Sigma \cup V_E)^*$ is an *existential term*.

If x and y are two existential terms, then $x \equiv y$ is an *existential atomic formula*. If x and y are two universal terms, then $x \equiv y$ is a *universal atomic formula*. Finally, if one of x and y is an existential term and the other is a universal term, then $x \equiv y$ is a *mixed atomic formula*.

An atomic formula is a *formula*. If F_1 and F_2 are formulas, then $(F_1 \wedge F_2)$ and $(F_1 \vee F_2)$ are *formulas*. A formula is called *linear* if no variable occurs twice in it.

If F is a formula with existential variables $v_1, \ldots, v_q$ and universal variables $u_1, \ldots, u_p$, then

$$\forall u_1 \forall u_2 \ldots \forall u_p \exists v_1 \exists v_2 \ldots \exists v_q F \text{ and } \exists v_1 \exists v_2 \ldots \exists v_q \forall u_1 \forall u_2 \ldots \forall u_p F$$

are sentences. By LINSEN(Σ) we denote the set of all sentences over Σ that contain only linear formulas.

Let S be a string-rewriting system on Σ. If φ is a sentence over Σ containing the variables $v_1, \ldots, v_p \in (V_E \cup V_U)$, and if $L_1, \ldots, L_p$ are subsets of Σ^*, then we obtain the following interpretation of φ:

(i) for each i, $1 \leq i \leq p$, the variable v_i takes values in the set L_i;
(ii) the symbol $\equiv$ is interpreted as the congruence $\leftrightarrow_S^*$;
(iii) the symbol $\wedge$ is interpreted as conjunction and the symbol $\vee$ is interpreted as disjunction.

Under this interpretation the sentence φ is either true or false as a statement about the congruence $\leftrightarrow^*_S$ and the sets $L_1, \ldots, L_p \subseteq \Sigma^*$, and hence about the monoid M_S.

For example, a string w is *left-divisible* by z if and only if w is congruent to a string with prefix z, that is, if the linear sentence $\exists v \ : \ w \equiv z \cdot v$ is true under the interpretation induced by S and the set Σ^*.

If S is a finite, monadic, and confluent system, then with each term x of a linear sentence we can associate a regular set $L(x)$ of irreducible strings based on the structure of the term and the regular sets serving as domains for the variables occurring in x. In this way the question of whether or not the linear sentence is true under the given interpretation is reduced to a question about regular languages. This yields the following decidability result.

Proposition 6. [3]
Let S be a string-rewriting system on Σ that is finite, monadic, and confluent. Then the following validity problem for linear sentences is decidable in polynomial space:

INSTANCE : *A sentence $\varphi \in \text{LINSEN}(\Sigma)$ containing variables $v_1, v_2, \ldots, v_m$, and regular sets $L_1, \ldots, L_m \subseteq \Sigma^*$ that are specified by finite-state acceptors.*

QUESTION : *Is φ true under the interpretation induced by S and $L_1, \ldots, L_m$?*

Actually, if the linear sentences φ considered do not contain mixed atomic formulas or if their quantifier prefixes are of the form $\exists^i \forall^j$, then the validity of these sentences is even decidable in polynomial time.

In fact, also some other decision problems, for which there does not seem to be a way of expressing them by linear sentences, can be solved for finite, monadic, and confluent string-rewriting systems in a similar way. An example for this is the property of *left-cancellativity*. Here the monoid M_S is called *left-cancellative* if, for all $u, v, w \in \Sigma^*$, $uv \leftrightarrow^*_S uw$ implies that $v \leftrightarrow^*_S w$ holds.

Proposition 7. [46]
Let S be a string-rewriting system that is length-reducing, interreduced, and confluent. Then the monoid M_S is not left-cancellative if and only if there exists a rule $(au, v) \in S$, where $a \in \Sigma$ and $u, v \in \text{IRR}(S)$, such that $\Delta^*_S(L_1) \cap \Delta^*_S(L_2) \neq \emptyset$, where $L_1 = \{auw \mid w \in \Sigma^* \text{ such that } uw \in \text{IRR}(S)\}$ and $L_2 = \{ax \mid x \in \text{IRR}(S), u \text{ is not a prefix of } x\}$.

If S is finite, then L_1 and L_2 are regular languages, finite-state acceptors for which can be constructed in polynomial time. Thus, if additionally S is monadic, then the condition stated in the proposition above can be verified in polynomial time. By considering various other regular languages associated with monadic string-rewriting systems, it can be shown that the (*left-, right-*) *conjugacy problem* and the *common left-* (*right-*) *multiplier problem* are decidable in polynomial time for each finite, monadic, and confluent string-rewriting system [47].

For a finite noetherian string-rewriting system S on Σ the test for confluence of S reduces to checking whether the intersection $\Delta^*_S(u) \cap \Delta^*_S(v)$ is non-empty for each of the finitely many critical pairs (u, v) of S. However, it is much more difficult in general to decide whether the system S is confluent on a certain congruence class $[w]_S$. Here S is called *confluent on* $[w]_S$ for some string $w \in \Sigma^*$ if, for all $u, v, x \in [w]_S$, $u \to_S^* v$ and $u \to_S^* x$ imply that $\Delta_S^*(v) \cap \Delta_S^*(x)$ is non-empty.

For $u \in \Sigma^*$ and $w \in \mathrm{IRR}(S)$, let $\mathrm{Con}_u(w) := \{x \# y \mid x, y \in \mathrm{IRR}(S)$ and $xuy \to_{\ell,S}^* w\}$ be the *set of contexts of u for w* modulo S, where $\to_{\ell,S}^*$ denotes the *left-most reduction* modulo S, and $\#$ is a new letter. Further, let $\mathrm{UCP}(S)$ denote the set of those critical pairs (y, z) of S for which the intersection $\Delta_S^*(y) \cap \Delta_S^*(z)$ is empty. Then we have the following characterization.

Proposition 8. [51]
Let S be a finite noetherian string-rewriting system on Σ, and let $w \in \mathrm{IRR}(S)$. Then S is confluent on $[w]_S$ if and only if $\mathrm{Con}_y(w) = \mathrm{Con}_z(w)$ holds for each pair $(y, z) \in \mathrm{UCP}(S)$.

Even for finite length-reducing systems confluence on a given congruence class is undecidable in general [51]. If, however, S is a finite monadic system, then each language of the form $\mathrm{Con}_u(w)$ is a deterministic one-turn language, and in fact, from S and the strings u and w, a deterministic one-turn pushdown automaton for $\mathrm{Con}_u(w)$ can be constructed effectively. Thus, we obtain the following decidability result due to the solvability of the equivalence problem for deterministic one-turn pushdown automata [70].

Corollary 1. [51]
For finite monadic string-rewriting systems confluence on a given congruence class is decidable in doubly exponential time.

However, for special systems this result can be improved considerably by analyzing the form of the generated reduction sequences in more detail.

Corollary 2. [53]
For finite special string-rewriting systems confluence on a given congruence class is decidable in polynomial time.

This result even extends to testing whether a finite monadic system S is *weakly confluent*, that is, whether S is confluent on $[a]_S$ for each $a \in \mathrm{range}(S)$ [44]. I would like to mention in passing that based on these confluence tests Knuth-Bendix like procedures for weak completion have been developed

1. for finite special systems [52], and
2. for finite monadic systems presenting groups [44].

Actually, for these two classes of string-rewriting systems further interesting results have been obtained that are based on language properties of certain associated sets.

For a string-rewriting system S on Σ and a language $L \subseteq \Sigma^*$, we denote by $I_S(L)$ the set $I_S(L) := [L]_S \cap \mathrm{IRR}(S)$ of irreducible strings that are congruent to some string from L.

Proposition 9. [60]
Let S be a finite special string-rewriting system that is confluent on $[\lambda]_S$, and let $L \subseteq \Sigma^*$ be a regular language. Then the set $I_S(L)$ is also regular, and a finite-state acceptor for $I_S(L)$ can be constructed in polynomial time from a finite-state acceptor for L.

Proposition 9 also holds for finite, monadic, and weakly confluent systems that present groups [44]. In particular, this implies that the result on linear sentences (Proposition 6) carries over to finite, special systems that are weakly confluent and to finite, monadic, and weakly confluent systems presenting groups.

3 Prefix-rewriting systems

In this section we take a look at prefix-rewriting systems and relate them to the subgroup problem of finitely presented groups.

A *prefix-rewriting system* on Σ is a subset of $\Sigma^* \times \Sigma^*$. Its elements are called *prefix-rules*. If P is a prefix-rewriting system, then $\operatorname{dom}(P)$ and $\operatorname{range}(P)$ are defined as for string-rewriting systems.

The *prefix-reduction relation* $\Rightarrow_P^*$ defined by P is the reflexive transitive closure of the *single-step prefix-reduction relation* $\Rightarrow_P := \{(\ell w, rw) \mid (\ell, r) \in P, w \in \Sigma^*\}$, and by $\Leftrightarrow_P^*$ we denote the reflexive, symmetric, and transitive closure of $\Rightarrow_P$. Obviously $\Leftrightarrow_P^*$ is a left-congruence on Σ^*. By $\operatorname{RED}(P)$ we denote the set of all reducible strings, and $\operatorname{IRR}(P)$ denotes the set of irreducible strings. Obviously, $\operatorname{RED}(P) = \operatorname{dom}(P) \cdot \Sigma^*$ and $\operatorname{IRR}(P) = \Sigma^* \setminus \operatorname{RED}(P)$. Hence, if $\operatorname{dom}(P)$ is a regular language, then $\operatorname{RED}(P)$ and $\operatorname{IRR}(P)$ are regular languages as well. In this situation the prefix-rewriting system P is called *left-regular*.

The prefix-rewriting system P is called *noetherian, confluent, convergent, λ-confluent, λ-convergent, interreduced,* or *canonical* if the corresponding condition is satisfied by $\Rightarrow_P$. It is interesting to observe that a prefix-rewriting system is convergent whenever it is interreduced, that is, it is canonical if and only if it is interreduced. This is an immediate consequence of the corresponding result for ground-term rewriting systems (Proposition 17), since a prefix-rewriting system P on Σ can be interpreted as a ground-term rewriting system on the signature F_Σ.

Next we will show how prefix-rewriting systems are related to the *subgroup problem*. Let G be a group that is given through a finite presentation $(\Sigma; S)$, and let $^{-1} : \Sigma^* \to \Sigma^*$ denote a function realizing the inverse function of G. Further, let $U \subseteq \Sigma^*$ be a finite set, where we assume without loss of generality that U is *closed under inverses*, that is, for each $u \in U$, there exists an element $v \in U$ such that $v \leftrightarrow_S^* u^{-1}$. Then a string $w \in \Sigma^*$ presents an element of the subgroup $\langle U \rangle$ of G that is generated by U if and only if there exist $u_1, \ldots, u_k \in U$ such that $w \leftrightarrow_S^* u_1 u_2 \cdots u_k$. The *subgroup problem* for G is the problem of deciding, given a finite set $U \subset \Sigma^*$ and a string $w \in \Sigma^*$, whether or not w belongs to the subgroup $\langle U \rangle$ of G.

With U we associate a binary relation $\sim_U$ on Σ^* as follows:

$$x \sim_U y \quad \text{iff} \quad \exists u \in \langle U \rangle : x \leftrightarrow_S^* uy.$$

Then $w \in \langle U \rangle$ if and only if $w \sim_U \lambda$.

With $(\Sigma; S)$ and U we now associate a prefix-rewriting system $P := P_U \cup P_S$, where

$$P_U := \{(u, \lambda) \mid u \in U\}$$

and

$$P_S := \{(x\ell, xr) \mid x \in \Sigma^* \text{ and } (\ell \to r) \in S\}.$$

Then P is a left-regular system, and the following property is easily verified.

Proposition 10. [41] *The left-congruences $\sim_U$ and $\Leftrightarrow_P^*$ coincide.*

Hence, if P is λ-confluent, then a string $w \in \Sigma^*$ belongs to $\langle U \rangle$ if and only if $w \Rightarrow_P^* \lambda$, and if P is convergent, then $\mathrm{IRR}(P)$ is a complete set of coset representatives for $\langle U \rangle$ in G.

If S is noetherian, then P is noetherian, but in general P will not be convergent even in case S is. However, as for string-rewriting systems confluence of the prefix-rewriting system P can be characterized through the convergence of finitely many critical pairs. Based on this confluence test a Knuth-Bendix style completion procedure for prefix-rewriting systems has been developed in [41] that applies to groups G that are given through finite convergent presentations.

Also confluence on $[\lambda]_{\sim_U}$ can be characterized as for string-rewriting systems (Proposition 8) [42]. However, there is another criterion for deciding this property that exploits automata-theoretical arguments.

Let $(\Sigma; S)$ be a finite convergent presentation of a group, let P_U be a set of prefix-rules on Σ, and let $P := P_U \cup P_S$, where we assume that the set $U := \{uv^{-1} \mid (u, v) \in P_U\}$ is closed under taking inverses and that P is noetherian. Then $[\lambda]_P = \langle U \rangle$, and hence, $w \in [\lambda]_P$ if and only if $w \Leftrightarrow_P^* z$ for some $z \in U^*$. Now P is confluent on $[\lambda]_{\sim_U}$ if and only if each string $w \in \langle U \rangle \smallsetminus \{\lambda\}$ is reducible by P, that is, if and only if $[\lambda]_P \cap \mathrm{IRR}(P) = \{\lambda\}$. However, since S is convergent, the latter equality is equivalent to the equality $(\Delta_S^*(U^*) \cap \mathrm{IRR}(S)) \cap \mathrm{IRR}(P) = \{\lambda\}$.

If S and P_U are both finite, then the sets $\mathrm{IRR}(S)$ amd $\mathrm{IRR}(P)$ are both regular, and finite-state acceptors for them can be constructed effectively. Also U^* is a regular set in this situation. Hence, this criterion becomes decidable whenever the set $\Delta_S^*(U^*)$ (or the set $\Delta_S^*(U^*) \cap \mathrm{IRR}(S)$) allows an effective specification for which the intersection with the regular set $\mathrm{IRR}(P)$ can be determined effectively.

If $(\Sigma; S)$ is a finite, weight-reducing, and confluent presentation of a group and $U \subseteq \Sigma^*$ is a finite set, then it is still an open problem whether or not the set $\Delta_S^*(U^*)$ is necessarily regular. However, if we restrict the set $\Delta_S^*(U^*)$ to only those strings that are obtained by left-most reductions, then this subset $\Delta_{L,S}^*(U^*)$ of $\Delta_S^*(U^*)$ can be shown to always be regular [40]. In fact, a finite-state acceptor for this language can be constructed effectively. Since $\Delta_{L,S}^*(U^*) \cap \mathrm{IRR}(S) = \Delta_S^*(U^*) \cap \mathrm{IRR}(S)$, we obtain a finite-state acceptor for the set $\Delta_S^*(U^*) \cap \mathrm{IRR}(P)$. This gives the following decidability result.

Proposition 11. [40]
Let $(\Sigma; S)$ be a finite, weight-reducing, and confluent presentation of a group,

and let P_U be a finite set of prefix-rules on Σ such that the set $U := \{uv^{-1} \mid (u,v) \in P_U\}$ is closed under taking inverses, and $P := P_U \cup P_S$ is noetherian. Then it is decidable whether the prefix-rewriting system P is λ-confluent.

Now assume that $(\Sigma; S)$ is a finite, weight-reducing, and confluent presentation of a group G, let $U \subseteq \Sigma^+$ be a finite set that is closed under taking inverses, and let $P_U := \{(u, \lambda) \mid u \in U\}$. From $(\Sigma; S)$ and U we can construct a finite-state acceptor $A = (Q, \Sigma, q_0, F, \delta)$ for the language $\Delta_S^*(U^*) \cap \mathrm{IRR}(S)$. From A we extract a finite set of prefix-rules P_U' as follows, where we identify A with its state graph in order to simplify the notation:

(i) For every simple path in A leading from the initial state q_0 to a final state $q_f \in F$, which does not pass through any final state, we put the rule (x, λ) into P_U', where x is the label along the path considered.

(ii) For every path p in A from q_0 to a final state $q_f \in F$, which does not pass through any final state, and which can be partitioned into three parts $p = p_1, p_2, p_3$ such that p_1 is a simple path, and p_2 is a simple loop, we put the rule $(x_1 x_2, x_1)$ into P_U', where x_i is the label along the subpath p_i, $i = 1, 2$.

Obviously, P_U' is a finite set of prefix-rules that can effectively be obtained from A. For $w \in \langle U \rangle$ there exists a unique string $w_0 \in \mathrm{IRR}(S)$ such that $w \to_S^* w_0$. Since $w \in \langle U \rangle$, $w_0 \in \Delta_S^*(U^*) \cap \mathrm{IRR}(S)$, and hence, w_0 is accepted by A. From the construction of P_U' it follows that $w \Rightarrow_{P'}^* \lambda$ holds, where $P' := P_U' \cup P_S$. Since $u \sim_U v$ holds for each rule $(u, v) \in P_U'$, it follows that $\Leftrightarrow_{P'}^* = \sim_U$, and P' is confluent on $[\lambda]_{\sim_U}$.

Proposition 12. [40]
Let $(\Sigma; S)$ be a finite, weight-reducing, and confluent presentation of a group G, and let $U \subseteq \Sigma^+$ be a finite set. Then a finite set of length-reducing prefix-rules P_U' can be determined effectively such that the prefix-rewriting system $P := P_U' \cup P_S$ is confluent on $[\lambda]_P$ and $\Leftrightarrow_P^* = \sim_U$.

Actually, this construction carries over to the case of groups that are presented through finite and monadic string-rewriting systems that are only confluent on the congruence class of the empty string [43].

Finally, we want to address *automatic structures* for monoids, which is a fairly recent development. An automatic structure for a monoid-presentation $(\Sigma; S)$ can be interpreted as a finite description of the multiplication table of the monoid M_S. Originally automatic structures were developed for groups (see [19] for a detailed presentation), but recently automatic structures have also been considered for semigroups and monoids [9].

In order to define automatic structures we need the following definition as we will be dealing with infinite sets of pairs of strings that are to be recognized by finite-state acceptors.

Let Σ be a finite alphabet, and let $\# \notin \Sigma$ be an additional "padding" symbol. Then by $\Sigma_\#$ we denote the following finite alphabet:

$$\Sigma_\# := ((\Sigma \cup \{\#\}) \times (\Sigma \cup \{\#\})) \setminus \{(\#, \#)\}.$$

This alphabet is called the *padded extension* of Σ. An encoding $\nu : \Sigma^* \times \Sigma^* \to \Sigma_\#^*$ is now defined as follows:
if $u := a_1 a_2 \cdots a_n$ and $v := b_1 b_2 \cdots b_m$, where $a_1, \ldots, a_n, b_1, \ldots, b_m \in \Sigma$, then

$$\nu(u,v) := \begin{cases} (a_1, b_1)(a_2, b_2) \cdots (a_m, b_m)(a_{m+1}, \#) \cdots (a_n, \#), & \text{if } m < n, \\ (a_1, b_1)(a_2, b_2) \cdots (a_m, b_m), & \text{if } m = n, \\ (a_1, b_1)(a_2, b_2) \cdots (a_n, b_n)(\#, b_{n+1}) \cdots (\#, b_m) & \text{if } m > n. \end{cases}$$

A *prefix-rewriting system* P on Σ is called *synchronously regular, s-regular* for short, if $\nu(P)$ is accepted by some finite-state acceptor over $\Sigma_\#$. Obviously, if P is s-regular, then $\text{dom}(P)$ and $\text{range}(P)$, and therewith also $\text{RED}(P)$ and $\text{IRR}(P)$, are regular languages.

An *automatic structure* for a finitely generated monoid-presentation $(\Sigma; S)$ consists of finite-state acceptors W over Σ and $M_=$ and M_a $(a \in \Sigma)$ over $\Sigma_\#$ satisfying the following conditions:

(0.) $L(W) \subseteq \Sigma^*$ is a complete set of (not necessarily unique) representatives for the monoid M_S, that is, $L(W) \cap [u]_S \neq \emptyset$ holds for each $u \in \Sigma^*$,
(1.) $L(M_=) = \{ \nu(u,v) \mid u, v \in L(W) \text{ and } u \leftrightarrow_S^* v \}$, and
(2.) for all $a \in \Sigma$, $L(M_a) = \{ \nu(u,v) \mid u, v \in L(W) \text{ and } ua \leftrightarrow_S^* v \}$.

Actually, one may require that the set $L(W)$ is a cross-section for M_S, in which case we say that we have an *automatic structure with uniqueness* [19]. In this situation the finite-state acceptor $M_=$ is trivial, and hence, it will not be mentioned explicitly.

A monoid-presentation is called *automatic* if it has an automatic structure, and a monoid is called *automatic* if it has an automatic presentation. Automatic monoids have word problems that are decidable in quadratic time based on the automatic structure. For automatic groups many additional nice properties have been obtained, while for automatic monoids in general the situation is not quite as nice [9, 54, 59]. Here we are interested in automatic structures with uniqueness, for which the set of representatives considered is in addition prefix-closed. It is an open problem whether or not every automatic group does have an automatic structure with this additional property. But at least the following characterization can be obtained.

Proposition 13. [55]
Let $(\Sigma; S)$ be a finitely generated monoid-presentation. Then the following two statements are equivalent:

(a) *There exists an automatic structure $(W, A_a(a \in \Sigma))$ with uniqueness for $(\Sigma; S)$ such that the set $L(W)$ is prefix-closed.*
(b) *There exists an s-regular canonical prefix-rewriting system P on Σ that is equivalent to S, that is, the left-congruence $\leftrightarrow_P^*$ coincides with the Thue congruence $\leftrightarrow_S^*$.*

There exists a group with a finite convergent presentation, which does not admit an automatic structure [24]. Hence, no finitely generated presentation of this group has an s-regular canonical prefix-rewriting system that defines the corresponding Thue congruence.

The monoid N of [59] has an automatic structure that is based on a regular cross-section that is the set of irreducible strings modulo some infinite left-regular convergent string-rewriting system. Hence, this set is certainly prefix-closed and so Proposition 13 shows that this presentation of N admits an s-regular canonical prefix-rewriting system. However, N does not admit any finite convergent presentation. These observations yield the following result.

Corollary 3. *The class of finitely presented monoids that admit a finite convergent presentation and the class of finitely presented monoids that admit an s-regular canonical prefix-rewriting system are incomparable under set inclusion.*

4 Church-Rosser languages

In the previous sections we have seen how techniques from automata theory have been used to establish properties for string-rewriting systems. Here we show that also rewriting theory has had some influence on formal language theory.

Definition 1. [45]

(a) *A language $L \subseteq \Sigma^*$ is a Church-Rosser language (CRL) if there exist an alphabet $\Gamma \supsetneq \Sigma$, a finite, length-reducing, confluent string-rewriting system R on Γ, two strings $t_1, t_2 \in (\Gamma \smallsetminus \Sigma)^* \cap \mathrm{IRR}(R)$, and a letter $Y \in (\Gamma \smallsetminus \Sigma) \cap \mathrm{IRR}(R)$ such that, for all $w \in \Sigma^*$, $t_1 w t_2 \to_R^* Y$ if and only if $w \in L$.*

(b) *A language $L \subseteq \Sigma^*$ is a Church-Rosser decidable language (CRDL) if there exist an alphabet $\Gamma \supsetneq \Sigma$, a finite, length-reducing, confluent string-rewriting system R on Γ, two strings $t_1, t_2 \in (\Gamma \smallsetminus \Sigma)^* \cap \mathrm{IRR}(R)$, and two distinct letters $Y, N \in (\Gamma \smallsetminus \Sigma) \cap \mathrm{IRR}(R)$ such that, for all $w \in \Sigma^*$, the following statements hold:*

- $t_1 w t_2 \to_R^* Y$ *if and only if $w \in L$, and*
- $t_1 w t_2 \to_R^* N$ *if and only if $w \notin L$.*

By admitting weight-reducing instead of length-reducing string-rewriting systems in the definition, we obtain the class $GCRL$ of *generalized Church-Rosser languages* [8]. Obviously, the membership problem for a $GCRL$ is decidable in linear time, and so $GCRL$ is contained in the class CSL of context-sensitive languages. Further, it is shown in [45] that each deterministic context-free language is a Church-Rosser decidable language, while there exist languages in $CRDL$ that are not context-free. Hence, we have the following sequence of inclusions:

$$DCFL \subset CRDL \subseteq CRL \subseteq GCRL \subset CSL.$$

However, while it was conjectured in [45] that the class CFL of context-free languages is not contained in CRL, this remained open at the time.

Another subclass of CSL that received quite some attention in the literature is the class $GCSL$ of growing context-sensitive languages. Here a language is called *growing context-sensitive* if it is generated by a *growing context-sensitive grammar*, that is, a grammar $G = (N, \Sigma, S, P)$ satisfying the following conditions:

1. the start symbol S does not occur on the right-hand side of any production, and
2. for each production $(\ell, r) \in P$, $|\ell| < |r|$ or $\ell = S$.

In [12] Dahlhaus and Warmuth proved that the membership problem for a growing context-sensitive language can be solved in polynomial time. In [7] Buntrock and Otto introduced the following type of automaton in order to characterize the class *GCSL* of growing context-sensitive languages.

Definition 2.

(a) A two-pushdown automaton (*TPDA*) *is a nondeterministic automaton with two pushdown stores. Formally, it is a 7-tuple* $M = (Q, \Sigma, \Gamma, \delta, q_0, \bot, F)$, *where*
   - Q *is the finite set of states,*
   - Σ *is the finite input alphabet,*
   - Γ *is the finite tape alphabet with* $\Gamma \supsetneq \Sigma$ *and* $\Gamma \cap Q = \emptyset$,
   - $q_0 \in Q$ *is the initial state,*
   - $\bot \in \Gamma \setminus \Sigma$ *is the bottom marker of the pushdown stores,*
   - $F \subseteq Q$ *is the set of final (or accepting) states, and*
   - $\delta : Q \times \Gamma \times \Gamma \to 2^{Q \times \Gamma^* \times \Gamma^*}$ *is the transition relation, where* $\delta(q, a, b)$ *is a finite set for each triple* $(q, a, b) \in Q \times \Gamma \times \Gamma$.

   M *is a* deterministic two-pushdown automaton (*DTPDA*), *if* δ *is a (partial) function from* $Q \times \Gamma \times \Gamma$ *into* $Q \times \Gamma^* \times \Gamma^*$.

(b) A (DTPDA) TPDA M *is called* shrinking *if there exists a weight function* $\varphi : Q \cup \Gamma \to \mathbb{N}_+$ *such that, for all* $q \in Q$ *and* $a, b \in \Gamma$, *if* $(p, u, v) \in \delta(q, a, b)$, *then* $\varphi(puv) < \varphi(qab)$. *By sTPDA and sDTPDA we denote the corresponding classes of shrinking automata.*

A *configuration* of a (DTPDA) TPDA M can be described as uqv with $q \in Q$ and $u, v \in \Gamma^*$, where u is the contents of the first pushdown store with the first letter of u at the bottom and the last letter of u at the top, q is the current state, and v is the contents of the second pushdown store with the last letter of v at the bottom and the first letter of v at the top. M induces a *computation relation* $\vdash_M^*$ on the set of configurations, which is the reflexive, transitive closure of the *single-step computation relation* $\vdash_M$ (see, e.g., [31]). For an input string $w \in \Sigma^*$, the corresponding *initial configuration* is $\bot q_0 w \bot$. M accepts by empty pushdown stores:

$$L(M) := \{w \in \Sigma^* \mid \exists q \in F : \bot q_0 w \bot \vdash_M^* q\}.$$

Buntrock and Otto established the following characterization for the classes of languages that are accepted by nondeterministic or deterministic shrinking *TPDAs*, respectively.

Proposition 14. [7, 8]

(a) A language is accepted by some shrinking TPDA if and only if it is growing context-sensitive.

(b) A language is accepted by some shrinking DTPDA if and only if it is a generalized Church-Rosser language.

Thus, the generalized Church-Rosser languages can be viewed as the deterministic variants of the growing context-sensitive languages. Further, it is observed in [8] that the language $L = \{ww \mid w \in \{a, b\}^+\}$ does not belong to the class $GCSL$. Since the class $GCRL$ is clearly closed under complement due to Proposition 14(b), it follows that the language $L^c = \{a, b\}^* \setminus L$ is a context-free language that is not generalized Church-Rosser. This finally settled the conjecture of [45] mentioned above.

Finally, Niemann and Otto showed that each $sDTPDA$ can be simulated by some finite length-reducing and confluent string-rewriting system [48], thus establishing the following equalities.

Proposition 15. [48] *The classes CRDL, CRL, and GCRL coincide.*

Thus, CRL is incomparable with the class CFL under set inclusion, it is closed under complement and under left and right quotient with a single string [45]. However, it is not closed under union or intersection [48], and it is not closed under homomorphisms, since CRL is a basis for the recursively enumerable languages [57, 58]. Since CFL is a full abstract family of languages [31], this indicates a certain duality between CRL and CFL. Based on a generalization of the so-called *restarting automata with rewriting* [35] this duality is further explored in [49].

5 Ground term-rewriting systems

Finally we turn to rewriting systems over terms. For the following considerations let F denote a finite *signature*, that is, F is a finite set of *function symbols*, each of which is associated with a fixed arity. For each $n \geq 0$, F_n is the subset of F consisting of the function symbols of arity n. The elements of F_0 are called *constants*. To avoid degenerate cases we will always assume that the set of constants is non-empty.

The set of *terms* $T(F)$ is defined inductively as follows:

(1.) Each constant is a term.

(2.) If $f \in F_n$ for some $n > 0$ and $t_1, \ldots, t_n \in T(F)$, then $f(t_1, \cdots, t_n)$ is a term.

Actually, since they do not contain any variables, the terms considered here are usually called *ground terms*. However, we call them simply terms here, as we will not consider terms with variables until the next section.

A term $t \in T(F)$ can be seen as a finite ordered tree, the leaves of which are labeled with constants and the internal nodes of which are labeled with function symbols of positive arity such that the outdegree of an internal node equals the arity of its label. Thus, a *position* within a term can be represented – in Dewey decimal notation – as the sequence of positive integers which describes the path from the root to that position. Accordingly, the set $O(t)$ of *occurrences* of the term t is the set of sequences of positive integers describing the positions in t. The length of the longest of these sequences is called the *depth* of the term t, which is denoted as $\mathrm{depth}(t)$, and the number of sequences in $O(t)$ is the *size* of t, denoted as $\mathrm{size}(t)$. For $p \in O(t)$, t/p denotes the subterm of t at occurrence p. If s is another term, then $t[p \leftarrow s]$ denotes the term that is obtained by replacing the subterm of t at occurrence p by the term s.

A *ground term-rewriting system* R is a subset of $T(F) \times T(F)$, the elements of which are called *(rewrite) rules*. The *reduction relation* associated with a ground term-rewriting system R is the reflexive and transitive closure $\to_R^*$ of the following *single-step reduction relation*: $s \to_R t$ if and only if there exist an occurrence $p \in O(s)$ and a rule $(\ell \to r) \in R$ such that $s/p = \ell$ and $t = s[p \leftarrow r]$. A term t is said to be in *normal form* or *irreducible* modulo the ground term-rewriting system R if no reduction can be applied to t. By $\mathrm{IRR}(R)$ we denote the set of all these irreducible terms, and $\mathrm{RED}(R) = T(F) \smallsetminus \mathrm{IRR}(R)$ is the set of reducible terms.

The equational theory that is associated with a ground term-rewriting system R is the congruence $=_R$ that is generated by the reduction relation $\to_R$, that is, it is the congruence $(\to_R \cup \leftarrow_R)^*$.

A ground term-rewriting system R is called *noetherian, (locally) confluent*, or *convergent* if the reduction relation $\to_R$ has the corresponding property. It is *depth-reducing* if $\mathrm{depth}(\ell) > \mathrm{depth}(r)$ holds for each rule $\ell \to r$ of R. Finally, R is called *interreduced* if $\mathrm{range}(R) \subseteq \mathrm{IRR}(R)$ and $\ell \in \mathrm{IRR}(R \smallsetminus \{\ell \to r\})$ for each rule $(\ell \to r) \in R$. If R is convergent and interreduced, then it is called *canonical*.

A *term language* over F is a subset of $T(F)$. As for strings term languages can be defined by formal grammars and by various types of automata. Here we are mainly interested in the class of *regular term languages* which can be defined as follows.

A *non-deterministic bottom-up tree automaton (NBUTA)* is given through a 4-tuple $A = (Q, F, R_A, Q_a)$, where Q is a finite set of states, F is a finite signature, $Q_a \subseteq Q$ is the set of accepting states, and R_A is a ground term-rewriting system on the signature $F \cup Q$, where each state symbol from Q is considered as a new constant. The rules of R_A are of the form

(i) $c \to q$, where $c \in F_0$ and $q \in Q$, and
(ii) $f(q_1, \ldots, q_n) \to q$, where $f \in F_n$ for some $n > 0$ and $q_1, \ldots, q_n, q \in Q$.

A is a *deterministic bottom-up tree automaton (BUTA)*, if R_A does not contain two rules with the same left-hand side. The language $L(A)$ accepted by A is defined as $L(A) = \{t \in T(F) \mid t \to_{R_A}^* q \text{ for some } q \in Q_a\}$. A language $L \subseteq T(F)$ is *regular* if and only if it is accepted by some *NBUTA*, and this is the case if and only if it is accepted by some *BUTA* [22].

For a finite ground term-rewriting system R, a *BUTA* A can easily be constructed such that $L(A) = \mathrm{RED}(R)$. Since the class of regular term languages is effectively closed under complement, we also obtain a *BUTA* for the set of irreducible terms IRR(R). Thus, RED(R) and IRR(R) are regular term languages.

In contrast to the situation for string-rewriting systems (or for that matter general term-rewriting systems) it is decidable whether or not a finite ground term-rewriting system is noetherian [32]. Further, even confluence is decidable for these systems [14, 15, 61]. Oyamaguchi's proof, which is combinatorically quite involved, reduces the confluence property of finite ground-term rewriting systems to the equivalence problem for non-deterministic top-down tree automata [61], while Dauchet and his co-authors invented a new kind of transducer to describe the confluence property [14, 15].

A *ground tree transducer* (*GTT*) consists of a pair (G, D) of NBUTAs $G = (Q_G, F, R_G, Q_G)$ and $D = (Q_D, F, R_D, Q_D)$ such that $Q_G \cap Q_D$ is non-empty. The relation $\rightarrow^{(G,D)} \leftarrow$ on $T(F)$ that is induced by (G, D) is defined as follows:

$$t \rightarrow^{(G,D)} \leftarrow t' \text{ iff } \exists s \in T(F \cup (Q_G \cap Q_D)): t \rightarrow^*_G s \leftarrow^*_D t'.$$

If a binary relation $\sim$ on $T(F)$ coincides with the relation $\rightarrow^{(G,D)} \leftarrow$ induced by a *GTT*, then $\sim$ is called a *GTT-relation*.

Proposition 16. [14, 15]

(1.) The inverse of a GTT-relation is a GTT-relation.
(2.) The semi-congruence closure of a GTT-relation is a GTT-relation.
(3.) The composition of two GTT-relations is a GTT-relation.

In fact, these closure properties are effective in that, given a *GTT* for a relation $\sim$, a *GTT* for the inverse relation $\sim^{-1}$ can be constructed effectively, and similar for the other two operations.

Now from a finite ground term-rewriting system R a *GTT* A_R can be constructed for the reduction relation $\rightarrow^*_R$ on $T(F)$. From A_R we obtain *GTTs* $A_{diverge}$ and $A_{converge}$, where $A_{diverge}$ realizes the relation $\leftarrow^*_R \circ \rightarrow^*_R$ and $A_{converge}$ realizes the relation $\rightarrow^*_R \circ \leftarrow^*_R$. Observe that R is confluent if and only if $\leftarrow^*_R \circ \rightarrow^*_R \subseteq \rightarrow^*_R \circ \leftarrow^*_R$. Hence, the test for confluence of R is reduced to the inclusion problem for two *GTT*-relations. Since the inclusion of *GTT*-relations is decidable [14, 15], this immediately yields the announced decidability result.

Corollary 4. [14, 15, 61]

The confluence property is decidable for finite ground term-rewriting systems.

Based on the same technique Dauchet and Tison even show that the first-order theory of a ground term-rewriting system is decidable [16].

We close this section with a remarkable observation concerning interreduced ground term-rewriting systems. Note that the following considerations also apply to prefix-rewriting systems, as a prefix-rewriting system on some alphabet Σ can be interpreted as a ground term-rewriting system on the signature F_Σ.

Let R be a ground term-rewriting system that is interreduced. Then range(R) $\subseteq$ IRR(R), and hence, it is easily seen that R is noetherian. Further, the left-hand side of no rule of R contains the left-hand side of another rule as a subterm. Therefore, R has no critical pairs at all, and hence, it is also confluent. Hence, we have the following characterization.

Proposition 17. [66]

A ground term-rewriting system is canonical if and only if it is interreduced.

Thus, by interreduction a finite ground-term rewriting system R that is noetherian can be transformed into an equivalent finite system R_0 that is canonical. By reorienting some of its rules if necessary, R can always be turned into an equivalent system that is noetherian. This yields the following result.

Corollary 5. *For each finite ground term-rewriting system an equivalent finite ground term-rewriting system can effectively be determined that is canonical.*

In fact this process can be performed in time $O(n \log n)$ [66,67] exploiting Shostak's congruence closure algorithm [65]. Also see [36] for a discussion of this algorithm and its relation to completion of ground term-rewriting systems.

6 Term-rewriting systems

In this section we will consider terms with variables, which we will again simply call terms. Accordingly, the terms without variables considered in the previous section will be called *ground terms* in the following.

Let F be a finite signature, and let V be a countable set of variables. Then $T(F, V)$ denotes the set of *terms* generated by F and V. As before $T(F)$ denotes the subset of ground terms of $T(F, V)$. For a term $t \in T(F, V)$, $\mathrm{Var}(t)$ denotes the set of variables that have occurrences in t. If no variable occurs more than once in t, then t is called a *linear* term.

A *substitution* is a mapping $\sigma : V \to T(F, V)$ such that $\sigma(v) = v$ holds for almost all variables v. It can uniquely be extended to a morphism $\sigma : T(F, V) \to T(F, V)$.

A *term-rewriting system* R is a (finite) set of *rules* $R = \{\ell_i \to r_i \mid i \in I\}$, where ℓ_i and r_i are terms from $T(F, V)$. While ground term-rewriting systems can be seen as a generalization of the prefix-rewriting systems considered in Section 3, term-rewriting systems are the corresponding generalization of string-rewriting systems to general finite signatures.

A term t is *reducible* modulo R if there is a rule $\ell \to r$ in R, an occurrence $p \in O(t)$, and a substitution σ such that $\sigma(\ell) = t/p$. The term $t[p \leftarrow \sigma(r)]$ is the result of *reducing* t by $\ell \to r$ at p, and this reduction is written as $t \to_R t[p \leftarrow \sigma(r)]$. The *reduction relation* associated with the term-rewriting system R is the reflexive and transitive closure $\to_R^*$ of this *single-step reduction relation* $\to_R$. A term t is said to be in *normal form* or *irreducible* modulo R if no reduction can be applied to t. By $\mathrm{IRR}(R)$ we denote the set of all those ground terms that are irreducible, and $\mathrm{RED}(R)$ is the set $T(F) \setminus \mathrm{IRR}(R)$ of reducible ground terms.

As for ground term-rewriting systems the equational theory that is associated with a term-rewriting system R is the congruence $=_R$ that is generated by the reduction relation $\to_R$, that is, it is the congruence $(\to_R \cup \leftarrow_R)^*$. Usually we are only interested in the restriction of this congruence to ground terms.

A term-rewriting system is called *noetherian*, (*locally*) *confluent, convergent* or *canonical* if the induced reduction relation has the corresponding property. It is called *left-linear* if the left-hand side of each rule of R is a linear term.

If R is a finite term-rewriting system that is left-linear, then a regular tree grammar can easily be constructed from R that generates the set $\text{RED}(R)$ of reducible ground terms. Hence, we have the following result.

Proposition 18. [21]
For a finite term-rewriting system that is left-linear the set of irreducible ground terms as well as the set of reducible ground terms is a regular term language.

The left-linearity of the term-rewriting system considered is a crucial hypothesis for Proposition 18, as a finite non-left-linear system can easily be constructed for which the set of irreducible ground terms is not regular. However, some finite systems yield regular sets of irreducible ground terms although they are not left-linear. An example in kind is the following system which is essentially taken from [38]:

$$
\begin{aligned}
eq(x,x) &\to s(0), & eq(0, s(x)) &\to 0, \\
eq(s(x), 0) &\to 0, & eq(s(x), s(y)) &\to eq(x,y), \\
eq(eq(x,y), z) &\to 0, & eq(x, eq(y,z)) &\to 0, \\
s(eq(x,y)) &\to 0.
\end{aligned}
$$

Thus, the question arises whether there are regular tree languages that occur as sets of irreducible ground terms for some finite term-rewriting systems that are not left-linear, but that do not occur as sets of irreducible ground terms for any finite left-linear systems. Surprisingly this is not the case.

Proposition 19. [38]
For a finite term-rewriting system R, if $\text{IRR}(R)$ is a regular term language, then there exists a finite left-linear system R_{lin} such that $\text{IRR}(R_{lin}) = \text{IRR}(R)$.

In fact, R_{lin} consists of linear instantiations of rules of R. By associating with each transition rule of a top-down tree automaton a regular set of ground terms governing the applicability of that rule, the class of *deterministic top-down tree automata with prefix look-ahead* is defined in [20]. It yields the following characterization.

Proposition 20. [20]
A term language $L \subseteq T(F)$ is recognized by a one-state deterministic top-down tree automata with prefix look-ahead if and only if there exists a finite term-rewriting system R satisfying $\text{IRR}(R) = L$.

Thus, the one-state deterministic top-down tree automata with prefix look-ahead, the finite left-linear term-rewriting systems, and the finite term-rewriting systems that are not left-linear all define the same subclass of the class of all regular term languages. In addition, the following decidability result holds.

Proposition 21. [29, 39, 69]
Given a finite term-rewriting system R, it is decidable whether or not $\text{IRR}(R)$ is a regular term language. If $\text{IRR}(R)$ is indeed a regular term language, then a linear instantiation R_{lin} of R can be constructed such that $\text{IRR}(R_{lin}) = \text{IRR}(R)$ holds.

A term-rewriting system R on a signature F is called *F-regularity preserving* if, for each regular term language $L \subseteq T(F)$, the set $\Delta_R^*(L)$ of all descendants is again regular. It is called *regularity preserving* if it is F-regularity preserving for each signature F containing all the function symbols that actually occur in the rules of R.

If $F := \{f, g, a\}$, where f and g are unary symbols and a is a symbol of arity 0 (a constant), then for $R := \{f(g(x)) \to f(f(g(g(x)))), f(a) \to a, g(a) \to a, a \to f(a), a \to g(a)\}$ it is easily seen that $\Delta_R^*(t) = T(F)$ holds for all ground terms $t \in T(F)$. However, if $F_1 := F \cup \{h\}$, where h is another unary function symbol, then $\Delta_R^*(f(g(h(a)))) = \{f^n(g^n(h(t))) \mid t \in T(F)\}$, which is not regular. Thus, R does preserve F-regularity, but not F_1-regularity [28]. Obviously the ground rules contained in R are responsible for this, since the subsystem $R' := \{f(g(x)) \to f(f(g(g(x))))\}$ of R does not even preserve F-regularity.

It is well-known that the property of being F-regularity preserving is undecidable in general [25, 26]. In fact, this property is even undecidable for finite string-rewriting systems [56]. On the other hand, while for term-rewriting systems the property of regularity preservation depends on the actually chosen signature as indicated by the example above, this is not true for string-rewriting systems [56]. Actually, we have the following result.

Proposition 22. [56]
Let S be a string-rewriting system on Σ, let $F_\Sigma = \Sigma \cup \{\natural\}$, where $\natural$ is a constant and each letter from Σ is interpreted as a unary function symbol, and let R_S be the term-rewriting system $R_S = \{\ell(x) \to r(x) \mid (\ell, r) \in S\}$ on F_Σ. Then R_S is regularity preserving if and only if it preserves F_Σ-regularity.

On the other hand it is known that certain restricted classes of term-rewriting systems are regularity preserving. This applies to those systems that contain only ground rules [6], to term-rewriting systems that are right-linear and monadic [63], that are linear and semi-monadic [11], or that are linear and generalized semi-monadic [28].

While we refer to the literature for the other notions mentioned above, we recall the definition of monadic term-rewriting systems. These systems were introduced by Book and Gallier as a direct generalization of the monadic string-rewriting systems [21]. A term-rewriting system R is called *monadic* if it is left-linear and if $\text{depth}(r) \leq 1$ holds for each rule $\ell \to r$ of R.

The process of reduction with respect to a finite monadic term-rewriting system that is noetherian can be realized by a *tree pushdown automaton* (*TreePDA*) [21]. For a *TreePDA* A, $L(A, B)$ denotes the set of all ground terms t for which there exists an accepting computation of A that, while processing t, produces a term from B.

Proposition 23. [21]
Let R be a finite monadic term-rewriting system that is convergent. Then for every regular tree language B, there exists a deterministic TreePDA A such that $L(A, B)$ coincides with the set of terms $[B]_R = \bigcup\{[t]_R \mid t \in B \cap \mathrm{IRR}(R)\}$.

As shown by K. Salomaa [63] the deterministic tree pushdown automata of Gallier and Book are more powerful than the corresponding automata of Schimpf [64]. An investigation of various classes of tree pushdown automata and a generalization of the results on monadic term-rewriting systems to semi-monadic systems can be found in [11].

The technique for deciding linear sentences (see Proposition 6) can obviously be lifted to those finite convergent term-rewriting systems which are effectively regularity preserving and for which the set of irreducible ground terms is regular. In particular, this has the following consequence.

Corollary 6. [25, 26]
The validity of linear sentences is decidable for finite convergent term-rewriting systems that are (1.) linear and monadic, or (2.) linear and semi-monadic, or (3.) linear generalized semi-monadic.

There are many more applications of tree automata to rewriting systems. For example, Comon shows that *strong sequentiality* [33] of left-linear rewriting systems and *NV-sequentiality* [62] of linear rewriting systems are definable in WSkS, the weak second-order monadic logic of k successor functions [10], by exploiting the correspondence between this logic and tree automata [68]. Following Comon's approach Jacquemard shows that *sequentiality* is decidable for each linear rewriting system that is *growing* [34].

Further, finite *test sets* have been found to be a useful tool for deciding the membership problem for the universal closure of a given tree language, that is, for deciding whether all the ground instances of a given term belong to the language considered. By relating test sets to tree automata and to appropriate congruences Hofbauer and Huber [30] obtain characterizations of ground and non-ground test sets, and they show how to compute and to minimize these test sets.

Finally, by introducing a class of more powerful bottom-up tree automata, called *reduction automata*, Dauchet et al prove that the *first-order theory of reduction* is decidable [13].

7 Conclusion

As we have seen automata theory provides essential tools for the study of rewriting systems and their properties. On the other hand, rewriting theory has influenced the theory of automata considerably in that motivated by problems encountered in rewriting theory new classes of automata have been developed. In fact, rewriting theory with its many applications to such diverse fields as automated theorem proving, functional and logic programming, and semigroup and group theory to mention just a few, can be seen as one of the main users of and contributors to automata theory.

Acknowledgement. I would like to thank the program committee of RTA'99 for inviting me to present this survey. Further, I want to thank Dieter Hofbauer and Klaus Madlener for their comments on a preliminary version of this paper.

References

1. R.V. Book. Confluent and other types of Thue systems. *Journal Association Computing Machinery*, 29:171–182, 1982.
2. R.V. Book. The power of the Church-Rosser property in string-rewriting systems. In D.W. Loveland, editor, *6th Conference on Automated Deduction*, Lecture Notes in Computer Science 138, pages 360–368. Springer-Verlag, Berlin, 1982.
3. R.V. Book. Decidable sentences of Church-Rosser congruences. *Theoretical Computer Science*, 24:301–312, 1983.
4. R.V. Book, M. Jantzen, and C. Wrathall. Monadic Thue systems. *Theoretical Computer Science*, 19:231–251, 1982.
5. R.V. Book and F. Otto. *String-Rewriting Systems*. Springer-Verlag, New York, 1993.
6. W.J. Brainerd. Tree generating regular systems. *Information and Control*, 14:217–231, 1969.
7. G. Buntrock and F. Otto. Growing context-sensitive languages and Church-Rosser languages. In E.W. Mayr and C. Puech, editors, *Proc. of STACS 95*, Lecture Notes in Computer Science 900, pages 313–324. Springer-Verlag, Berlin, 1995.
8. G. Buntrock and F. Otto. Growing context-sensitive languages and Church-Rosser languages. *Information and Computation*, 141:1–36, 1998.
9. C.M. Campbell, E.F. Robertson, N. Ruškuc, and R.M. Thomas. Automatic semigroups. Technical Report No. 1997/29, Dep. of Mathematics and Computer Science, University of Leicester, 1997.
10. H. Comon. Sequentiality, second order monadic logic and tree automata. In *Proceedings 10th Symposium on Logic in Computer Science*, pages 508–517. IEEE Computer Society Press, San Diego, 1995.
11. J.-L. Coquidé, M. Dauchet, R. Gilleron, and S. Vágvölgyi. Bottom-up tree pushdown automata: classification and connection with rewrite systems. *Theoretical Computer Science*, 127:69–98, 1994.
12. E. Dahlhaus and M. Warmuth. Membership for growing context-sensitive grammars is polynomial. *Journal Computer System Sciences*, 33:456–472, 1986.
13. M. Dauchet, A.-C. Caron, and J.-L. Coquidé. Automata for reduction properties solving. *Journal of Symbolic Computation*, 20:215–233, 1995.
14. M. Dauchet, T. Heuillard, P. Lescanne, and S. Tison. Decidability of the confluence of finite ground term rewrite systems and of other related term rewrite systems. *Information and Computation*, 88:187–201, 1990.
15. M. Dauchet and S. Tison. Decidability of confluence for ground term rewriting systems. In *Fundamentals of Comp. Theory, Cottbus 1985*, Lecture Notes in Computer Science 199, pages 80–89. Springer-Verlag, Berlin, 1985.
16. M. Dauchet and S. Tison. The theory of ground rewriting systems is decidable. In J.C. Mitchell, editor, *Proc. of 5th LICS*, pages 242–248. IEEE Computer Society Press, Los Alamitos, CA, 1990.

17. M.D. Davis and E.J. Weyuker. *Computability, Complexity, and Languages*. Academic Press, New York, 1983.

18. N. Dershowitz and J.P. Jouannaud. Rewrite systems. In J. van Leeuwen, editor, *Handbook of Theoretical Computer Science, Vol. B.: Formal Models and Semantics*, pages 243–320. Elsevier, Amsterdam, 1990.

19. D.B.A. Epstein, J.W. Cannon, D.F. Holt, S.V.F. Levy, M.S. Paterson, and W.P. Thurston. *Word Processing In Groups*. Jones and Bartlett Publishers, Boston, 1992.

20. Z. Fülöp and S. Vágvölgyi. A characterization of irreducible sets modulo left-linear term rewriting systems by tree automata. *Fundamenta Informaticae*, 13:211–226, 1990.

21. J. Gallier and R.V. Book. Reductions in tree replacement systems. *Theoretical Computer Science*, 37:123–150, 1985.

22. F. Gécseg and M. Steinby. *Tree Automata*. Akadémiai Kiadó, Budapest, 1984.

23. F. Gécseg and M. Steinby. Tree languages. In G. Rozenberg and A. Salomaa, editors, *Handbook of Formal Languages, Vol. III*, pages 1–68. Springer-Verlag, Berlin, 1997.

24. S.M. Gersten. Dehn functions and ll-norms of finite presentations. In G. Baumslag and C.F. Miller III, editors, *Algorithms and Classification in Combinatorial Group Theory*, Math. Sciences Research Institute Publ. 23, pages 195–224. Springer-Verlag, New York, 1992.

25. R. Gilleron. Decision problems for term rewriting systems and recognizable tree languages. In C. Choffrut and M. Jantzen, editors, *Proc. of STACS'91*, Lecture Notes in Computer Science 480, pages 148–159. Springer-Verlag, 1991.

26. R. Gilleron and S. Tison. Regular tree languages and rewrite systems. *Fundamenta Informaticae*, 24:157–175, 1995.

27. R. Gilman. Presentations of groups and monoids. *Journal of Algebra*, 57:544–554, 1979.

28. P. Gyenizse and S. Vágvölgyi. Linear generalized semi-monadic rewrite systems effectively preserve recognizability. *Theoretical Computer Science*, 194:87–122, 1998.

29. D. Hofbauer and M. Huber. Linearizing term rewriting systems using test sets. *Journal of Symbolic Computation*, 17:91–129, 1994.

30. D. Hofbauer and M. Huber. Test sets for the universal and existential closure of regular tree languages. *This volume.*

31. J.E. Hopcroft and J.D. Ullman. *Introduction to Automata Theory, Languages, and Computation*. Addison-Wesley, Reading, M.A., 1979.

32. G. Huet and D. Lankford. *On the uniform halting problem for term rewriting systems*. Lab. Report No. 283, INRIA, Le Chesnay, France, March 1978.

33. G. Huet and J.J. Lévy. Computations in orthogonal rewriting systems I and II. In J.L. Lassez and G. Plotkin, editors, *Computational Logic: Essays in Honor of Alan Robinson*, pages 395–443. MIT Press, 1991. This paper was written in 1979.

34. F. Jacquemard. Decidable approximations of term rewriting systems. In H. Ganzinger, editor, *Proc. of RTA'96*, Lecture Notes in Computer Science 1103, pages 362–376. Springer-Verlag, Berlin, 1996.

35. P. Jančar, F. Mráz, M. Plátek, and J. Vogel. On restarting automata with rewriting. In G. Păun and A. Salomaa, editors, *New Trends in Formal Languages*, Lecture Notes in Computer Science 1218, pages 119–136. Springer-Verlag, Berlin, 1997.

36. D. Kapur. Shostak's congruence closure as completion. In H. Comon, editor, *Rewriting Techniques and Applications, Proc. of RTA'97*, Lecture Notes in Computer Science 1232, pages 23–37. Springer-Verlag, Berlin, 1997.

37. T. Kretschmer. A closure property of regular languages. *Theoretical Computer Science*, 61:283–287, 1988.

38. G. Kucherov. On relationship between term rewriting systems and regular tree languages. In R.V. Book, editor, *Rewriting Techniques and Applications, Proceedings RTA '91*, Lecture Notes in Computer Science 488, pages 299–311. Springer-Verlag, Berlin, 1991.

39. G. Kucherov and M. Tajine. Decidability of regularity and related properties of ground normal form languages. *Information and Computation*, 118:91–100, 1995.

40. N. Kuhn. *Zur Entscheidbarkeit des Untergruppenproblems für Gruppen mit kanonischen Darstellungen*. Dissertation, Universität Kaiserslautern, Fachbereich Informatik, 1991.

41. N. Kuhn and K. Madlener. A method for enumerating cosets of a group presented by a canonical system. In *Proc. ISSAC'89*, pages 338–350. ACM Press, New York, 1989.

42. N. Kuhn, K. Madlener, and F. Otto. A test for λ-confluence for certain prefix rewriting systems with applications to the generalized word problem. In S. Watanabe and M. Nagata, editors, *Proceedings ISSAC'90*, pages 8–15. ACM, New York, 1990.

43. N. Kuhn, K. Madlener, and F. Otto. Computing presentations for subgroups of polycyclic groups and of context-free groups. *Applicable Algebra in Engineering, Communication and Computing*, 5:287–316, 1994.

44. K. Madlener, P. Narendran, F. Otto, and L. Zhang. On weakly confluent monadic string-rewriting systems. *Theoretical Computer Science*, 113:119–165, 1993.

45. R. McNaughton, P. Narendran, and F. Otto. Church-Rosser Thue systems and formal languages. *Journal Association Computing Machinery*, 35:324–344, 1988.

46. P. Narendran and C. Ó'Dúnlaing. Cancellativity in finitely presented semigroups. *Journal of Symbolic Computation*, 7:457–472, 1989.

47. P. Narendran and F. Otto. The problems of cyclic equality and conjugacy for finite complete rewriting systems. *Theoretical Computer Science*, 47:27–38, 1986.

48. G. Niemann and F. Otto. The Church-Rosser languages are the deterministic variants of the growing context-sensitive languages. In M. Nivat, editor, *Foundations of Software Science and Computation Structures, Proceedings FoSSaCS'98*, Lecture Notes in Computer Science 1378, pages 243–257. Springer-Verlag, Berlin, 1998.

49. G. Niemann and F. Otto. *Restarting automata, Church-Rosser languages, and confluent internal contextual languages*. Mathematische Schriften Kassel 4/99, Universität Kassel, March 1999.

50. F. Otto. Some undecidability results for non-monadic Church-Rosser Thue systems. *Theoretical Computer Science*, 33:261–278, 1984.

51. F. Otto. On deciding the confluence of a finite string-rewriting system on a given congruence class. *Journal Computer System Sciences*, 35:285–310, 1987.

52. F. Otto. Completing a finite special string-rewriting system on the congruence class of the empty word. *Applicable Algebra in Engineering, Communication and Computing*, 2:257–274, 1992.

53. F. Otto. The problem of deciding confluence on a given congruence class is tractable for finite special string-rewriting systems. *Mathematical Systems Theory*, 25:241–251, 1992.

54. F. Otto. *On Dehn functions of finitely presented bi-automatic monoids*. Mathematische Schriften Kassel 8/98, Universität Kassel, July 1998.

55. F. Otto. *On s-regular prefix-rewriting systems and automatic structures*. Mathematische Schriften Kassel 9/98, Universität Kassel, September 1998.

56. F. Otto. Some undecidability results concerning the property of preserving regularity. *Theoretical Computer Science*, 207:43–72, 1998.
57. F. Otto, M. Katsura, and Y. Kobayashi. Cross-sections for finitely presented monoids with decidable word problems. In H. Comon, editor, *Rewriting Techniques and Applications, Proceedings RTA '97*, Lecture Notes in Computer Science 1232, pages 53–67. Springer-Verlag, Berlin, 1997.
58. F. Otto, M. Katsura, and Y. Kobayashi. Infinite convergent string-rewriting systems and cross-sections for finitely presented monoids. *Journal of Symbolic Computation*, 26:621–648, 1998.
59. F. Otto, A. Sattler-Klein, and K. Madlener. Automatic monoids versus monoids with finite convergent presentations. In T. Nipkow, editor, *Rewriting Techniques and Applications, Proceedings RTA '98*, Lecture Notes in Computer Science 1379, pages 32–46. Springer-Verlag, Berlin, 1998.
60. F. Otto and L. Zhang. Decision problems for finite special string-rewriting systems that are confluent on some congruence class. *Acta Informatica*, 28:477–510, 1991.
61. M. Oyamaguchi. The Church-Rosser property for ground term-rewriting systems is decidable. *Theoretical Computer Science*, 49:43–79, 1987.
62. M. Oyamaguchi. NV-sequentiality: a decidable condition for call-by-need computations in term-rewriting systems. *SIAM Journal on Computing*, 22:114–135, 1993.
63. K. Salomaa. Deterministic tree pushdown automata and monadic tree rewriting systems. *Journal Computer System Sciences*, 37:367–394, 1988.
64. K.M. Schimpf and J.H. Gallier. Tree pushdown automata. *Journal Computer System Sciences*, 30:25–40, 1985.
65. R.E. Shostak. An algorithm for reasoning about equality. *Communications of the Association for Computing Machinery*, 21:583–585, 1978.
66. W. Snyder. Efficient ground completion: an $O(n \log n)$ algorithm for generating reduced sets of ground rewrite rules equivalent to a set of ground equations E. In N. Deshowitz, editor, *Rewriting Techniques and Applications, Proceedings RTA '89*, Lecture Notes in Computer Science 355, pages 419–433. Springer-Verlag, Berlin, 1989.
67. W. Snyder. A fast algorithm for generating reduced ground rewriting systems from a set of ground equations. *Journal of Symbolic Computation*, 15:415–450, 1993.
68. J. Thatcher and J. Wright. Generalized finite automata with an application to a decision problem of second-order logic. *Mathematical Systems Theory*, 2:57–82, 1968.
69. S. Vágvölgyi and R. Gilleron. For a rewrite system it is decidable whether the set of irreducible ground terms is recognizable. *Bulletin of the EATCS*, 48:197–209, 1992.
70. L.G. Valiant. The equivalence problem for deterministic finite-turn pushdown automata. *Information and Control*, 25:123–133, 1974.

A Rewrite System
Associated with Quadratic Pisot Units

Christiane Frougny[1] and Jacques Sakarovitch[2]

[1] Université Paris 8 and
LIAFA, Université Paris 7 and C. N. R. S.
[2] Laboratoire Traitement et Communication de l'Information (C. N. R. S. URA 820),
E. N. S. T.,   46, rue Barrault   –   75 634 Paris Cedex 13 (France).
Email: sakarovitch@enst.fr, WWW home page: http://www.enst.fr/~jsaka

Dedicated to the memory of David Klarner

Abstract. In a previous work, we have investigated an automata-theoretic property of numeration systems associated with quadratic Pisot units that yields, for every such number θ, a certain group G_θ.
In this paper, we characterize a cross-section of a congruence γ_θ of $\mathbb{Z}^4$ that had arisen when constructing G_θ. In spite of the algebraic connections and implications of that characterization, the proof is combinatorial, and based upon rewriting techniques.
The main point is to show that the rewrite system made up by the relations that generate γ_θ, though non-confluent, behaves as if it were confluent.

Dans un article précédent, nous avions associé à chaque nombre de Pisot quadratique unitaire θ un certain groupe G_θ par le biais de la construction d'un automate qui réalise le passage entre les représentations des entiers dans deux systèmes de numération naturellement attachés à θ.
Dans cet article, nous donnons une caractérisation d'un ensemble de représentants pour une congruence γ_θ de $\mathbb{Z}^4$ qui avait été utilisée pour la définition de G_θ. Bien que les motivations, le cadre, et les implications de cette caractérisation soient algébriques, la preuve est combinatoire et utilise les techniques des systèmes de réécriture.
Le point crucial consiste à montrer que le système de réécriture formé par les relations qui engendrent γ_θ se comporte comme un système confluent bien qu'il ne le soit pas.

We describe here a rewrite system that we have been led to consider in order to characterize a group that is associated with quadratic Pisot units, *via* numeration systems.

It is straightforward to associate numeration systems to Pisot numbers and recent publications have shown spectacular appearances of these systems in different questions, putting an emphasis on quadratic Pisot units. For instance, they are involved in the mathematical description of quasicrystals; and every quasicrystal observed so far in the real world is indeed defined by a quadratic

Pisot unit (see [1]). As another example, these numeration systems are also present in the realization of arithmetic codings of hyperbolic automorphisms of the torus (see [7]).

There are indeed *two* numeration systems associated with every Pisot number. In a previous work of ours, we showed that, in the case of a *quadratic Pisot unit*, there exists a finite two-tape automaton that translates the representation of integers in one system into the representation of the same integer in the other system ([3]). This automaton is fairly complicated (several hundreds of states in the simplest case where θ is the golden mean) but, thanks to a series of decompositions, its description boils down to the computation of a relatively small group ($\mathbb{Z}/5\mathbb{Z}$ in the case of the golden mean). This group appears as *a certain subgroup G_θ* of the quotient H_θ of $\mathbb{Z}^4$ by a certain congruence γ_θ, and has been fully described in [3].

The purpose of this paper is a characterization of a cross-section of that congruence γ_θ — Theorem 1 — the exact statement of which requires some more definitions and notation.

We found Theorem 1 of interest for several reasons. First, it gives the key to the computation in H_θ, hence to the determination of H_θ itself. Second, the definition of γ_θ provided by Theorem 1 bears a remarkable similarity with the description of the symbolic dynamical system associated, by a theorem of Parry, with Pisot numbers (see [6]). Due to space limitation, these two aspects will not be developed here (and are presented in [4]). Finally, the only proof we know for Theorem 1, and which is the subject of this paper, is purely combinatorial and relies on rewriting techniques; this is remarkable too, for similar results, the theorem of Parry quoted above for instance, have algebraic, and even analytical, proofs.

1 The context

We first give a glimpse of the automatic conversion between numeration systems, based on the example of the Fibonacci system. The reader is referred to [3] for a complete presentation of the subject and of the result.

Let $F = \{F_n \mid n \in \mathbb{N}\}$ be the sequence of Fibonacci numbers, defined by the recurrence relation $F_{n+2} = F_{n+1} + F_n$ and by the "initial conditions" [1] $F_0 = 1$, $F_1 = 2$. It is well-known[2] that every positive integer can be written as a sum of Fibonacci numbers; the sequence F together with the two-digit alphabet $A = \{0, 1\}$ defines thus the *Fibonacci numeration system, i.e.* every integer is represented by a sequence of 0's and 1's. Every integer can be given a *normal representation*, which is unique and characterized by the fact it does not contain two consecutive 1's

Let φ be the *golden mean i.e.* the larger zero of

$$P(X) = X^2 - X - 1 ,$$

---

[1] These are *not* the "usual" initial conditions but they happen to be the "good" ones when one wants to turn the Fibonacci sequence into a numeration system.

[2] and usually credited to Zeckendorf [8]; *cf.* also the Exercise 1.2.8.34 in [5].

which is the characteristic polynomial of the above recurrence relation. It is known (*cf.* [5, Exercise 1.2.8.35]) that every number x can be written as a sum of (positive and negative) powers of φ and thus can be represented as a sequence — possibly infinite — of 0's and 1's together with a radix point. Every real number can be given such a sequence, called its φ-*expansion*, which is unique and characterized by the fact it does not contain two adjacent 1's and does not terminate by the infinite factor $101010\ldots$.

Table 1 below gives the Fibonacci normal representation of some integers together with their φ-expansion as well as the same φ-expansion written in the convenient folded form.

N	Fibonacci representations	φ-expansions	Folded φ-expansions
1	1	1.	1 0
5	1000	1000.1001	1 0 0 0 1 0 0 1
10	10010	10100.0101	1 0 1 0 0 0 1 0 1 0
15	100010	100101.001001	1 0 0 1 0 1 1 0 0 1 0 0

Table 1. Fibonacci representations and φ-expansions of some integers

The result proved in [3] is the following:

THEOREM A   *There exists a letter-to-letter finite two-tape automaton $\mathcal{A}_\varphi$ that maps the Fibonacci representation of any integer onto its folded φ-expansion.*

The automaton $\mathcal{A}_\varphi$ is not constructed directly. Rather, its construction is broken up into several steps. The main step in proving Theorem A is the construction of an automaton $\mathcal{T}_\varphi$ that reads words where the letters have been *grouped into blocks of length* 4, and with the property that there is *at most one digit* 1 *in every block*. In contrast with $\mathcal{A}_\varphi$, this automaton $\mathcal{T}_\varphi$ is remarkably simple (*cf.* Figure 1). Its transition monoid is the group $\mathbb{Z}/5\mathbb{Z}$.

Indeed, the result proved in [3] is a generalization of Theorem A, proving the property not only for the golden mean φ but for *any quadratic Pisot unit* θ. And the corresponding automaton $\mathcal{T}_\theta$ is computed *via* its transition monoid G_θ, a group obtained from $\mathbb{Z}^4$ by a certain congruence γ_θ which is the subject of the present paper.

2   The result

A *quadratic Pisot unit* θ is the root greater than 1 of a polynomial

$$P_\theta(X) = X^2 - rX - \varepsilon \ ,$$

with either: $\varepsilon = +1$ and $r \geq 1$ — this will be referred to as CASE 1,
or $\varepsilon = -1$ and $r \geq 3$ — this will be referred to as CASE 2.

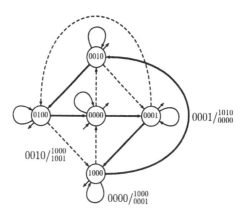

$0001/^{1010}_{0000}$

$0010/^{1000}_{1001}$

$0000/^{1000}_{0001}$

Fig. 1. The automaton $\mathcal{T}_\varphi$ (partial view: the only transitions represented are those labelled by 0 0 0 1 (bold arrows), by 0 0 1 0 (dashed arrows) and by 0 0 0 0 (loops) on the first coordinate; the full label is given for three transitions only)

The elements of the commutative group $\mathbb{Z}^4$ are seen as "words" of length 4 over the alphabet $\mathbb{Z}$ and γ_θ is the congruence generated by the following equalities[3]: $1\ \bar{r}\ \bar{\varepsilon}\ 0 = \bar{r}\ \bar{\varepsilon}\ 0\ 1 = \bar{\varepsilon}\ 0\ 1\ \bar{r} = 0\ 1\ \bar{r}\ \bar{\varepsilon} = 0\ 0\ 0\ 0$.

Two words in $\mathbb{Z}^4$ are said to be *conjugate*[4] if there exists a circular permutation of their digits that sends one onto the other.

The definition of the set R_θ of *reduced words* depends then upon the case we consider:

CASE 1. $r \geq 1$ and $\varepsilon = +1$. R_θ is the set of words of $\mathbb{N}^4$ with the property that they, and all their conjugates, are strictly smaller than $r\ 0\ r\ 0$ in the lexicographic order.

CASE 2. $r \geq 3$ and $\varepsilon = -1$. R_θ is the set of words of $\mathbb{N}^4$ with the property that they, and all their conjugates, are different from $r_{-2}\ r_{-2}\ r_{-2}\ r_{-2}$ and strictly smaller than $r_{-1}\ r_{-2}\ r_{-2}\ r_{-2}$ in the lexicographic order.

Theorem 1 *Every class of $\mathbb{Z}^4$ modulo γ_θ contains exactly one element in R_θ.*

The proof of Theorem 1[5] is quite different in CASE 1 and in CASE 2, much simpler in the latter case. For CASE 1, it is first easily established that every class contains at least one element of R_θ (Part A). The proof of uniqueness is more involved. An element, or word, of $\mathbb{Z}^4$ is said to be *positive* if all its digits belong to $\mathbb{N}$. We first consider only positive words and we give *an orientation* to the defining relations of γ_θ. If the *rewrite system* obtained that way were *confluent* — that is to say, if "no matter how one diverges from a common ancestor, there are paths joining at a common descendent" [2] — the uniqueness of a reduced positive word would follow from a standard argument. What is developed in Part B through a detailled analysis is that this reduction "behaves" as if the

---

[3] With the convention that if n is an integer, $\bar{n}$ denotes $-n$.

[4] Though this is *not* the conjugacy relation in the group $\mathbb{Z}^4$ (which is the identity since $\mathbb{Z}^4$ is commutative).

[5] announced, without proof, as Proposition 14 in [3].

system were confluent, though it is not. The last (and easy) step amounts to verify that reduction pathes through non positive words do not bring any further possibilities of equivalence between words (Part C). For CASE 2, we directly derive a *confluent rewrite system* from the defining relations of γ_θ.

3 Proof for Case 1

Notation and conventions. By definition, γ_θ is generated by the following relations:

$$1\,\bar{r}\,\bar{1}\,0 \;=\; 0\,0\,0\,0 \qquad (1) \qquad\qquad \bar{1}\,0\,1\,\bar{r} \;=\; 0\,0\,0\,0 \qquad (3)$$

$$\bar{r}\,\bar{1}\,0\,1 \;=\; 0\,0\,0\,0 \qquad (2) \qquad\qquad 0\,1\,\bar{r}\,\bar{1} \;=\; 0\,0\,0\,0 \qquad (4)$$

Any linear combination of these relations gives rise to another relation that is also satisfied by the congruence γ_θ. In particular (1)+(3), and (2)+(4), yield respectively:

$$0\,\bar{r}\,0\,\bar{r} \;=\; 0\,0\,0\,0 \qquad (5) \qquad\qquad \bar{r}\,0\,\bar{r}\,0 \;=\; 0\,0\,0\,0 \qquad (6)$$

The opposite of a relation (α) is another relation, denoted by $\overline{(\alpha)}$; *e.g.*

$$\bar{1}\,r\,1\,0 \;=\; 0\,0\,0\,0 \;. \qquad\qquad\qquad \overline{(1)}$$

By abuse, we denote as the sum $w + (\alpha)$ the digit-addition of w and the non-zero member of the relation (α), $1 \le \alpha \le 4$; *e.g.*

$$\text{if}\quad w = x\;y\;u\;t\,, \quad\text{then}\quad w + (1) = x_{+1}\,y_{-r}\,u_{-1}\,t \;.$$

The notation extends to subtraction: $w - (\alpha) = w + \overline{(\alpha)}$.

If w' is obtained from w by adding one of the four defining relations of γ_θ or of their opposite, we write $w \leftrightarrow w'$; if moreover w and w' are both positive, we write $w \Leftrightarrow w'$. If w' is obtained by a sequence of such additions we write $w \overset{*}{\leftrightarrow} w'$ and such a sequence is called a *path* from w to w'. If moreover every word encountered on the path is positive, we write $w \overset{*}{\Leftrightarrow} w'$. By definition, w and w' are equivalent modulo γ_θ if, and only if, $w \overset{*}{\leftrightarrow} w'$.

The four relations (1) to (4) will also be considered as reductions and written as such:

$$w \overset{(\alpha)}{\rightarrow} w + (\alpha) \;.$$

If *both w and $w + (\alpha)$ are positive*, we write

$$w \overset{(\alpha)}{\Rightarrow} w + (\alpha)$$

and we say that w is *positively reducible* by (α). If it is not the case, w, *supposed to be positive*, is said to be (α)-*irreducible*. A positive word is called *positively irreducible*, or *p-irreducible*, if no such reduction is possible.

Every word in R_θ is p-irreducible but the converse obviously does not hold; *e.g.* $0\,r_{+1}\,0\,0$ is p-irreducible but not in R_θ.

In addition to the cases described by these general conventions, we shall also write

$$0 \; r \; 0 \; r \; \overset{(5)}{\Rrightarrow} \; 0 \; 0 \; 0 \; 0 \qquad \text{and} \qquad r \; 0 \; r \; 0 \; \overset{(6)}{\Rrightarrow} \; 0 \; 0 \; 0 \; 0 \;,$$

which state that $0 \; r \; 0 \; r$, and $r \; 0 \; r \; 0$, are *positively* reducible to $0 \; 0 \; 0 \; 0$. But this word $w = 0 \; r \; 0 \; r$ (resp. $w = r \; 0 \; r \; 0$) is the only one to which the reduction (5), (resp. (6)) may be applied, since they would be otherwise even more cases to be analysed later.

A positive path between two positive words f and g is thus a sequence of positive reductions following each other either in the direct or in the reverse direction; *e.g.*

$$f \overset{(\alpha)}{\Rightarrow} f + (\alpha) \overset{(\beta)}{\Leftarrow} f + (\alpha) + \overline{(\beta)} = g \;.$$

A last definition: the sum of the digits of an element w of $\mathbb{Z}^4$ is called the *weight* of w, and is denoted by $W(w)$. A positive word has positive weight but the converse does not hold. For any w in $\mathbb{Z}^4$ and for any reductions α, β, and γ it holds:

$$W(w + (\alpha)) = W(w + (\alpha) + \overline{(\beta)} + (\gamma)) = W(w) - r \;,$$

$$\text{and} \qquad W(w + (\alpha) + \overline{(\beta)}) = W(w) \;.$$

Part A. Every class modulo γ_θ contains positive words, by adding $\overline{(1)} + \overline{(3)}$ and $\overline{(2)} + \overline{(4)}$ a sufficient number of times to any word of $\mathbb{Z}^4$.

We show, by case examination, that for any positive word w not in R_θ, it is possible to find a positive path (of length 1, 2 or 3) that leads to a word w' which is either in R_θ or has a weight reduced by r. Hence, from any positive word there exists a positive path reaching R_θ, for otherwise it would be possible to build a positive path reaching a word of non positive weight, and thus, non positive, a contradiction. And then, every class modulo γ_θ contains at least one element in R_θ.

Let $w = x \; y \; u \; t$ be in $\mathbb{N}^4$ and not in R_θ. Without loss of generality, one can suppose that x and u on one hand, and y and t on the other hand, are not both greater than r, for otherwise a sequence of reductions $(2) + (4)$ or $(1) + (3)$ could be used[6]. Similarly, one can suppose that no digit greater than r is followed by a strictly positive digit for otherwise one of the reductions (1) to (4) could obviously be used.

Since γ_θ commutes with the circular shift, one can suppose, without loss of generality, that $y \; u \; t \; x$ is the largest circular factor (in the lexicographic order) of $w = x \; y \; u \; t$. Since we suppose that w is not in R_θ, *i.e.* $y \; u \; t \; x$ is greater than $r \; 0 \; r \; 0$ in the lexicographic order, it implies from the above remark that

$$x \leq r - 1, \qquad y \geq r + 1, \qquad u = 0 \quad \text{and} \quad t \leq r - 1.$$

---

[6] in the case where the other two digits (y and t, or x and u) are 0, this sequence is prefixed by $\overline{(1)}$ (resp. $\overline{(2)}$) and suffixed by (1) (resp. (2))

We then apply $\overline{(4)}$ followed by (1):

$$w = x\ y\ 0\ t\ \overset{(4)}{\Leftarrow}\ x\ y{-}1\ r\ t{+}1\ \overset{(1)}{\Rightarrow}\ x{+}1\ y{-}r{-}1\ r{-}1\ t{+}1\ = w'$$

We have $W(w) = W(w')$. Which relation can be further applied to w' then depends on the actual values of x, y, and t, and we have to examine the different possible cases.

1   If $t = r - 1$,   then $w' = x{+}1\ y{-}r{-}1\ r{-}1\ r\ \overset{(3)}{\Rightarrow}\ x\ y{-}r{-}1\ r\ 0\ = w''$.

2   If $t \le r - 2$ (and thus $r \ge 2$)

    2.1   If $x = r - 1$

       2.1.1 If $y = r + 1$,   then $w' = r\ 0\ r{-}1\ t{+}1$ is in R_θ .

       2.1.2 If $y \ge r + 2$,   then

$$w' = r\ y{-}r{-}1\ r{-}1\ t{+}1\ \overset{(2)}{\Rightarrow}\ 0\ y{-}r{-}2\ r{-}1\ t{+}2\ = w''$$

    2.2   If $x \le r - 2$

       2.2.1 If $y \le 2r$,   then $w' = x{+}1\ y{-}r{-}1\ r{-}1\ t{+}1$ is in R_θ .

       2.2.2 If $y \ge 2r + 1$

$$w' = x{+}1\ y{-}r{-}1\ r{-}1\ t{+}1\ \overset{(1)}{\Rightarrow}\ x{+}2\ y{-}2r{-}1\ r{-}2\ t{+}1\ = w''$$

is a positive word since $r \ge 2$.

Thus, as announced, in any case, a positive word w is lead either to a word of strictly smaller weight by a positive path of length 1 or 3 or into R_θ by a positive path of length 2.

Part B. The system defined by the relations (1) to (4) — oriented, as in Part A, from left to right — is not confluent when it is restricted to positive words.

It is easy to verify, by inspection on the possible values of the digits of w, the following claims[7].

Claim 1 *Let w be a positive word; then,*   i) $w \overset{(1)}{\Rightarrow} w'$ *and* $w \overset{(3)}{\Rightarrow} w''$ *imply that there exists a (positive) word v such that* $w' \overset{(3)}{\Rightarrow} v$ *and* $w'' \overset{(1)}{\Rightarrow} v$. *Similarly,* ii) $w \overset{(2)}{\Rightarrow} w'$ *and* $w \overset{(4)}{\Rightarrow} w''$ *imply that there exists a (positive) word v such that* $w' \overset{(4)}{\Rightarrow} v$ *and* $w'' \overset{(2)}{\Rightarrow} v$.

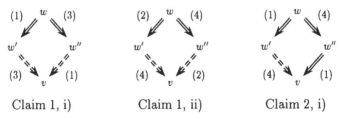

Claim 1, i)   Claim 1, ii)   Claim 2, i)

---

[7] The diagrams express the claims: the reductions that hold by hypothesis are drawn with solid arrows, the reductions that are deduced from the claims are drawn with dashed arrows.

Claim 2 *Let w be a positive word and suppose that $w \overset{(1)}{\Rightarrow} w'$ and $w \overset{(4)}{\Rightarrow} w''$ hold. Then:* i) w'' *(1)-reducible implies that w' is (4)-reducible and vice versa;*
ii) w'' *(4)-reducible implies that w'' is (1)-reducible as well (and thus w' is (4)-reducible);*
iii) w'' *(3)-reducible implies that both w and w' are (3)-reducible;*
iv) w'' *(2)-reducible implies that w is (2)-reducible.*

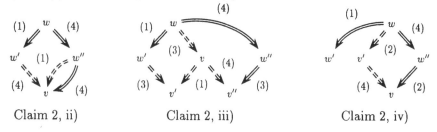

Claim 2, ii)   Claim 2, iii)   Claim 2, iv)

These first two claims deal with the cases where the reduction behaves as if it were confluent; the next one describes in detail the case where the reduction is not confluent: one of the branch happens to be a deadend from where one cannot escape by another derivation than the branch itself.

Claim 3 *Let w be a positive word. Suppose that $w \overset{(1)}{\Rightarrow} w'$ and $w \overset{(4)}{\Rightarrow} w''$ hold and that w'' is p-irreducible. If there exists two (positive) words h and k and two distincts reductions (α) and (β) such that $h \overset{(\alpha)}{\Rightarrow} w''$ and $h \overset{(\beta)}{\Rightarrow} k$, then necessarily* i) $h = w$ *[and $(\alpha) = (4)$];*   ii) *if $k \neq w'$ then $(\beta) = (3)$.*

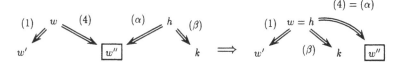

Proof. Let $w = x\ y\ u\ t$; thus $w'' = x\ y+1\ u-r\ t-1$.
   The hypothesis implies that:

$y \geq r$ since w is (1)-reducible; $u = r$,   since w'' is (1)-irreducible and $y \geq r$;

$x < r$,   since w'' is (2)-irreducible; and $t < r+1$,   since w'' is (3)-irreducible.
   Now, $h \overset{(3)}{\Rightarrow} w''$ is impossible since $w'' = x\ y+1\ 0\ t-1$.
   Suppose $h \overset{(2)}{\Rightarrow} w''$; then $h = x+r\ y+2\ 0\ t-2$ which makes $h \overset{(4)}{\Rightarrow} k$ or $h \overset{(1)}{\Rightarrow} k$ impossible; $h \overset{(3)}{\Rightarrow} k$ is impossible as well since $t < r+1$.
   Suppose $h \overset{(1)}{\Rightarrow} w''$; then $h = x-1\ y+r+1\ 1\ t-1$ which implies $r > 1$ since $x < r$. Now $h \overset{(4)}{\Rightarrow} k$ is impossible since $r > 1$; $h \overset{(3)}{\Rightarrow} k$ is impossible since $t < r+1$ and $h \overset{(2)}{\Rightarrow} k$ is impossible since $x < r$.
   The only possibility left by $(\alpha) \neq (\beta)$ is thus $(\alpha) = (4)$.
   As before, $(\beta) = (2)$ is impossible since $x < r$ and the claim is established. ∎

A simple verification leads to the following claim.

Claim 4 *Let w be a positive word; $w \overset{(1)}{\Rightarrow} w'$ and $w \overset{(4)}{\Rightarrow} w''$ imply that w'' does not belong to R_θ. Similarly, $w \overset{(4)}{\Rightarrow} w'$ and $w \overset{(3)}{\Rightarrow} w''$, or $w \overset{(3)}{\Rightarrow} w'$ and $w \overset{(2)}{\Rightarrow} w''$, or $w \overset{(2)}{\Rightarrow} w'$ and $w \overset{(1)}{\Rightarrow} w''$ imply $w'' \notin R_\theta$.* ■

The collection of claims we have just established allows us to adapt the classical scheme of the demonstration of the uniqueness of reduced words in a class modulo a confluent relation. Let us suppose, by way of contradiction, that there exist distinct f and g in R_θ with the property that there exists a positive path between them. [Recall that a positive path is a sequence of reductions (1), (2), (3) or (4) between positive words — together with the possible occurrence of $0\,r\,0\,r \overset{(5)}{\Rightarrow} 0\,0\,0\,0$ and of $r\,0\,r\,0 \overset{(6)}{\Rightarrow} 0\,0\,0\,0$ — in either directions.]

Since f and g are both p-irreducible, such a path Π must contain a "*peak*", that is a factor

$$w' \Leftrightarrow w \Leftrightarrow w'' \qquad \text{of the form} \qquad w' \overset{(\alpha)}{\Leftarrow} w \overset{(\beta)}{\Rightarrow} w'' \ .$$

The path Π thus contains a peak of *maximal weight*; the weight of such peak will be *the weight of the path Π*. Pathes are then *ordered by weight* and pathes of equal weight are *ordered by the number of peaks* of maximal weight.

In the set of all positive pathes between f and g — non empty by hypothesis — let us choose a minimal path Π_0, *i.e.* a path of minimal weight with a minimal number of peaks of maximal weight. Notice that by a circular permutation of every word of a path, and a possible exchange of the extremities, the weight of a path is unchanged and thus the effect of such a transformation on Π_0 gives a minimal path.

Let w be one of the peaks of maximal weight in Π_0 and let $w \overset{(\alpha)}{\Rightarrow} w'$ and $w \overset{(\beta)}{\Rightarrow} w''$ be the two reductions that go out of w on Π_0, which can thus be written in the following form:

$$f \overset{*}{\Leftrightarrow} w' \overset{(\alpha)}{\Leftarrow} w \overset{(\beta)}{\Rightarrow} w'' \overset{*}{\Leftrightarrow} g$$

By Claim 1, it is not possible to have $\alpha = 1$ and $\beta = 3$; for otherwise we would have a word v such that $w' \overset{(3)}{\Rightarrow} v$ and $w'' \overset{(1)}{\Rightarrow} v$ and thus the path

$$f \overset{*}{\Leftrightarrow} w' \overset{(3)}{\Rightarrow} v \overset{(1)}{\Leftarrow} w'' \overset{*}{\Leftrightarrow} g$$

is smaller than Π_0, a contradiction. For the same reason it is not possible to have $\alpha = 2$ and $\beta = 4$.

Up to a circular permutation of every word on Π_0, and a possible exchange of f and g, we can assume that $\alpha = 1$ and $\beta = 4$. By Claim 2 i) and ii), and with same argument as just above, w'' is neither (1)- nor (4)-reducible. By Claim 2 iii), w'' is not (3)-reducible for otherwise we would have $w' \overset{(3)}{\Rightarrow} v'$, $w \overset{(3)}{\Rightarrow} v$ and $w'' \overset{(3)}{\Rightarrow} v''$ and since reductions commute we would get the path

$$f \overset{*}{\Leftrightarrow} w' \overset{(3)}{\Rightarrow} v' \overset{(1)}{\Leftarrow} v \overset{(4)}{\Rightarrow} v'' \overset{(3)}{\Leftarrow} w'' \overset{*}{\Leftrightarrow} g$$

which again is smaller than Π_0. We have now to consider the remaining two cases: **case a)** w'' *is not* (2)-reducible (and thus w'' is p-irreducible), or **case b)** w'' *is* (2)-reducible. Note that w'' is neither (5) nor (6)-reducible.

case a.– w'' is p-irreducible. By Claim 4, w'' is not in R_θ and thus not equal to g. The path Π_0 factorizes into

$$f \overset{*}{\Leftrightarrow} w' \overset{(1)}{\Leftarrow} w \overset{(4)}{\Rightarrow} w'' \overset{(\alpha)}{\Leftarrow} h \Leftrightarrow k \overset{*}{\Leftrightarrow} g \,.$$

The only possibilities left by Claim 3 for the reductions $w'' \overset{(\alpha)}{\Leftarrow} h \Leftrightarrow k$ are either

case a.1.– $w'' \overset{(4)}{\Leftarrow} w = h \overset{(1)}{\Rightarrow} k = w'$ in which case w'' is indeed a dead end in Π_0 and

$$f \overset{*}{\Leftrightarrow} w' \overset{*}{\Leftrightarrow} g$$

is a path smaller than Π_0, a contradiction; or

case a.2.– $w'' \overset{(4)}{\Leftarrow} w = h \overset{(3)}{\Rightarrow} k$ in which case w'' is again a dead end in Π_0 which reads

$$f \overset{*}{\Leftrightarrow} w' \overset{(1)}{\Leftarrow} w = h \overset{(3)}{\Rightarrow} k \overset{*}{\Leftrightarrow} g \,,$$

a case that is ruled out by Claim 1; or

case a.3.– $w'' \overset{(\alpha)}{\Leftarrow} h \overset{(\gamma)}{\Leftarrow} k$ in which case k has a greater weight than w, another contradiction.

case b.– w'' is (2)-reducible. By Claim 2 iv), w is (2)-reducible as well. If w' is also (2)-reducible the situation is the same as if w'' were (3)-reducible and leads to a contradiction. Let us suppose then that w' is (2)-irreducible and let us sum up the constraints on the digits of $w = x\ y\ u\ t$ given by all hypothesis we have made up to that point.

$$w' = x{+}1\ y{-}r\ u{-}1\ t \overset{(1)}{\Leftarrow} w = x\ y\ u\ t \overset{(4)}{\Rightarrow}$$

$$w'' = x\ y{+}1\ u{-}r\ t{-}1 \overset{(2)}{\Rightarrow} x{-}r\ y\ u{-}r\ t$$

w'' (1)-irreducible implies $u = r$ and thus w' is (4)-irreducible; w' (2)-irreducible implies $y = r$ (and thus w' is (1)-irreducible). There are thus two possibilities: case b.1), w' is (3)-reducible or, case b.2), w' is p-irreducible. Note that w' is neither (5) nor (6)-reducible.

case b.1.– w' is (3)-reducible; w'' (3)-irreducible implies $t = r$. The path Π_0 reads then

$$f \overset{*}{\Leftrightarrow} x{+}1\ 0\ r{-}1\ r \overset{(1)}{\Leftarrow} w = x\ r\ r\ r \overset{(4)}{\Rightarrow} x\ r{+}1\ 0\ r{-}1 \overset{*}{\Leftrightarrow} g \,.$$

If x is greater than r, the path

$$f \overset{*}{\Leftrightarrow} x{+}1\ 0\ r{-}1\ r \overset{(3)}{\Rightarrow} x\ 0\ r\ 0 \overset{(1)}{\Leftarrow} x{-}1\ r\ r{+}1\ 0 \overset{(2)}{\Rightarrow} x{-}r{-}1\ r{-}1\ r{+}1\ 1$$

$$\overset{(3)}{\Leftarrow} x{-}r\ r{-}1\ r\ r{+}1 \overset{(4)}{\Rightarrow} x{-}r\ r\ 0\ r \overset{(2)}{\Leftarrow} x\ r{+}1\ 0\ r{-}1 \overset{*}{\Leftrightarrow} g$$

is smaller than Π_0. If $x = r$, the path

$$f \overset{*}{\leftrightarrow} x{+}1\ 0\ r{-}1\ r \overset{(3)}{\Rightarrow} r\ 0\ r\ 0 \overset{(6)}{\Rightarrow} 0\ 0\ 0\ 0 \overset{(5)}{\Leftarrow} 0\ r\ 0\ r \overset{(2)}{\Leftarrow} x\ r{+}1\ 0\ r{-}1 \overset{*}{\leftrightarrow} g$$

is again smaller than Π_0, and of a form consistent with the hypothesis on a positive path. Contradiction for any possible value of x.

case b.2.– w' is p-irreducible. Since w is (2)-reducible and using Claim 1, we can transform the path Π_0 into a path Π_0'

$$f \overset{*}{\leftrightarrow} w' \overset{(1)}{\Leftarrow} w \overset{(2)}{\Rightarrow} w''' \overset{*}{\leftrightarrow} g$$

which is also minimal. The image of Π_0' by the circular permutation[8] of the digits σ^{-1} then reads

$$k \overset{*}{\leftrightarrow} m'' \overset{(4)}{\Leftarrow} m \overset{(1)}{\Rightarrow} m' \overset{*}{\leftrightarrow} l$$

with $k = \sigma^{-1}(f)$, $l = \sigma^{-1}(g)$, $m = \sigma^{-1}(w)$, $m'' = \sigma^{-1}(w')$ and $m' = \sigma^{-1}(w''')$. As w', m'' is p-irreducible and we are back to case a), that leads to a contradiction.

This terminates the proof of the fact that no two distinct elements of R_θ can be joined by a positive path.

Part C. It remains to show that if two elements f and g of R_θ are congruent modulo γ_θ, they are equal. Indeed, f and g are congruent modulo γ_θ if, and only if, $f \overset{*}{\leftrightarrow} g$, that is, if, and only if, there exists a path Σ between f and g. The idea is to "lift" the path Σ into a *positive path* Π between f and g (as sketched on Figure 2) and the conclusion follows from Part B.

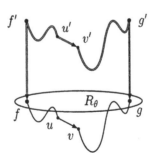

Fig. 2. Σ lifts to Π .

The lifting relies on a lemma and a remark.

Lemma 1 *Let n be any positive integer and let h_n be the word*

$$h_n = nr\ nr\ nr\ nr\ .$$

For any f in R_θ there exists a positive path $f + h_n \overset{*}{\Rightarrow} f$.

---

[8] Where the last digit becomes the first one; we denote it by σ^{-1}, and not, more simply, by σ to be consistent with the previous paper ([3])

Proof. If $f = 0\,0\,0\,0$, the path may begin with

$$0\,0\,0\,0 \overset{(6)}{\Leftarrow} r\,0\,r\,0$$

and thus one can assume that at least one digit of $f = x\,y\,u\,t$ is strictly positive. Up to a circular permutation, we suppose that this digit is x. One has the sequence:

$$f = x\,y\,u\,t \overset{(1)}{\Leftarrow} x{-}1\ y{+}r\ u{+}1\ t \overset{(3)}{\Leftarrow} x\ y{+}r\ u\ t{+}r$$

$$\overset{(2)}{\Leftarrow} x{+}r\ y{+}r{+}1\ u\ t{+}r{-}1 \overset{(4)}{\Leftarrow} x{+}r\ y{+}r\ u{+}r\ t{+}r = f + h_1$$

Without any further care on the order of the rewriting, one has

$$f + h_i \overset{(1)+(2)+(3)+(4)}{\Longleftarrow} f + h_{i+1}$$

for any positive i and then $f \overset{*}{\Leftarrow} f + h_n$ for any positive n. ∎

Remark 1   Let u and v be two non positive elements of $\mathbb{Z}^4$ and k the lower bound of the digits in u and v; let n be a positive integer such that $nr > -k$. If $u \overset{(\alpha)}{\to} v$, then $u + h_n \overset{(\alpha)}{\to} v + h_n$, for any reduction (α).

Let Σ be any path of reductions that links f and g, which we write $f \overset{\Sigma}{\longleftrightarrow} g$. It is clear now how to lift Σ: let k be the lower bound of the digits of the words that appear in Σ and let n as above, i.e. such that $nr > -k$. Let $f' = f + h_n$ and $g' = g + h_n$. We have, by Lemma 1,

$$f \overset{*}{\Leftarrow} f' \overset{\Sigma+h_n}{\longleftrightarrow} g' \overset{*}{\Rightarrow} g$$

and by the remark " $\Sigma + h_n$ " is a positive path. Thus $f \overset{*}{\leftrightarrow} g$ implies $f \overset{*}{\Leftrightarrow} g$ which has been shown impossible and this complete the proof of the proposition in CASE 1. ∎

4   Proof for Case 2

Unless otherwise stated, all notation and conventions described in the previous section are still valid. The congruence γ_θ is now generated by the following relations:

$1\,\bar{r}\,1\,0 = 0\,0\,0\,0$	(1')		$1\,0\,1\,\bar{r} = 0\,0\,0\,0$	(3')
$\bar{r}\,1\,0\,1 = 0\,0\,0\,0$	(2')		$0\,1\,\bar{r}\,1 = 0\,0\,0\,0$	(4')

which can be turned into a rewrite system $(\mathcal{S})$ by giving the orientation from left to right:

$$1\,\bar{r}\,1\,0 \to 0\,0\,0\,0 \qquad\qquad (1')\ etc.$$

Let us recall also that R_θ is now the set of words of $\mathbb{N}^4$ with the property that they, and all their conjugates, are different from $r-2\ r-2\ r-2\ r-2$ and strictly smaller than $r-1\ r-2\ r-2\ r-2$ in the lexicographic order.

It is immediate to check that for any positive word w and any two distinct reductions (α) and (β) in $(\mathcal{S})$, if $w \overset{(\alpha)}{\Rightarrow} w'$ and $w \overset{(\beta)}{\Rightarrow} w''$ hold, then $w' \overset{(\beta)}{\Rightarrow} v$ and $w'' \overset{(\alpha)}{\Rightarrow} v$ hold as well.

This fundamental difference with CASE 1 can be stated as follows:

Claim 5 $(\mathcal{S})$ *is confluent on the set of positive words.* ∎

We are not yet done, for the rewrite system $(\mathcal{S})$ is not equivalent to γ_θ on the set of positive words. But the solution is at hand and will be reached by the construction of a richer system.

Let $(\mathcal{T})$ be the rewrite system obtained by adding any subset of relations in $(\mathcal{S})$. We get then the following relations.

$$\overline{r-1}\ \overline{r-1}\ 1\ 1 \to 0\ 0\ 0\ 0 \quad (5') \qquad 1\ 1\ \overline{r-1}\ \overline{r-1} \to 0\ 0\ 0\ 0 \quad (7')$$
$$\overline{r-1}\ 1\ 1\ \overline{r-1} \to 0\ 0\ 0\ 0 \quad (6') \qquad 1\ \overline{r-1}\ \overline{r-1}\ 1 \to 0\ 0\ 0\ 0 \quad (8')$$

that is: $(5') = (1') + (2')$, $(6') = (2') + (3')$, $(7') = (3') + (4')$ and $(8') = (4') + (1')$.

$$\overline{r-1}\ 2\ \overline{r-1}\ \overline{r-2} \to 0\ 0\ 0\ 0 \quad (9') \qquad \overline{r-1}\ \overline{r-2}\ \overline{r-1}\ 2 \to 0\ 0\ 0\ 0 \quad (11')$$
$$2\ \overline{r-1}\ \overline{r-2}\ \overline{r-1} \to 0\ 0\ 0\ 0 \quad (10') \qquad \overline{r-2}\ \overline{r-1}\ 2\ \overline{r-1} \to 0\ 0\ 0\ 0 \quad (12')$$

that is : $(9') = (2') + (3') + (4')$, $(10') = (3') + (4') + (1')$, $(11') = (4') + (1') + (2')$ and $(12') = (1') + (2') + (3')$. And finally $(13') = (1') + (2') + (3') + (4')$:

$$\overline{r-2}\ \overline{r-2}\ \overline{r-2}\ \overline{r-2} \to 0\ 0\ 0\ 0 \quad (13')$$

A simple case inspection shows that

Claim 6 R_θ *is the set of irreducible words for the system* $(\mathcal{S} + \mathcal{T})$. ∎

The core of the proof lies then in the following:

Claim 7 $(\mathcal{S} + \mathcal{T})$ *is confluent on the set of positive words.*

Proof. Since we have not spared him the slightest detail yet, the reader may be scared by the prospect of checking the 78 critical pairs of the system $(\mathcal{S} + \mathcal{T})$. Hopefully, thanks to the symmetries and the very specific form of the relations, the number of cases to be examined boils down to 11, of which we shall make only 3 explicit.

Let $w = x\ y\ u\ t$ be a positive word and suppose that $w \overset{(\alpha)}{\Rightarrow} w'$ and $w \overset{(\beta)}{\Rightarrow} w''$ hold. By Claim 5, one can assume that (α) and (β) are not both in $(\mathcal{S})$. Up to an exchange of (α) and (β), we suppose that β is in $(\mathcal{T})$.

1   (α) is in $(\mathcal{S})$. Up to a circular shift, one can assume that $(\alpha) = (1')$.

1.1 (β) "contains" (α), *i.e.* it exists (γ) in $(\mathcal{S} + \mathcal{T})$ such that $(\beta) = (\alpha) + (\gamma)$.

Then, obviously, $w' \overset{(\gamma)}{\Rightarrow} w''$.

1.2 (β) does not "contain" (α); the only possibilities are $(\beta) = (6')$, $(7')$, or $(9')$. Immediate computations show that $w' \overset{(\beta)}{\Rightarrow} v$ and $w'' \overset{(\alpha)}{\Rightarrow} v$ hold as well.

For instance, let $(\beta) = (9')$; it comes:

$$x\ y\ u\ t \overset{(1')}{\Longrightarrow} x{+}1\ y{-}r\ u{+}1\ t\ = w' \tag{14'}$$

and
$$x\ y\ u\ t \overset{(9')}{\Longrightarrow} x{-}r{+}1\ y{+}2\ u{-}r{+}1\ t{-}r{+}2\ = w'' \ . \tag{15'}$$

From (14') follows $y \geq r$ and thus $w'' \overset{(1')}{\Longrightarrow} v$; from (15') follows x, $u \geq r-1$, $t \geq r-2$ and thus $w' \overset{(9')}{\Longrightarrow} v$.

2 (α) is not in $(\mathcal{S})$. By symmetry, one can assume that (β) is as "large" as (α), *i.e.* is the sum of as many relations from $(\mathcal{S})$ as (α). Up to a circular permutation, one can assume that $(\alpha) = (5')$ or $(9')$ (since $(\alpha) = (13')$ implies $(\beta) = (13')$).

2.1 (β) "contains" (α). Same solution as in 1.1.

2.2 Case 2.1 does not hold. Then there exist (γ), (δ) and (ε) in $(\mathcal{S}+\mathcal{T})$ such that $(\alpha) = (\gamma) + (\delta)$ and $(\beta) = (\gamma) + (\varepsilon)$. It is then a matter af immediate computation to verify that, in every case, $w' \overset{(\varepsilon)}{\Rightarrow} v$ and $w'' \overset{(\delta)}{\Rightarrow} v$ hold. The only possibilities are:

2.2.1 if $(\alpha) = (5')$, then $(\beta) = (6')$, $(7')$, $(8')$,$(9')$ or $(10')$.

For instance, let $(\beta) = (6')$; it comes:

$$x\ y\ u\ t \overset{(5')}{\Longrightarrow} x{-}r{+}1\ y{-}r{+}1\ u{+}1\ t{+}1\ = w' \tag{16'}$$

and
$$x\ y\ u\ t \overset{(6')}{\Longrightarrow} x{-}r{+}1\ y{+}1\ u{+}1\ t{-}r{+}1\ = w'' \ . \tag{17'}$$

From (16') follows $y \geq r-1$, thus $y+1 \geq r$ and thus $w'' \overset{(1')}{\Longrightarrow} v$; from (17') follows $t \geq r-1$, thus $t+1 \geq r$ and thus $w' \overset{(3')}{\Longrightarrow} v$.

2.2.2 if $(\alpha) = (9')$, then $(\beta) = (10')$, $(11')$, or $(12')$.

For instance, let $(\beta) = (10')$; it comes:

$$x\ y\ u\ t \overset{(9')}{\Longrightarrow} x{-}r{+}1\ y{+}2\ u{-}r{+}1\ t{-}r{+}2\ = w' \tag{18'}$$

and
$$x\ y\ u\ t \overset{(10')}{\Longrightarrow} x{+}2\ y{-}r{+}1\ u{-}r{+}2\ t{-}r{+}1\ = w'' \ . \tag{19'}$$

From (18') follows $x \geq r-1$, thus $x+2 > r$ and thus $w'' \overset{(2')}{\Longrightarrow} v$; from (19') follows $y \geq r-1$, thus $y+2 > r$ and thus $w' \overset{(1')}{\Longrightarrow} v$.

The claim is established. ∎

CASE 2 is now immediately settled. Every class modulo γ_θ contains positive words, by adding $\overline{(13')}$ to any word of $\mathbb{Z}^4$, a sufficient number of times. And any positive word reduces to a unique word in R_θ, using reductions in $(\mathcal{S}+\mathcal{T})$.

Two words f and g of R_θ are congruent modulo γ_θ if, and only if, $f \overset{*}{\leftrightarrow} g$, that is, if, and only if, there exists between f and g a path Σ consisting in reductions of $(\mathcal{S})$ taken in either directions. As in CASE 1, this path Σ can be "lifted" into a *positive path* Π by using of the same reduction (13') as before. Since we can use that reduction, the actual value of f has not to be taken into consideration — in other words, Lemma 1 becomes trivial — and the lifting is even simpler than in CASE 1.

By construction, the path Π consists in reductions of $(\mathcal{S} + \mathcal{T})$. By Claims 6 and 7, two distincts words of R_θ cannot be joinned by such a path, hence $f = g$.

And this completes the proof of Theorem 1. ∎

Acknowledgements

We are grateful to the two meticulous referees who have not been disheartened by the technicalities of the paper. They have checked every detail of the proof and corrected an error in one of the claims and a number of misprints.

We were fixing the last details of this paper when we learn the passing away of our friend David Klarner. David was the editor who read the first version of our work on the translation from the Fibonacci system to the golden mean base by a finite automaton. His advice and enthusiastic encouragement were of great help to us for giving the final shape to our result. We sorrowfully dedicate this subsequent work to his memory.

References

1. Č. Burdík, Ch. Frougny, J.P. Gazeau, and R. Krejcar, Beta-integers as natural counting systems for quasicrystals. *J. of Physics A: Math. Gen.* **31**, 1998, 6449–6472.
2. N. Dershowitz and J. P. Jouannaud, Rewrite Systems, in *Handbook of Theoretical Computer Science* (J. van Leeuwen, ed.), vol. B, Elsevier, 1990, 243–320.
3. Ch. Frougny and J. Sakarovitch, Automatic conversion from Fibonacci representation to representation in base φ, and a generalization, to appear in *Int. J. of Alg. and Comput.* Available at http://www.enst.fr/~jsaka
4. Ch. Frougny and J. Sakarovitch, Two groups associated with quadratic Pisot units, to appear. Available at http://www.enst.fr/~jsaka .
5. D. E. Knuth, The Art of Computer Programming, Vol. 1, Addison-Wesley, 1968.
6. W. Parry, On the β-expansions of real numbers. *Acta Math. Acad. Sci. Hungar.* **11**, 1960, 401–416.
7. N. Sidorov and A. Vershik, Bijective arithmetic codings of hyperbolic automorphisms of the 2-torus, and binary quadratic forms. *J. Dynam. Control Systems* **4**, 1998, 365–399.
8. E. Zeckendorf, Représentation des nombres entiers par une somme de nombres de Fibonacci ou de nombre de Lucas, *Bull. Soc. Roy. Liège* **41**, 1972, 179–182.

Fast Rewriting of Symmetric Polynomials

Manfred Göbel

Deutsches Fernerkundungsdatenzentrum
Algorithmen und Prozessoren
Deutsches Forschungszentrum für Luft- und Raumfahrt e.V.
DLR Oberpfaffenhofen, 82234 Weßling, Germany
Manfred.Goebel@dlr.de

Abstract. This note presents a fast version of the classical algorithm to represent any symmetric function in a unique way as a polynomial in the elementary symmetric polynomials by using power sums of variables. We analyze the worst case complexity for both algorithms, the original and the fast version, and confirm our results by empirical run-time experiments. Our main result is a fast algorithm with a polynomial worst case complexity w.r.t. the total degree of the input polynomial compared to the classical algorithm with its exponential worst case complexity.
Keywords. Symmetric polynomials, dynamic power sum representation, fast rewriting techniques

1 Introduction

Symmetric polynomials play an important rôle in algebra and its applications. In particular, many proofs in the theory of algebraic numbers refer explicitly to them. Resultants [15] and Sturm sequences [12] can be generated as symmetric polynomials. The unique representation of a symmetric polynomial as a polynomial in the elementary symmetric polynomials [15, §33] is an important and central method in algebra, which occurs, e.g., in Gauß' (second) proof of the *Fundamentalsatz* of the algebra in 1812 [3, section 3.1], or in Galois Theory [4]. It is therefore reasonable to search for fast algorithms and reduction techniques [11] to compute this representation for any symmetric polynomial and analyze their complexity.

A comprehensive and very detailed analysis of the complexity of the classical algorithm was done by Lauer and Loos [9, 10]. In addition, they have shown that a special representation of symmetric polynomials and its corresponding arithmetic lead to a significant empirical run-time improvement. In the following we call this method the *Lauer-Loos algorithm*.

Another approach was studied in [2]. Their method expresses the orbits of a symmetric polynomial as a sum of determinants in the elementary symmetric polynomials: The orbits are first represented in terms of Schur polynomials, which are then evaluated as determinants in the elementary symmetric polynomials.

In this paper, we present a fast reduction technique using a *dynamic power sum representation* of symmetric polynomials to compute the representation in terms of the elementary symmetric polynomials, i.e. in the first phase of the algorithm all power sums up to the total degree of the given polynomial are used for the representation. In a second phase, a post-processing step follows to compute the representation in terms of the elementary symmetric polynomials. In a simplified worst case complexity analysis, we show how this approach reduces the *exponential* run-time of the classical algorithm to a *polynomial* run-time, and we confirm our results by a set of benchmark experiments. Our empirical results will show, e.g., that our fast rewriting technique is superior to the classical algorithm in the average case, and moreover, that it *outperforms* the Lauer-Loos algorithm.

We proceed as follows: Section 2 recalls the basic definitions and the classical algorithm. Section 3 presents a power sum algorithm which leads to the fast reduction algorithm developed in Section 4. Furthermore, we explain the necessary pre- and post-processing steps for the fast version, present empirical run-times, and compare our results with other work. In addition, Section 2, 3 and 4 contain a simplified worst case complexity analysis w.r.t. the number of monomial multiplications during the reduction. Section 5 presents our conclusion.

2 The Classical Rewriting Technique

Notation: Let $\mathbb{Q}$ be the set of rational numbers, let $\mathbb{Q}[X_1, \ldots, X_n]$ be the commutative polynomial ring over $\mathbb{Q}$ in the indeterminates X_i, let T be the set of terms (= power-products of the X_i) in $\mathbb{Q}[X_1, \ldots, X_n]$, and let $T(f)$ be the set of terms occurring in $f \in \mathbb{Q}[X_1, \ldots, X_n]$ with non-zero coefficients. Maximal variable degree and total degree of $t \in T$ and $f \in \mathbb{Q}[X_1, \ldots, X_n]$ are defined as usual. $X_1^{e_1} \ldots X_n^{e_n}$ is called descending, if $e_1 \geq \ldots \geq e_n$.

S_n denotes the symmetric group, $\mathbb{Q}[X_1, \ldots, X_n]^{S_n}$ the set of symmetric polynomials in $\mathbb{Q}[X_1, \ldots, X_n]$, and

$$orbit_{S_n}(t) = \sum_{s \in \{\pi(t) | \pi \in S_n\}} s$$

the symmetric orbit of t.

We assume in the following that the multiplication of two monomials in n variables can be done in a constant amount of time, and that the run-time of an algorithm is linear w.r.t. the number of necessary monomial multiplications.

We first recall the basic facts about the classical algorithm that represents any symmetric function as a polynomial in the elementary symmetric polynomials

$$\sigma_1 = X_1 + X_2 + \ldots + X_n,$$
$$\sigma_2 = X_1 X_2 + X_1 X_3 + \ldots + X_{n-1} X_n,$$
$$\ldots$$
$$\sigma_n = X_1 X_2 \ldots X_n.$$

The i-th elementary symmetric polynomial has $\binom{n}{i}$ monomials and $\sigma_1, \ldots, \sigma_n$ are algebraically independent.

Theorem 1. *Let $f \in \mathbb{Q}[X_1, \ldots, X_n]^{S_n}$. Then f has a computable representation as $f = p(\sigma_1, \ldots, \sigma_n)$, where $p \in \mathbb{Q}[X_1, \ldots, X_n]$ is uniquely determined by f.*

Proof. See [1, section 10.7]. Let max-desc(f) be the maximal descending monomial in f w.r.t. the lexicographical term order. Then the polynomial $p \in \mathbb{Q}[X_1, \ldots, X_n]$ can be computed as follows:

Algorithm 1.   **Classical Algorithm**

1. *INPUT $f \in \mathbb{Q}[X_1, \ldots, X_n]^{S_n}$; lex. term order;*
2. *$\hat{f} := f; p := 0;$*
3. *WHILE $\hat{f} \neq 0$ DO*
4.   *$aX_1^{e_1} \ldots X_n^{e_n} := max\text{-}desc(\hat{f});$*
5.   *$p := p + a \cdot X_1^{e_1 - e_2} \ldots X_{n-1}^{e_{n-1} - e_n} X_n^{e_n};$*
6.   *$\hat{f} := \hat{f} - a \cdot \sigma_1^{e_1 - e_2} \ldots \sigma_{n-1}^{e_{n-1} - e_n} \sigma_n^{e_n};$*
7.   *ENDWHILE;*
8. *OUTPUT $f = p(\sigma_1, \ldots, \sigma_n)$ with $p \in \mathbb{Q}[X_1, \ldots, X_n]$;*

Corollary 1. *Let $f = p(\sigma_1, \ldots, \sigma_n) \in \mathbb{Q}[X_1, \ldots, X_n]^{S_n}$ as stated in Theorem 1 and assume that the maximal variable degree of f is at most d. Then p has a maximal variable degree of at most d and at most $\binom{n+d}{n}$ monomials.*

Proof. The maximal variable degree bound for p is a consequence of Algorithm 1; the number of descending terms with a maximal variable degree of at most d is $\binom{n+d}{n}$. This is a bound for the number of loop-runs in Algorithm 1 for a given polynomial f and therefore for the number of monomials in p.

Lemma 1. *Let $f \in \mathbb{Q}[X_1, \ldots, X_n]^{S_n}$ with a maximal variable degree d. Then the classical algorithm needs at most $O(\binom{n+d}{n} r(n)^d)$ multiplications of monomials to compute the representation, where*

$$r(n) = \max\left\{ \binom{n}{i} \mid 1 \leq i \leq n \right\} = \binom{n}{\lceil \frac{n}{2} \rceil}.$$

Proof. The number of descending terms with a maximal variable degree of at most d is $\binom{n+d}{n}$, and any term has to be reduced at most once. Each elementary symmetric polynomial has at most $r(n)$ monomials and we have to multiply at most d elementary symmetric polynomials during a single loop-run.

The number of monomial multiplications for the classical algorithm is — from the point of view of worst case complexity — *exponential* in the maximal variable degree d of the given polynomial $f \in \mathbb{Q}[X_1, \ldots, X_n]^{S_n}$.

The number of monomials of the input polynomial $f \in \mathbb{Q}[X_1, \ldots, X_n]^{S_n}$ and the output polynomial $p \in \mathbb{Q}[X_1, \ldots, X_n]$ of Algorithm 1 is always polynomial in the maximal variable degree d and also in the total degree D.

3 The Power Sum Rewriting Technique

It is well known, that any symmetric polynomial can be also represented as a polynomial in the first n power sums $s_j = X_1^j + \ldots + X_n^j$ with $1 \le j \le n$ [13].[1] Any polynomial s_j has n monomials and $s_1, \ldots, s_n$ are algebraically independent. In contrast to this result, our next algorithm uses dynamically all power sums s_j for $1 \le j \le D$ up to the total degree D of the given polynomial $f \in \mathbb{Q}[X_1, \ldots, X_n]^{S_n}$.

Theorem 2. *Let* $f \in \mathbb{Q}[X_1, \ldots, X_n]^{S_n}$ *be a polynomial with a total degree of at most* D. *Then* f *has a computable representation as* $f = p(s_1, \ldots, s_D)$ *with* $p \in \mathbb{Q}[X_1, \ldots, X_D]$.

Proof. Let min-desc(f) be the minimal descending monomial in f w.r.t. the lexicographical term order. Then the polynomial $p \in \mathbb{Q}[X_1, \ldots, X_D]$ can be computed as follows:

Algorithm 2. **Power Sum Algorithm**

1. *INPUT* $f \in \mathbb{Q}[X_1, \ldots, X_n]^{S_n}$; *lex. term order;*
2. $\hat{f} := f$; $p := 0$; $X_0 := 1$; $s_0 := 1$;
3. *WHILE* $\hat{f} \ne 0$ *DO*
4. $\quad aX_1^{e_1} \ldots X_n^{e_n} := $ *min-desc*$(\hat{f})$; $c := $ *coeff. of* $X_1^{e_1} \ldots X_n^{e_n}$ *in* $s_{e_1} \ldots s_{e_n}$;
5. $\quad p := p + \frac{a}{c} \cdot X_{e_1} \ldots X_{e_n}$;
6. $\quad \hat{f} := \hat{f} - \frac{a}{c} \cdot s_{e_1} \ldots s_{e_n}$;
7. *ENDWHILE*;
8. *OUTPUT* $f = p(s_1, \ldots, s_D)$ *with* $p \in \mathbb{Q}[X_1, \ldots, X_D]$;

The loop invariant is $f = \hat{f} + p(s_1, \ldots, s_D)$. Every pass through the while-loop removes at least the symmetric orbit containing $aX_1^{e_1} \ldots X_n^{e_n}$ from $\hat{f}$ and adds only symmetric orbits to $\hat{f}$ which have a higher descending head term w.r.t. the lexicographical term order. The number of descending terms with a maximal variable degree of at most D is finite, i.e. $\hat{f} = 0$ will be reached after finitely many cycles.

Corollary 2. *Let* $f = p(s_1, \ldots, s_D) \in \mathbb{Q}[X_1, \ldots, X_n]^{S_n}$ *as stated in Theorem 2. Furthermore, let* $k = \max\{\sum_{i=1}^n \delta(e_i) \mid X_1^{e_1} \ldots X_n^{e_n} \in T(f)\}$ *with* $\delta(e) = 0$, *if* $e = 0$, *and* $\delta(e) = 1$, *if* $e \ne 0$. *Then* p *has a total degree of at most* $k \le n$ *and at most* $\binom{D+k}{k}$ *monomials.*

Proof. The total degree bound for p is a consequence of Algorithm 2; any polynomial in D variables with a total degree of at most k has at most $\binom{D+k}{k}$ monomials [14].

---

[1] Another basis for $\mathbb{Q}[X_1, \ldots, X_n]^{S_n}$ are, e.g., complete symmetric functions.

Lemma 2. *Let* $f \in \mathbb{Q}[X_1, \ldots, X_n]^{S_n}$ *with a total degree* D*. Then the power sum algorithm needs at most* $O\left(\binom{n+D}{n} n^n\right)$ *multiplications of monomials to compute the representation.*

Proof. The number of terms with a total degree of at most D is $\binom{n+D}{n}$, and any term has to be reduced at most once. Each power sum consists of n monomials and we have to multiply at most n power sums during a single loop-run.

The number of monomial multiplications for the power sum algorithm is — from the point of view of worst case complexity — *polynomial* in the total degree D of the given polynomial $f \in \mathbb{Q}[X_1, \ldots, X_n]^{S_n}$. In a single loop-run, the amount of work to do depends only on the number of variables n.

4 The Fast Rewriting Technique

The basic idea for a fast reduction of symmetric polynomials is to use Algorithm 2 instead of Algorithm 1 and perform some pre- and post-processing steps. This notion is carried out in this section.

Lemma 3. *The following relations between the elementary symmetric polynomials and the power sum polynomials are satisfied:*

1. $s_j - \sigma_1 s_{j-1} + \sigma_2 s_{j-2} - \ldots + (-1)^{j-1}\sigma_{j-1}s_1 + (-1)^j \sigma_j j = 0$ *for* $1 \le j \le n$.
2. $s_j - \sigma_1 s_{j-1} + \sigma_2 s_{j-2} - \ldots + (-1)^{n-1}\sigma_{n-1}s_{j-n+1} + (-1)^n \sigma_n s_{j-n} = 0$ *for* $j > n$.

Proof. The relations in (1) are the *Newton's identities* and a proof can be found in [4]. For the proof of (2), we observe that

$$\sigma_k s_{j-k} = orbit_{S_n}(X_1^{j-k+1} X_2 \ldots X_k) + orbit_{S_n}(X_1^{j-k} X_2 \ldots X_{k+1})$$

for $1 \le k \le n-1$ and $\sigma_n s_{j-n} = orbit_{S_n}(X_1^{j-n+1} X_2 \ldots X_n)$. Hence, the alternating sum $\sum_{k=1}^{n}(-1)^k \sigma_k s_{j-k}$ is equal to $-s_j$, and therefore (2) is valid.

4.1 Using Only Post-Processing

The change of the representation from the power sum algorithm to the representation of the classical algorithm can be achieved as follows:

Algorithm 3. **Post-Processing Step**

1. *INPUT* $f = p(s_1, \ldots, s_D)$ *with* $p \in \mathbb{Q}[X_1, \ldots, X_D]$;
2. *Consider* p *as elem. of* $\mathbb{Q}[\sigma_1, \ldots, \sigma_n][X_1, \ldots, X_D]$;
3. *FOR* $j := D$ *TO* $n+1$ *BY* -1 *DO*   /* *Lemma 3 (2)* */
    $p(s_1, \ldots, s_{j-1}) := p(s_1, \ldots, s_{j-1}, \sigma_1 s_{j-1} - \ldots - (-1)^n \sigma_n s_{j-n})$;

4. FOR $j := n$ TO 1 BY -1 DO /* Lemma 3 (1) */
  $p(s_1,\ldots,s_{j-1}) := p(s_1,\ldots,s_{j-1},\sigma_1 s_{j-1} - \ldots - (-1)^j \sigma_j j)$;
5. OUTPUT $f = p(\sigma_1,\ldots,\sigma_n)$ with $p \in \mathbb{Q}[X_1,\ldots,X_n]$;

Termination of Algorithm 3 is obvious. Correctness is due to Lemma 3 and the fact that the representation computed by the classical algorithm is unique.

We have implemented the classical algorithm and the fast algorithm, i.e. the power sum algorithm and the post-processing step in the invariant package of MAS [6, 8]. The distributive polynomial representation over the rational numbers was used. All run-times of the benchmark test in Table 1 were obtained on a SUN Ultra-Sparc running Solaris 2.5.1. Each entry in the table shows the run-time in seconds for (i) the classical algorithm and (ii) the fast algorithm using power sums and the post-processing step for the reduction of all symmetric orbits in n variables with a total degree D for different values of n and D. In addition, the table contains (iii) the percentage how often the fast algorithm was superior to the classical algorithm.

Tot.	Class. Alg. /		Pow. Sum Alg. + Post-Proc. /				%	
Deg. D	$n = 5$		$n = 6$		$n = 7$		$n = 8$	
5	1.6	1.2 58	4.4	2.8 72	13.0	8.3 72	40.9	29.6 72
6	4.3	2.1 70	14.3	6.3 64	43.0	16.6 73	125.8	54.6 73
7	9.8	3.9 62	37.7	10.8 65	125.5	33.5 60	374.1	100.6 67
8	25.3	7.9 62	112.6	22.2 65	416.8	63.9 77		
9	59.4	15.8 61	304.8	43.8 66	1265.1	131.8 72		
10	139.2	32.5 57	827.8	92.0 66	3858.7	269.7 69		
12	650.7	151.7 56	4989.8	413.0 66				
14	2658.6	708.7 50						

Table 1. Empirical run-time results [sec.] (using only post-processing)

Additional experiment information is given in Table 2: The entries indicate the absolute number of all symmetric orbits in n variables with a total degree D for different values of n and D, which where used in our experiments. For example, for $n = D = 5$ we have the 7 symmetric orbits $orbit_{S_5}(X_1^5)$, $orbit_{S_5}(X_1^4 X_2)$, $orbit_{S_5}(X_1^3 X_2^2)$, $orbit_{S_5}(X_1^3 X_2 X_3)$, $orbit_{S_5}(X_1^2 X_2 X_3 X_4)$, $orbit_{S_5}(X_1^2 X_2^2 X_3)$, and $orbit_{S_5}(X_1 X_2 X_3 X_4 X_5)$ (cf. Table 2). The total time for the reduction of these polynomials was 1.6 sec. with the classical algorithm and 1.2 sec. with the fast algorithm. In 58% of the cases the fast algorithm was superior to the classical algorithm (cf. Table 1).

The results shown in Table 1 are promising: In all cases, the fast algorithm was superior to the classical algorithm in more than 50% of the reduction tasks.[2] And when looking at the average run-times of all reductions the fast version shows speed-ups of up to 12 and more compared to the classical algorithm.

---

[2] Note that the fast version has no chance to beat the classical algorithm when computing representations for polynomials like, e.g., $orbit_{S_n}(X_1^e \ldots X_n^e) = \sigma_n^e$.

Tot.	Number of Symmetric Orbits			
Deg. D	$n = 5$	$n = 6$	$n = 7$	$n = 8$
5	7	7	7	7
6	10	11	11	11
7	13	14	15	15
8	18	20	21	
9	23	26	28	
10	30	35	38	
12	47	58		
14	70			

Table 2. Additional experiment information

We omit here to state and prove a bound for the number of monomial multiplications, because it doesn't seem to lead to a theoretical result which confirms our run-time experiments. A similar phenomenon occurred in the work of Lauer and Loos [9, 10]: They could show that a special representation of symmetric polynomials speed-up the classical algorithm but the theoretical analysis of their approach did not confirm this. The next section shows that we can handle this problem by introducing a pre-processing step and a modified post-processing step.

4.2 Using Pre- and Post-Processing

For a fixed number of variables and a given upper total degree D, we can simply pre-process the power sums s_i, $i = 1, \ldots, D$ and compute their representation as polynomials in the elementary symmetric polynomials by using Lemma 3 or the classical algorithm. Once this is done, we can come up with the following result for a modified post-processing step.

Lemma 4. *Let $f = p(s_1, \ldots, s_D) \in \mathbb{Q}[X_1, \ldots, X_n]^{S_n}$ be a representation computed by the power sum algorithm, and let $s_1 = p_1(\sigma_1, \ldots, \sigma_n)$, ..., $s_D = p_D(\sigma_1, \ldots, \sigma_n)$. Then we need at most $O\!\left(nD \binom{n+D}{n}^{n+1}\right)$ multiplications of monomials to compute the representation of Algorithm 1.*

Proof. The change of the representation from the power sum algorithm to the representation of the classical algorithm can be achieved as follows:

Algorithm 4.   **Modified Post-Processing Step**

1. *INPUT $f = p(s_1, \ldots, s_D)$ with $p \in \mathbb{Q}[X_1, \ldots, X_D]$;*
   $s_1 = p_1(\sigma_1, \ldots, \sigma_n)$, ..., $s_D = p_D(\sigma_1, \ldots, \sigma_n)$;
2. *Consider p as elem. of $\mathbb{Q}[\sigma_1, \ldots, \sigma_n][X_1, \ldots, X_D]$;*
3. *FOR $j := 1$ TO D DO $p(s_{j+1}, \ldots, s_D) := p(p_j(\sigma_1, \ldots, \sigma_n), s_{j+1}, \ldots, s_D)$;*
4. *OUTPUT $f = p(\sigma_1, \ldots, \sigma_n)$ with $p \in \mathbb{Q}[X_1, \ldots, X_n]$;*

Termination and correctness of Algorithm 4 is obvious. Any p_j has at most $\binom{n+D}{n}$ monomials (cf. Corollary 1) and any term in p has a total degree of at most n (cf. Corollary 2). In the worst case, we have to compute $p_j^2, \ldots, p_j^n$ during step j of the loop in Algorithm 4, which needs at most $O(n\binom{n+D}{n}^n)$ multiplications of monomials. Furthermore, we have to multiply at most $\binom{n+D}{n}$ monomials of p (cf. Corollary 2) with at most $O(\binom{n+D}{n}^n)$ monomials of p_j^k for some $1 \le k \le n$. Putting all together and considering the number of loop-runs D implies at most $O(nD\binom{n+D}{n}^{n+1})$ multiplications of monomials during the execution of Algorithm 4.

The number of monomial multiplications for the modified post-processing step is — from the point of view of worst case complexity — polynomial in the total degree D of the given polynomial $f \in \mathbb{Q}[X_1, \ldots, X_n]^{S_n}$. Therefore, the fast version consisting of Algorithm 2 followed by Algorithm 4 has a worst case complexity which is polynomial in D, and so also polynomial in the maximal variable degree d of f. This is a significant improvement compared to the classical algorithm, which has an exponential worst case complexity in d w.r.t. the number of monomial multiplications.

The run-times presented in Table 3 confirm this theoretical analysis: The setup for this experiment was precisely the same as for Table 1 (see also Table 2). We have assumed that the representations of $s_1, \ldots, s_D$ have been computed before in a pre-processing step. Each entry in the table shows the run-time in seconds for (i) the classical algorithm and (ii) the fast algorithm using power sums and the modified post-processing step for the reduction of all symmetric orbits in n variables with a total degree D for different values of n and D. In addition, the table contains (iii) the percentage how often the fast algorithm was superior to the classical algorithm.

Tot. Deg. D	Class. Alg. / Pow. Sum Alg. + Mod. Post-Proc. / %			
	$n = 5$	$n = 6$	$n = 7$	$n = 8$
5	1.6 1.0 72	4.4 2.7 72	13.0 7.8 72	40.9 28.8 72
6	4.3 1.8 70	14.3 5.7 64	43.0 15.1 73	125.8 49.6 82
7	9.8 3.1 62	37.7 9.4 65	125.5 29.4 74	374.1 85.7 74
8	25.3 5.2 67	112.6 17.7 70	416.8 53.1 77	
9	59.4 8.3 70	304.8 31.0 74	1265.1 104.7 75	
10	139.2 13.0 74	827.8 55.4 72	3858.7 198.1 72	
12	650.7 30.0 75	4989.8 151.3 75		
14	2658.6 64.4 76			

Table 3. Empirical run-time results [sec.] (using pre- and post-processing)

The results shown in Table 3 are again very promising: In all cases, the fast algorithm was superior to the classical algorithm in more than 62% of the reduction tasks. And when looking at the average run-times of all reductions, the fast version leads to speed-ups of up to 15 and more compared to the classical algorithm. Furthermore, this fast version shows speed-ups in all cases compared to the previous fast version.

4.3 Comparison with Existing Algorithms

In order to obtain an empirical comparison with existing fast reduction methods we have implemented the *Lauer-Loos algorithm* in MAS. This algorithm is known to be very fast [9, table 3]. It works precisely in the same way as the classical algorithm, but uses a special, space economical component representation for symmetric polynomials together with its corresponding arithmetic during the reduction process. Simply speaking, the more variables are involved in the reduction problem, the larger the benefit due to the component representation and arithmetic, and the more powerful and swift is this algorithm.

Figure 1 displays the results of our benchmark experiments: Each run-time (y-axis, logarithmic scaling) is, once again, the result of the reduction of all symmetric orbits in n variables with a total degree D for different values of D (x-axis) and n (see also Table 2). There are four pairs of curves (Lauer-Loos

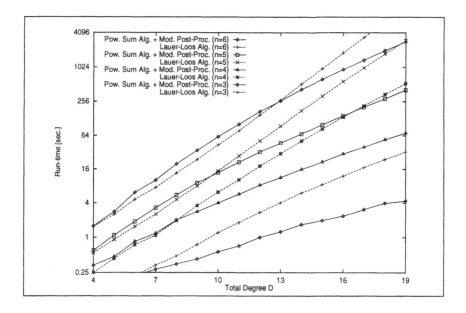

Fig. 1. Empirical run-time comparison [sec.]

algorithm (dotted line); power sum algorithm and modified post-processing step (solid line)) showing from above the run-time results for $n = 6$, 5, 4, and 3 variables. We see that the empirical asymptotic behaviour of the fast algorithm using power sums and the modified post-processing step is better than the Lauer-Loos algorithm in our average case run-time experiments. Whenever the total degree D is greater than approximately $2n$, our method *out-performs* the Lauer-Loos algorithm. For smaller and medium sized problems ($D < 2n$) both algorithms perform almost the same.

It would be of interest to know, whether a combination of both our *fast rewriting technique* and the *component representation and arithmetic* of Lauer and Loos is possible, and if so, if it leads to additional speed-ups during the reduction of symmetric functions.

5   Conclusion

We have seen by which means a *fast rewriting of symmetric polynomials* can be achieved. Our methods can compete with other well-known fast rewriting techniques without making use of a special data structure and polynomial arithmetic for symmetric polynomials.

A generalization of the classical algorithm to arbitrary permutations groups G [5] leads to a representation of any $f \in \mathbb{Q}[X_1, \ldots, X_n]^G$ as a finite linear combination of special G-invariant orbits with symmetric polynomials as coefficients, i.e.

$$f = \sum_{t \text{ spec.}} p_t(\sigma_1, \ldots, \sigma_n) \cdot orbit_G(t)$$

with $p_t \in \mathbb{Q}[X_1, \ldots, X_n]$. It should not be too difficult — at least for permutation groups with a small index $|S_n|/|G|$ — to speed-up this general algorithm by using a dynamic power sum representation for the coefficients of the linear combination. The Lauer-Loos algorithm for the generalization is already described and evaluated in [7].

Acknowledgments

This work was supported by the Deutsche Forschungsgemeinschaft (DFG), grant Go 789/2-1. It was done at the International Computer Science Institute (ICSI), 1947 Center Street, Berkeley, CA 94704, USA. The author would like to thank Dr. Andreas Weber (Tübingen) and several anonymous referees for their comments and remarks.

References

1. Becker, T., Weispfenning, V., in Cooperation with Kredel, H. (1993). Gröbner Bases: A Computational Approach to Commutative Algebra. Springer
2. Bratley, P., McKay, J. (1967, 1968). Algorithm 305, 450: Symmetric Polynomials. Communications of the ACM, Vol. 10, 450, Vol. 11, 272

3. Eisenbud, D. (1994). Commutative Algebra with a View Towards Algebraic Geometry. Springer

4. Edwards, H. M. (1993). Galois Theory. Springer

5. Göbel, M. (1995). Computing Bases for Permutation-Invariant Polynomials. Journal of Symbolic Computation 19, 285–291

6. Göbel, M. (1997). The Invariant Package of MAS. In: Comon, H., (ed.), Rewriting Techniques and Applications, 8th Intl. Conf., RTA-97, volume 1232 of *LNCS*, Springer, 327–330

7. Göbel, M. (1998). On the Reduction of G-invariant Polynomials for Arbitrary Permutation Groups G. In: Bronstein, M., Grabmeier, J., Weispfenning, V., (eds.), Symbolic Rewriting Techniques, Progress in Computer Science and Applied Logic (PCS 15), 71–92, Monte Verita, Schwitzerland. Birkhäuser

8. Kredel, H. (1990). MAS: Modula-2 Algebra System. In: Gerdt, V. P., Rostovtsev, V. A., Shirkov, D. V., (eds.), IV International Conference on Computer Algebra in Physical Research. World Scientific Publishing Co., Singapore, 31–34

9. Lauer, E. (1976a). Algorithms for Symmetrical Polynomials. In: Jenks, R. D., (ed.), ACM Symposium on Symbolic and Algebraic Computation. ACM Press, New York, 242–247

10. Lauer, E. (1976b). Algorithmen für symmetrische Polynome. Diplomarbeit. Universität Kaiserslautern

11. Loos, R. (1981). Term Reduction Systems and Algebraic Algorithms. In: Siekmann, J. H. (ed.), GWAI-81, German Workshop on Artificial Intelligence, Bad Honnef. Informatik-Fachberichte, Herausgegeben von W. Brauer im Auftrag der Gesellschaft für Informatik (GI), 47. Springer, 214–234

12. Netto, E. (1896). Vorlesungen über Algebra. Leipzig, Teubner

13. Macdonald, I. G. (1995). Symmetric Functions and Hall Polynomials. Oxford Mathematical Monographs. Oxford University Press.

14. Sturmfels, B. (1993). Algorithms in Invariant Theory. Springer

15. van der Waerden, B. L. (1971). Algebra I. Springer

On Implementation of Tree Synchronized Languages

Frédéric Saubion and Igor Stéphan

LERIA, Université d'Angers
2 Bd Lavoisier
F-49045 Angers Cedex 01
{Frederic.Saubion,Igor.Stephan} @univ-angers.fr

Abstract. Tree languages have been extensively studied and have many applications related to the rewriting framework such as order sorted specifications, higher order matching or unification. In this paper, we focus on the implementation of such languages and, inspired by the Definite Clause Grammars that allows to write word grammars as Horn clauses in a Prolog environment, we propose to build a similar framework for particular tree languages (TTSG) which introduces a notion of synchronization between production rules. Our main idea is to define a proof theoretical semantics for grammars and thus to change from syntactical tree manipulations to logical deduction. This is achieved by a sequent calculus proof system which can be refined and translated into Prolog Horn clauses. This work provides a scheme to build goal directed procedures for the recognition of tree languages.

Keywords : Tree Synchronized Grammars - Linear Logic - Proof Systems - Prolog Implementation

1 Introduction

Tree languages have been extensively studied and have many applications related to the rewriting framework such as order sorted specifications, higher order matching or unification, term schematization [3, 10, 8]. These languages can be handled either from the generation (i.e. grammars) or from the recognition point of view (i.e. automata). In this paper, we focus on the implementation of such languages. Of course, tree automata [3] appear to be easily implementable tools for recognition, but, inspired by the implementation of Definite Clause Grammars [2] that allows to write word grammars as Horn clauses in a Prolog environment, our aim is to propose a similar approach for particular tree languages. This method is especially interesting concerning languages for which the notion of automaton does not exist.

This paper does not address the implementation of all types of tree grammars but focuses on particular tree languages that have been developed to handle E-unification problems. E-unification [15] is known to be undecidable in general, but some decidable classes can be characterized by using tree languages [10, 12]. Moreover, this approach has been extended to disunification problems [11] (presented at RTA'98).

Basically, the idea is to describe solutions of E-unification problems by tree grammars instead of usual first order term substitutions. This allows to get finite representations (schematizations) of infinite sets of terms. But, the remaining point is to provide a framework to use these solution languages in practice.

Tree synchronized grammars $TTSG$[1] have been introduced in [10] to extend decidability results for E-unification. Their main particularity is their notion of synchronization between production rules (i.e. some productions can be applied at the same time). For this reason, they can describe context sensitive languages such as $\{f(s^n(z), s^{2n}(z)) \mid n \in \mathbb{N}\}$: this simple example will be developed along this paper.

Example 1 *We want to define the language corresponding to $\{f(s^n(z), s^{2n}(z)) \mid n \in \mathbb{N}\}$ which is obviously not a regular language. The corresponding $TTSG$ consists of the following production rules:*

$$R_0 : I \Rightarrow f(X, Y)$$
$$R_1 : X \Rightarrow z$$
$$R_2 : Y \Rightarrow z$$
$$P_1 : \{R_3 : X \Rightarrow s(X), R_4 : Y \Rightarrow s(s(Y))\}$$

where X, Y are non terminal symbols and f, s, z terminal symbols. $P_1 : \{R_3, R_4\}$ denotes that, using the pack of production P_1, R_3 and R_4 have to be applied at the same time (synchronization). The axiom[2] is $(I, \#)$. One possible derivation of this grammar is thus (where a is a new symbol):

$$(I, \#) \Rightarrow_{R_0} f((X, \#), (Y, \#)) \Rightarrow_{P_1} f(s((X, a)), s(s((Y, a))))$$

$$\Rightarrow_{R_1} f(s(z), s(s(Y, a))) \Rightarrow_{R_2} f(s(z), s(s(z)))$$

The second step of this derivation clearly illustrates the notion of synchronized production rules.

It should be noticed that grammar derivations can only be used to generate elements of the language but not for the recognition of such elements. Moreover, there does not exist (as far as we know) any notion of automaton related to

---

[1] Formally, $TTSG$ means tree tuple synchronized grammars since they are defined over tuples of trees. In this paper, we have decided to consider only trees instead of tuple of trees. The two points of view are indeed equivalent and this will be discussed in section 2.1.

[2] Note that, the axiom and the derivation are in fact a bit sophisticated due to the control of synchronizations. Basically, counters are associated to non terminal symbols to insure the integrity of the application of synchronized productions. Intuitively, only non terminals having the same level of synchronization can be used for the application of a pack of productions. This mechanism will be explained in section 2.1 and we refer the reader to [10] for further details.

this class of tree grammars and thus there is no practical framework for the recognition of such languages. For this reason, we propose an approach that can be viewed as a general method which could be extended to various tree languages. Another important point is that our translation allows us to transform the initial grammar problem into an applicative known framework (based on Forum and Prolog).

The key idea of this paper is to define a proof framework to describe the behavior of tree grammars and therefore to provide a proof theoretical semantics for grammar derivations. In this context, we change from syntactic generation of tree to logical deduction (i.e. productions rules are translated into logical formulas and grammar derivations into logical inferences). Then, the membership test, for a language described by a grammar, just consists in proving a particular formula in a sequent calculus using a proof system. Now, we briefly describe the method.

First, production rules are translated into logical formulas built over linear logic [5] to handle the problem of synchronization (this idea has been investigated in [6]). Then, a sequent calculus proof system, inspired by the proof system of [14], allows us to translate $TTSG$ derivations into proof searches.

Example 1 (following)

We translate the previous $TTSG$ into the following set of logical formulas:

$$F_0 : \forall c(\forall X \forall Y(i(f(X,Y),c) \circ\!\!-x(X,c) \otimes y(Y,c)))$$
$$F_1 : \forall c(x(z,c) \circ\!\!-1)$$
$$F_2 : \forall c(y(z,c) \circ\!\!-1)$$
$$F_3 : \forall c \exists nc((\forall X(x(s(X),c) \circ\!\!-x(X,nc))) \otimes (\forall Y(y(s(s(Y)),c) \circ\!\!-y(Y,nc))))$$

where x,y are predicate symbols, X,Y variables and s,z function symbols. Actually, terminal symbols of the initial grammar are translated into predicate symbols and logical variables. The connectors $\circ\!\!-$ and $\otimes$ respectively denote the linear implication and multiplicative conjunction (see [5]). The counters are handled thanks to quantified variables (c and nc). A pack of synchronized productions is transformed into a conjunction of formulas corresponding to the rules of the pack. A sequent calculus presented in section 3.2, defines the semantics of such a set of formulas. Thus, we are able to prove that the goal $f(s(z),s(s(z)))$ can be deduced from the set of formulas F_0, F_1, F_2 and F_3 (this means that this term belongs to the language).

Note that this example could be transformed into linear logic formulas without counters. We keep this very simple example to give the intuition of our approach but more complex synchronized grammars issued from unification problems [10] cannot be treated without these counters variables.

The equivalence between the notion of grammar computation and our notion of proof is established by a correctness and completeness result. The last part of our work consists in refining this initial proof system in order to introduce notion of strategy in the proof search and thus to get a goal directed procedure. At this step, both transformation function and logical inference system can be easily translated into Prolog Horn clauses.

This paper is organized as follows : in section 2, we recall some basic definitions and notions related to tree synchronized grammars, first order logic and linear logic formalisms. The proof theoretical semantics for $TTSG$ is introduced in section 3. The transformation of grammar productions into linear logic formulas is defined as well as its associated sequent calculus proof system. Section 4 describes how this proof system can be refined to get a more operational and easily implementable equivalent system. The implementation of this system is discussed in section 5 illustrated by an execution example.

2 Preliminaries

In this paper, we introduce different notions related to tree grammars and linear logic proof systems. We refer the reader to [3, 10, 5, 14, 4] for more details.

2.1 Tree synchronized grammars

Tree grammars are indeed similar to word grammars except that basic objects are trees. Note that we slightly differ from the initial definition given in [10] where the axiom of the grammar is given as a tuple of pairs of non terminals and counters. But obviously, tuples of size n can be constructed using a particular terminal symbol $(\_, \_, ..., \_)$ with arity n. Thus, the two points of view can be considered as equivalent. For the sake of history, we keep the name $TTSG$ but we adapt the different definitions.

Let $\mathcal{L}$ be a finite set of symbols (a signature), $T(\mathcal{L})$ denotes the first-order algebra of ground terms over $\mathcal{L}$. $\mathcal{L}$ is partitioned in two parts : the set $\mathcal{F}$ of terminal symbols, and the set $\mathcal{N}$ of non terminal symbols. Upper-case letters denote elements of $\mathcal{N}$. $t|u$ denotes the subterm of t at occurrence u and $t[u \leftarrow v]$ denotes the term constructed by replacing in t the subterm $t|u$ by v. $C(X_1, ..., X_n)$ denotes a term with occurences $\{1...n\}$ such that $C(X_1, ..., X_n)|i = X_i$. C will be called a context. We first define the notion of productions for $TTSG$. We require that $\mathcal{F} \cap \mathcal{N} = \emptyset$ and that each element of $\mathcal{F} \cup \mathcal{N}$ has a fixed arity.

Definition 1. Productions

A production is a rule of the form $X \Rightarrow t$ where $X \in \mathcal{N}$ and $t \in T(\mathcal{L})$. A pack of productions is a set of productions denoted $\{X_1 \Rightarrow t_1, \ldots, X_n \Rightarrow t_n\}$.

- *When the pack is a singleton of the form $\{X_1 \Rightarrow C(Y_1, \ldots, Y_n)\}$ where C is a context of terminal symbols and $Y_1, \ldots, Y_n$ non terminals. The production is said free, and is written more simply $X_1 \Rightarrow C(Y_1, \ldots, Y_n)$.*
- *When the pack is of the form $\{X_1 \Rightarrow Y_1, \ldots, X_n \Rightarrow Y_n\}$ where $Y_1, \ldots, Y_n$ are terminals or non terminals. The productions of the pack are said synchronized.*

We can then define the notion of $TTSG$.

Definition 2. *$TTSG$*

A $TTSG$ is defined by a 4-tuple $(\mathcal{F}, \mathcal{N}, PP, TI)$ where

- *$\mathcal{F}$ is the set of terminals,*
- *$\mathcal{N}$ is the finite set of non terminals,*
- *PP is a finite set of packs of productions,*
- *TI is the axiom of the $TTSG$ denoted $(I, \#)$ where I is a non terminal and $\#$ a new symbol added to the signature.*

Definition 3. Computations of a $TTSG$

The set of computations of a $TTSG$
$Gr = (\mathcal{F}, \mathcal{N}, PP, TI)$, denoted $Comp(Gr)$, is the smallest set defined by:

- *TI is in $Comp(Gr)$,*
- *if t is in $Comp(Gr)$, $t|_u = (X, c)$ and the free production $X \Rightarrow C(Y_1, \ldots, Y_n)$ is in PP then $t[u \leftarrow C((Y_1, c), \ldots, (Y_n, c))]$ is in $Comp(Gr)$,*
- *if t is in $Comp(Gr)$ and there exists n pairwise different occurrences $u_1, \ldots, u_n$ of t such that $\forall i \in [1, n]$ $t|_{u_i} = (X_i, c_i)$ and $c_i = a$ and the pack of productions $\{X_1 \Rightarrow Y_1, \ldots, X_n \Rightarrow Y_n\} \in PP$, then $t[u_1 \leftarrow (Y_1, b)] \ldots [u_n \leftarrow (Y_n, b)]$ (where b is a new symbol) is in $Comp(Gr)$.*

The symbol $\Rightarrow$ denoting also the above two deduction steps, a derivation of Gr is a sequence of computations $TI \Rightarrow t_1 \Rightarrow \ldots \Rightarrow t_n$.

As mentioned in the introduction, counters are introduced to control the application of the synchronized production rules. The previous definition imposes that only non terminals having the same control symbol can be used in a synchronized production. It should be noticed that $TTSG$ were originally defined using tuple of counters. Here, as we already did, concerning tuples of terms in the definition of the axiom, we consider a single counter to control synchronization in order to simplify the presentation of the basic concepts.

Definition 4. Recognized language

The language recognized by a $TTSG$ Gr, denoted $Rec(Gr)$, is the set of tree composed of terminal symbols $Comp(Gr) \cap T(\mathcal{F})$.

2.2 Linear logic and sequent calculus

We recall here some basic notations and notions related to first order and linear logics [5, 4], and proof systems in sequent calculus [14].
Let us consider a first order logic signature Σ with V a countable set of variables, $\Sigma_{\mathcal{F}}$ a finite set of function symbols with fixed arity and Σ_N a finite set of predicate symbols. $T(\Sigma_{\mathcal{F}}, V)$ denotes the first order term algebra built over

$\Sigma_{\mathcal{F}}$ and V and $\mathcal{T}(\Sigma_{\mathcal{F}})$ denotes the set of ground terms. Atoms are, as usual, built over predicates symbols and terms. A substitution is a mapping from V to $T(\Sigma_{\mathcal{F}}, V)$ which is extended to $T(\Sigma_{\mathcal{F}}, V)$. A substitution assigning t to a variable x will be denoted $\{x \leftarrow t\}$. We introduce some linear logic[3] notations : ∘– denotes the linear implication and $\otimes$ denotes the multiplicative conjunction (see [5, 4] for the precise definitions of these connectives). A formula $A \circ\!\!-\!B_1 \otimes ... \otimes B_n$ will be called a clause. The set $\Sigma_{formula}$ is the set of Σ-formulas built using atoms and the logic connectives.

A sequent will be written as $\Sigma : \Delta \to G$ where Σ is the signature, Δ a multiset of Σ-formulas and G a Σ-formula. A proof system consists of a set of inference rules between sequents. An inference rule is presented here as :

$$\frac{\Sigma : \Delta \to G}{\Sigma' : \Delta' \to G'}$$

We use the classical notion of proof (i.e. there is a proof of a sequent in a proof system if this sequent is the root of a proof tree constructed using inference rules with empty leaves).

3 From *TTSG* Computation to Proof Search

In this section, we define the translation of grammar production rules into linear logic formulas. Then, designing a particular proof system, we show that usual grammar computations can be reduced to proof searches in sequent calculus using this system.

3.1 Transforming *TTSG* into Linear Logic Formulas

Given a *TTSG* $(\mathcal{F}, \mathcal{N}, PP, TI)$, the set PP of production rules is partitioned into PP_{free} the set of free production rules and PP_{sync} the set of packs of synchronized production rules. We define now a transformation function Ψ which is decomposed into two different mappings (corresponding respectively to the transformation of free and synchronized production rules).

Definition 5. Transformation function Ψ

Let $\sigma_{\mathcal{N}} : \mathcal{N} \to \Sigma_{\mathcal{N}}$ be the mapping that translates every non terminal symbol into a predicate symbol (to simplify, we will write $\sigma_{\mathcal{N}}(N) = n$). Let $\sigma_{\mathcal{V}} : \mathcal{N} \to \mathcal{V}$ be the mapping that transforms every non terminal symbol into a logical variable (we will write $\sigma_{\mathcal{V}}(N) = N$). For the sake of readability, universal quantifications have been omitted.

---

[3] Intuitively, the key idea of linear logic we use here is that, when performing logical inferences, some hypothesis can be consumed. This means that, along a proof, some formulas are persistent (as in usual first order logic) but some formulas can only be used once.

Let $\sigma_{PP_{free}} : PP_{free} \to \Sigma_{formula}$ be the mapping that translate every free production rule into a Σ-formula and $\sigma_{PP_{sync}} : PP_{sync} \to \Sigma_{formula}$ the mapping that translates every pack of synchronized production rules into a Σ-formula.

Free productions

Let $N \to g(N_1, \ldots, N_p)$ and $N \to t$ in PP_{free}.

$$\sigma_{PP_{free}}(N \to g(N_1, \ldots, N_p))$$
$$= n(g(N_1, \ldots, N_p), c) \circ\!\!-n_1(N_1, c) \otimes \ldots \otimes n_p(N_p, c)$$

$$\sigma_{PP_{free}}(N \to t) = n(t, c) \circ\!\!-\mathbf{1}$$

Synchronized Productions

Let $S = \{R \mid P \in PP_{sync} \wedge R \in P\}$ and $\{R_1; \ldots; R_n\} \in PP_{sync}$:
$$\sigma_{PP_{sync}}(\{R_1, \ldots, R_n\}) = \exists nc(\sigma_S(R_1, c, nc) \otimes \ldots \otimes \sigma_S(R_n, c, nc))$$

where $\sigma_S : S \times C \times C \to \Sigma_{formula}$ is defined[4], for $N \to g(N_1, \ldots, N_p)$ and $N \to t$ in S, as :

$$\sigma_S(N \to g(N_1, \ldots, N_p), c, nc)$$
$$= n(g(N_1, \ldots, N_p), c) \circ\!\!-n_1(N_1, nc) \otimes \ldots \otimes n_p(N_p, nc)$$

$$\sigma_S(N \to t, c, nc) = n(t, c) \circ\!\!-\mathbf{1}$$

Let $\Sigma_\Psi = \Sigma_\mathcal{F} \cup \Sigma_\mathcal{N}$ and $\Psi((\mathcal{F}, \mathcal{N}, PP, TI)) = \sigma_{PP_{free}}(PP_{free}) \cup \sigma_{PP_{sync}}(PP_{sync})$ (in the following, Ψ will denote a set of formulas obtained from a $TTSG$ and will be called a $TTSG$ program). At this step, a particular sequent calculus defines the semantics of Ψ.

3.2 The Proof System FG

In this section, we define a sequent calculus proof system inspired by the system Forum of D. Miller [14] (as a consequence, our system FG is proved correct w.r.t. Forum (proof can be found in [16])). This initial system defines the basic proof theoretical semantics of a $TTSG$ program Ψ.

Our proof system FG (**F**orum for **G**rammars) is defined by the following inference rules.

Definition 6. The FG system

[1] $$\frac{}{\Sigma :\to \mathbf{1}}$$

---

[4] The set C denotes a set of counters (i.e. new symbols which are not in the current signature).

[*Sync*]

$$\frac{\Sigma, \alpha : C_1\theta, \ldots, C_r\theta, \Delta \to G}{\Sigma : \Delta \to G}$$

where $\forall c \exists nc(C_1 \otimes \ldots \otimes C_r) \in \Psi$, $\theta = \{c \leftarrow \beta, nc \leftarrow \alpha\}$, $\beta \in \Sigma$, $\alpha \notin \Sigma$.

[*Back* !]

$$\frac{\Sigma : \Delta_1 \to A_1\sigma \ldots \Sigma : \Delta_p \to A_p\sigma}{\Sigma : \Delta_1, \ldots, \Delta_p \to G}$$

where $C = (H \circ\!\!-\, A_1 \otimes \ldots \otimes A_p) \in \Psi$ and $H\sigma = G$.

[*Back* ?]

$$\frac{\Sigma : \Delta_1 \to A_1\sigma \ldots \Sigma : \Delta_p \to A_p\sigma}{\Sigma : (H \circ\!\!-\, A_1 \otimes \ldots \otimes A_p), \Delta_1, \ldots, \Delta_p \to G}$$

where $H\sigma = G$.

Comments : It should be noticed that we distinguish free production rules and synchronized production rules (see the definition of the transformation function Ψ in section 3.1). Clearly free productions appear as clauses $H \circ\!\!-\, A_1 \otimes \ldots \otimes A_n$ and packs of synchronized productions appear as formulas $\forall c \exists nc(C_1 \otimes \ldots \otimes C_n)$ where the C_i's are $H \circ\!\!-\, A_1 \otimes \ldots \otimes A_n$ and are called linear clauses (in the following linear clause will always refer to a clause generated by synchronization rules). Clauses corresponding to free productions in Ψ are persistent along the proof and are used in the inference [*Back* !] (this corresponds to a step of the grammar derivation using this free production). The treatment of a pack of synchronization is performed thanks to the rules [*Sync*] and [*Back* ?]. The first step consists in generating the formula corresponding to this pack in Δ (rule [*Sync*]). The control is insured by the instantiation of the counter variables with new symbols added to the signature Σ in rule [*Sync*]. Since Δ is the linear logic part of our system, the linear clauses will be consumed when used by rule [*Back* ?] (in the philosophy of linear logic). This insures the integrity of the synchronization and the control of the simultaneous application of the different productions of this pack. [*Back*?] and [*Back*!] lead to branching in the search tree.

The correctness and completeness of FG w.r.t. the notion of $TTSG$ computation is insured by the following theorem.

Theorem 1. Correctness and completeness of FG

Given a $TTSG$ $(\mathcal{F}, \mathcal{N}, PP, (I, \#))$, the corresponding $TTSG$ program Ψ and $t \in T(\Sigma_{\mathcal{F}})$, $(\Sigma_\Psi, \# :\to \sigma_{\mathcal{N}}(I)(t, \#))$ has a proof in FG w.r.t. the $TTSG$ program Ψ if and only if $((I, \#) \overset{*}{\Rightarrow} t)$.

Proof. Due to a lack of space, we refer the reader to [16] for the proof. □

Example 1 (following)

We build the derivation tree for the TTSG program Ψ (Σ_Ψ is omitted in the left hand side of the sequents of the following proof in order to simplify the notation, we only mention the new symbols added to the signature along the proof and universal quantifiers are omitted in the sequents) :

$\{ F_0 : \forall c (\forall X \forall Y (i(f(X,Y),c) \circ\!\!-x(X,c) \otimes y(Y,c))),$

$\quad F_1 : \forall c(x(z,c) \circ\!\!-1),$

$\quad F_2 : \forall c(y(z,c) \circ\!\!-1),$

$\quad F_3 : \forall c \exists nc((\forall X(x(s(X),c) \circ\!\!-x(X,nc))) \otimes (\forall Y(y(s(s(Y)),c) \circ\!\!-y(Y,nc))))\}$

$$
\cfrac{
\cfrac{
\delta_1 \qquad\qquad\qquad\qquad\qquad\qquad \delta_2
}{
\#,\beta : x(s(X),\#) \circ\!\!-x(X,\beta), y(s(s(Y)),\#) \circ\!\!-y(Y,\beta) \to i(f(s(z),s(s(z))),\#)
}\text{[Back !] } F_0
}{
\# :\to i(f(s(z),s(s(z))),\#)
}\text{[Sync] } F_3
$$

where

$\delta_1 : \begin{cases} \\ \cfrac{\cfrac{\cfrac{1}{\#,\beta :\to x(z,\beta)}\text{[Back !] } F_1}{\#,\beta : x(s(X),\#) \circ\!\!-x(X,\beta) \to x(s(z),\#)}\text{[Back ?]}}{} \\ \end{cases}$

and

$\delta_2 : \begin{cases} \\ \cfrac{\cfrac{\cfrac{1}{\#,\beta :\to y(z,\beta)}\text{[Back !] } F_2}{\#,\beta : y(s(s(Y)),\#) \circ\!\!-y(Y,\beta) \to y(s(s(z)),\#)}\text{[Back ?]}}{} \\ \end{cases}$

4   From Linear Logic to Prolog

From an operational point of view, it appears clearly that the previous system FG does not provide any strategy for a proof search (especially concerning the use of the rule [Sync]). Furthermore, a refinement will help us to get the implementation of the system. Therefore, we define a goal directed proof system FG^{dir} with the following inference rules[5].

Definition 7. The FG^{dir} system

[Back !^{dir}]

$$
\cfrac{
\begin{array}{c}
\Sigma : ((\Delta_1 \to G_1),\ldots,(\Delta_{i-1} \to G_{i-1}), \\
(\Delta_1' \to A_1\sigma),\ldots,(\Delta_p' \to A_p\sigma), \\
(\Delta_{i+1} \to G_{i+1}),\ldots,(\Delta_n \to G_n))
\end{array}
}{
\Sigma : ((\Delta_1 \to G_1),\ldots,(\Delta_n \to G_n))
}
$$

---

[5] We define the set operator $\uplus$ as $\uplus_{1 \le i \le n} A_i = \cup_{1 \le i \le n} A_i$ such that $\forall 1 \le i,j \le n, i \ne j, A_i \cap A_j = \emptyset$.

if $H \circ\!\!-A_1 \otimes \ldots \otimes A_p \in \Psi$, *there exists a substitution* σ *such that* $H\sigma = G_i$ *and* $\Delta_i = \uplus_{1 \le k \le p} \Delta'_k$.

[*Back* $?^{dir}$]

$$\frac{\begin{array}{c} \Sigma : ((\Delta_1 \to G_1), \ldots, (\Delta_{i-1} \to G_{i-1}), \\ (\Delta'_1 \to A_1\sigma), \ldots, (\Delta'_p \to A_p\sigma), \\ (\Delta_{i+1} \to G_{i+1}), \ldots, (\Delta_n \to G_n)) \end{array}}{\Sigma : ((\Delta_1 \to G_1), \ldots, (\Delta_n \to G_n))}$$

if $C = (H \circ\!\!-A_1 \otimes \ldots \otimes A_p) \in \Delta_i$, *there exists a substitution* σ *such that* $H\sigma = G_i$ *and* $\Delta_i \setminus \{C\} = \uplus_{1 \le k \le p} \Delta'_k$.

[*Sync+* ^{dir}]

$$\frac{\begin{array}{c} \Sigma, \alpha : ((\Delta_1 \cup \Delta''_1 \to G_1), \ldots, (\Delta_{i-1} \cup \Delta''_{i-1} \to G_{i-1}), \\ (\Delta'_1 \cup \Delta'''_1 \to A_1\sigma), \ldots, (\Delta'_p \cup \Delta'''_p \to A_p\sigma), \\ (\Delta_{i+1} \cup \Delta''_{i+1} \to G_{i+1}), \ldots, (\Delta_n \cup \Delta''_n \to G_n)) \end{array}}{\Sigma : \qquad ((\Delta_1 \to G_1), \ldots, (\Delta_n \to G_n))}$$

if $\forall c \exists nc(C_1 \otimes \ldots \otimes C_r) \in \Psi$, $1 \le j \le r$, $C_j\theta = (H \circ\!\!-A_1 \otimes \ldots \otimes A_p)$, $H\sigma = G_i$, $\Delta_i = \uplus_{1 \le k \le p} \Delta'_k$, $\{C_1\theta, \ldots, C_{j-1}\theta, C_{j+1}\theta, \ldots, C_r\theta\} = (\uplus_{1 \le k(\neq i) \le n} \Delta''_k) \uplus (\uplus_{1 \le k \le p} \Delta'''_k)$, $\beta \in \Sigma$, $\alpha \notin \Sigma$ *and* $\theta = \{c \leftarrow \beta, nc \leftarrow \alpha\}$.

[*Sync1* ^{dir}]

$$\frac{\begin{array}{c} \Sigma, \alpha : ((\Delta_1 \cup \Delta'_1 \to G_1), \ldots, (\Delta_{i-1} \cup \Delta'_{i-1} \to G_{i-1}) \\ (\Delta_{i+1} \cup \Delta'_{i+1} \to G_{i+1}), \ldots, (\Delta_n \cup \Delta'_n \to G_n)) \end{array}}{\Sigma : \qquad ((\Delta_1 \to G_1), \ldots, (\Delta_n \to G_n))}$$

if $\forall c \exists nc(C_1 \otimes \ldots \otimes C_r) \in \Psi$, $1 \le j \le r$, $C_j\theta = (H \circ\!\!-1)$, $H\sigma = G_i$, $\Delta_i = \emptyset$, $\{C_1\theta, \ldots, C_{j-1}\theta, C_{j+1}\theta, \ldots, C_r\theta\} = (\uplus_{1 \le k(\neq i) \le n} \Delta'_k)$, $\beta \in \Sigma$, $\alpha \notin \Sigma$ *and* $\theta = \{c \leftarrow \beta, nc \leftarrow \alpha\}$.

[*Axiom* $!^{dir}$]

$$\frac{\begin{array}{c} \Sigma : ((\Delta_1 \to G_1), \ldots, (\Delta_{i-1} \to G_{i-1}), \\ (\Delta_{i+1} \to G_{i+1}), \ldots, (\Delta_n \to G_n)) \end{array}}{\Sigma : \quad ((\Delta_1 \to G_1), \ldots, (\Delta_n \to G_n))}$$

if $H \circ\!\!-1 \in \Psi$, $\Delta_i = \emptyset$ *and there exists a substitution* σ *such that* $H\sigma = G_i$.

[*Axiom* $?^{dir}$]

$$\frac{\begin{array}{c} \Sigma : (\Delta_1 \to G_1), \ldots, (\Delta_{i-1} \to G_{i-1}), \\ (\Delta_{i+1} \to G_{i+1}), \ldots, (\Delta_n \to G_n) \end{array}}{\Sigma : \quad (\Delta_1 \to G_1), \ldots, (\Delta_n \to G_n)}$$

if $\Delta_i = \{H \circ\!\!-1\}$, *there exists a substitution* σ *such that* $H\sigma = G_i$.

The equivalence of the two systems is insured by the following theorem :

Theorem 2. $FG^{dir} \Leftrightarrow FG$

There exists a proof of a sequent in the system FG if and only if there exists a proof of this sequent in the system FG^{dir}.

Proof. This proof is inspired by [17]. The first step of the proof is to define a notion of restricted proof : a proof is said to be restricted if all the instances of the rule [*Sync*] are at the root of the proof tree. This amounts to generate all the synchronizations before using them. We can prove that if there exists an FG proof then there exists a restricted FG proof (see figure below).

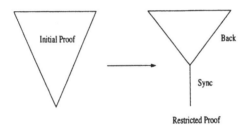

From proof to restricted proof

The second step of the proof is to define a notion of ordered linear proof : a linear proof is ordered if the order of synchronizations corresponds to the order of use of the clauses generated by the corresponding synchronizations. The transformation of the proof tree into a linear proof tree is achieved by recombining the different branches of the proof tree into a single derivation thanks to a reformulation of the sequent calculus proof system (see figure below). The intermediate system FG^{lin} is proved equivalent to the initial system FG by induction.

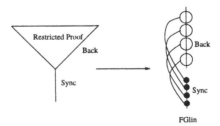

From restricted proof to linear proof

The last step of this proof consists in giving the strategy for the use of the synchronization rule. In fact, synchronizations will be only generated when needed (see figure below). Of course, FG^{dir} is proved equivalent to FG^{lin} by induction $\square$.

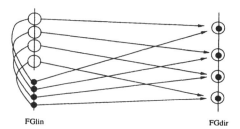

FGlin   FGdir

From linear proof to goal directed linear proof

We have to mention here that this system provides a goal directed approach for the recognition of an element of a tree language defined by a particular $TTSG$. This kind of operation is not possible using the initial definition of a $TTSG$ derivation which gives a method to produce elements of the language but does not give any strategy to recognize an element.

The introduction of unification and variable renaming in the system FG^{dir} would also insure the generation of the element of the language. This has to be formally proved thanks to a lifting lemma (as it is done for the SLD resolution of Prolog [13]). As a consequence, the generation is available in the Prolog implementation of our system described in the next section.

Concerning the decidability of our approach, it is clear that problems encountered with DCG's appear here (mainly due to left recursions and empty transitions). Since Prolog is the underlying framework, the problems related to its depth left first strategy also occur. Thus, termination of our method depends on this search strategy.

5 Implementation Issue

This section briefly describes how the implementation can be achieved from the previous inference system. At this time, a library is able, taking a $TTSG$ as input, to provide, as output, a predicate `phrase_ttsg` which can recognize or generate a term for this $TTSG$ from the axiom. The implementation of the method described along this paper can be divided in two parts : the translation of the grammar into linear logic formulas (described in section 3.1) and the implementation of the proof system FG^{dir} in Prolog. The following presentation of this implementation is related to the example 1.

5.1 Management of signature and linear context

The extension of the signature (i.e. new symbols introduced by existential quantifiers of synchronization) is inspired by the management of essentially universal quantifier (quantifier `pi` in the body of clauses) of λProlog [1] : only the cardinality of this extension is needed.

The management of linear clauses is meta-programmed by a linear program continuation LPC (i.e. a set containing the remaining linear clauses introduced by instances of $[Sync+^{dir}]$ or $[Sync1]$ which have to be used later). This technique is inspired by the management of intuitionistic implication of λProlog [9, 1].

5.2 Transformation Function

The transformation is illustrated here by the implementation of the clause F_0 which corresponds to the free production R_0.

```
i(C, f(X, Y), SigmaIn, SigmaOut, DeltaIn, DeltaOut) :-
      x(C, X, SigmaIn, SigmaInter, DeltaIn, DeltaInter),
      y(C, Y, SigmaInter, SigmaOut, DeltaInter, DeltaOut).
```

SigmaIn, SigmaInter and SigmaOut represent the cardinality of the extension of the signature. DeltaIn, DeltaInter and DeltaOut represent the LPC.

5.3 Implementation of the Proof System FG^{dir} in Prolog

The inference rules $[Back\ !]$ and $[Axiom\ !]$ are handled by the reduction mechanism of Prolog.

The inference rules $[Back\ ?]$ and $[Axiom\ ?]$ are implemented by a meta-programming of Prolog by the predicate linear over the linear program continuation.

```
linear(PredicatSymbol, C, Term, SigmaIn, SigmaOut,
      [C_j|Cj_plus_1_Cm], C1_Cj_minus_1, DeltaOut) :-
  C_j = (Head :- Body),
  Head =.. [PredicatSymbol, C, Term, SigmaIn, SigmaOut,
            C1_Cm_but_Cj, DeltaOut],
  append(C1_Cj_minus_1, Cj_plus_1_Cm, C1_Cm_but_Cj),
  call(Body).
```

The linear clause C_j is meta-interpreted and then discarded from the linear program continuation : C1_Cm_but_Cj = C1_Cj_minus_1 ⊎ Cj_plus_1_Cm (see append predicate). A recursive prolog clause is added for the program continuation traversal.

The inference rules $[Sync+^{dir}]$ and $[Sync1]$ are implemented in a similar way as respectively $[Back\ !]$ and $[Axiom\ !]$ but with an increase of the linear program continuation by the linear clauses introduced by the synchronization.
The prolog clause below corresponds to the implementation of x^6 for the translation of the pack of production $\{X \Rightarrow s(X), Y \Rightarrow s(s(Y))\}$:

$$(\forall c \exists nc((\forall X(x(s(X),c) \circ -x(X,nc))) \otimes (\forall Y(y(s(s(Y)),c) \circ -y(Y,nc))))).$$

---

[6] There is of course a symmetrical clause for y.

```
x(C, s(X), SigmaIn, SigmaOut, DeltaIn, DeltaOut) :-
  NC is SigmaIn + 1,
  Cy = (y(C, s(s(Y)), SI, SO, DI, DO) :- y(NC, Y, SI, SO, DI, DO)),
  x(NC, X, NC, SigmaOut, [Cy|DeltaIn], DeltaOut).
```

The linear clause $\sigma_S(Y \Rightarrow s(s(Y)), 0, 1) = (\forall Y(y(s(s(Y)), 0) \circ -y(Y, 1)))$ is implemented by `Cy` added to the linear program continuation.

5.4 Reduction strategy

Due to the leftmost selection rule and the depth-first search strategy of Prolog the reduction strategy is a left-outermost strategy. This is illustrated by the trace below for the goal `phrase_ttsg(f(s(z),s(s(z))),i)` (the clause `(y(0, s(s(Y)), SI, SO, DI, DO) :- y(1, Y, SI, SO, DI, DO))` is denoted Cy). The first column is the (simplified [7]) prolog trace, the second one contains the linear program continuation and the third one the term already recognized.

```
phrase_ttsg(f(s(z),s(s(z))),i)    LPC
i(0,f(s(z),s(s(z))))              {}     (I,0)
x(0,s(z)),y(0,s(s(z)))            {}     f((X, 0), (Y, 0))
x(1,z),y(0,s(s(z)))               {Cy}   f(s((X, 1)), (Y, 0))
y(0,s(s(z)))                      {Cy}   f(s(z), (Y, 0))
y(1,z)                            {}     f(s(z), s(s((Y, 1))))
                                  {}     f(s(z), s(s(z)))
```

The source code of this implementation is available at :
`http://www.info.univ-angers.fr/pub/stephan/Research/Download.html`.

6 Conclusion and Future Work

In this paper, we describe an implementation scheme for a particular type of tree language including synchronization features (TTSG). The main idea is to provide a framework in order to compute over tree grammars. As it has been done for word grammars with DCG [2], we propose a transformation method that allows us to get a set of Prolog Horn clauses from a grammatical definition of a tree language. This method consists in translating grammar derivation into proof search using a sequent calculus proof system based on linear logic. By successive refinements, we get a goal directed procedure implemented in Prolog. Since such tree languages appear as powerful tools for the schematization of sets of terms to represent solutions of symbolic computation problems, it seemed necessary to define a method to use these representations. Moreover, our approach provides an uniform framework for the implementation and combination of different tree languages.

---

[7] In the real system, variables are denoted as in Prolog (i.e. of the form _1234). For the sake of readability, explicit names have been introduced and meta-programming arguments have been discarded.

This general approach is only illustrated here for TTSG but could be extended to other particular tree languages such as primal grammars [8] or [7] for which the notion automaton does not exist (as far as we know) and does not seem to be obvious to deduce from the grammatical definition.

Concerning basic operations on languages (such as intersection) it will be interesting to study how these operations can be achieved through combination of proofs.

Acknowledgements. We would like to thank the anonymous reviewers for their helpful comments.

References

1. P. Brisset. *Compilation de λProlog*. PhD thesis, Thèse de doctorat de l'université de Rennes, 1992.
2. J. Cohen and T.J. Hickey. Parsing and compiling using prolog. *ACM Transactions on Programming Languages and Systems*, 9(2):125–163, 1987.
3. H. Comon, M. Dauchet, R. Gilleron, D. Lugiez, S. Tison, and M. Tommasi. *Tree Automata Techniques and Applications*. 1997.
4. J.-Y. Girard. Linear logic, its syntax and semantics. In Regnier Girard, Lafont, editor, *Advances in Linear Logic*, number 222 in London Mathematical Society Lecture Notes Series, pages 355–419. Cambridge University Press, 1993.
5. Jean-Yves Girard. Linear Logic. *Theoretical Computer Science*, (50):1–102, 1987.
6. P. De Groote and G. Perrier. A Note on Kobayashi's and Yonezawa's "Asynchronous Communication Model Based on Linear Logic ". *Formal Aspects of Computing*, 10, 1998.
7. Y. Guan, G. Hotz, and A. Reichert. Tree Grammars with Multilinear Interpretation. Technical Report FB14-S2-01, 1992.
8. M. Hermann and R. Galbavy. Unification of infinite sets of terms schematized by primal grammars. *Theoretical Computer Science*, 176, 1997.
9. J. S. Hodas and D. Miller. Logic Programming in a Fragment of Intuitionistic Linear Logic. In *Proceedings of LICS'91*, pages 32–42, 1991.
10. S. Limet and P. Réty. E-Unification by Means of Tree Tuple Synchronized Grammars. *Discrete Mathematics and Theoretical Computer Science*, 1:69–98, 1997.
11. S. Limet and P. Réty. Solving Disequations modulo some Class of Rewrite System. In T. Nipkow, editor, *Proceedings of 9th Conference on Rewriting Techniques and Applications*, volume 1379 of *LNCS*, pages 121–135. Springer Verlag, 1998.
12. S. Limet and F. Saubion. Primal Grammars for R-unification. In *PLILP/ALP'98*, number 1490 in LNCS. Springer-Verlag, 1998.
13. J.W. Lloyd. *Foundations of Logic Programming*. Symbolic Computation series. Springer Verlag, 1987.
14. Dale Miller. A Multiple-Conclusion Meta-Logic. In *LICS 1994*, pages 272–281, 1994.
15. G. Plotkin. Building-in equational theories. *Machine Intelligence*, 7:73–90, 1972.
16. F. Saubion and I. Stephan. Grammaires TTSG et Système FG. Technical report, 1998. Available at http://www.info.univ-angers.fr/pub/saubion/Research/Download.html.
17. I. Stéphan. *Nouvelles fondations pour la programmation en logique disjonctive*. PhD thesis, Thèse de doctorat de l'université de Rennes, 1995.

Author Index

Springer
and the
environment

At Springer we firmly believe that an international science publisher has a special obligation to the environment, and our corporate policies consistently reflect this conviction.
We also expect our business partners – paper mills, printers, packaging manufacturers, etc. – to commit themselves to using materials and production processes that do not harm the environment. The paper in this book is made from low- or no-chlorine pulp and is acid free, in conformance with international standards for paper permanency.

 Springer

Lecture Notes in Computer Science

For information about Vols. 1–1548
please contact your bookseller or Springer-Verlag